Theoretical Physics:
The First Problem

JRBreton

Theoretical Physics:
The First Problem
by JRBreton

Published by:
Foundation for Theoretical Physics
3 Apple Tree Lane
Walpole, MA 02081–2301

Web address: FoundationForTheoreticalPhysics.org
email address: Theoretical.Physics.books@gmail.com

Copies of this book and other offerings of the Foundation may be obtained preferably online from tp.vendevor.com, the Foundation's website and also from amazon.com and similar sites.

ISBN, print ed. 978–0–9844299–1–2

Second printing 2019

Printed in the United States of America

Library of Congress Control Number 2010921214

Preface

Reality is created and thus defined by the one God:
Father, Son, and Spirit.

Physics is the study of reality observable as:
extended, moving, or forcing.

The first problem of Theoretical Physics: How to transform Mathematics into Theoretical Physics.

Physics defined (page 7); Theoretical Physics defined (11); Statement of the First Problem (12); Symbols (13) .

Mathematical foundations (17); Real numbers (17); Topology (17); Measurable sets (17); Limits (18); Functions (18); Continuity (21); Differentiation and Integration (22): Derivatives (22), Integrals (28); The Fundamental Theorem of Integral Calculus (34); Step Functions (36); Three Dimensional set of vectors––V3 (51); Vector Algebra (51); Basic Operations (51); the Origin (58); Operations referred to the Origin (60); Reciprocal Vectors (63); Solutions of Vector Equations (65); Matrix functions (73); Solutions of Matrix Equations (78); Functions of V3 (70); Straight Maps in V3 (87); Vector Calculus (87); Topology of V3 (89); Limits in V3 (92); Curves in V3 (90); Process in V3 (103); Differentiation in V3 (107); Directional Derivatives (108); Derivatives along a curve (110); Directional Divergences, Curls, and Gradients (116); First Quadrant Divergences, Curls, and Gradients (122); Gradient of a Vector (126) ; First Quadrant Derivatives of Compound Functions (129); Local Differentiation (131); Sectional Gradients (135); Relationships between Gradients (138); Rank of a Transformation (145); Local Differentiation of Sums and Products (146); Integration in V3 (149); Integration of Processes (150); Convergences, Incurls, and Ingradients along curves (157); Step Functions over curves (162); Vector Measures in V3 (174); Step Functions in V3 (189).

Theoretical Physics: Rules for Physical Units (221); the Idea of a Particle (226); Material and Local References (229); the Principle of Non–Collocation (234); Derivatives of Theoretical Physics (235); Derivatives of Primary Functions (237); Derivatives of Generalized Functions (284); Derivatives: Mathematical versus those of Theoretical Physics (293); Relationships between the Derivatives of Theoretical Physics (297); Results for Primary Functions (316); Step functions of Theoretical Physics (320); Integrals of Theoretical Physics (331); Directional Integration of Directional gradients (344); Sectional Integration of sectional gradients (375).

The bridge theorem (409); Fallacies related to Partial Derivatives (433).

Summary (439).

Epilog (443).

End–notes (444).

Indices (471).

Reflection. Physics is reductionist: reality which is not observable as extended, moving, or forcing lies outside its purview. Physics does not study, nor has it anything to say about, peace, justice, morality, religion, or indeed about many matters which substantially engage the human spirit.

Physicists as persons, however, are engaged in a specifically human, and thus in a moral activity. By the very fact of engaging in Physics, physicists testify to and acknowledge, at least existentially, a reality larger than Physics. Even more, they depend on this larger reality to validate their methods and personal Integrity.

Moderns have made a substantial error by Inverting this order. It is not difficult to uncover claims by modern physicists and by their popularizers which assert that reality consists only of what is physically observable––and conversely that what is not physically observable does not exist.

This modern attitude amounts to a kind of idolatry, since only God defines reality. For the post–modern world, the modern claim amounts to nothing more than arrogant nonsense.

First Definitions

Definition of Physics

Physics is the study of reality observable as extended, moving, or forcing.

First Definitions

Reflection. Physics as science, unified knowledge, is thus a kind of adoration. In their efforts to explain and unify, physicists testify existentially to the unity in reality which reflects its creation by the one God.

First Definitions

Physics has an inherent structure which needs to be contemplated.

Physics is:

the study of
reality observable as
extended objects which may be
moving
or forcing.

Study
An ordering of observations through ideas,
rather than a mere taxonomy.

Observable reality
The process of observation may be either direct
or Indirect through instruments;
ultimately a human observer is Implicated.
The observation itself involves a physical process.
The observation has both a minimum
and a maximum scope
whose range can be enlarged by instruments.
Observation implies the observer can Identify
the observed object as it changes.

Extended objects
Physical objects have parts, and thus are measurable.
Physical objects are localized and oriented.
Separate objects cannot be collocated.
Extension, as an attribute of objects,
is logically prior to motion or force.

Moving or forcing
Thus changing and mutable.

Upon reflection, we note that physical objects are always observed as extended, but not necessarily as moving or as forcing. Sometimes physical objects are observed as moving without forcing or contrariwise forcing without moving. These relationships are clearly of fundamental importance to the science of Physics and require profound reflection.

Reality, observable as extended, moving, or forcing, is overwhelming in its extent and complexity. Physics as a study, has as goal the analysis and unification of physical observations. Analysis and unification make the transmission of physical knowledge more efficient and furthers the elaboration of Physics as unified knowledge, that is to say, of Physics as science.

First Definitions

Reflection. The logical consistency of Theoretical Physics is an act of faith. By this act of faith the physicist professes that the result of a consistent, logical elaboration from valid premises will be confirmed in reality. In effect, reality mirrors God in not contradicting itself.

Reflection. It is apparent the ideas of Physics are not themselves physical realities. The idea of velocity, for example, does not itself move and may be considered apart from a moving object, whereas physical objects are observed only in certain locations, moving with a certain motion and being subjected to and exerting certain forces. The idea of velocity is nevertheless real. It is clearly not observable as extended, moving, or forcing. The idea of velocity, and indeed all the ideas of Theoretical Physics, are non–physical realities.

Reflection. By its requirement for logical consistency, Theoretical Physics differs from engineering. Engineers seek the utilitarian exploitation of physical reality for which purpose they accept approximations. Approximations imply logical inconsistency. The boundaries of Theoretical Physics are its axioms (the validity of its results comes from its logical consistency) whereas the boundaries of engineering are the utility of its approximations. While approximations cannot be true logically and thus cannot be understood, they can be useful.

Much of what commonly passed for science and Physics in the modern era was, in fact, engineering.

In this book the word "theory" is used for a logically consistent elaboration of a set of axioms. The theory of numbers or the mathematical theory of measure, provide Illustrations. In contrast, an engineering model is a constructlon, often mathematical, to suit a utilitarian purpose.

When in the Bible the circumference of a column (an easy measurement) is said to measure three times its diameter (a hard measurement), a model useful for constructlon and transporting columns of a temple is Indicated. Referring to the ratio of the circumference of a circle to its diameter as pi is theoretical.

First Definitions

Definition of Theoretical Physics

Theoretical Physics is a set of coherent ideas related in a framework of logically consistent statements useful for understanding physical observations.

Ideas are intellectual instruments. Theoretical Physics is needed for the study of Physics, because while extended, moving, or forcing objects can be sensed, only ideas can be understood.

Theoretical Physics is constrained to be logically consistent[1]. Contradictions can be stated, believed, acted upon, and even given the appearance of logically true statements. Contradictions arise from inconsistent pseudo–ideas or false statements. Contradictions cannot be understood. Only truth can be understood. Theoretical Physics cannot accept contradictions as instruments for studying reality because reality is what it is and is not what it is not.

Accepting as true any *one* contradiction in a set of otherwise logically consistent propositions makes all possible propositions *logically* true.

For example, if the proposition {1=0} in arithmetic is accepted as true, then

$$(n*1)+m = (n*0)+m = m.$$

Consequently any integer can then be "proven" to equal any another integer using the operations of arithmetic logically and consistently. The science of arithmetic is thus annihilated.

Similarly, accepting any contradiction in Theoretical Physics radically annihilates Physics as science.

Unfortunately, such is the condition of modern Physics. It is widely conceded that modern Physics is "badly posed" mathematically. The corollary to this concession is not so widely appreciated, namely that given sufficient mathematical manipulation, *any* proposition of Physics can be "proven" true.

In practice, this license is restrained by recourse to experiment which, in effect, reattaches a given proposition of Physics to reality. Experiments, however, carry their own burden of error. Thus where experiment is guide, modern Physics is reduced to engineering. Where experiment is not possible, modern Physics remains vulnerable to wild speculation.

The ultimate purpose of this book is to help physicists reestablish Physics as science.

First Definitions

How, then, can the study of physical observations be brought into a framework of ideas which

guarantees logical consistency,
has some hope for achieving intellectual unity, and
relates to our observations of measurable and mutable objects?

Fortunately logically consistent systems of statements with their embedded ideas have been elaborated by the intellectual toil of previous generations of mankind. The structures of Mathematics are particularly helpful because they consist of simple, axiomatic foundations with well–developed, logically consistent, elaborations.

If a way can be found to relate physical observations to mathematical axioms, the mathematical elaborations may relate to physical reality. Even better, the understanding of Mathematics will guide the understanding of Physics. The logical unity of Mathematics can thus help create the science of Physics.

Statement of the First Problem.

With this hope in mind, we now pose the first problem of Theoretical Physics:

How can mathematical ideas and expressions be transformed into Theoretical Physics?

To illustrate the problem on an elementary level, consider the mathematical relationship:

$$\exp(x) = 1 + x/1! + x^2/2! + \ldots$$

To use this relationship for Theoretical Physics, a physical unit should be attached to the variable, x.

The first term on the right is 1 which is a simple number and has no physical unit. The next term is x, which should have the physical unit of x. The next term is $x^2/2$ which presumably has a different physical unit than the physical unit of x. Etc.

So what can be the physical unit of exp(x)? Clearly, exp(x) is a mathematical idea. For Theoretical Physics it must be transformed into a new and different, but related idea.

This book answers the first problem by first developing aspects of mathematics suitable for the transformation. Then it explains how the mathematical ideas are transformed into Theoretical Physics. The process results not only in a clear appreciation of their differences, but ends by giving the reader a solid and comprehensive set of intellectual tools by which to explain physical observations.

But how may ideas, which cannot be sensed, whether of mathematics or of Theoretical Physics, be nevertheless expressed or communicated? To answer this question we next discuss symbology.

Symbols

A **symbol** is a physical object which is considered not as a physical object but rather as a pointer to an idea. In contrast to ideas, which must be consistent, symbols are arbitrary.

Symbols, as notational shorthand for ideas, are thus tools for thinking. Their efficient and appropriate use reflects on our ability to communicate.

There is an inescapable tension between the use of a large number of symbols (appropriate to subtlety of thought) and efficiency (which favors small number of symbols). To symbolize numbers, for instance, the broad human culture has evolved ten special symbols (for decimal notation) repeated positionally as superior to alphabetical symbols.

Theoretical Physics uses for symbols only the Latin alphabet, certain punctuations and typefaces, arithmetic operators like plus and minus, set operators like union and intersections, and the Arabic numerals.

The symbols are governed by the following rules (with occasional exceptions).

First Definitions

1 *Lower case letters denote variables, quantities, parameters, vectors.*
  1.1 The first part of the alphabet is often reserved for symbols whose definition is given locally.
  1.2 The middle part of the alphabet is often used for indexing.
  1.3 The later part of the alphabet is often reserved for symbols defined in larger context, tending to standardization.
  1.4 Vectors are denoted by bold typeface.

2 *Definitions are indicated by the symbol* ≡.

3 *Propositions (statements) are indicated by the symbol* =. For relevant contexts propositions may also be indicated by ≥, ≤, <, >, ⟶, ⟵, or ⟷.

4 *Upper case letters denote operators or sets.* The corresponding lower case is then taken as an individual of the set.

5 *Arithmetical operations are always denoted explicitly.* Therefore, characters run together will denote one symbol, for example, pi or exp.

6 *Wherever possible, a bias is exerted toward mnemonics* (for example, **f** for force.)

Rules for Symbols

Example: a, **b**, f are usually defined locally. They may refer to functions, parameters, variables, vectors, etc. variously in any given location.

Example: i,m,n are often used as indices.

Example: q, **x**, **r** often stand for charge, particle, location.

Example: Given the set R, r is an element of R.

Example: the ratio of the circumference of a circle to its diameter is symbolized as pi. (π is not used.)

The local context should always be consulted for the symbol's meaning, because the local definition always prevails.

Words, too, are symbols for ideas. Words as symbols differ from the symbols of Theoretical Physics not as symbols but rather in their tolerance for ambiguity[2]. The symbols of Theoretical Physics tolerate no ambiguity. Each must relate clearly to a single idea derived ultimately from observations of extension, motion, or force. Words, in contrast, shoulder the much heavier burden of referring to a much larger range of human experience, not only intellectual but emotional, volitional, and religious experience as well. Clarity is often desirable with words, but sometimes also ambiguity. Not so with Theoretical Physics.

Since both are symbolic, the symbols and statements of Theoretical Physics may be mixed with words when useful. In particular, the rules for punctuation apply to the symbolic statements of Theoretical Physics.

Example: Let there be given a, a real variable.

Example: The function,

$$\exp(x) = 1 + x/1! + x^2/2! + \ldots,$$

is a mathematical function.

Theoretical Physics, as a coherent set of ideas, faces the task of creating new ideas from the given, basic ideas associated with extension, motion, or force. An arbitrary creation, contradictory to a previously elaborated framework, will annihilate the previous framework. Such a pseudo–idea and the process used to create it are called **invalid** with respect to the previous framework. Theoretical Physics must deal only with **valid** ideas created by a **valid** creative process, that is, ones which are logically coherent with the basic ideas and the previous framework built upon them.

Moreover symbols referring to the new ideas must be unambiguous.

Definitions exhibit the process by which new ideas are created. Definitions may be created informally or formally.

A simple, informal definition may be made by placing a symbol in apposition to a word signifying the idea. Or the idea may be described and given a name. The simple naming of an idea as above is usually made merely by emboldening the symbolic word.

First Definitions

The symbol ≡ is used generally to exhibit the creation of a new idea from basic ideas or previously defined ones, sometimes in a local context. Formal definitions, often involved, are usually named and numbered.

Example: At location **r1**

Example:

$2 \equiv 1 + 1$ defines the symbol "2";

$4 = 2 + 2$ is a mathematical statement.

Given the above

$4 \equiv 2+3$ is an invalid definition.

Example of a formal definition:

Definition (title)

Given x,

for y=x,

then

definition.

end of definition

Given these rules, the text eschews special alphabets or other special symbols including the usual ones for integration and differentiation.

Mathematical Foundations

The starting point being Mathematics, some useful mathematical structures are now invoked, of which, from this point on, the reader[3] is presumed knowledgeable. For the most part commonly known mathematical results will not be derived, formal proofs being reserved for less well–known results. Many of the mathematical results are revisited in sections devoted to Theoretical Physics where the development becomes more formal. The missing mathematical proofs will then become apparent.

The text does contain some new mathematical material, notably
division in the set of vectors
a detailed discussion of gradients
new types of integrals.

Preliminaries

Some elementary mathematical ideas are first described.

An **algebra** is a set which supports addition, subtraction, multiplication and division.

We first invoke the **real number system**, R. R consists of infinitely many mathematical objects called real numbers which are **ordered**, **complete** with and in a given t**opology** with an **algebra** which is a **field**[4].

Topologies are defined in terms of subsets of a given set. Consider the set, RT, of subsets of R which are arbitrary unions of sets

S = {r|abs(r–r1)<e}

for all positive e of R and any r1 of R. Then
R and the null set are subsets of RT
and RT is closed with respect to the intersection
of a finite number of its subsets,
and RT is closed with respect to the union
of an arbitrary collection of its members.

RT is called a **topology** in R. The members of RT are called the **open** subsets of R.

Call RM the smallest set of subsets of R which includes all its open sets, their complements, and all denumerable unions of the such subsets. The sets of RM are then called the **measurable** sets of R.

Mathematics

A **calculus** is a set which supports differentiable functions.

The mathematical theory of measurable sets is the basis upon which the concepts of integral calculus are built.

The **absolute value** function, abs, serves as a **positive metric** in R.

The idea of a **limit** stems from topology. An element *x* of a set *X* is said to be a **limit point of X** if any open set containing *x* also contains an element *y* in *X* which is not *x*.

Functions

A **function** is a rule describing a specific relationship from a set *X* to a set *R* wherein each element of *X* is related to one and only one element of *R*. Functions are symbolized as

$$f(x) = r$$

or to emphasize the sets involved,

$$f: X \longrightarrow R.$$

The set *X* is called the **domain** of *f*, the set *R* is called the **range** of *f*, the subset of *R* which is related to *X* is called the **image** of *X* by *f*. The image of *f* is denoted by f(X).

Different functions may describe the same relationship between between *X* and *R*; that is, a function is *not* uniquely defined by a given relationship[5]. Two different functions having the same values are called **look–alike functions**.

A **real function** is a function whose range are real numbers, including their topology.

A real function f :X⟶R has a **limit** f1 as x ⟶x1 if x1 is a limit point in X and given any er>0 there can be found an ex>0 such that

$$0<\text{abs}(x-x1) <\text{ex implies abs}(f(x)-f1)<\text{er}.$$

Note f1 need not equal f(x1) nor be a limit point of f(x).
The distinction between f(x1) and lim f as x⟶x1 as ideas is useful for the study of discontinuities.

Taking limits of functional values is thus a process symbolized as

$$\lim f = f1 \text{ as } x \longrightarrow x1. \quad (1)$$

The real number system because of its property of order allows a further distinction.

A real function f :X⟶R has a **limit f0 from below** as *x* approaches *x1*

if

x1 is a limit point of X

x<x1

given any er>0 there can be found an ex>0 such that

0<abs(x–x1)<ex whenever abs(f(x)–f0)<er.

Similarly, a real function f:X⟶R has a **limit f1 from above** as *x* approaches *x1*

if

x1 is a limit point of X

x>x1

given any er>0 there can be found an ex>0 such that

0<abs(x–x1)<ex whenever abs(f(x)–f1)<er.

Note f(x1) may not exist at all, may equal either *f1* or *f0* or any other value in the range of f. Moreover, the limit from below need not equal the limit from above.

A **linear** function is one for which, given *x1* and *x2* of *X*,

$$f(x1+x2) = f(x1) + f(x2). \qquad (2)$$

Two functions may interact provided they are defined appropriately.

For

f:X⟶R and g:X⟶R

the sum f+g and product f✱g are defined as f(x) + g(x) and f(x)✱g(x) respectively.

Furthermore two functions may interact depending on the arrangement of their domains and images.

Definition 1 (restricted function)
Given
$f:X \longrightarrow R$;
for
$Y \subset X$, Y a subset of X
then

$$f|Y \equiv f:Y \longrightarrow R$$

end of definition

Definition 2 (compound function)
Given
$f:X \longrightarrow R$;
$g:Y \longrightarrow X$;
then

$$f(g\ ) \equiv f(g(y)) : Y \longrightarrow R$$

end of definition

The function g of the compound function f(g) is called **a compound variable**.

A restricted function is the same rule as the original function but for its restricted domain. The restriction is often to a constant value.

One the other hand, a compound function is generally a different rule from the original.

Example: Let
$f(x) = 2 + 3 * x$;
$x = g^2$
Then
$f(g) = 2 + 3 * g^2$.

Thus f(x) is linear, but f(g) is parabolic.

Clearly, f(x) and f(g) are two different rules of the same relationship. When confusion threatens, the functions will be labeled explicitly as fx or fg.

Continuity

Continuity is an attribute of certain functions, f:X⟶R with metric domains and ranges, by which points close in the function's domain remain close in the function's image. Formally,

---

Definition 3 (continuity of a real function)

Given

sets X and R with metric domains and ranges

f:X⟶R

x1, a limit point of X

for

any real number ef>0

if there exists a real number ex>0 such that

whenever abs(x–x1)<ex, abs(f(x)–f(x1))<ef

then

f **is continuous at** x1

end of definition

Definition 3 is called an **operational** definition because it envisages an interplay between choosing *ef* and finding *ex*.

Continuity and limits are closely related. Indeed, if the limit of *f* at *x1* from above equals the limit of *f* at *x1* from below and both equal f(x1), the function is continuous at x1.

The idea of continuity applies to the real number system but may also be extended to any *topological* set with a metric. In this larger context the definition becomes:

A function f:X⟶R is continuous if given any open set S of R, and f(Y) = S, then Y is an open set in X.

Mathematics

The idea of continuity is tied to a relationship rather than to a specific function expressing that relationship. That is, given *F*, the set of all look–alike functions describing the relationship, if any *f* in *F* is continuous, then all *f* in *F* are similarly continuous.

Measurability is an attribute of certain functions, f:X⟶R.

Definition 4 (measurability of a real function)

Given

f(X) ⟶ R, a real function;
U, an open subset of the image of f;
V, a subset of the domain of f;

if for

every U such that f(V) = U, implies V is a measurable set of X

then

f **is measurable over** X.

end of definition

Differentiation and Integration

The idea of a derivative, unlike that of continuity, pertains to functions rather than to relationships. That is, given *F*, the set of all look–alike functions describing a relationship, if *f1* in *F* has a derivative, it does *not* follow that any other *f* in *F* has the same derivative, or indeed any derivative at all. Thus derivatives are attributes of functions which do not necessarily arise from functional values.[6]

Differentiation is a limiting process which yields a **derivative**. The symbol *D* is used for the process and by extension for the result. For functions of a real variable *f:X⟶R*, the syntax for derivatives is,

$$D[x1](f(t)|u;dt) \tag{3}$$

to denote the derivative of the function *f* at the real domain value *x1* (called the **reference**) with respect to the real variable *t* (called the **index**) constrained by the **restriction** *u*.

The real derivative of a real function thus involves three distinguishable sets, X, R1, R2

$$D:(f:X\longrightarrow R1)\longrightarrow R2$$

The derivative is either a range value (for fixed domain value *x1*) or a function of x1 (for variable *x1*). In the latter case, *D* is considered an **operator** transforming the function *f:X⟶R1* into another function, *z:R1⟶R2*

$$D:f(x)\longrightarrow z(x) \equiv D[x](f(t)|u;dt).$$

To emphasize *D* as an operator, a contrasting symbol is often used to distinguish the reference as a variable from the index, as for example,

$$D[x(a)](f(t)|u;dt) = z(x(a)).$$

On the other hand, when the distinction between the index and the reference is clear, especially in complicated expressions focused on other considerations, the symbology is sometimes confused as:

$$D[x](f(t)|u;dt) = z(x).$$

Now consider functions continuous at *x1.*

For any function *f* with an *arbitrary*, bounded derivative at *x1,* it is always true for *any* value of the derivative that

$$\lim f(x1+dt) = f(x1) + \lim (dt * D[x1](f(t)|u;dt)) \quad \text{as } dt\longrightarrow 0. \quad (4)$$

From this fact, however, it neither follows necessarily that

$$D[x1](f(t)|u;dt) = \lim (f(x1+dx) - f(x1))/dx \quad \text{as } dx\longrightarrow 0$$

nor

$$f(x1+dt) \approx f(x1) + dt * D[x1](f(t)|u;dt).$$

Only specialized functions have these attributes. Functions continuous at x1 **with a basic derivative** are those whose derivative do satisfy[7]

$$D[x1](f(x)|u;dx) = \lim ((f(x1+dx) - f(x1))/dx, \quad \text{as } dx\longrightarrow 0. \quad (5)$$

Functions with a basic derivative are called **differentiable functions**; they form a subset of the larger class of functions with a derivative.

Only for functions with a basic derivative does,

$$f(x1+dx) \approx f(x1) + dx * D[x1](f(x)|u;dx). \tag{6}$$

Differentiable functions need not be continuously differentiable.

Inasmuch as the idea of limit applies more generally than the idea of continuity, the idea of a derivative may be extended to functions which though mostly continuous have nevertheless a finite number of finite discontinuities.

Definition 5 (basic derivative from above/below)

Given

f(x), a function continuous

except for a finite discontinuity at x1;

fa, the limit of f from above at x1;

fb, the limit of f from below at x1;

for

$dx > 0$

then

$$D[x1](f(x)|u;d_f x) \equiv \lim (f(x1+dx) - fa)/dx \ \text{ as } dx \longrightarrow 0$$

is the **basic derivative of f at x1 from above**; (7)

$$D[x1](f(x)|u;d_b x) \equiv \lim (fb - f(x1-dx))/dx \ \text{ as } dx \longrightarrow 0$$

is the **basic derivative of f at x1 from below**. (8)

end of definition

With definition 5 comes the idea of **direction**. One might have defined

$$\lim (fa - f(x1+dx))/dx \ \text{ as } dx \longrightarrow 0$$

or

$$\lim (f(x1-dx) - fb)/dx \ \text{ as } dx \longrightarrow 0$$

as derivatives in the **negative** direction. The derivatives of the definition are called derivatives in the **positive** direction.

Thus, in the real number system, there are four basic derivatives which may defined for *f* at x1.

1. lim (f(x1+dx)–fa)/dx   as dx⟶ 0, dx>0
2. lim (fa–f(x1+dx))/dx   as dx⟶ 0, dx>0
3. lim (fb–f(x1–dx))/dx   as dx⟶ 0, dx>0
4. lim (f(x1–dx)–fb)/dx   as dx⟶ 0, dx>0.

They are called respectively:

1. **forward derivative in the positive direction**
2. **forward derivative in the negative direction**
3. **backward derivative in the positive direction**
4. **backward derivative in the negative direction**.

In the real number system the direction may be made explicit by assigning:

positive direction as +1
negative direction as –1

and setting the increment as

+1 * dx, dx>0 as a positive increment,

and –1 * dx, dx>0 as a negative increment.

Then the derivatives may be rewritten in terms of direction as

1. lim (f(x1+(+1* dx)) - fa)/(+1*dx) as dx⟶ 0
2. lim (f(x1+(+1*dx)) - fa)/(–1*dx) as dx⟶ 0
3. lim (fb+f(x1 + (–1*dx)))/(+1* dx) as dx⟶ 0
4. lim (fb+f(x1 + (–1*dx)))/(–1*dx) as dx⟶ 0.

The symbolism pointing to these four distinct, but related, mathematical ideas must be capable of distinguishing between them.

Two conventions for symbolizing derivatives, one called **positive definite**, and a second called **basic**, are given in the following table.

type	limit	positive definite	basic
Forward, positive	(f(x1+dx) – fa)/dx	D[x1](f(x)$\vert$u;d_fx)	D[x1,x1+dx](f(x)$\vert$u;dx)
Forward, negative	(fa – f(x1+dx))/dx	–D[x1](f(x)$\vert$u;d_fx)	D[x1+dx,x1](f(x)$\vert$u;dx)
Backward, positive	(fb – f(x1–dx))/dx	D[x1](f(x)$\vert$u;d_bx)	D[x1–dx,x1](f(x)$\vert$u;dx)
Backward, negative	(f(x1–dx)– fb)/dx	–D[x1](f(x)$\vert$u;d_bx)	D[x1,x1–dx](f(x)$\vert$u;dx)

Positive Definite and Basic Conventions

The symbolism thus expresses which convention is employed.[8] The **basic** convention first fixes the direction and so distinguishes in the reference; the **positive definite** convention first fixes the increment and so indicates the direction with a sign[9].

The symbolism of the syntax defined in equation (3) is called a **generic symbol.** In contrast, the symbols and definitions of the table are called **specific**. Propositions using generic terms are proposed generically for application to specific instances.

The derivative, as a function, has all the attributes of functions in addition to a special relationship its parent function. It, too, may have a derivative, which in turn may have a derivative, etc. With reference to the parent function, the first derivative is called the **derivative of the first order**; the nth derivative **the derivative of the nth order**.

Now consider four possible attributes of both functions and their derivatives:

1. functional continuity at all points of the domain
2. functional continuity at all points of the domain except for a finite number of points
3. functional discontinuity on a denumerable set of of points of the domain
4. functional discontinuity on a non–denumerable set of points of the domain.

Each of these attributes may apply independently to a function or its derivative. While all sixteen combinations of functions and derivatives are properly the subject of Mathematics, for the purposes of Theoretical Physics only the first two[10] attributes will be considered.

Consequently, four categories of real functions and their derivatives are considered:

1. continuous functions with continuous derivatives;
2. continuous functions with continuous derivatives except at a finite number of points;
3. continuous functions except at a finite number of points with continuous derivatives;
4. continuous functions except at a finite number of points with continuous derivatives except at a finite number of points.

A basic function may have forward and backward basic derivatives. If either of these derivatives exist at the referenced point of interest, the function is called **differentiable** there. If the forward and backward derivatives exist and are equal, the function is said to have a **continuous** derivative at that point. If such is the case with all points, the function is said to have a **continuous derivative everywhere**.

Continuous functions need not be differentiable; differentiable functions need not have a continuous derivative.

While continuous, differentiable functions are prized in Mathematics, it is rather functions with abrupt changes which hold special interest in Theoretical Physics.

The derivative of a restricted function is the derivative of the function itself as restricted. The restriction is often taken as constant; however when the restriction itself varies a derivative may be defined:

$$D[x1](f(x)|g(x);dx) \equiv \lim (f(x1+dx)|g(x1+dx) - f(x1)|g(x1))/dx, \quad \text{as } dx \longrightarrow 0. \qquad (9)$$

The derivative of a compound function may be indexed by either the base variable or by the compound variable, and so two generic derivatives arise:

$$D[x1](f(g(x))|u;dx)$$

and

$$D[g(x1)](f(g)|u;dg).$$

Since f(g(x)) = f(x) the first reduces to D[x1](f(x)|u;dx).

The second, however, differs, but is related generically.[11] For a compound variable *g(x) with a continuous, non–zero derivative D[x(a)](g(x)|u;dx)*

$$D[x(a)](f(x)|u;dx) = D[g(a)](f(g)|u;dg) * D[x(a)](g(x)|u;dx), \quad (10)$$

a proposition called **the chain rule**.

For g(x) = c0 + c1*x, with c0 and c1 constant

$$\begin{aligned} D[x(a](f(x)|u;dx) &= D[g(a)](f(g)|u;dg) * D[x(a]((c0+c1*x)|u;dx) \\ &= c1 * D[g(a)](f(g)|u;dg). \end{aligned} \quad (11)$$

With c0=0 and c1=1, g=x, the **chain rule** reduces to equation (3). In this latter case, the variable g is called a **dummy variable** because

$$D[a](f(g)|u;dg) = D[a](f(x)|u;dx). \quad (12)$$

Whenever the restriction *u* is simply stated, *u* is meant to be held constant. If *u* is omitted, the differentiation is unconstrained.

Integration, like differentiation, is a limiting process which yields an **integral**. The symbol I is used for the process and by extension for the result. For a function *f* defined on a measurable set *M*, the generic syntax is

$$I[M](f(m)|u;dm) \quad (13)$$

mapping the function *f(m)* into the value *g(M)*, with *M* either a fixed set or a set of subsets. Analogous to differentiation, integration may be considered an operator mapping functions over a set of subsets, M, in the domain of *f* into other functions over a set of subsets, E, in the image of *f*, symbolized as:

$$I:f(M) \longrightarrow g(E) \quad (14)$$

where again a contrasting symbol is used to distinguish the domains.

Many specific kinds of integration are thus allowable, each classified according to an elementary kind of integration called a measure, i.e.

$$m(M) = I[M](1;dm),$$

with M restricted to the measurable sets of the function's domain.

The consequence, namely that if M1and M2 are disjoint, then

{m(M1∪M2) = m(M1)+m(M2)}

is often used in the definition of a measure.

A **directed** measure allows m(M) to take both positive and negative values; a **positive** measure allows m(M) to take on only non–negative values. Positive measures are useful when the results of the integration are to be interpreted non–negatively (such as the "area under a curve"); directed measures are useful when the results are to be interpreted in more than one direction (such as "balance").

For ordered sets the measurable set *M* is often specified to an interval. Intervals may be open or closed at either extremity and are symbolized as:

(x0,xm)	open lower, open upper
(x0,xm]	open lower, closed upper
[x0,xm)	closed lower, open upper
[x0,xm]	closed lower,closed upper.

For real functions defined over an interval *[x0,xn]* of a real variable, the syntax is written specifically as,

I[x0,xn](f(t)|u;dt) (15)

to denote the integral of the function *f* over the interval [x0,xn] in the function's domain with respect to the real variable *t* (called the **index**) constrained by the **restriction** *u*.

The real integral of a real function thus involves three distinguishable sets,

I:(f:R1⟶R2)⟶R3.

The integral is either a range value (for fixed interval [x0,xn]) or a function of a varying interval (denoted [x0,x]). In the latter case, I is an **operator** transforming the function *f:R⟶R* into another function, *z:R⟶R*

I:f(x)⟶z(x).

Mathematics

The integration of a real continuous function *f* over an interval [x0,xn] is defined as a limiting process

$$
\begin{aligned}
\mathrm{I}[x0,xn](f(t)|u;dt) \\
&\equiv \lim dx * (f(x0) \\
&\qquad + f(x0 + dx) \\
&\qquad + f(x0 + 2 * dx) \\
&\qquad + \ldots \\
&\qquad + f(xn)) \qquad \text{as } dx \rightarrow 0. \qquad (16)
\end{aligned}
$$

The integral as a function is written

$$g([x0,x]) = \mathrm{I}[x0,x](f(t)|u;dt) \qquad (17)$$

which is is often shortened, with accompanying ambiguity, to g(x).

Two different functions may possibly integrate over the same interval to the same value. Similarly, two different derivatives may also possibly integrate to the same value.

If *f1* and *f2* are two look–alike functions then

$$\mathrm{I}[x0,x](f1(t)|u;dt) = \mathrm{I}[x0,x](f2(t)|u;dt). \qquad (18)$$

However, since their derivatives may differ it does not follow that the integral of their derivatives are equal.

Conversely if *f1* and *f2* are functions with different values over [x0,xn] but with identical derivatives, their integrals may differ while the integrals of their derivatives are equal.

The categories of differentiation and integration involving possible look–alike functions are given in the following table with

f1 or f2 functions, possibly look-alike
Df symbolizing D[x](f(t)|u;dt)
If symbolizing I[x0,x](f(t)|u;dt)
IDf symbolizing I[x0,x](D[x](f(t)|u;dt)|u;dx).

Condition		Possible Result	
f	Df	I_f	I_{Df}
f1=f2	Df1=Df2	$I_{f1}=I_{f2}$	$I_{Df1}=I_{Df2}$
f1=f2	Df1≠Df2	$I_{f1}=I_{f2}$	$I_{Df1}=I_{Df2}$
f1=f2	Df1≠Df2	$I_{f1}=I_{f2}$	$I_{Df1}\neq I_{Df2}$
f1≠f2	Df1=Df2	$I_{f1}=I_{f2}$	$I_{Df1}=I_{Df2}$
f1≠f2	Df1≠Df2	$I_{f1}=I_{f2}$	$I_{Df1}=I_{Df2}$
f1≠f2	Df1≠Df2	$I_{f} =I_{f2}$	$I_{Df1}\neq I_{Df2}$
f1≠f2	Df1≠Df2	$I_{f1}\neq I_{f2}$	$I_{Df1}=I_{Df2}$
f1≠f2	Df1≠Df2	$I_{f1}\neq I_{f2}$	$I_{Df1}\neq I_{Df2}$

Categories of Integration

For functions with a finite number of finite discontinuities in the interval [x0,xn], the definition of equation (16) remains unambiguous.

Also, because adding a set of numbers in any order yields the same result, the notion of forward or backward integrals is not needed.

The idea of direction remains, however, and so also the basic and positive definite conventions. In the positive definite convention the integral in the negative direction is written,

$$\lim -dx*(f(xn) + f(xn - dx) + f(xn - 2*dx) + \ldots + f(x0)) \quad \text{as } dx\rightarrow 0$$
$$\equiv I[x0,xn](f(x)|u;d_b x) \qquad (19)$$
$$\equiv -I[x0,xn](f(x)|u;d_f x). \qquad (20)$$

In the basic convention the same limit is written as

$$\lim dx*(f(xn) + f(xn - dx) + f(xn - 2*dx) + \ldots + f(x0)) \quad \text{as } dx\rightarrow 0$$
$$\equiv I[xn,x0](f(x)|u;dx). \qquad (21)$$

Mathematics

The integral, as a function, has all the attributes of functions in addition to a special relationship its parent function. It too may have an integral, which in turn may have an integral, etc. With reference to the parent function, the first integral is called the **integral of the first order**; the nth integral **the integral of the nth order**.

Since integrals are summations, they participate in the algebra of their domains and ranges. For I:f=g and *c* a fixed element of R,

$$\mathrm{I}:c*f = c*g \tag{22}$$

and if I*:f1=g1* and if I*:f2=g2*, then

$$\mathrm{I}:(f1+f2) = g1+g2. \tag{23}$$

Moreover, similar to differentiation, integration may be defined to include restricted and compound functions.

The integral of a restricted function is the integral of the function itself as restricted. The restriction is most often taken as constant; however when the restriction itself varies an integral may be defined:

$$\begin{aligned} &\mathrm{I}[x0,xn](f(x)|u(x);dx) \\ &\qquad \equiv \lim\ (dx*(f(x0)|u(x0) \\ &\qquad\qquad + f(x0+dx)|u(x0+dx) \\ &\qquad\qquad + f(x0+2*dx)|u(x0+2*dx) \\ &\qquad\qquad + \ldots \\ &\qquad\qquad + f(xn)|u(xn))),\ \text{as } dx \longrightarrow 0. \end{aligned} \tag{24}$$

The integral of a compound function may be indexed by either the base variable or by the compound variable and so two generic integrals arise:

$$\mathrm{I}[x0,xn](f(g(x))|u;dx)$$

and

$$\mathrm{I}[g(x0),g(xn)](f(g)|u;dg).$$

Since f(g(x)) = f(x) the first reduces to I[x0,xn](f(x))|u;dx.

The second, however, differs but is related generically.[12] For a compound variable g(x) with a continuous non–zero derivative D[x(a)](g(x)|u;dx)

I[g(x0),g(xn)](f(g)|u;dg)
= I[x0,xn](f(x)∗D[x](g(x)|u;dx)|u;dx). (25)

I[x0,xn](f(x)|u;dx)
= I[g(x0),g(xn)](f(g)∗D[g](x(g)|u;dg)|u;dg). (26)

For g(x) = c0 + c1∗x, with c0 and c1 constant

I[g(x0),g(xn)](f(g)|u;dg)
= c1∗I[x0,xn](f(x)∗D[x](g(x)|u;dx)|u;dx). (27)

With c1=1,

I[g(x0),g(xn)](f(g)|u;dg)
=I[x0,xn](f(x)∗D[x](g(x)|u;dx)|u;dx), (28)

in which case the compound variable becomes a dummy variable.

If *x0* or *xn* is not expressed, the domain of integration is unbounded. For example, I[,xn] means that the integration is to proceed from negative infinity to *xn*. Whether or not this integral exists depends on the function, *f*, to which the integral operator is applied.

Whenever the condition *u* is simply stated, *u* is to be held constant. If *u* is omitted, it means that the integration is unconstrained.

Summation, considered as discrete integration, is symbolized analogously to integration, that is

S[m,n](f_t|u) (29)

denotes the summation of f_t from t=m to t=n subject to the condition u.

We start with two positive definite measures:

- that of H. Lebesgue for real functions (for the calculus)
- the positive discrete measure for integers (for summation).

Differentiation and integration will be broadened below beyond the real number system.

Mathematics

To complete the appropriation of mathematical concepts we finally invoke the concept of a three dimensional set of vectors, *V3*, over *R*.

Although *R*, its measurable subsets, and the extension to a three–dimensional set of vectors *V3,* give the mathematician a rich world of intellectual instruments, not one of these ideas conceptualizes the physical observation of objects extended, moving, or forcing.

Enabled by this symbology to communicate mathematical ideas, let us now collect[13] some mathematical results.

The Fundamental Theorem of Integral Calculus.

An inverse relationship between integration and differentiation can sometimes be captured in an analytical context. Suppose, f(x) continuous in [x1,x2] with a basic derivative in the interval.

Then

$$\text{I}[x1,x2](D[x](f(t);dt);dx) = f(x2) - f(x1), \quad x1 \le x2. \tag{30}$$

This result follow easily from the definitions.

$$\begin{aligned}
\text{I}[x1,x2](D[x](f(t);dt);dx) \\
&= \lim dx*(D[x1](f(t);dt) + D[x1+dx](f(t);dt) \\
&\qquad + D[x1+2dx](f(t);dt)+\cdots+D[x2](f(t);dt) \\
&= \lim dx*((f(x1 +dx) - f(x1)/dx) \\
&\qquad + (f(x1 +2*dx) - f(x1+dx)/dx) \\
&\qquad + \cdots \\
&\qquad + (f(x2 +dx) - f(x2)/dx) \\
&= \lim (dx/dx)*(fx1+dx) - f(x1) \\
&\qquad + f(x1 +2*dx) - f(x1+dx) \\
&\qquad + \cdots \\
&\qquad + (f(x2 +dx) - f(x2)) \\
&= \lim (-f(x1) +f(x2 +dx)) \text{ as } dx \longrightarrow 0 \\
&= f(x2) - f(x1).
\end{aligned}$$

In effect, the interior contributions to the summation cancel.

Equation 30 is called **the fundamental theorem of integral calculus**.

Mathematics

Although propositions to follow may often pertain to a wider class, consideration hereafter, in pursuit of an answer to the first problem of Theoretical Physics, will be centered on functions with basic derivatives.

Mathematics

Step Functions

The restrictions of continuity can be greatly eased with step functions. The set of bounded functions with a finite number of steps but otherwise continuous is called the set of **generalized functions**. Derivatives, perhaps unbounded, exist for generalized functions everywhere even at points of discontinuity.

With the goal of Theoretical Physics in mind, the functions considered hereafter will embrace generalized functions.

Consider[14] first the unit step functions defined as

$$\begin{aligned} u(x) &\equiv 1, \quad x>0 \\ &\equiv 0, \quad x\le 0 \end{aligned} \tag{31}$$

and

$$\begin{aligned} v(x) &\equiv 1, \quad x\ge 0 \\ &\equiv 0, \quad x<0. \end{aligned} \tag{32}$$

The unit step functions are constant everywhere with a discontinuity at *x=0*. The discontinuity can be translated to *x=xi* to form

$$\begin{aligned} u(x-xi) &= 1, \quad x-xi>0 \\ &= 0, \quad x-xi\le 0 \end{aligned} \tag{33}$$

$$\begin{aligned} v(x-xi) &= 1, \quad x-xi\ge 0 \\ &= 0, \quad x-xi<0. \end{aligned} \tag{34}$$

The two unit step functions are related topologically by

$$\lim\, (u(x+dx-xi)) = v(x-xi) \tag{35}$$
$$\lim\, (u(x-dx-xi)) = v(x-xi) \tag{36}$$
$$\lim\, (v(x+dx-xi)) = v(x-xi) \tag{37}$$
$$\lim\, (v(x-dx-xi)) = u(x-xi) \tag{38}$$
$$\text{as } dx \longrightarrow 0.$$

The derivatives of the unit step functions yield a unique functional called the impulse function, *i(t)*, defined as

$$
\begin{aligned}
D[x](u(t);d_f t) &\equiv \lim((u(x+dt) - u(x))/dt) \\
&= 0, \qquad x>0 \\
&= 0, \qquad x<0 \\
&= \lim\ (1-0)/dt, \qquad x=0. \qquad (39)
\end{aligned}
$$

$$
\begin{aligned}
D[x](v(t);d_b t) &\equiv \lim((v(x) - v(x-dt))/dt) \\
&= 0, \qquad x>0 \\
&= 0, \qquad x<0 \\
&= \lim\ (1-0)/dt, \qquad x=0. \qquad (40)
\end{aligned}
$$

Consequently,

$$
\begin{aligned}
D[x](u(t);d_f t) &= D[x](v(t);d_b t) \qquad (41) \\
&\equiv i(x). \qquad (42)
\end{aligned}
$$

However,

$$
\begin{aligned}
D[x](u(t-xi);d_b t) &= 0, \qquad \text{for all } x \\
D[x](v(t-xi);d_f t) &= 0, \qquad \text{for all } x. \qquad (43)
\end{aligned}
$$

The impulse function always arises from a specific step function and a specific direction.

For a generalized function *f*[15]

$$
\begin{aligned}
&D[x](u(t-xi)*f(t);d_f t) \\
&\quad = \lim\ (f(x+dx)*u(x+dx-xi) - f(x)*u(x-xi))/dx \\
&\quad = \lim\ (f(x+dx)*u(x+dx-xi) \\
&\qquad\quad -f(x+dx)*u(x-xi) \\
&\qquad\quad + f(x+dx)*u(x-xi) \\
&\qquad\quad -f(x)*u(x-xi))/dx \\
&\quad = \lim\ (f(x+dx)*D[x](u(t-xi);d_f t) \\
&\qquad\qquad + u(x-xi)*(f(x+dx)-f(x))/dx) \qquad (44) \\
&\quad = D[x](f(t);d_f t), \qquad x>xi \qquad (45) \\
&\quad = 0, \qquad x<xi. \qquad (46)
\end{aligned}
$$

At xi however

$$
\begin{aligned}
&\lim\ (f(xi+dx)*D[xi](u(t-xi);d_f t) + u(xi-xi)*(f(xi+dx)-f(xi))/dx \\
&\quad = f1*D[xi](u(t-xi);d_f t) \qquad (47)
\end{aligned}
$$

where[16] *f1* is the limit of *f* from above at *xi*.

Thus,

$$\begin{aligned} &D[x](f(t) \ast u(t-xi);d_f t) \\ &\qquad = 0, && x<xi; \\ &\qquad = D[x](f(t);d_f t), && x>xi; \\ &\qquad = f1 \ast D[xi](u(t-xi);d_f t), && x=xi. \end{aligned} \tag{48}$$

Furthermore,

$$\begin{aligned} &f(x) \ast D[x](u(t-xi);d_f t) \\ &\qquad = 0, && x<xi; \\ &\qquad = 0, && x>xi; \\ &\qquad = f(xi) \ast D[xi](u(t-xi);d_f t), && x=xi; \end{aligned} \tag{49}$$

and

$$\begin{aligned} &u(x-xi) \ast D[x](f(t);d_f x) \\ &\qquad = 0, && x<xi; \\ &\qquad = D[x](f(t);d_f x), && x>xi; \\ &\qquad = 0, && x=xi. \end{aligned} \tag{50}$$

Except at x=xi

$$\begin{aligned} &D[x](f(t) \ast u(t-xi);d_f t) \\ &\qquad = (f(x) \ast D[x]u(t-xi);d_f t) + u(x-xi) \ast D[x](f(t);d_f t), \qquad x \neq xi. \end{aligned} \tag{51}$$

Equation 51 is an example of the **differential product rule**.

At x=xi however, if f(xi) ≠ f1

$$D[x](f(t) \ast u(t-xi);d_f t)$$

does *not* obey the differential product rule.

If *f* is continuous at *xi*, then equation 51 holds even at xi.

For *f* continuous at *xi*, equation 51 is sufficiently written as

$$\begin{aligned} &D[x](f(t) \ast u(t-xi);d_f t) \\ &\qquad = f(x) \ast D[xi](u(t-xi);d_f t) + v(x-xi) \ast D[x](f(t);dt) \end{aligned} \tag{52}$$

since u(x–xi) = v(x–xi), except at xi.

Further,

$D[x](f(t)*u(t-xi);d_bt)$
$= 0, \quad x<xi;$
$= D[x](f(t);d_bt), \quad x>xi;$
$= 0, \quad x=xi;$
(53)

$(f(x)*D[x]u(t-xi);d_bt)$
$= 0, \quad x<xi;$
$= 0, \quad x>xi;$
$= 0, \quad x=xi;$
(54)

and

$u(x-xi)*D[x](f(t);d_bt)$
$= 0, \quad x<xi;$
$= D[x](f(t);d_bt), \quad x>xi;$
$= 0, \quad x=xi.$
(55)

Thus, the product rule holds for $D[x](f(t)*u(t-xi);d_bt)$.

Similarly,

$D[x](f(t)*v(t-xi);d_ft)$
$= 0, \quad x<xi;$
$= D[x](f(t);d_ft), \quad x>xi;$
$= D[xi](f(t);d_ft), \quad x=xi;$
(56)

$f(x)*D[x](v(t-xi);d_ft)$
$= 0, \quad x<xi;$
$= 0, \quad x>xi;$
$= 0, \quad x=xi;$
(57)

and

$v(x-xi)*D[x](f(t);d_ft)$
$= 0, \quad x<xi;$
$= D[x](f(t);d_ft), \quad x\geq xi.$
(58)

Thus, the differential product rule holds for $D[x](f(t)*v(t-xi);d_ft)$.

Also,

D[x](v(t–xi)∗f(t);d_bt)
    = 0, $x<xi$;
    = D[x](f(t);d_bt), $x>xi$;
    = D[xi](f(t);d_bt) + f0∗D[xi](v(t–xi);d_bt), $x=xi$;

(59)

where f0 is the lower limit of f at xi.

f(x)∗D[x](v(t–xi);d_bt)
    = 0, $x<xi$;
    = 0, $x>xi$;
    = f(xi)∗D[xi](v(t–xi);d_bt), $x=xi$;

(60)

and

v(x–xi)∗D[x](f(t);d_bt)
    = 0, $x<xi$;
    = D[x](f(t);d_bt), $x>xi$;
    = D[x](f(t);d_bt), $x=xi$.

(61)

Thus, the differential product rule holds for D[x](f(t)∗v(t–xi);d_bt) everywhere except at x=xi, and even there when *f* is continuous from below at xi.

The previous results are summarized in the following table.

Derivative	x<xi	x=xi	x>xi
D[x](u(t−xi)∗f(t);d_ft)	0	f1∗D[xi](u(t−xi);d_ft)	D[x](f(t);d_ft)
f(x)∗D[x]u(t−xi);d_ft)	0	f(xi)∗D[xi](u(t−xi);d_ft)	0
u(x−xi)∗D[x](f(t);d_fx)	0	0	D[x](f(t);d_fx)
D[x](u(t−xi)∗f(t);d_bt)	0	0	D[x](f(t);d_bt)
f(x)∗D[x]u(t−xi);d_bt)	0	0	0
u(x−xi)∗D[x](f(t);d_bx)	0	0	D[x](f(t);d_bx)
D[x](v(t−xi)∗f(t);d_ft)	0	f0∗D[xi](v(t−xi);d_bt)	D[x](f(t);d_ft)
f(x)∗D[x]v(t−xi);d_ft)	0	0	0
v(x−xi)∗D[x](f(t);d_fx)	0	f(xi)∗D[xi](v(t−xi);d_bt)	D[x](f(t);d_fx)
D[x](v(t−xi)∗f(t);d_bt)	0	f(xi)/dt	D[x](f(t);d_bt)
f(x)∗D[x]v(t−xi);d_bt)	0	lim f(xi)/dx	0
v(x−xi)∗D[x](f(t);d_fx)	0	0	D[x](f(t);d_bx)

Derivatives involving Step Functions

Consider now for *f* a generalized function[17],

I[](f(x)∗D[x](u(t−xi);d_ft);dx)

= lim dx∗(...

+ f(xi−dx)∗(u(xi) − u(xi−dt))/dt

+ f(xi)∗(u(xi+dt) − u(xi))/dt

+ f(xi+dx)∗(u(xi+dx+dt) − u(xi+dx))/dt

+ ...), dx=dt>0

= f(xi) (62)

≡I[](f(x)∗i(x);dx) (63)

while

I[](f(x)∗D[x](u(t−xi);d_bt);dx) = 0. (64)

Mathematics

Likewise,

$$I[](f(x)*D[x](v(t-xi);d_f t);dx) = 0 \quad (65)$$

$$I[](f(x)*D[x](v(t-xi);d_b t);dx) = f(xi). \quad (66)$$

The ability of the impulse function to select one specific value in a function's image is equivalent to invoking the **axiom of choice**.

Further, for *f*

with a basic derivative
continuous[18]
bounded, f(x)<∞ for any x
asymptotic, f(x)⟶0 as x⟶∞

$$\begin{aligned}
&I[](u(x-xi)*D[x](f(t);d_f t);dx) \\
&\quad = \lim \lim dx*(\ldots \\
&\qquad + (f(xi-dx+dt) - f(xi-dx))*u(xi-xi-dx) \\
&\qquad + (f(xi+dt) - f(xi))*u(xi-xi) \\
&\qquad + (f(xi+dx+dt) - f(xi+dx))*u(xi-xi+dx) \\
&\qquad + (f(xi+2*dx+dt) - f(xi+2*dx))*u(xi-xi+2*dx) \\
&\qquad + \ldots)/dt \\
&\quad = \lim \lim (-f(xi+dx)*u(dx) \\
&\qquad\quad + f(xi+dx+dt)*u(dx) \\
&\qquad\quad -f(xi+2*dx)*u(2*dx) \\
&\qquad\quad + \ldots) \\
&\quad = \lim (-f(xi+dx)) \qquad \text{as } dx \longrightarrow 0 \\
&\quad = -f(xi). \qquad (67)
\end{aligned}$$

Again

$$\begin{aligned}
&I[](u(x-xi)*D[x](f(t);d_b t);dx) \\
&\quad = \lim \lim dx*(\ldots \\
&\qquad + (f(xi+dx)-f(xi+dx-dt))*u(xi-xi+dx) \\
&\qquad + (f(xi+2*dx) -f(xi+2*dx-dt))*u(xi-xi+2*dx) \\
&\qquad + \ldots)/dt \\
&\quad = \lim (-f(xi+dx-dt)) \\
&\quad = -f(x1). \qquad (68)
\end{aligned}$$

Likewise

$$I[](v(x-xi)*D[x](f(t);d_b t);dx) = -f(x1) \quad (69)$$

$$I[](v(x-xi)*D[x](f(t);d_f t);dx) = -f(x1). \quad (70)$$

Further, for f
continuous
bounded f(x)<∞for any x
asymptotic f(x)⟶0 as x⟶∞

I[](D[x](f(t)*u(t–xi);d_ft);dx)
= lim dx*(...
+ f(xi–dx+dt)*u(xi–dx+dt)
– f(xi–dx)*u(xi–dx)
+ f(xi+dt)*u(xi+dt)
– f(xi)*u(xi)
+ f(xi+dx+dt)*u(xi+dx+dt)
– f(xi+dx)*u(xi+dx)
+ f(xi+2*dx+dt)*u(xi+2*dx+dt)
– f(xi+dx+dt)*u(xi+dx+dt)
+ ...)/dt
= lim dx*(...
+ f(xi+dt)*u(xi+dt)
– f(xi+dx)*u(xi+dx)
+ f(xi+dx+dt)*u(xi+dx+dt)
– f(xi+dx+dt)*u(xi+dx+dt)
+ f(xi+2*dx+dt)*u(xi+2*dx+dt)
+ ...)/dt
= lim dx*(f(xi+n*dx+dt)*u(xi+n*dx+dt))/dt as n⟶∞
= 0 (71)

since *f* is asymptotic.

Also,

I[](D[x](f(t)*u(t–xi);d_bt);dx)= 0. (72)

Similarly,

I[](D[x](f(t)*v(t–xi);d_ft);dx)= 0 (73)
I[](D[x](f(t)*v(t–xi);d_bt);dx)= 0. (74)

Mathematics

Consequently for *f*

continuous at $x = xi$

whose integral over the unbounded domain is bounded,

$$I[](D[x](f(t)*u(t-xi);d_{f}t);dx)$$
$$= I[](f(x)*D[x](u(t-xi);d_{f}t);dx)$$
$$+ I[](u(x-xi)*D[x](f(t);dt);dx) \qquad (75)$$

and

$$I[](D[x](f(t)*v(t-xi);d_{b}t);dx)$$
$$= I[](f(x)*D[x](v(t-xi);d_{b}t);dx)$$
$$+ I[](v(x-xi)*D[x](f(t);dt);dx). \qquad (76)$$

The latter two equations are examples of a mutual attribute between two functions called the **integral product rule**.[19]

Since for continuous functions with bounded integrals,

$$I[](D[x](f(t)*u(t-xi);d_{f}t);dx) = 0,$$

it follows for these functions

$$I[](f(x)*D[t](u(t-xi);d_{f}t);dx) = -\ I[](D[x](f(t);dt)*u(x-xi);dx) \qquad (77)$$

and similarly,

$$I[](f(x)*D[t](v(t-xi);d_{b}t);dx) = -\ I[](D[x](f(t);dt)*v(x-xi);dx). \qquad (78)$$

The following table summarizes the above results.

Integral	**Value**	**Conditions on *f***
I[](f(x)∗D[x](u(t–xi);d_ft);dx)	f(xi)	generalized
I[](f(x)∗D[x](u(t–xi);d_bt);dx)	0	generalized
I[](f(x)∗D[x](v(t–xi);d_ft);dx)	0	generalized
I[](f(x)∗D[x](v(t–xi);d_bt);dx)	f(xi)	generalized
I[](u(x–xi)∗D[x](f(t);d_ft);dx)	–f(xi)	continuous, bounded, asymptotic
I[](u(x–xi)∗D[x](f(t);d_bt);dx)	–f(xi)	continuous, bounded, asymptotic
I[](v(x–xi)∗D[x](f(t);d_ft);dx)	–f(xi)	continuous, bounded, asymptotic
I[](v(x–xi)∗D[x](f(t);d_bt);dx)	–f(xi)	continuous, bounded, asymptotic
I[](D[x](f(t)∗u(t–xi);d_ft);dx)	0	continuous, bounded, asymptotic
I[](D[x](f(t)∗u(t–xi);d_bt);dx)	0	continuous, bounded, asymptotic
I[](D[x](f(t)∗v(t–xi);d_ft);dx)	0	continuous, bounded, asymptotic
I[](D[x](f(t)∗v(t–xi);d_bt);dx)	0	continuous, bounded, asymptotic

Integrals Involving Step Functions

To see the utility of the impulse function, consider for f continuous at x,

f(x) = I[](f(t)∗i(t–x);dt) = I[](f(t+x)∗i(t);dt) (79)

where t is a dummy variable.

Equation 79 may be interpreted to mean that the continuous parts of *f* can be decomposed into weighted impulse functions.

Now let

f(x) = f1(x) + c*D[x](u(t–x1);d_ft)

where f1(x) is continuous in the interval [x1,xn] containing x1 and c is constant.

Then

I[x0,xn](f(t);dt) = I[x0,xn](f1(t);dt) + c. (80)

Although the two functions *f* and *f1* look–alike everywhere except at *x1* the value of their integrals need not correspond at all.

For *c* a non–zero constant,

u(c*(x–xi)) = u(x–xi), c>0 (81)
u(c*(x–xi)) = 1–v(x–xi), c<0 (82)
v(c*(x–xi)) = v(x–xi), c>0 (83)
v(c*(x–xi)) = 1–u(x–xi), c<0 (84)

which are instances of different functions with the same functional values.

The pairs, u(x–xi) and 1–v(x–xi), and v(x–xi) and 1–u(x–xi), are called **mirror functions**.

Now for c>0
D[x](u(c*(t–xi));d_ft)
= lim (u(c*(t+dt–xi))–u(c*(t–xi)))/dt
= lim (u(c*t+c*dt–c*xi)–u(c*t+–c*xi))/dt
=abs(c)*lim (u(c*t+c*dt–c*xi)–u(c*t–c*xi))
/(abs(c)*dt)
=abs(c)*lim (u(t+dt–xi)–u(t–xi))/dt
=abs(c)*D[x](u((t–xi));d_ft) (85)

while for c<0
D[x](u(c*(t–xi));d_ft)
= lim (u(c*(t+dt–xi))–u(c*(t–xi)))/dt
= lim (1–v(c*t+c*dt –c*xi)–1+v(c*t–c*xi))/dt
= lim (v(c*t–c*xi)–v(c*t+c*dt–c*xi))/dt
= abs(c)*lim (v(c*t–c*xi)–v(c*t+c*dt–c*xi))
/(abs(c)*dt)
= abs(c)*lim ((v(t–xi)–v(t–dt–xi)))/dt
= abs(c)*D[x](v((t–xi));d_bt). (86)

Similarly,

D[x](v(c∗(t–xi));d_bt) = abs(c)∗D[x](v((t–xi));d_bt), c>0 (87)
D[x](v(c∗(t–xi));d_bt) = abs(c)∗D[x](u((t–xi));d_ft), c<0 (88)

and specifically for c=–1

D[x](u(–(t–xi));d_ft) = D[x](v((t–xi));d_bt), (89)
D[x](v(–(t–xi));d_bt) = D[x](u((t–xi));d_ft). (90)

Consider now the integral of i(x–xi)
I[,x](D[x](u(t–xi);d_ft);dx) = 0, x<xi
= 1, x≥xi
= v(x–xi). (91)

Likewise,
I[,x](D[x](v(t–xi);d_bt);dx) = 0, x<xi
= 1, x≥xi
= v(x–xi). (92)

Consider next,

D[x](D[t](u(z–xi);d_fz);d_ft)
= lim lim (u(x–xi+dt+dx)–u(x–xi+dt)
–u(x–xi+dx)+u(x–xi))/(dt∗dx) as dx>dt>0
= 0, x≠xi
= –∞, x=xi. (93)
D[x](D[t](u(z–xi);d_fz);d_bt)
= lim lim (u(x–xi+dt) – u(x–xi+dt–dx)
– u(x–xi) + u(x–xi–dx))/(dt∗dx) as dx>dt>0
= 0. (94)
D[x](D[t](v(z–xi);d_bz);d_ft)
= lim lim (v(x–xi+dx)–v(x–xi–dt+dx)
–v(x–xi)+v(x–xi–dt))/(dt∗dx)
= 0. (95)
D[x](D[t](v(z–xi);d_bz);d_bt)
= lim lim (v(x–xi)–v(x–xi–dx)
– v(x–xi–dt)+v(x–xi–dt–dx))/(dt∗dx)
= 0, x ≠xi
= ∞, x=xi. (96)

Mathematics

Somewhat ambiguously, the derivatives of i(t) are called **the doublet**.

The action of the doublet is seen for generalized functions by

I[](f(x)*D[x](D[y](u(z–xi);d_fz);d_fy);dx)
    = lim dx*(...
        + f(xi–dx)*(u(dz) –2*u(0) + u(–dx))
        + f(xi)*(u(2*dz) – 2*u(dx) + u(0))
        + f(xi+dz)* (u(3*dx) – 2*u(2*dz) + u(dx))
        + ...)/(dz*dy)
    = lim (–f(xi) + f(xi–dx))/dy
    = –D[xi](f(x);d_bx). (97)

Likewise

I[](f(x)*D[x](D[y](v(z–xi);d_bz);d_by);dx)= –D[xi](f(x);d_fx). (98)

In contrast

I[](D[x](f(t);d_ft)*D[x](u(t–xi);d_ft);dx) = D[xi](f(t);d_ft)
I[](D[x](f(t);d_bt)*D[x](u(t–xi);d_ft);dx) = D[xi](f(t);d_bt)
I[](D[x](f(t);d_ft)*D[x](v(t–xi);d_bt);dx) = D[xi](f(t);d_ft)
I[](D[x](f(t);d_bt)*D[x](v(t–xi);d_bt);dx) = D[xi](f(t);d_bt).

Consequently,

I[](D[x](f(t);d_ft)*D[x](u(t–xi);d_ft);dx)
    = –I[](f(x)*D[x](D[y](v(z–xi);d_bz);d_by);dx)
    = I[](D[x](f(t);d_ft)*D[x](v(t–xi);d_bt);dx)
    = D[xi](f(t);d_ft) (99)
I[](D[x](f(t);d_bt)*D[x](u(t–xi);d_ft);dx)
    = –I[](f(x)*D[x](D[y](u(z–xi);d_fz);d_fy);dx)
    = I[](D[x](f(t);d_bt)*D[x](v(t–xi);d_bt);dx)
    = D[xi](f(t);d_bt). (100)

The restriction of a bounded integral can be somewhat alleviated since[20] for *f* with integral bounded merely in the interval [x0,x2] with $x0<xi<x2$

$$\mathrm{I}[x0,x2](f(t);dt) = \mathrm{I}[]((v(t-x0)-u(t-x2))*f(t);dt) \tag{101}$$

Thus,

$$\mathrm{I}[x0,x2](f(x)*D[x](u(t-xi);d_f t);dx) = \mathrm{I}[]((v(x-x0) - u(x-x2))*f(x)*D[x](u(t-xi);d_f t);dx) \tag{102}$$

$$\mathrm{I}[x0,x2](f(x)*D[x](u(t-xi);d_b t);dx) = \mathrm{I}[]((v(x-x0) - u(x-x2))*f(x)*D[x](u(t-xi);d_b t);dx) \tag{103}$$

$$\mathrm{I}[x0,x2](f(x)*D[x](v(t-xi);d_f t);dx) = \mathrm{I}[]((v(x-x0) - u(x-x2))*f(x)*D[x](v(t-xi);d_f);dx) \tag{104}$$

$$\mathrm{I}[x0,x2](f(x)*D[x](v(t-xi);d_b t);dx) = \mathrm{I}[]((v(x-x0) - u(x-x2))*f(x)*D[x](v(t-xi);d_b t);dx). \tag{105}$$

Example: For constant *c*,

$$\begin{aligned}\mathrm{I}[](D[x](c*u(t);d_f t);dx) &= \mathrm{I}[](c*D[x](u(t);d_f);dx)\\ &\quad+\mathrm{I}[](D[x](c;dt)*u(x);dx)\\ &= c, \text{ not 0 as in the table above.}\end{aligned}$$

Further,

$$D[x](\mathrm{I}[](c*u(t);dt);d_f x) \text{ is indeterminate.}$$

But then $\mathrm{I}[](c;dx)$ is unbounded.

Example. Let f(x) = u(x−5) + u(x−10) − 2*u(x−20), where u is the step function defined above. Then, after bounding the interval from equation 101 on page 49,

$$\begin{aligned}\mathrm{I}[](f;dx) &= -\mathrm{I}[](x*(i(x-5)+i(x-10)-2*i(x-20));dx)\\ &= -5-10+2*20\\ &= 25.\end{aligned}$$

More generally, if

f1(x) is continuous in [x0,x2];
g(x), the anti–derivative of f, that is, D[x](g(t);dt) = f(x);
x0<xi<x2;
d constant;

then the function

f(x) = f1(x) + d*u(x–xi) (106)

is integrated as

I[x0,x2]((f1(x) + d*u(x–xi));dx)
= I[](f1(x)*(v(x–x0)–u(x–x2))
+ d*(u(x–xi)–u(x–x2));dx)
= I[](g(x)*(–D[x](v(t–x0);d_bt)
+ D[x](u(t–x2);d_ft))
+ d*(–D[x](u(t–xi);d_ft))
+ D[x](u(t–x2);d_ft))
= g(x2) – g(x0) + d*(x2–xi). (107)

In this way the fundamental theorem of integral calculus is extended to functions with a finite number of bounded discontinuities.

Example. Let f(g) = u(g–5); g(x) = 3*x. Then
f(x) = u(3*x–5)
D[xi](f;d_fx) = 3*D[xi](u(3*x–5);d_fx)
= D[xi](u(x–5/3);d_fx)
D[g(xi)](f;d_fg) = D[g(xi)](u(g–5);d_fg);
D[xi](g;dx) = 3,
so that
D[g(xi)](f;d_fg)*D[xi](g;dx) = D[g(xi)](u(g–5);d_fg)*3
= D[xi](f;d_fx).

In contrast let f(g) = 3*g; g(x) = u(x–5). Then
f(x) = 3*u(x–5)
D[xi](f;d_fx) = 3*D[xi](u(x–5);d_fx)
D[(xi)](f;d_fg) = 3;
D[xi](g;d_fx) = D[xi](u(x–5);d_fx)
so that
D[g(xi)](f;d_fg)*D[xi](g;d_fx) = 3*D[xi](u(x–5);d_fx).

The set of vectors, V3

We observe physical reality as extended and mutable. To advance the process of transforming mathematical ideas into those of Theoretical Physics, particularly those associated with extension, consider next the three dimensional set of vectors, *V3*.

> Reflection. God has revealed himself as a triple unity. The special suitability of three dimensions to describe his created glory is a profound act of natural adoration by that creation of its creator.

Each vector in V3 is characterized by a length and a direction; the entire set can be endowed with both an algebra and a topology.

Vector Algebra

The mathematical set of real vectors[21], *V3*, presupposes (overlies) the field of real numbers, *R*, used to describe the magnitude and direction for each vector. The algebra of *V3* enables combinations of the vectors.

Algebraic Operations

V3 is endowed with two algebraic operations, **addition** and **multiplication**.

The **sum** of any two vectors is defined as a vector whose direction and length are those of the major diagonal of the parallelogram formed from the magnitudes and directions of the two vectors; their **difference** is a vector which takes its direction and magnitude from the minor diagonal of the parallelogram in a manner consistent with vectorial addition.

For any three vectors, **r1**, **r2**, and **r3** and any real number, *c*, the following statements are taken as true axiomatically:

r1 + r2 is also a vector
there is a **0** vector such that **r1 + 0** = **r1**
c∗**r1** is also a vector
$$\mathbf{r1} + \mathbf{r2} = \mathbf{r2} + \mathbf{r1} \quad (108)$$
$$(\mathbf{r1} + \mathbf{r2}) + \mathbf{r3} = \mathbf{r1} + (\mathbf{r2} + \mathbf{r3}) \quad (109)$$
$$c*(\mathbf{r1} + \mathbf{r2}) = (\mathbf{r1} + \mathbf{r2})*c = c*\mathbf{r1} + c*\mathbf{r2}. \quad (110)$$

The vector **0** is called the **origin** of V3. The origin of the set of vectors has special relevance to the first problem of Theoretical Physics because of its potential application in referencing the process of observation.

The operation symbolized by the multiplication of a real number with a vector, c∗**r1**, is called **scalar multiplication**.

While arithmetical notation is used for addition and multiplication in *V3*, the reader should note that the use of identical symbols for different operations (vector addition and arithmetical addition) is fundamentally confusing. Some justification for the use of arithmetical symbols in a vectorial context is given below. Their correct interpretation, however, requires attention to context. As help, vectorial operations are usually high–lighted with bold–face type.

Two vectors **r1** and **r2** are said to be **parallel** if there is a real constant, c, such that

$$\mathbf{r1} = c*\mathbf{r2}. \qquad (111)$$

If c>0, **r1** and **r2** are called **positively parallel**; If c<0, **r1** and **r2** are called **negatively parallel**.

The vector **0** is thus parallel to all vectors.

Two elementary functions are associated with each vector: its **length** or **magnitude** and its **direction**.

The **length** of vector is a function from the set of vectors to the underlying field of real numbers and is denoted as abs(**r**).

The following properties pertain to abs(**r**).

$$\begin{aligned} \text{abs}(\mathbf{r}) &\geq 0 \\ \text{abs}(c*\mathbf{r}) &= \text{abs}(c)*\text{abs}(\mathbf{r}) \\ \text{abs}(\mathbf{r1}+\mathbf{r2}) &\leq \text{abs}(\mathbf{r1}) + \text{abs}(\mathbf{r2}) \end{aligned}$$

for any real c where abs(c) denotes its absolute value. Abs(**r**) is defined specifically for V3 below in equation (127).

The **direction** of **r**, symbolized as **u**(**r**) or often merely as **ur**, is itself a vector of unit length parallel to **r**. The relationship

$$\mathbf{r} \equiv \text{abs}(\mathbf{r})*\mathbf{u}(\mathbf{r}) \qquad (112)$$

defines **u**(**r**).

Since the length of a direction is

$$\text{abs}(\mathbf{u}(\mathbf{r})) = \text{abs}(\mathbf{r}/\text{abs}(\mathbf{r})) = \text{abs}(\mathbf{r})/\text{abs}(\mathbf{r}) = 1 \qquad (113)$$

directions are sometimes called **unit vectors**.

Scalar multiplication merely changes the vector's length, not its direction. Because scalar multiplication is analogous to arithmetic multiplication, a common symbol is used for both distinctly different operations. For example, the equation

$$c*\mathbf{r} = c*\text{abs}(\mathbf{r})*\mathbf{u}(\mathbf{r})$$

embodies the two different kinds of multiplication. The first and third apply to the set of vectors; the second to real numbers.

Length and direction, referenced to a specific vector, decompose the vector as,

$$\text{abs}(\mathbf{r}) = \text{abs}(\text{abs}(\mathbf{r})*\mathbf{ur}) = \text{abs}(\mathbf{r}) \qquad (114)$$

and

$$\mathbf{u}(\mathbf{r}) = \mathbf{u}(\text{abs}(\mathbf{r})*\mathbf{u}(\mathbf{r})) = \mathbf{u}(\mathbf{r}). \qquad (115)$$

The length of a vector is often abbreviated to $r \equiv \text{abs}(\mathbf{r})$.

There is a certain ambiguity in the decomposition

$$\mathbf{r} = r*\mathbf{u}(\mathbf{r})$$

which may also be written

$$\mathbf{r} = -r*\mathbf{u}(-\mathbf{r}).$$

The decomposition of $-\mathbf{r}$ is then

$$-\mathbf{r} = r*\mathbf{u}(-\mathbf{r})$$

or

$$-\mathbf{r} = -(r*\mathbf{u}(\mathbf{r})) = -r*\mathbf{u}(\mathbf{r}).$$

The designation fixing $r>0$ is called the **positive definite convention** and consequently **u**(**r**) is always the direction of **r**. In the positive definite convention *r* is called the vector's **length**.

The designation fixing the direction **u**(**r**) first is called the **basic convention**. In the basic convention the sign of *r*, which may be positive or negative, indicates the direction of ***r*** relative to the referenced direction **u**(**r**). In the basic convention *r* is called the vector's **magnitude**.

The positive definite convention is most useful when a discussion emphasizes distances. The basic convention is most useful for discussions emphasizing direction.

In some circumstances the convention for the decomposition of the vector may need to be made explicit.

Direction is symbolized more generally by

$$\mathbf{u}(\mathbf{r}-\mathbf{r0})$$

as the direction from **r0** to **r**.

Two vectors may also be multiplied.

Three such vectorial multiplications are possible depending on their product. The product may reside either in:

the underlying field of real numbers, or
the set of vectors, or
the set of vectorial transformations.

The first of these multiplications, symbolized by " • ", is called the **inner product** and is defined for any two vectors, **r1** and **r2**, as

$$\mathbf{r1}\bullet\mathbf{r2} \equiv \text{abs}(\mathbf{r1})*\text{abs}(\mathbf{r2})*\cos(a) \tag{116}$$

where *a* is the angle formed between **r1** and **r2**. Thus,

$$\cos(a) = \mathbf{ur1}\bullet\mathbf{ur2} \tag{117}$$

and

$$\mathbf{r1}\bullet\mathbf{r2} \equiv \text{abs}(\mathbf{r1})*\text{abs}(\mathbf{r2})*\mathbf{ur1}\bullet\mathbf{ur2}. \tag{118}$$

The inequality

$$\mathbf{r1}\bullet\mathbf{r2}*\mathbf{r1}\bullet\mathbf{r2} \leq (\text{r1}*\text{r1})*(\text{r2}*\text{r2}) \tag{119}$$

is often useful.

The value of the inner product varies from **–r1•r2** to **r1•r2** as a function of the angle *a* and can thus be used as a measure of co–directionality. If the value of **r1•r2** = 0, the two vectors are said to be **orthogonal** (a=pi/2).

With this definition, **0** is orthogonal to all other vectors.

The second of these vectorial multiplications is called the **cross product**, whose operator is symbolized by "$\wedge$." The cross product is semi–defined as

$$\mathbf{r1}\wedge\mathbf{r2} \equiv \mathrm{abs}(\mathbf{r1})*\mathrm{abs}(\mathbf{r2})*\sin(a)*\mathbf{un} \quad (120)$$

$$\equiv \mathrm{abs}(\mathbf{r1})*\mathrm{abs}(\mathbf{r2})*\mathbf{ur1}\wedge\mathbf{ur2} \quad (121)$$

where *a*, as before, is the angle formed between **r1** and **r2** and ***un*** is a unit vector orthogonal to both **r1** and **r2**. There are two such orthogonal vectors, ***un*** and −***un***, so one is assigned to $\mathbf{r1}\wedge\mathbf{r2}$ and the other to $\mathbf{r2}\wedge\mathbf{r1}$ in a consistent way. The length of the result also varies from $-\mathbf{r1}\bullet\mathbf{r2}$ to $\mathbf{r1}\bullet\mathbf{r2}$ as a function of *a* and can be used as a measure of orthogonality. The final definition of the cross product is given in equation (146).

For three vectors **r1**, **r2**, **r3**, the real product,

$$(\mathbf{r1}\wedge\mathbf{r2})\bullet\mathbf{r3} = \mathbf{r1}\bullet(\mathbf{r2}\wedge\mathbf{r3}) = \mathbf{r3}\bullet(\mathbf{r1}\wedge\mathbf{r2}) \quad (122)$$

is the "volume" of the parallelepiped defined by the three vectors. The product is called the **triple product**. Note the triple product need not be positive.

The third of these multiplications is called the **outer product**, whose operation is symbolized by " $*$ " and is defined from

$$[\mathbf{r1}*\mathbf{r2}] \equiv \mathrm{abs}(\mathbf{r1})*\mathrm{abs}(\mathbf{r2})*[\mathbf{ur1}*\mathbf{ur2}] \quad (123)$$

where $[\mathbf{ur1}*\mathbf{ur2}]$ is a transformation such that for any direction **ur0**

$$\mathbf{ur0}\bullet[\mathbf{ur1}*\mathbf{ur2}] \equiv (\mathbf{ur0}\bullet\mathbf{ur1})*\mathbf{ur2} \quad (124)$$

that is, $[\mathbf{r1}*\mathbf{r2}]$ transforms **r0** into a vector parallel to **r2**. Note that $[\mathbf{r1}*\mathbf{r2}]$ differs from $[\mathbf{r2}*\mathbf{r1}]$.

The operation of the cross product may also be regarded as a transformation. To emphasize this transformation we write

$$\mathbf{r1}\wedge\mathbf{r2} \equiv \mathbf{r1}\bullet\mathbf{C}(\mathbf{r2}). \quad (125)$$

The explicit form of the transformation **C** is given in equation (233).

Mathematics

With *a*, *b* scalars and **r1**, **r2**, **r3**, **r4** vectors, the following equations hold for these multiplications.

$$\mathbf{r1}\bullet\mathbf{r2} = \mathbf{r2}\bullet\mathbf{r1}$$
$$b*\mathbf{r1}\bullet c*\mathbf{r2} = b*c*(\mathbf{r1}\bullet\mathbf{r2})$$
$$\mathrm{abs}(\mathbf{r1}\bullet\mathbf{r2}) \le \mathrm{abs}(\mathbf{r1}) * \mathrm{abs}(\mathbf{r2})$$

$$\mathbf{r1}\wedge\mathbf{r2} = -(\mathbf{r2}\wedge\mathbf{r1})$$
$$= ((-\mathbf{r2})\wedge\mathbf{r1})$$
$$= (\mathbf{r2}\wedge(-\mathbf{r1}))$$
$$\mathbf{r1}\wedge\mathbf{r1} = \mathbf{0}$$
$$\mathbf{r1}\bullet(\mathbf{r1}\wedge\mathbf{r2}) = \mathbf{r2}\bullet(\mathbf{r1}\wedge\mathbf{r2}) = 0$$
$$(b*\mathbf{r1})\wedge(c*\mathbf{r2}) = b*c*(\mathbf{r1}\wedge\mathbf{r2})$$
$$\mathrm{abs}(\mathbf{r1}\wedge\mathbf{r2}) \le \mathrm{abs}(\mathbf{r1}) * \mathrm{abs}(\mathbf{r2})$$
$$(\mathrm{abs}(\mathbf{r1})*\mathrm{abs}(\mathbf{r2}))^2 = (\mathrm{abs}(\mathbf{r1}\wedge\mathbf{r2}))^2 + (\mathrm{abs}(\mathbf{r1}\bullet\mathbf{r2}))^2$$

$$\mathbf{r1}*\mathbf{r2} = \mathbf{T}[\mathbf{r2}*\mathbf{r1}]$$
$$(b*\mathbf{r1})*(c*\mathbf{r2}) = b*c*(\mathbf{r1}*\mathbf{r2})$$
$$\mathrm{abs}(\mathbf{r1}\bullet(\mathbf{r2}*\mathbf{r3})) = \mathrm{abs}(\mathbf{r1}\bullet\mathbf{r2})*\mathrm{abs}(\mathbf{r3})$$

$$\mathbf{r1}\bullet(\mathbf{r2}+\mathbf{r3}) = \mathbf{r1}\bullet\mathbf{r2} + \mathbf{r1}\bullet\mathbf{r3}$$
$$\mathbf{r1}\wedge(\mathbf{r2}+\mathbf{r3}) = \mathbf{r1}\wedge\mathbf{r2} + \mathbf{r1}\wedge\mathbf{r3}$$
$$\mathbf{r1}*(\mathbf{r2}+\mathbf{r3}) = \mathbf{r1}*\mathbf{r2} + \mathbf{r1}*\mathbf{r3}$$

$$\mathbf{r1}\bullet(\mathbf{r2}\wedge\mathbf{r3}) = \mathbf{r2}\bullet(\mathbf{r3}\wedge\mathbf{r1})$$
$$= \mathbf{r3}\bullet(\mathbf{r1}\wedge\mathbf{r2})$$
$$= (\mathbf{r1}\wedge\mathbf{r2})\bullet\mathbf{r3}$$
$$= (\mathbf{r2}\wedge\mathbf{r3})\bullet\mathbf{r1}$$
$$= (\mathbf{r3}\wedge\mathbf{r1})\bullet\mathbf{r2}$$
$$(\mathbf{r1}\bullet\mathbf{r2})*\mathbf{r3} = \mathbf{r1}\bullet(\mathbf{r2}*\mathbf{r3})$$
$$= (\mathbf{r2}\bullet\mathbf{r1}*\mathbf{r3}$$
$$= \mathbf{r2}\bullet(\mathbf{r1}*\mathbf{r3})$$

$$\mathbf{r1}\wedge(\mathbf{r2}\wedge\mathbf{r3}) = (\mathbf{r1}\bullet\mathbf{r3})*\mathbf{r2} - (\mathbf{r1}\bullet\mathbf{r2})*\mathbf{r3}$$
$$= \mathbf{r3}\bullet(\mathbf{r1}*\mathbf{r2}) - \mathbf{r3}*(\mathbf{r1}\bullet\mathbf{r2})$$
$$= \mathbf{r1}\bullet(\mathbf{r3}*\mathbf{r2} - \mathbf{r2}*\mathbf{r3})$$
$$(\mathbf{r1}\wedge\mathbf{r2})\wedge\mathbf{r3} = (\mathbf{r1}\bullet\mathbf{r3})*\mathbf{r2} - (\mathbf{r2}\bullet\mathbf{r3})*\mathbf{r1}$$
$$= \mathbf{r1}\bullet(\mathbf{r3}*\mathbf{r2}) - \mathbf{r1}*(\mathbf{r3}\bullet\mathbf{r2})$$
$$= \mathbf{r3}\bullet(\mathbf{r1}*\mathbf{r2} - \mathbf{r2}*\mathbf{r1})$$
$$\mathbf{r1}*(\mathbf{r2}\wedge\mathbf{r3}) = (\mathbf{r1}*\mathbf{r2})\bullet\mathbf{C}(\mathbf{r3})$$
$$= -(\mathbf{r1}*\mathbf{r3})\bullet\mathbf{C}(\mathbf{r2})$$

$$(r1 \wedge r2) * r3 = C(r1) \bullet (r2 * r3)$$
$$= -C(r2) \bullet (r1 * r3)$$
$$r1 \wedge (r2 \wedge r3) - (r1 \wedge r2) \wedge r3 = r2 \bullet (r3 * r1 - r1 * r3)$$
$$r1 \wedge (r2 \wedge r3) + r2 \wedge (r3 \wedge r1) + r3 \wedge (r1 \wedge r2) = 0$$

$$(r1+r2) \bullet (r3+r4) = r1 \bullet r3 + r1 \bullet r4 + r2 \bullet r3 + r2 \bullet r4$$
$$(r1+r2) \wedge (r3+r4) = r1 \wedge r3 + r1 \wedge r4 + r2 \wedge r3 + r2 \wedge r4$$
$$(r1+r2) * (r3+r4) = r1 * r3 + r1 * r4 + r2 * r3 + r2 * r4$$

$$(r1 \wedge r2) \bullet (r3 \wedge r4) = (r1 \bullet r3) * (r2 \bullet r4) - (r1 \bullet r4) * (r2 \bullet r3)$$
$$= r1 \bullet (r2 \wedge (r3 \wedge r4))$$
$$r1 \bullet ((r2 \wedge r3) \wedge r4) = (r1 \bullet r3) * (r2 \bullet r4) - (r1 \bullet r2) * (r3 \bullet r4)$$
$$= (r1 \wedge r4) \bullet (r3 \wedge r2)$$
$$(r1 \wedge r2) \bullet (r3 \wedge r4) + (r1 \wedge r3) \bullet (r4 \wedge r2) + (r1 \wedge r4) \bullet (r2 \wedge r3) = 0$$

$$(r1 \wedge r2) \wedge (r3 \wedge r4) = (r4 \bullet (r1 \wedge r2)) * r3 - (r3 \bullet (r1 \wedge r2)) * r4$$
$$= (r1 \bullet (r3 \wedge r4)) * r2 - (r2 \bullet (r3 \wedge r4)) * r1$$
$$r1 \wedge (r2 \wedge (r3 \wedge r4)) = (r2 \bullet r4) * (r1 \wedge r3) - (r2 \bullet r3) * (r1 \wedge r4)$$
$$= r1 \wedge ((r2 \wedge r3) \wedge r4) + r1 \wedge (r3 \wedge (r2 \wedge r4))$$
$$((r1 \wedge r2) \wedge r3) \wedge r4 = (r1 \bullet r3) * (r2 \wedge r4) - (r2 \bullet r3) * (r1 \wedge r4)$$
$$= (r1 \wedge (r2 \wedge r3)) \wedge r4 + ((r1 \wedge r3) \wedge r2) \wedge r4$$
$$r1 \wedge (r2 \wedge (r3 \wedge r4)) - ((r1 \wedge r2) \wedge r3) \wedge r4$$
$$= (r1 \bullet r3) * (r2 \wedge r4) - (r2 \bullet r4) * (r1 \wedge r3)$$

$$r1 \bullet (r2 \bullet r3 * r4) = (r1 \bullet r4) * (r2 \bullet r3)$$
$$r1 \bullet (r2 * r3) \bullet r4 = (r1 \bullet r2) * (r3 \bullet r4)$$
$$= r1 \bullet (r2 * r4) \bullet r3$$
$$= r2 \bullet (r1 * r3) \bullet r4$$
$$= r2 \bullet (r1 * r4) \bullet r3$$

$$(r1 \bullet r2) * (r3 * r4) = (r1 \bullet (r2 * r3)) * r4$$
$$= r3 * (r1 \bullet r2 * r4$$
$$= (r3 * r1) \bullet (r2 * r4)$$
$$= r3 * (r2 \bullet r1) * r4$$
$$= (r3 * r2) \bullet (r1 * r4)$$
$$(r1 * r2) \bullet (r3 * r4) = (r2 \bullet r3) * (r1 * r4)$$

$$(r1 \wedge r2) * (r3 \wedge r4) = -C(r2) \bullet (r1 * r3) \bullet C(r4)$$
$$(r1 * r2) \bullet (r3 \wedge r4) = r1 * ((r2 \wedge r3) \bullet r4)$$
$$(r1 \wedge r2) \bullet (r3 * r4) = (r1 \bullet (r2 \wedge r3)) * r4$$

$$
\begin{aligned}
\mathbf{r1}\wedge(\mathbf{r2}\bullet(\mathbf{r3}*\mathbf{r4})) &= (\mathbf{r2}\bullet\mathbf{r3})*(\mathbf{r1}\wedge\mathbf{r4})\\
(\mathbf{r1}\bullet\mathbf{r2})*(\mathbf{r3}\wedge\mathbf{r4}) &= (\mathbf{r1}\wedge\mathbf{r3})\bullet(\mathbf{r4}*\mathbf{r2}) + (\mathbf{r1}\wedge\mathbf{r3})\wedge(\mathbf{r4}\wedge\mathbf{r2})\\
\mathbf{r1}\bullet((\mathbf{r2}\wedge\mathbf{r3})*\mathbf{r4}) &= (\mathbf{r1}\bullet(\mathbf{r2}\wedge\mathbf{r3}))*\mathbf{r4}\\
&= (\mathbf{r2}\bullet(\mathbf{r3}\wedge\mathbf{r1}))*\mathbf{r4}\\
&= (\mathbf{r3}\bullet(\mathbf{r1}\wedge\mathbf{r2}))*\mathbf{r4}\\
&= ((\mathbf{r1}\wedge\mathbf{r2})\bullet\mathbf{r3})*\mathbf{r4}\\
&= ((\mathbf{r2}\wedge\mathbf{r3})\bullet\mathbf{r1})*\mathbf{r4}\\
&= ((\mathbf{r3}\wedge\mathbf{r1})\bullet\mathbf{r2})*\mathbf{r4}\\
&= \mathbf{r2}\bullet((\mathbf{r3}\wedge\mathbf{r1})*\mathbf{r4})\\
&= \mathbf{r3}\bullet((\mathbf{r1}\wedge\mathbf{r2})*\mathbf{r4})\\
(\mathbf{r1}*(\mathbf{r2}\wedge\mathbf{r3}))\bullet\mathbf{r4} &= \mathbf{r1}*((\mathbf{r2}\wedge\mathbf{r3})\bullet\mathbf{r4})\\
&= \mathbf{r1}*((\mathbf{r3}\wedge\mathbf{r4})\bullet\mathbf{r2})\\
&= \mathbf{r1}*((\mathbf{r4}\wedge\mathbf{r2})\bullet\mathbf{r3})\\
&= \mathbf{r1}*(\mathbf{r2}\bullet(\mathbf{r3}\wedge\mathbf{r4}))\\
&= \mathbf{r1}*(\mathbf{r3}\bullet(\mathbf{r4}\wedge\mathbf{r2}))\\
&= \mathbf{r1}*(\mathbf{r4}\bullet(\mathbf{r2}\wedge\mathbf{r3}))
\end{aligned}
\qquad (126)
$$

Note that $\mathbf{r1}\bullet(\mathbf{r2}*\mathbf{r3})-(\mathbf{r2}*\mathbf{r3})\bullet\mathbf{r1} = (\mathbf{r2}\wedge\mathbf{r3})\wedge\mathbf{r1}$, not $\mathbf{0}$ generally, and that $(\mathbf{r1}\wedge\mathbf{r2})\bullet(\mathbf{r3}*\mathbf{r4})$ does not equal $\mathbf{r1}\wedge(\mathbf{r2}\bullet(\mathbf{r3}*\mathbf{r4}))$.

The magnitude of a vector may now be defined as

$$\mathrm{abs}(\mathbf{r}) \equiv \mathrm{sqrt}(\mathbf{r}\bullet\mathbf{r}) \equiv r, \qquad (127)$$

where sqrt is the square root of the non–negative real number $\mathbf{r}\bullet\mathbf{r}$.

The origin.

We may now clarify the meaning of "origin." The origin of V3 is $\mathbf{0}$, the zero vector.

Since any vector,

$$\mathbf{r} = \mathbf{r} - \mathbf{0} \qquad (128)$$

we see that $\boldsymbol{0}$ is simply a reference for other vectors.

Now any vector, including $\boldsymbol{0}$ has both a length and a direction.

To be a reference for the length of any vector, $\boldsymbol{0}$ must itself have zero length. Indeed, since $0 \le \mathrm{abs}(\mathbf{0}) = \mathrm{abs}(\mathbf{r}-\mathbf{r}) \le 2*\mathrm{abs}(\mathbf{r})$, for any $\mathbf{r}$, it follows that $\mathrm{abs}(\mathbf{0})=0$.

But what is the direction of ***0***? As for any other vector,

$$\mathbf{0} = \text{abs}(\mathbf{0})*\mathbf{u0} = 0*\mathbf{u0}, \qquad (129)$$

where ***u0*** is the direction of the origin, a unit vector. The direction ***u0*** may thus be chosen arbitrarily from a unit sphere of vectors centered on ***0***.

Further precision may be made.

Choose an arbitrary direction, ***u01***. Then consider all unit vectors, ***un***, orthogonal to ***u01***. The vectors ***un*** form a great circle of the unit sphere. Choose one of these orthogonal vectors, ***un2***. These two arbitrary choices of orthogonal unit vectors define yet a third unit vector, **ub3** = **u01∧un2**. In *V3*, ***ub3*** is unique.

In terms of these vectors, any direction, **u**(**r**), including the direction of the origin, may be written

$$\mathbf{u}(\mathbf{r}) = c1*\mathbf{u01} + c2*\mathbf{un2} + c3*\mathbf{ub3} \qquad (130)$$

for some constants *ci* such that $\text{sqrt}(c1^2 + c2^2 + c3^2) = 1$.

The origin, then, implies a choice of an arbitrary but specific minimum number of mutually orthogonal directions in terms of which the direction of any vector may be written. This number[22] is called the **dimension** of the set of vectors. The dimension of V3 is 3.

Thus clarified, the origin becomes the reference not only for the length of any vector, but also an orientation for its direction.

The origin of the set of vectors is completely arbitrary. To show this, label the origin as **r0**. Then for another fixed vector **r1** and any other vector **r**

$$\mathbf{r} - \mathbf{r1} = (\mathbf{r}-\mathbf{r0}) - (\mathbf{r1}-\mathbf{r0})$$
$$\mathbf{r} - \mathbf{r0} = (\mathbf{r}-\mathbf{r1}) - (\mathbf{r0}-\mathbf{r1}).$$

Let any vector referenced to **r0** be labeled **r|r0** = **r** – **r0**; let any vector referenced to **r1** be labeled **r|r1** = **r** – **r1**. Then

$$\mathbf{r|r1} = \mathbf{r|r0} - \mathbf{r1|r0}$$

and

$$\mathbf{r|r0} = \mathbf{r|r1} - \mathbf{r0|r1}.$$

Thus either **r0** or **r1** may serve as a reference for location.

What about the origin's orientation? Let

$$\mathbf{A} = \mathbf{u1}*\mathbf{ur1} + \mathbf{u2}*\mathbf{ur2} + \mathbf{u3}*\mathbf{ur3} \quad (131)$$

where **u1**, **u2**, and **u3** are the orientation of **r0** and **ur1**, **ur2**, and **ur3** are any other three mutually orthogonal directions.

Then **A** has an inverse

$$\mathbf{A}^{-1} = \mathbf{ur1}*\mathbf{u1} + \mathbf{ur2}*\mathbf{u2} + \mathbf{ur3}*\mathbf{u3} \quad (132)$$

Notice that, with ***I*** the identity transformation,

$$\begin{aligned}(\mathbf{r}-\mathbf{r0})\bullet\mathbf{I} &= ((\mathbf{r}-\mathbf{r1}) - (\mathbf{r0}-\mathbf{r1}))\bullet\mathbf{A}^{-1}\bullet\mathbf{A} \\ &= (\mathbf{r}-\mathbf{r1})\bullet\mathbf{A}^{-1}\bullet\mathbf{A} - (\mathbf{r0}-\mathbf{r1})\bullet\mathbf{A}^{-1}\bullet\mathbf{A}\end{aligned}$$

Now let **r|r1** be defined as

$$\mathbf{r}|\mathbf{r1} \equiv (\mathbf{r}-\mathbf{r1})\bullet\mathbf{A}^{-1} \quad (133)$$

to indicate another origin changed in location and orientation. Then

$$\mathbf{r}|\mathbf{r0} = (\mathbf{r}|\mathbf{r1} - \mathbf{r0}|\mathbf{r1})\bullet\mathbf{A} \quad (134)$$

$$\mathbf{r}|\mathbf{r1} = (\mathbf{r}|\mathbf{r0} - \mathbf{r1}|\mathbf{r0})\bullet\mathbf{A}^{-1}. \quad (135)$$

Thus **r1** is as suitable as **r0** to serve as the reference and thus the origin for any vector **r**.

Operations on vectors suppose a common origin. Thus any vector may be used as origin ***provided it is used consistently***.

Operations referred to the origin

For simplicity the arbitrary orientation of the origin[23] is symbolized in terms of three mutually orthogonal unit vectors, **u1**, **u2**, and **u3**.

Because any vector,

$$\begin{aligned}\mathbf{r1} &= \text{abs}(\mathbf{r1})*\mathbf{u}(\mathbf{r1}) \\ &\equiv r1 * (c1*\mathbf{u1} + c2*\mathbf{u2} + c3*\mathbf{u3}) \\ &\equiv r11*\mathbf{u1} + r12*\mathbf{u2} + r13*\mathbf{u3} \\ &= \mathbf{r1}\bullet\mathbf{u1}*\mathbf{u1} + \mathbf{r1}\bullet\mathbf{u2}*\mathbf{u2} + \mathbf{r1}\bullet\mathbf{u3}*\mathbf{u3} \\ &= \mathbf{r1}\bullet(\mathbf{u1}*\mathbf{u1} + \mathbf{u2}*\mathbf{u2} + \mathbf{u3}*\mathbf{u3}),\end{aligned} \quad (136)$$

it follows that:

$$r1i = \mathbf{r1 \bullet ui} \quad (137)$$
$$r1 = sqrt(r11^2 + r12^2 + r13^2) \quad (138)$$
$$ci = r1i/r1 \quad (139)$$
$$\mathbf{I = u1 * u1 + u2 * u2 + u3 * u3}. \quad (140)$$

I is the **identity transformation**. The *ci* are called **directional cosines**.

For

$$\mathbf{r1} = r1 * \mathbf{ur1}$$
$$\equiv r1 * (c11 * \mathbf{u1} + c12 * \mathbf{u2} + c13 * \mathbf{u3})$$

and

$$\mathbf{r2} = r2 * \mathbf{ur2}$$
$$\equiv r2 * (c21 * \mathbf{u1} + c22 * \mathbf{u2} + c23 * \mathbf{u3}),$$

$$\mathbf{r1+r2} = (r1 * c11 + r2 * c21) * \mathbf{u1} + (r1 * c12 + r2 * c22) * \mathbf{u2} + (r1 * c13 + r2 * c23) * \mathbf{u3} \quad (141)$$
$$\mathbf{r1 \bullet r2} = r1 * r2 * (c11 * c21 + c12 * c22 + c13 * c23) \quad (142)$$
$$\mathbf{r1 \wedge r2} = r1 * r2 * ((c12 * c23 - c13 * c22 * \mathbf{u1} + (c13 * c21 - c11 * c23) * \mathbf{u2} + (c11 * c22 - c12 * c21) * \mathbf{u3}) \quad (143)$$
$$\mathbf{r1 * r2} = r1 * r2 * \mathbf{U}, \quad (144)$$
$$\mathbf{U} \equiv \mathbf{ur1} * \mathbf{ur2} \equiv \begin{bmatrix} c11*c21 & c11*c22 & c11*c23 \\ c12*c21 & c12*c22 & c12*c23 \\ c13*c21 & c13*c22 & c13*c23 \end{bmatrix} \quad (145)$$

With these definitions[24]

$$\mathbf{u1 \wedge u2 = u3}$$
$$\mathbf{u2 \wedge u3 = u1}$$
$$\mathbf{u3 \wedge u1 = u2} \quad (146)$$

These equations finally specify **r1∧r2** uniquely.

If either **r1=0** or **r2=0**,

r1•r2 = 0,

but not necessarily the reverse

r1∧r2 = **0**,

but not necessarily the reverse

r1∗r2 = [**0**];

however **r1∗r2** = [**0**] does imply either **r1=0** or **r2=0**.

(147)

The transformations **r1∗r2** and **r2∗r1** can be seen to be related as an interchange of rows and columns.

Given two vectors **a** = a***ua** and **b** = b***ub**, it is always true that

ua•ua ≥ ua•ub.

It does not follow, however, that **a•a ≥ a•b**. The relative values of **a•a** and **a•b** for various conditions are given in the following table.

	a<b	a=b	a>b
ua•ub = 1	**a•b**>**a•a**>0>–**a•a**	**a•a**=**a•b**>0>–**a•a**	**a•a**>**a•b**>0>–**a•a**
0<**ua•ub**<1 and **ua•ub**>a/b	**a•b**>**a•a**>0>–**a•a**	NA	NA
0<**ua•ub**<1 and **ua•ub**=a/b	**a•a**=**a•b**>0>–**a•a**	NA	NA
0<**ua•ub**<1 and **ua•ub**<a/b	**a•a**>**a•b**>0>–**a•a**	**a•a**>**a•b**>0>–**a•a**	**a•a**>**a•b**>0>–**a•a**
ua•ub = 0	**a•a**>**a•b**=0>–**a•a**	**a•a**>**a•b**=0>–**a•a**	**a•a**>**a•b**=0>–**a•a**
–1<**ua•ub** < 0 and **ua•ub**>–a/b	**a•a**>0>**a•b**>–**a•a**	**a•a**>0>**a•b**>–**a•a**	**a•a**>0>**a•b**>–**a•a**
ua•ub < 0 and **ua•ub=a/b**	**a•a**>0>**a•b**=–**a•a**	NA	NA
–1<**ua•ub** < 0 and **ua•ub**<–a/b	**a•a**>0>–**a•a**>**a•b**	NA	NA
ua•ub = –1	**a•a**>0>–**a•a**>**a•b**	**a•a**>0>**a•b**= –**a•a**	**a•a**>0>**a•b**>–**a•a**

Inequalities of inner products

Reciprocal vectors

Just as the field properties of *R* require a reciprocal for every non–zero real number, so V3 over *R* contains reciprocal vectors which enable division in the set of vectors.

Definition 6 (reciprocal vectors)

Given

$$\mathbf{r} = r*\mathbf{ur} = r*(c1*\mathbf{u1} + c2*\mathbf{u2} + c3*\mathbf{u3});$$

the following vectors are called **reciprocal vectors of r**:

$$\mathbf{q11}(\mathbf{r}) \equiv \mathbf{u1}/(c1*r) \quad (148)$$
$$\mathbf{q12}(\mathbf{r}) \equiv \mathbf{u2}/(c2*r) \quad (149)$$
$$\mathbf{q13}(\mathbf{r}) \equiv \mathbf{u3}/(c3*r) \quad (150)$$
$$\mathbf{qd}(\mathbf{r}) \equiv \mathbf{ur}/r \quad (151)$$
$$= (c1*\mathbf{u1} + c2*\mathbf{u2} + c3*\mathbf{u3})/r \quad (152)$$
$$\mathbf{q21}(\mathbf{r}) \equiv ((\mathbf{u2}/c2) + (\mathbf{u3}/c3))/r \quad (153)$$
$$\mathbf{q22}(\mathbf{r}) \equiv ((\mathbf{u1}/c1) + (\mathbf{u3}/c3))/r \quad (154)$$
$$\mathbf{q23}(\mathbf{r}) \equiv ((\mathbf{u1}/c1) + (\mathbf{u2}/c2))/r \quad (155)$$
$$\mathbf{q}(\mathbf{r}) \equiv ((\mathbf{u1}/c1) + (\mathbf{u2}/c2) + (\mathbf{u3}/c3))/r \quad (156)$$
$$= \mathbf{q}(\mathbf{ur})/r \quad (157)$$

end of definition

It follows that

$$\mathbf{q}(\mathbf{r}) = \mathbf{q1i}(\mathbf{r}) + \mathbf{q2i}(\mathbf{r}),\ i = 1,2,3 \quad (158)$$
$$= \mathbf{q11}(\mathbf{r}) + \mathbf{q12}(\mathbf{r}) + \mathbf{q13}(\mathbf{r}). \quad (159)$$

Note that these definitions may be applied to vector functions. For example

$$\mathbf{qd((r1 \wedge r2)} = \mathbf{un}/(r1*r2*\sin(\text{angle}(\mathbf{r1},\mathbf{r2}))) \quad (160)$$

since $\mathbf{r1 \wedge r2} = r1*r2*\sin(\text{angle}(\mathbf{r1},\mathbf{r2}))*\mathbf{un}$.

Definition 7 (orthogonal reciprocal vectors.)

Given

$$\mathbf{r} = r*\mathbf{ur} = r*(c1*\mathbf{u1} + c2*\mathbf{u2} + c3*\mathbf{u3})$$

the following vectors are called **orthogonal reciprocal vectors of r:**

$$\mathbf{qn1}(\mathbf{ur}) = (\mathbf{u1}\wedge\mathbf{u2}/c3 + \mathbf{u3}\wedge\mathbf{u1}/c2)/r = (\mathbf{u2}/c2 + \mathbf{u3}/c3)/r \quad (161)$$

$$\mathbf{qn2}(\mathbf{ur}) = (\mathbf{u2}\wedge\mathbf{u3}/c1 + \mathbf{u1}\wedge\mathbf{u2}/c3)/r = (\mathbf{u1}/c1 + \mathbf{u3}/c3)/r \quad (162)$$

$$\mathbf{qn3}(\mathbf{ur}) = (\mathbf{u3}\wedge\mathbf{u1}/c2 + \mathbf{u2}\wedge\mathbf{u3}/c1)/r = (\mathbf{u1}/c1 + \mathbf{u2}/c2)/r \quad (163)$$

end of definition

Example:

For **r** = **u1**

$$\mathbf{q}(\mathbf{r}) = \mathbf{u1} + \mathbf{u2}/0 + \mathbf{u3}/0$$
$$\mathbf{qd}(\mathbf{r}) = \mathbf{u1}$$
$$\mathbf{qn1}(\mathbf{r}) = \mathbf{u2}/0 + \mathbf{u3}/0$$
$$\mathbf{qn2}(\mathbf{r}) = \mathbf{u1} + \mathbf{u3}/0$$
$$\mathbf{qn3}(\mathbf{r}) = \mathbf{u1} + \mathbf{u2}/0$$

For **r** = **u1** + **u2** + **u3**

$$\mathbf{q}(\mathbf{r}) = \mathbf{u1} + \mathbf{u2} + \mathbf{u3}$$
$$\mathbf{qd}(\mathbf{r}) = \mathbf{u1} + \mathbf{u2} + \mathbf{u3}$$
$$\mathbf{qn1}(\mathbf{r}) = (\mathbf{u2} + \mathbf{u3})$$
$$\mathbf{qn2}(\mathbf{r}) = (\mathbf{u1} + \mathbf{u3})$$
$$\mathbf{qn3}(\mathbf{r}) = (\mathbf{u1} + \mathbf{u2})$$

Solutions of Vector Equations

With the above definitions, vectors in V3 acquire the ability to be added, subtracted, multiplied and divided and so constitute an algebra. They may thus be assembled into equations whose solutions may be sought.

The solutions for the vector **r** in the equation **r+a=b**, where **a** and **b** are fixed vectors are unique, namely

$$\mathbf{r} = \mathbf{b} - \mathbf{a} \tag{164}$$

since if a different solution, say **r1**, existed, then

$$\mathbf{r} - \mathbf{r1} = \mathbf{b} - \mathbf{a} - (\mathbf{b} - \mathbf{a}) = \mathbf{0}.$$

The solutions for the vector **r** in the equation **r•a**=b, where **a** is a fixed vector and *b* a real constant, are *not* unique.

For **a**=a∗**ua** and **r**=r∗**ur**

$$r*(\mathbf{ua}\cdot\mathbf{ur}) = b/a \tag{165}$$

describes an infinite set of solution vectors.

For any given **ur**

$$r = b/(a*\mathbf{ua}\cdot\mathbf{ur}) \tag{166}$$

$$\mathbf{r} = b*\mathbf{ur}/(a*\mathbf{ua}\cdot\mathbf{ur}) \tag{167}$$

solves **r•a**=b;

for any given r

$$\mathbf{ur}\cdot\mathbf{ua} = b/(a*r) \tag{168}$$

so that any direction **ur** taken from from a circle on the unit sphere, the cosine of whose angle with **ua** is b/(a∗r), also solves **r•a**=b.

Reciprocal vectors provide specific solutions to **r•a**=b.

For **r** = b∗**qd**(**a**)

$$b*\mathbf{qd}(\mathbf{a})\cdot\mathbf{a}=b \tag{169}$$

so **r** =b∗**qd**(**a**) is a solution of **r•a**=b.

Likewise,

$$b*\mathbf{q}(\mathbf{a})/3 \tag{170}$$

$$b*\mathbf{q11}(\mathbf{a}) \tag{171}$$

$$b*\mathbf{q12}(\mathbf{a}) \tag{172}$$

$$b*\mathbf{q13}(\mathbf{a}) \tag{173}$$

$$b*\mathbf{q21}(\mathbf{a})/2 \tag{174}$$

$$\mathrm{b}*\mathbf{q22}(\mathbf{a})/2 \quad (175)$$
$$\mathrm{b}*\mathbf{q23}(\mathbf{a})/2 \quad (176)$$

solve $\mathbf{r}\bullet\mathbf{a}=\mathrm{b}$.

There exists, however, a unique solution which minimizes *r*, namely the direction which maximizes $\mathbf{ua}\bullet\mathbf{ur}$, namely $\mathbf{ur} = \mathbf{ua}$. The unique minimum solution[25] is thus,

$$\mathbf{r} = \mathrm{b}*\mathbf{ua}/\mathrm{a} \quad (177)$$
$$= \mathrm{b}*\mathbf{qd}(\mathbf{a}) \quad (178)$$

The difference between any two specific solutions, ***r1*** and ***r2*** is orthogonal to ***a*** since

$$(\mathbf{r1}-\mathbf{r2})\bullet\mathbf{a} = \mathbf{r1}\bullet\mathbf{a} - \mathbf{r2}\bullet\mathbf{a} = \mathrm{b} - \mathrm{b} = 0. \quad (179)$$

In particular for any solution ***r***, let

$$\mathbf{r} - \mathrm{b}*\mathbf{ua}/\mathrm{a} = \mathrm{c}*\mathbf{un}.$$

The entire set of solutions may thus be written as

$$\mathbf{r} = \mathrm{b}*\mathbf{ua}/\mathrm{a} + \mathrm{c}*\mathbf{un} \quad (180)$$

where c is any real number and $\mathbf{ua}\bullet\mathbf{un} = 0$. Consequently the entire set of solutions is a plane orthogonal to ***ua*** offset from the origin by $\mathrm{b}*\mathbf{ua}/\mathrm{a}$.

The solutions for the vector **r** in the equation $\mathbf{r}\wedge\mathbf{a}=\mathbf{b}$, where **a** and **b** are fixed vectors are *not* unique.

First note the equation implies that **a** and **b** may not be arbitrary choices. For the equation to hold, consider

$$\begin{aligned}(\mathbf{r}\wedge\mathbf{a})\wedge\mathbf{b} &= (\mathbf{r}\bullet\mathbf{b})*\mathbf{a} - (\mathbf{a}\bullet\mathbf{b})*\mathbf{r}\\ &= \mathbf{b}\wedge\mathbf{b}\\ &= \mathbf{0}\end{aligned}$$

so that

$$(\mathbf{r}\bullet\mathbf{b})*\mathbf{a} = (\mathbf{a}\bullet\mathbf{b})*\mathbf{r}.$$

Now this condition cannot be met unless

$$\mathbf{r} = \mathbf{a}$$

or

$$\mathbf{a} \bullet \mathbf{b} = 0.$$

If $\mathbf{r} = \mathbf{a}$, then $\mathbf{b} = \mathbf{0}$, a trivial solution.
If $\mathbf{a} \bullet \mathbf{b} = 0$, then **a** and **b** are orthogonal and $(\mathbf{r} \bullet \mathbf{b}) * \mathbf{a} = \mathbf{0}$, so either

$$\mathbf{a} = \mathbf{0}$$

or

$$\mathbf{r} \bullet \mathbf{b} = 0.$$

If $\mathbf{a} = \mathbf{0}$, then $\mathbf{b} = \mathbf{0}$, another trivial solution.

If $\mathbf{r} \bullet \mathbf{b} = 0$, then **r** and **b** must be orthogonal.

Consequently the set of solutions consists of the trivial solutions,

$$\mathbf{a} = \mathbf{0} = \mathbf{b}$$

and non-trivial solutions where **r** and **a** are both orthogonal to **b**.

For the non-trivial solutions, let the vectors **a** and **b** be orthogonal, **r** orthogonal to **b**, but not to **a**.

Then

$$\mathbf{r} \wedge \mathbf{a} = a * r * (\sin(\text{angle}(\mathbf{a},\mathbf{r}))) * \mathbf{ub}.$$

Let $\mathbf{ur} = \cos(\text{angle}(\mathbf{a},\mathbf{r}))) * \mathbf{ua} + \sin(\text{angle}(\mathbf{a},\mathbf{r}))) * (\mathbf{ua} \wedge \mathbf{ub})$.

Then

$$\begin{aligned}
\mathbf{ur} \wedge \mathbf{ua} &= \cos(\text{angle}(\mathbf{a},\mathbf{r})) * \mathbf{ua} + \sin(\text{angle}(\mathbf{a},\mathbf{r})) * (\mathbf{ua} \wedge \mathbf{ub}) \wedge \mathbf{ua} \\
&= \sin(\text{angle}(\mathbf{a},\mathbf{r})) * (\mathbf{ua} \wedge \mathbf{ub}) \wedge \mathbf{ua} \\
&= \sin(\text{angle}(\mathbf{a},\mathbf{r})) * ((\mathbf{ua} \bullet \mathbf{ua}) * \mathbf{ub} - (\mathbf{ub} \bullet \mathbf{ua}) * \mathbf{ua}) \\
&= \sin(\text{angle}(\mathbf{a},\mathbf{r})) * ((\mathbf{ua} \bullet \mathbf{ua}) * \mathbf{ub} \\
&= \sin(\text{angle}(\mathbf{a},\mathbf{r})) * \mathbf{ub}.
\end{aligned}$$

Now[26] for

$$\mathbf{r} = b * \mathbf{ur} / (a * \sin(\text{angle}(\mathbf{a},\mathbf{r}))) \qquad (181)$$

$$\begin{aligned}
\mathbf{r} \wedge \mathbf{a} &= a * b * (\sin(\text{angle}(\mathbf{a},\mathbf{r}))) * \mathbf{ub} / (a * (\sin(\text{angle}(\mathbf{a},\mathbf{r})))) \\
&= \mathbf{b}
\end{aligned}$$

which solves $\mathbf{r} \wedge \mathbf{a} = \mathbf{b}$ for any angle$(\mathbf{a},\mathbf{r})$.

There is, however, an unique solution which minimizes *r*, namely the direction which maximizes sin(angle(**a**,**r**)). For this minimum solution **r** is orthogonal to both **a** and **b**. For this condition, let

$$\mathbf{r} = b*\mathbf{ur}/a. \tag{182}$$

Then

$$\begin{aligned}\mathbf{r}\wedge\mathbf{a} &= b*\mathbf{ur}\wedge\mathbf{a}/a \\ &= b*a*\mathbf{ub}/a \\ &= b*\mathbf{ub} \\ &= \mathbf{b}.\end{aligned}$$

The minimum solution may be rewritten in terms of quotient vectors. Since **qd**(**b∧a**) = **ur**/(a∗b)

$$\mathbf{r} = b^2*\mathbf{qd}(\mathbf{r}\wedge\mathbf{a}). \tag{183}$$

For any two solutions of **r∧a=b**, say **r1** and **r2**,

$$\mathbf{r1}\wedge\mathbf{a} = \mathbf{r2}\wedge\mathbf{a}$$

since

$$\mathbf{r1}\wedge\mathbf{a} - \mathbf{r2}\wedge\mathbf{a} = \mathbf{b} - \mathbf{b} = \mathbf{0}.$$

The solutions for the vector **r** in the equation **r∗a=B**, where **a** is a fixed vector and **B** a fixed transformation are unique.

First note the equation implies that **B** must be an outer product, say

$$\mathbf{B} = \mathbf{b1}*\mathbf{b2}.$$

Further, if **a** = **0**, then **B** = **b1∗0**; if **B** = **0∗b2** then **r** = **0**.

For **r**=r∗**ur**, **a**=a∗**ua**, **b1**=b1∗**ub1**, **b2**=b2∗**ub2**

$$\begin{aligned}\mathbf{r}*\mathbf{a} &= r*a*\mathbf{ur}*\mathbf{ua} \\ &= \mathbf{b1}*\mathbf{b2} \\ &= b1*b2*\mathbf{ub1}*\mathbf{ub2}\end{aligned}$$

Consequently,

$$\begin{aligned}&\mathbf{ur} = \mathbf{ub1} \\ &\mathbf{ua} = \mathbf{ub2} \\ &r*a = b1*b2\end{aligned}$$

Therefore

$$\mathbf{r} = b2 * \mathbf{b1}/a \tag{184}$$

where $\mathbf{a} = a * \mathbf{ub2}$ solves $\mathbf{r} * \mathbf{a} = \mathbf{B}$ uniquely.

The solutions for vector equations of vector multiplication are summarized in the following table.

Solutions	$\mathbf{r} \bullet \mathbf{a} = b$	$\mathbf{r} \wedge \mathbf{a} = \mathbf{b}$	$\mathbf{r} * \mathbf{a} = \mathbf{B}$
general	$\mathbf{r} = b * \mathbf{ur} / (a * (\mathbf{ua} \bullet \mathbf{ur}))$	$\mathbf{r} = b * \mathbf{ur} / (a * (\mathbf{ur} \wedge \mathbf{ua}) \bullet \mathbf{ub})$	$\mathbf{r} = b2 * \mathbf{b1}/a$
minimum	$\mathbf{r} = b * \mathbf{qd}(\mathbf{a})$	$\mathbf{r} = b^2 * \mathbf{qd}(\mathbf{b} \wedge \mathbf{a})$	$\mathbf{r} = b2 * \mathbf{b1}/a$
entire set	$\mathbf{r} = b * \mathbf{ua}/a + c * \mathbf{un}(\mathbf{a})$	$\mathbf{r} = b * \mathbf{ur} / (a * \sin(\text{angle}(\mathbf{a}, \mathbf{r})))$	$\mathbf{r} = b2 * \mathbf{b1}/a$

Solutions of algebraic vector equations

Mathematics: Functions of V3

Functions of V3

Functions of the set of vectors may have either *V3* or its underlying field *R* as domain or range. Three categories of functions may thus be defined:

f:V3 ⟶R written as f(**r**)
f:V3⟶V3 written as **f**(**r**)
f:R ⟶V3 written as **f**(r)

(185)

As with real functions f:R⟶R, different look-alike functions of V3 may describe the same relationship between domain and range.

Functions are called by their ranges. Those symbolized as f(**r**) are called **scalar functions**; both **f**(**r**) and **f**(r) are called **vector functions**.

Functions of V3 participate in the algebra defined on their ranges.

If f1(**r**) and f2(**r**) are scalar functions then so too are

f1(**r**) + f2(**r**)

and

f1(**r**) * f2(**r**).

For vector functions **f1**(**r**) and **f2**(**r**) or **f1**(r) and **f2**(r) and scalar *c*

f1(**r**) + **f2**(**r**),
f1(**r**) ∧ **f2**(**r**),
c***f**(**r**),
f1(r) + **f2**(r),
f1(r) ∧ **f2**(r),
c***f**(r)

are likewise vector functions, while

f1(**r**) • **f2**(**r**),
f1(r) • **f2**(r)

are scalar functions, and

f1(**r**) * **f2**(**r**),
f1(r) * **f2**(r)

are transformations.

Mathematics: Functions of V3

Functions of V3 are also susceptible to restrictions in their domains by either scalar or vector functions.

Definition 8 (restricted functions of V3)

Given
$\mathbf{f}$:V3⟶V3;
$\mathbf{g}$:Y⟶V3;
or g:Y⟶R;
for
Y ⊂ V3, Y a subset of V3
then

$$\mathbf{f}|\mathbf{g} \equiv \mathbf{f}:Y \longrightarrow V3$$
$$\mathbf{f}|g \equiv \mathbf{f}:Y \longrightarrow V3.$$

Given
$\mathbf{f}$:R⟶V3;
$\mathbf{g}$:Y⟶V3;
or g:Y⟶V3;
for
Y ⊂ R, Y a subset of R
then

$$\mathbf{f}|\mathbf{g} \equiv \mathbf{f}:Y \longrightarrow V3$$
$$\mathbf{f}|g \equiv \mathbf{f}:Y \longrightarrow V3.$$

Given
f:V3⟶R;
$\mathbf{g}$:Y⟶V3;
or g:Y⟶V3;
for
Y ⊂ V3, Y a subset of V3
then

$$\mathbf{f}|\mathbf{g} \equiv \mathbf{f}:Y \longrightarrow R$$
$$\mathbf{f}|g \equiv \mathbf{f}:Y \longrightarrow R.$$

end of definition

Mathematics: Functions of V3

Likewise functions of V3 may be compounded.

Definition 9 (compound functions of V3)

Given

$\mathbf{f}$:V3⟶V3;

$\mathbf{g}(\mathbf{r})$:V3⟶V3;

or $\mathbf{g}(x)$:R⟶V3;

then

$$\mathbf{f}(\mathbf{g}) \equiv \mathbf{f}(\mathbf{g}(\mathbf{r})) \equiv \mathbf{f}\text{:V3}\longrightarrow\text{V3}$$

or

$$\mathbf{f}(\mathbf{g}) \equiv \mathbf{f}(\mathbf{g}(x)) \equiv \mathbf{f}\text{:R}\longrightarrow\text{V3.}$$

Given

$\mathbf{f}$:R⟶V3;

$g(\mathbf{r})$:V3⟶R;

or $g(x)$:R⟶R;

then

$$\mathbf{f}(g) \equiv \mathbf{f}(g(\mathbf{r})) \equiv \mathbf{f}\text{:R}\longrightarrow\text{V3}$$

or

$$\mathbf{f}(g) \equiv \mathbf{f}(g(x)) \equiv \mathbf{f}\text{:R}\longrightarrow\text{V3.}$$

Given

f:V3⟶R;

$\mathbf{g}(\mathbf{r})$:V3⟶V3;

or $\mathbf{g}(x)$:R⟶V3

then

$$f(\mathbf{g}) \equiv f(\mathbf{g}(\mathbf{r})) \equiv \mathbf{f}\text{:V3}\longrightarrow\text{R}$$

or

$$f(\mathbf{g}) \equiv f(\mathbf{g}(x)) \equiv \mathbf{f}\text{:R}\longrightarrow\text{R.}$$

end of definition

Matrix functions

Because of their utility we introduce a few functions of an arbitrary transformation, called a **matrix**,

$$\mathbf{A}=\begin{bmatrix} a11 & a12 & a13 \\ a21 & a22 & a23 \\ a31 & a32 & a33 \end{bmatrix}$$
$$\equiv \mathbf{u1*a1+u2*a2+u3*a3}. \quad (186)$$

$$\mathrm{tr}[\mathbf{A}] \equiv \mathbf{u1\bullet A\bullet u1 + u2\bullet A\bullet u2 + u3\bullet A\bullet u3} \quad (187)$$
$$\mathbf{g}[\mathbf{A}] \equiv a11*\mathbf{u1} + a22*\mathbf{u2} + a33*\mathbf{u3} \quad (188)$$
$$\mathbf{c}[\mathbf{A}] \equiv (a23-a32)*\mathbf{u1} + (a31-a13)*\mathbf{u2} + (a12-a21)*\mathbf{u3} \quad (189)$$
$$\mathbf{T}[\mathbf{A}] \equiv \mathbf{a1*u1+a2*u2+a3*u3}$$
$$= \begin{bmatrix} a11 & a21 & a33 \\ a12 & a22 & a32 \\ a13 & a23 & a33 \end{bmatrix} \quad (190)$$
$$\det[\mathbf{A}] \equiv \mathbf{a1\bullet(a2\wedge a3)} = \mathbf{a2\bullet(a3\wedge a1)} = \mathbf{a3\bullet(a1\wedge a2)} \quad (191)$$
$$= a11*a22*a33 + a12*a23*a31 + a13*a21*a32$$
$$-a11*a23*a32 - a12*a21*a33 - a13*a22*a31 \quad (192)$$

$$\det[\mathbf{A}]*\mathbf{A}^{-1}=\begin{bmatrix} a22*a33-a23*a32 & a13*a32-a12*a33 & a12*a23-a13*a22 \\ a23*a31-a21*a33 & a11*a33-a13*a31 & a13*a21-a11*a23 \\ a21*a32-a22*a31 & a12*a31-a11*a32 & a11*a22-a12*a21 \end{bmatrix}$$
$$\equiv \mathbf{(a2\wedge a3)*u1 + (a3\wedge a1)*u2 + (a1\wedge a2)*u3} \quad (193)$$

where the ***ai*** are the row vectors of ***A***; tr[**A**], called the **trace** of **A**, is the sum of the diagonal elements of ***A***; **g**[**A**], called the **diagonal matrix operator**, transforms the matrix into a vector specified by its diagonal elements; **c**[**A**], called the **curl matrix operator**, transforms the matrix into a vector specified by its off–diagonal elements; **T**[**A**], called the **transpose**[27] of **A**, interchanges the rows and columns of **A**; det[**A**], called the **determinant** of **A**, is the "volume" of the constituent vectors of **A**; $\mathbf{A}^{-1}$ is called the **inverse** of **A** since

$$\mathbf{A}\bullet[\mathbf{A}^{-1}]=[\mathbf{A}^{-1}]\bullet\mathbf{A}=\mathbf{I}.$$

Only matrices with non–zero determinants have inverses.

For such matrices it is easy to see that

$$\mathbf{T}[\mathbf{A}^{-1}] = [\mathbf{T}[\mathbf{A}]]^{-1}. \quad (194)$$

Mathematics: Functions of V3

For any matrix **A**

$$\det[\mathbf{T}[\mathbf{A}]] = \det[\mathbf{A}] \tag{195}$$

$$\mathbf{c}[\mathbf{A}] = -\mathbf{c}[\mathbf{T}[\mathbf{A}]]. \tag{196}$$

Some specific results are given below.

$$(\mathbf{r1} \wedge \mathbf{r2}) \ast \mathbf{r3} + (\mathbf{r3} \wedge \mathbf{r1}) \ast \mathbf{r2} + (\mathbf{r2} \wedge \mathbf{r3}) \ast \mathbf{r1} = ((\mathbf{r1} \wedge \mathbf{r2}) \bullet \mathbf{r3}) \ast \mathbf{I} \tag{197}$$

$$\mathbf{r1} \bullet (\mathbf{r2} \bullet \mathbf{A}) = r21 \ast \mathbf{r1} \bullet \mathbf{a1} + r22 \ast \mathbf{r1} \bullet \mathbf{a2} + r23 \ast \mathbf{r1} \bullet \mathbf{a3} \tag{198}$$

$$\mathbf{r1} \wedge (\mathbf{r2} \bullet \mathbf{A}) = r21 \ast \mathbf{r1} \wedge \mathbf{a1} + r22 \ast \mathbf{r1} \wedge \mathbf{a2} + r23 \ast \mathbf{r1} \wedge \mathbf{a3} \tag{199}$$

$$\mathbf{r1} \ast (\mathbf{r2} \bullet \mathbf{A}) = r21 \ast \mathbf{r1} \ast \mathbf{a1} + r22 \ast \mathbf{r1} \ast \mathbf{a2} + r23 \ast \mathbf{r1} \ast \mathbf{a3}. \tag{200}$$

For two vectors **r1** and **r2**,

$$\mathrm{tr}[\mathbf{r1} \ast \mathbf{r2}] = \mathbf{r1} \bullet \mathbf{r2} \tag{201}$$

$$\mathbf{g}[\mathbf{r1} \ast \mathbf{r2}] = \mathbf{g}[\mathbf{r2} \ast \mathbf{r1}] \tag{202}$$

$$\mathbf{c}[\mathbf{r1} \ast \mathbf{r2}] = \mathbf{r1} \wedge \mathbf{r2} \tag{203}$$

$$\det[\mathbf{r1} \ast \mathbf{r2}] = 0 \tag{204}$$

$$\mathbf{T}[\mathbf{r1} \ast \mathbf{r2}] = \mathbf{r2} \ast \mathbf{r1}. \tag{205}$$

For two transformations

$$\mathbf{A} \equiv \mathbf{u1} \ast \mathbf{a1} + \mathbf{u2} \ast \mathbf{a2} + \mathbf{u3} \ast \mathbf{a3}$$

and

$$\mathbf{B} \equiv \mathbf{u1} \ast \mathbf{b1} + \mathbf{u2} \ast \mathbf{b2} + \mathbf{u3} \ast \mathbf{b3}$$

$$\mathrm{tr}[\mathbf{A}+\mathbf{B}] = \mathrm{tr}[\mathbf{A}] + \mathrm{tr}[\mathbf{B}] \tag{206}$$

$$\mathbf{g}[\mathbf{A}+\mathbf{B}] = \mathbf{g}[\mathbf{A}] + \mathbf{g}[\mathbf{B}] \tag{207}$$

$$\mathbf{c}[\mathbf{A}+\mathbf{B}] = \mathbf{c}[\mathbf{A}] + \mathbf{c}[\mathbf{B}] \tag{208}$$

$$\mathbf{T}[\mathbf{A}+\mathbf{B}] = \mathbf{T}[\mathbf{A}] + \mathbf{T}[\mathbf{B}] \tag{209}$$

$$\mathrm{tr}[\mathbf{T}[\mathbf{A}] \bullet \mathbf{B}] = \mathrm{tr}[\mathbf{T}[\mathbf{B}] \bullet \mathbf{A}] \tag{210}$$

$$= \mathbf{a1} \bullet \mathbf{b1} + \mathbf{a2} \bullet \mathbf{b2} + \mathbf{a3} \bullet \mathbf{b3} \tag{211}$$

$$\mathbf{g}[\mathbf{T}[\mathbf{A}] \bullet \mathbf{B}] = \mathbf{g}[\mathbf{a1} \ast \mathbf{b1}] + \mathbf{g}[\mathbf{a2} \ast \mathbf{b2}] + \mathbf{g}[\mathbf{a3} \ast \mathbf{b3}] \tag{212}$$

$$\mathbf{c}[\mathbf{T}[\mathbf{A}] \bullet \mathbf{B}] = \mathbf{a1} \wedge \mathbf{b1} + \mathbf{a2} \wedge \mathbf{b2} + \mathbf{a3} \wedge \mathbf{b3} \tag{213}$$

$$\det[\mathbf{A} \bullet \mathbf{B}] = \det[\mathbf{A}] \ast \det[\mathbf{B}] \tag{214}$$

$$\mathbf{T}[\mathbf{A} \bullet \mathbf{B}] = \mathbf{T}[\mathbf{B}] \bullet \mathbf{T}[\mathbf{A}]. \tag{215}$$

If ***A*** and ***B*** have inverses,

$$[\mathbf{A} \bullet \mathbf{B}]^{-1} = [\mathbf{B}^{-1}] \bullet [\mathbf{A}^{-1}] \tag{216}$$

$$[\mathbf{T}[\mathbf{A} \bullet \mathbf{B}]]^{-1} = \mathbf{T}[\mathbf{A}^{-1}] \bullet \mathbf{T}[\mathbf{B}^{-1}] \tag{217}$$

$$= [\mathbf{T}[\mathbf{A}]]^{-1} \bullet [\mathbf{T}[\mathbf{B}]]^{-1} \tag{218}$$

$$= \mathbf{T}[[\mathbf{A} \bullet \mathbf{B}]^{-1}]. \tag{219}$$

The scalar

$$\begin{aligned}(\mathbf{r}\cdot\mathbf{A})\cdot(\mathbf{r}\cdot\mathbf{A}) &= (\mathbf{r}\cdot\mathbf{A})\cdot\mathbf{T}[\mathbf{A}]\cdot\mathbf{r}\\ &= r1*r1*\mathbf{a1}\cdot\mathbf{a1} + r1*r2*\mathbf{a1}\cdot\mathbf{a2} + r1*r3*\mathbf{a1}\cdot\mathbf{a3}\\ &= r2*r1*\mathbf{a2}\cdot\mathbf{a1} + r2*r2*\mathbf{a2}\cdot\mathbf{a2} + r2*r3*\mathbf{a2}\cdot\mathbf{a3}\\ &= r3*r1*\mathbf{a3}\cdot\mathbf{a1} + r3*r2*\mathbf{a3}\cdot\mathbf{a2} + r3*r3*\mathbf{a3}\cdot\mathbf{a3}\end{aligned} \qquad (220)$$

whereas,

$$\begin{aligned}(\mathbf{r}\cdot\mathbf{T}[\mathbf{A}])\cdot(\mathbf{r}\cdot\mathbf{T}[\mathbf{A}]) &= (\mathbf{r}\cdot\mathbf{T}[\mathbf{A}])\cdot\mathbf{A}\cdot\mathbf{r}\\ &= (\mathbf{r}\cdot\mathbf{a1})^2 + (\mathbf{r}\cdot\mathbf{a2})^2 + (\mathbf{r}\cdot\mathbf{a3})^2.\end{aligned} \qquad (221)$$

Note that the scalar

$$\mathbf{r1}\cdot\mathbf{A}\cdot\mathbf{r2} = \mathbf{r2}\cdot(\mathbf{r1}\cdot\mathbf{A}) = \mathbf{r1}\cdot(\mathbf{r2}\cdot\mathbf{T}[\mathbf{A}])$$

does not equal

$$\mathbf{r2}\cdot\mathbf{A}\cdot\mathbf{r1} = \mathbf{r1}\cdot(\mathbf{r2}\cdot\mathbf{A}) = \mathbf{r2}\cdot(\mathbf{r1}\cdot\mathbf{T}[\mathbf{A}])$$

generally.

The outer product of two non–zero vectors cannot produce a diagonal matrix. Therefore the following transformation, called the **diagonal vector operator**, is defined for any vector **r1**:

$$\begin{aligned}\mathbf{G}(\mathbf{r1}) &\equiv r1*(c1*\mathbf{u1}*\mathbf{u1} + c2*\mathbf{u2}*\mathbf{u2} + c3*\mathbf{u3}*\mathbf{u3}) \qquad (222)\\ &= r1*\begin{bmatrix} c1 & 0 & 0\\ 0 & c2 & 0\\ 0 & 0 & c3\end{bmatrix}\end{aligned}$$

Then

$$\mathbf{r1}\cdot\mathbf{G}(\mathbf{r2}) = \mathbf{r2}\cdot\mathbf{G}(\mathbf{r1}) \qquad (223)$$

$$= r1*r2 *(c11*c21*\mathbf{u1}+c12*c22*\mathbf{u2}+c13*c23*\mathbf{u3}) \qquad (224)$$

$$[\mathbf{r1}*\mathbf{r2}]\cdot\mathbf{G}(\mathbf{r3}) = [\mathbf{r1}*\mathbf{r3}]\cdot\mathbf{G}(\mathbf{r2}) \qquad (225)$$

$$\mathbf{G}(\mathbf{r1})\cdot[\mathbf{r2}*\mathbf{r3}] = \mathbf{G}(\mathbf{r2})\cdot[\mathbf{r1}*\mathbf{r3}]. \qquad (226)$$

The outer product of **r2*r1** transforms any other vector into another parallel to **r1**; **G**(**r1**), however, does not generally transform another vector into the direction of **r1**.

For any matrix **A**

$$\mathbf{T}[\mathbf{G}(\mathbf{r})\cdot\mathbf{A}] = \mathbf{T}[\mathbf{A}]\cdot\mathbf{G}(\mathbf{r}) \qquad (227)$$

$$\mathbf{T}[\mathbf{A}\cdot\mathbf{G}(\mathbf{r})] = \mathbf{G}(\mathbf{r})\cdot\mathbf{T}[\mathbf{A}] \qquad (228)$$

since $\mathbf{T}[\mathbf{G}(\mathbf{r})] = \mathbf{G}(\mathbf{r})$.

Mathematics: Functions of V3

The diagonal matrix operator is an operational inverse for the diagonal vector operator since for any **r**,

$$\mathbf{r} = \mathbf{g}[\mathbf{G}(\mathbf{r})]. \tag{229}$$

Also,

$$\mathbf{r1}\cdot\mathbf{r2} = \mathbf{g}[\mathbf{G}(\mathbf{r1})]\cdot\mathbf{g}[\mathbf{G}(\mathbf{r2})] \tag{230}$$

$$= \mathrm{tr}[\mathbf{G}(\mathbf{r1})\cdot\mathbf{G}(\mathbf{r2})]. \tag{231}$$

Cross products may also be described in terms of a transformation, **C**, called the **curl vector operator**. Specifically, for two vectors

$$\mathbf{r1} = \mathrm{r1}*(\mathrm{c11}*\mathbf{u1} + \mathrm{c12}*\mathbf{u2} + \mathrm{c13}*\mathbf{u3})$$

and

$$\mathbf{r2} = \mathrm{r2}*(\mathrm{c21}*\mathbf{u1} + \mathrm{c22}*\mathbf{u2} + \mathrm{c33}*\mathbf{u3})$$

$$\mathbf{r1}\wedge\mathbf{r2} \equiv \mathrm{r1}*\mathrm{r2}*(\mathrm{c11},\mathrm{c12},\mathrm{c13})\bullet\begin{bmatrix} 0 & -\mathrm{c23} & \mathrm{c22} \\ \mathrm{c23} & 0 & -\mathrm{c21} \\ -\mathrm{c22} & \mathrm{c21} & 0 \end{bmatrix} \tag{232}$$

$$\equiv \mathbf{r1}\bullet\mathbf{C}(\mathbf{r2}) = \mathbf{r2}\bullet\mathbf{T}[\mathbf{C}(\mathbf{r1})] = -\mathbf{r2}\wedge\mathbf{r1}. \tag{233}$$

Thus

$$\mathbf{r1}\wedge\mathbf{r2} = \mathbf{c}[\mathbf{r1}*\mathbf{r2}] = \mathbf{r1}\bullet\mathbf{C}(\mathbf{r2}). \tag{234}$$

The matrix **C** does not have an inverse, but neither can it be expressed as an outer product.

Given a direction **ur1** = c1∗**u1** + c2∗**u2** + c3∗**u3**, we define the symmetrical matrix

$$\mathbf{COS}(\mathbf{ur1}) \equiv \mathbf{ur1}*\mathbf{ur1} \equiv \begin{bmatrix} \mathrm{c1}*\mathrm{c1} & \mathrm{c1}*\mathrm{c2} & \mathrm{c1}*\mathrm{c3} \\ \mathrm{c2}*\mathrm{c1} & \mathrm{c2}*\mathrm{c2} & \mathrm{c2}*\mathrm{c3} \\ \mathrm{c3}*\mathrm{c1} & \mathrm{c3}*\mathrm{c2} & \mathrm{c3}*\mathrm{c3} \end{bmatrix} \tag{235}$$

called the **cosine matrix of ur1** since ci = **ur1**•**ui** is the cosine of the angle between **ur1** and the given orthogonal direction **ui**.

It is easily seen that

$$\mathrm{tr}[\mathbf{G}(\mathbf{r1})] = \mathrm{r1}*(\mathrm{c1}+\mathrm{c2}+\mathrm{c3}) \quad (236)$$
$$\mathrm{tr}[\mathbf{C}(\mathbf{r1})] = 0 \quad (237)$$
$$\mathrm{tr}[\mathbf{COS}(\mathbf{ur1})] = 1 \quad (238)$$
$$\mathbf{c}[\mathbf{G}(\mathbf{r1})] = \mathbf{0} \quad (239)$$
$$\mathbf{c}[\mathbf{C}(\mathbf{r1})] = -2*\mathbf{r1} \quad (240)$$
$$\mathbf{c}[\mathbf{COS}(\mathbf{ur1})] = \mathbf{0} \quad (241)$$
$$\mathrm{det}[\mathbf{G}(\mathbf{r1})] = \mathrm{r1}*\mathrm{r1}*\mathrm{r1}*\mathrm{c1}*\mathrm{c2}*\mathrm{c3} \quad (242)$$
$$\mathrm{det}[\mathbf{C}(\mathbf{r1})] = 0 \quad (243)$$
$$\mathrm{det}[\mathbf{COS}(\mathbf{ur1})] = 0 \quad (244)$$

and

$$\mathbf{C}(\mathbf{r1}+\mathbf{r2}) = \mathbf{C}(\mathbf{r1}) + \mathbf{C}(\mathbf{r2}) \quad (245)$$
$$\mathbf{C}(\mathbf{r1}\wedge\mathbf{r2}) = \mathbf{r2}*\mathbf{r1} - \mathbf{r1}*\mathbf{r2} \quad (246)$$
$$\mathbf{C}(\mathbf{c}[\mathbf{r1}*\mathbf{r2}]) = \mathbf{r2}*\mathbf{r1} - \mathbf{r1}*\mathbf{r2} \quad (247)$$
$$\mathbf{c}[\mathbf{C}(\mathbf{r1}\wedge\mathbf{r2})] = 2*\mathbf{r2}\wedge\mathbf{r1} = -2*\mathbf{r1}\wedge\mathbf{r2}. \quad (248)$$

Mathematics: Vector Algebra

Vector Solutions of Matrix Equations

Given a matrix **A** = **u1*a1+u2*a2+u3*a3** , the transformation V3⟶V3 represented by **r•A** may have as image either V3 entire or some proper subset.

Definition 10 (zero subset of a matrix)

Given

A, a matrix

N(**A**) = {**r**|**r•A** = **0**} (249)

is called the **zero subset** of **A**.

end of definition

Clearly, **N**(**A**) is a subset of vectors in V3 with its own algebra.

If a non–zero vector **r1** is in **N**(**A**), then, for any *x* in *R*, the line {x***r1**} is also in **N**(**A**). If these are the only vectors in **N**(**A**), its dimension is 1.

If the non–zero and non–parallel vectors **r1** and **r2** are in **N**(**A**), then, for any *x1* and *x2* in *R*, the plane x1***r1+** x2***r2** is also in **N**(**A**). If these are the only vectors in **N**(**A**), its dimension is 2.

If the non–zero and non–parallel vectors **r1**, **r2**, and **r3** are in **N**(**A**), then V3 = **N**(**A**) and **A** = [**0**]. The dimension of **N**(**A**) is thus 3.

If **N**(**A**) consists of the single vector **0**, then the transformation **A** is 1–1 since for the non–zero, distinct vectors **r1** and **r2**,

if **r1•A** = **b** and **r2•A** = **b**, then (**r1–r2**)**•A** = **0**, a contradiction.

Under **A** then, V3 is decomposed into two subsets of vectors:

N(**A**) and {**A – N(A) + 0**}

each with its own algebra.

The **rank** of **A**, rank(**A**), is the dimension of the image of V3 under **A**.

A matrix of rank 3 is a 1–1 transformation of V3 into V3. A matrix of rank 2 transforms V3 into a surface. A matrix of rank 1 transforms V3 into a line. A matrix of rank 0 transforms V3 into the origin.

In V3,

rank(**A**) + dimension(**N**(**A**)) = 3. (250)

What is the rank of an outer product? For **r•r1∗r2** = **0** we need only

r•r1= 0.

If **r1** or **r2** equals **0,** then the matrix also equals [**0**] and any **r** is in the zero subset. Otherwise **r** belongs to the plane of vectors orthogonal to **r1**. Thus the dimension of the zero subset is 2. The rank of the outer product is then 1. Indeed **r•**[**r1∗r2**] = (**r•r1**)**∗r2** which is a line in the **ur2** direction.

The curl matrix operator is an example of a matrix of rank 2.

The solution of **r•T**[**A**] = **b** for ***r***, where ***A*** and ***b*** are given, *may not* be unique.

r•T[**A**] = **r•a1∗u1** + **r•a2∗u2** + **r•a3∗u3**
≡ b1∗**u1** + b2∗**u2** + b3∗**u3**. (251)

The solution then devolves into the simultaneous solution of

r•a1 = b1
r•a2 = b2
r•a3 = b3

each of whose set of solutions is a plane orthogonal to bi∗**qd**(**ai**), i=1,2,3.

Now three planes may intersect at a point, in a line or be parallel. Thus if rank(**T**[**A**]) = 3, the solution is unique; if rank(**T**[**A**]) = 2, the solution set is a line of vectors; if rank(**T**[**A**]) = 1, the solution set is a plane of vectors. If rank(**T**[**A**]) = 0, there are no solutions unless **b=0** in which case every vector in V3 is a solution.

For rank(**T**[**A**])=3, the unique solution is

r = **b•T**[**A**]$^{-1}$
= (b1∗(**a2∧a3**) + b2∗(**a3∧a1**) + b3∗(**a1∧a2**))/det[**A**] (252)

since for this case **T**[**A**] is invertible and **N**(**T**[**A**]) = **0**.

Mathematics: Vector Algebra

For rank(**T**[**A**])=2, the dimension of **N**(**T**[**A**]) is 1.

Let **r1**, **r2**, and ***n*** be three non–parallel vectors with ***n*** in **N**(**T**[**A**]). Any vector in V3 may then be represented as

$$\mathbf{r} = t1 * \mathbf{r1} + t2 * \mathbf{r2} + t3 * \mathbf{n} \tag{253}$$

for some t1, t2, t3.

Then for the solution of **r•T**[**A**] = **b**

$$\mathbf{r} \bullet \mathbf{T}[\mathbf{A}] = (t1 * \mathbf{r1} + t2 * \mathbf{r2}) \bullet \mathbf{T}[\mathbf{A}] = \mathbf{b} \tag{254}$$

devolves into a solution for t1 and t2.

Now, however, the vector

$$\mathbf{b} = b1 * \mathbf{u1} + b2 * \mathbf{u2} + b3 * \mathbf{u3}$$

is constrained to lie outside **N**(**T**[**A**]), since otherwise only **r** = **0** is a solution. Thus

$$\mathbf{b} \bullet \mathbf{T}[\mathbf{A}] \neq \mathbf{0}. \tag{255}$$

Thus also

$$\begin{aligned} (t1 * \mathbf{r1} + t2 * \mathbf{r2}) \bullet \mathbf{a1} &= b1 \\ (t1 * \mathbf{r1} + t2 * \mathbf{r2}) \bullet \mathbf{a2} &= b2 \\ (t1 * \mathbf{r1} + t2 * \mathbf{r2}) \bullet \mathbf{a3} &= b3 \\ b1 * \mathbf{a1} + b2 * \mathbf{a2} + b3 * \mathbf{a3} &\neq \mathbf{0}. \end{aligned} \tag{256}$$

Then

$$\begin{aligned} t1 &= (b1 * \mathbf{r2} \bullet \mathbf{a2} - b2 * \mathbf{r2} \bullet \mathbf{a1})/(\mathbf{r1} \bullet \mathbf{a1} * \mathbf{r2} \bullet \mathbf{a2} - \mathbf{r1} \bullet \mathbf{a2} * \mathbf{r2} \bullet \mathbf{a1}) \\ &= (b1 * \mathbf{r2} \bullet \mathbf{a2} - b2 * \mathbf{r2} \bullet \mathbf{a1})/(\mathbf{r1} \bullet (\mathbf{a1} * \mathbf{a2} - \mathbf{a2} * \mathbf{a1}) \bullet \mathbf{r2}) \qquad (257) \\ t2 &= (b1 * \mathbf{r1} \bullet \mathbf{a2} - b2 * \mathbf{r1} \bullet \mathbf{a1})/(\mathbf{r2} \bullet \mathbf{a1} * \mathbf{r1} \bullet \mathbf{a2} - \mathbf{r2} \bullet \mathbf{a2} * \mathbf{r1} \bullet \mathbf{a1}) \\ &= (b2 * \mathbf{r1} \bullet \mathbf{a1} - b1 * \mathbf{r1} \bullet \mathbf{a2})/(\mathbf{r1} \bullet (\mathbf{a1} * \mathbf{a2} - \mathbf{a2} * \mathbf{a1}) \bullet \mathbf{r2}). \qquad (258) \end{aligned}$$

The line of solutions for **r•T**[**A**] = **b** is thus

$$\begin{aligned} \mathbf{r} = &(b1 * \mathbf{r2} \bullet \mathbf{a2} - b2 * \mathbf{r2} \bullet \mathbf{a1}) * \mathbf{r1}/(\mathbf{r1} \bullet (\mathbf{a1} * \mathbf{a2} - \mathbf{a2} * \mathbf{a1}) \bullet \mathbf{r2}) \\ &+ (b2 * \mathbf{r1} \bullet \mathbf{a1} - b1 * \mathbf{r1} \bullet \mathbf{a2}) * \mathbf{r2} \\ &\qquad /(\mathbf{r1} \bullet (\mathbf{a1} * \mathbf{a2} - \mathbf{a2} * \mathbf{a1}) \bullet \mathbf{r2}) \\ &+ t3 * \mathbf{n}. \end{aligned} \tag{259}$$

$$= (b1*\mathbf{a2}\cdot(\mathbf{r2}*\mathbf{r1} - \mathbf{r1}*\mathbf{r2}) + b1*\mathbf{a1}\cdot(\mathbf{r1}*\mathbf{r2} - \mathbf{r2}*\mathbf{r2}) / (\mathbf{r1}\cdot(\mathbf{a1}*\mathbf{a2} - \mathbf{a2}*\mathbf{a1})\cdot\mathbf{r2}) + (b2*\mathbf{r1}\cdot\mathbf{a1} - b1*\mathbf{r1}\cdot\mathbf{a2})*\mathbf{r2} / (\mathbf{r1}\cdot(\mathbf{a1}*\mathbf{a2} - \mathbf{a2}*\mathbf{a1})\cdot\mathbf{r2}) + t3*\mathbf{n} \quad (260)$$

$$= (b1*\mathbf{a2} - b2*\mathbf{a1})\cdot(\mathbf{r2}*\mathbf{r1} - \mathbf{r1}*\mathbf{r2}) / (\mathbf{r1}\cdot(\mathbf{a1}*\mathbf{a2} - \mathbf{a2}*\mathbf{a1})\cdot\mathbf{r2}) + t3*\mathbf{n}. \quad (261)$$

The image of V3 under **T**[**A**] is the plane

$$\{c1*\mathbf{r1}\cdot\mathbf{T}[\mathbf{A}] + c2*\mathbf{r2}\cdot\mathbf{T}[\mathbf{A}]\}$$

with c1 and c2 variable.

In the line of solutions there exists a unique solution **rn** orthogonal to **N**(**T**[**A**]) given by

$$\mathbf{rn} = t1*\mathbf{r1} + t2*\mathbf{r2} - (t1*\mathbf{r1} + t2*\mathbf{r2})\cdot\mathbf{n}*\mathbf{n}/(\mathbf{n}\cdot\mathbf{n}) \quad (262)$$

$$= ((b1*\mathbf{r2}\cdot\mathbf{a2} - b2*\mathbf{r2}\cdot\mathbf{a1})*\mathbf{r1} + (b2*\mathbf{r1}\cdot\mathbf{a1} - b1*\mathbf{r1}\cdot\mathbf{a2})*\mathbf{r2}) / (\mathbf{r1}\cdot(\mathbf{a1}*\mathbf{a2} - \mathbf{a2}*\mathbf{a1})\cdot\mathbf{r2})) \cdot[\mathbf{I} - \mathbf{n}*\mathbf{n}/\mathbf{n}\cdot\mathbf{n}]. \quad (263)$$

where **n** may be any vector in **N**(**T**[**A**]).

For rank(**T**[**A**])=1, the dimension of **N**(**T**[**A**]) is 2. Suppose non–parallel **n2** and **n3** in **N**(**T**[**A**]) while **r1** is not.

In this case ***T[A]*** maps V3 into a line which imposes a constraint on **b**, that is

$$\mathbf{b}\cdot\mathbf{T}[\mathbf{A}] \neq \mathbf{0} \quad (264)$$

Thus **b** is constrained to the line mapped by **T**[**A**]

Then for

$$\mathbf{r} = t1*\mathbf{r1} + t2*\mathbf{n2} + t3*\mathbf{n3}. \quad (265)$$

$$\mathbf{r}\cdot\mathbf{T}[\mathbf{A}] = t1*\mathbf{r1}\cdot\mathbf{T}[\mathbf{A}] = \mathbf{b}. \quad (266)$$

$$\begin{aligned} &t1 * \mathbf{r1} \bullet \mathbf{a1} = b1 \\ &t1 * \mathbf{r1} \bullet \mathbf{a2} = b2 \\ &t1 * \mathbf{r1} \bullet \mathbf{a3} = b3 \\ &b1 * \mathbf{a1} + b2 * \mathbf{a2} + b3 * \mathbf{a3} \neq \mathbf{0}. \end{aligned} \tag{267}$$

Thus

$$t1 = b1/\mathbf{r1} \bullet \mathbf{a1} = b2/\mathbf{r1} \bullet \mathbf{a2} = b3/\mathbf{r1} \bullet \mathbf{a3}. \tag{268}$$

The plane of solutions for $\mathbf{r} \bullet \mathbf{T}[\mathbf{A}] = \mathbf{b}$ is

$$P = \{\mathbf{r} | \mathbf{r} = b1 * \mathbf{r1}/\mathbf{a1} \bullet \mathbf{r1} + t2 * \mathbf{n2} + t3 * \mathbf{n3}\} \tag{269}$$

where now t2 and t3 vary.

The image of V3 under $\mathbf{T}[\mathbf{A}]$ is the line $\{t1 * \mathbf{b}\}$.

In the plane of solutions there exist a unique solution $\mathbf{rn}$ orthogonal to $\mathbf{N}(\mathbf{T}[\mathbf{A}])$ given[28] by

$$\begin{aligned} \mathbf{rn} = (b1/\mathbf{r1} \bullet \mathbf{a1}) * \mathbf{r1} \\ \bullet [\mathbf{I} - ((\mathbf{n2} * \mathbf{n3} - \mathbf{n3} * \mathbf{n2}) \bullet \mathbf{n3} * \mathbf{n2} \\ + (\mathbf{n3} * \mathbf{n2} - \mathbf{n2} * \mathbf{n3}) \bullet \mathbf{n2} * \mathbf{n3}) \\ /(\mathbf{n2} \bullet \mathbf{n2} * \mathbf{n3} \bullet \mathbf{n3} - \mathbf{n2} \bullet \mathbf{n3} * \mathbf{n2} \bullet \mathbf{n3}). \end{aligned} \tag{270}$$

The solutions of $\mathbf{r0} \bullet \mathbf{T}[\mathbf{A}] = \mathbf{r1}$ for $\mathbf{A}$, where $\mathbf{r0}$ and $\mathbf{r1}$ are given vectors are *not* generally unique since the equation is a vectored form of three equations in nine unknowns. The equation

$$\mathbf{r0} \bullet \mathbf{T}[\mathbf{A}] = \mathbf{r0} \bullet \mathbf{a1} * \mathbf{u1} + \mathbf{r0} \bullet \mathbf{a2} * \mathbf{u2} + \mathbf{r0} \bullet \mathbf{a3} * \mathbf{u3} \tag{271}$$

$$\equiv r11 * \mathbf{u1} + r12 * \mathbf{u2} + r13 * \mathbf{u3} \tag{272}$$

devolves into

$$\mathbf{r0} \bullet \mathbf{a1} = r11 \tag{273}$$

$$\mathbf{r0} \bullet \mathbf{a2} = r12 \tag{274}$$

$$\mathbf{r0} \bullet \mathbf{a3} = r13 \tag{275}$$

each with its plane of solutions.

Solutions for rank(**A**) = 3 must satisfy

$$\mathbf{r0} = \mathbf{r1}\bullet\mathbf{T}[\mathbf{A}]^{-1} \tag{276}$$
$$= (r11*(\mathbf{a2}\wedge\mathbf{a3}) + r12*(\mathbf{a3}\wedge\mathbf{a1}) + r13*(\mathbf{a1}\wedge\mathbf{a2}))/\det[\mathbf{A}] \tag{277}$$
$$\equiv r0*(c1*\mathbf{u1} + c2*\mathbf{u2} + c3*\mathbf{u3}). \tag{278}$$

As one solution, consider the solutions based on the **q1i**(**r0**) reciprocal vectors,

$$\mathbf{a1} = r11*\mathbf{u1}/(r0*c1) \tag{279}$$
$$\mathbf{a2} = r12*\mathbf{u2}/(r0*c2) \tag{280}$$
$$\mathbf{a3} = r13*\mathbf{u3}/(r0*c3). \tag{281}$$

Then

$$(\mathbf{a1}\wedge\mathbf{a2}) = r11*\mathbf{u1}/(r0*c1) \wedge r12*\mathbf{u2}/(r0*c2) \tag{282}$$
$$= r11*r12*\mathbf{u3}/(r0^2*c1*c2) \tag{283}$$
$$(\mathbf{a2}\wedge\mathbf{a3}) = r12*r13*\mathbf{u1}/(r0^2*c2*c3) \tag{284}$$
$$(\mathbf{a3}\wedge\mathbf{a1}) = r13*r11*\mathbf{u2}/(r0^2*c3*c1) \tag{285}$$
$$\det[\mathbf{A}] = \mathbf{a1}\bullet(\mathbf{a2}\wedge\mathbf{a3}) \tag{286}$$
$$= r11*r12*r13/(r0^3*c1*c2*c3). \tag{287}$$

For this choice

$$\mathbf{r0}\bullet\mathbf{T}[\mathbf{A}] = r11*\mathbf{u1} + r12*\mathbf{u2} + r12*\mathbf{u3}$$
$$= \mathbf{r1} \tag{288}$$

and

$$\mathbf{r1}\bullet\mathbf{T}[\mathbf{A}]^{-1} = (r11*r12*r13*\mathbf{u1}/(r0^2*c2*c3)$$
$$+ r11*r12*r13*\mathbf{u2}/(r0^2*c1*c3)$$
$$+ r11*r12*r13*\mathbf{u3}/(r0^2*c1*c2))$$
$$/(r11*r12*r13/(r0^3*c1*c2*c3))$$
$$= r0*(c1*\mathbf{u1} + c2*\mathbf{u2} + c3*\mathbf{u3})$$
$$= \mathbf{r0}. \tag{289}$$

Thus

$$\mathbf{A} = r11*\mathbf{u1}*\mathbf{u1}/(r0*c1)$$
$$+ r12*\mathbf{u2}*\mathbf{u2}/(r0*c2)$$
$$+ r13*\mathbf{u3}*\mathbf{u3}/(r0*c3) \tag{290}$$

is a solution[29] of rank 3 for **r0**•**T**[**A**] = **r1**.

Six similar solutions of rank 3 exist, namely

$$\mathbf{a1} = r11*\mathbf{u1}/(r0*c1) \quad (291)$$
$$\mathbf{a2} = r12*\mathbf{u2}/(r0*c2) \quad (292)$$
$$\mathbf{a3} = r13*\mathbf{u3}/(r0*c3) \quad (293)$$

$$\mathbf{a1} = r12*\mathbf{u2}/(r0*c1) \quad (294)$$
$$\mathbf{a2} = r11*\mathbf{u1}/(r0*c2) \quad (295)$$
$$\mathbf{a3} = r13*\mathbf{u3}/(r0*c3) \quad (296)$$

$$\mathbf{a1} = r13*\mathbf{u3}/(r0*c1) \quad (297)$$
$$\mathbf{a2} = r12*\mathbf{u2}/(r0*c2) \quad (298)$$
$$\mathbf{a3} = r11*\mathbf{u1}/(r0*c3) \quad (299)$$

$$\mathbf{a1} = r11*\mathbf{u1}/(r0*c1) \quad (300)$$
$$\mathbf{a2} = r13*\mathbf{u3}/(r0*c2) \quad (301)$$
$$\mathbf{a3} = r12*\mathbf{u2}/(r0*c3) \quad (302)$$

$$\mathbf{a1} = r12*\mathbf{u2}/(r0*c1) \quad (303)$$
$$\mathbf{a2} = r13*\mathbf{u3}/(r0*c2) \quad (304)$$
$$\mathbf{a3} = r11*\mathbf{u1}/(r0*c3) \quad (305)$$

$$\mathbf{a1} = r13*\mathbf{u3}/(r0*c1) \quad (306)$$
$$\mathbf{a2} = r11*\mathbf{u1}/(r0*c2) \quad (307)$$
$$\mathbf{a3} = r12*\mathbf{u2}/(r0*c3). \quad (308)$$

One solution is thus diagonal; three solutions have one element on the diagonal and finally two solutions have no diagonal elements. These solutions are called **sparse** solutions.

To find the solutions of **r0•T[A]** = **r1** of rank 3 in general, chose three distinct directions, **ua1**, **ua2**, and **ua3**. Next, define

$$t1 \equiv \mathbf{r1}\bullet\mathbf{u1}/(\mathbf{r0}\bullet\mathbf{ua1}) \quad (309)$$
$$t2 \equiv \mathbf{r1}\bullet\mathbf{u2}/(\mathbf{r0}\bullet\mathbf{ua2}) \quad (310)$$
$$t3 \equiv \mathbf{r1}\bullet\mathbf{u3}/(\mathbf{r0}\bullet\mathbf{ua3}). \quad (311)$$

Then

$$\mathbf{A} = t1 \ast \mathbf{u1} \ast \mathbf{ua1} + t2 \ast \mathbf{u2} \ast \mathbf{ua2} + t3 \ast \mathbf{u3} \ast \mathbf{ua3}$$

is a solution, since

$$\begin{aligned}
\mathbf{r0} \bullet \mathbf{T}[\mathbf{A}] &= \mathbf{r0} \bullet [t1 \ast \mathbf{ua1} \ast \mathbf{u1} + t2 \ast \mathbf{ua2} \ast \mathbf{u2} + t3 \ast \mathbf{ua3} \ast \mathbf{u3}] \\
&= t1 \ast \mathbf{r0} \bullet \mathbf{ua1} \ast \mathbf{u1} \\
&\quad + t2 \ast \mathbf{r0} \bullet \mathbf{ua2} \ast \mathbf{u2} \\
&\quad + t3 \ast \mathbf{r0} \bullet \mathbf{ua3} \ast \mathbf{u3} \\
&= (\mathbf{r1} \bullet \mathbf{u1} / \mathbf{r0} \bullet \mathbf{ua1}) \ast \mathbf{r0} \bullet \mathbf{ua1} \ast \mathbf{u1} \\
&\quad + (\mathbf{r1} \bullet \mathbf{u2} / \mathbf{r0} \bullet \mathbf{ua2}) \ast \mathbf{r0} \bullet \mathbf{ua2} \ast \mathbf{u2} \\
&\quad + (\mathbf{r1} \bullet \mathbf{u3} / \mathbf{r0} \bullet \mathbf{ua3}) \ast \mathbf{r0} \bullet \mathbf{ua3} \ast \mathbf{u3} \\
&= \mathbf{r1} \bullet [\mathbf{u1} \ast \mathbf{u1} + \mathbf{u2} \ast \mathbf{u2} + \mathbf{u3} \ast \mathbf{u3}] \\
&= \mathbf{r1}
\end{aligned} \tag{312}$$

To find the rank 2 solutions of **r0•T[A]** = **r1** chose two distinct directions, **ua1**, **ua2**. Next define

$$t1 \equiv \mathbf{r1} \bullet \mathbf{u1} / \mathbf{r0} \bullet \mathbf{ua1} \tag{313}$$

$$t2 \equiv \mathbf{r1} \bullet \mathbf{u2} / \mathbf{r0} \bullet \mathbf{ua2} \tag{314}$$

$$t3 \equiv \mathbf{r1} \bullet \mathbf{u3} / \mathbf{r0} \bullet \mathbf{ua2} \tag{315}$$

Then

$$\mathbf{A} = t1 \ast \mathbf{u1} \ast \mathbf{ua1} + t2\mathbf{u2} \ast \mathbf{ua2} + t3 \ast \mathbf{u3} \ast \mathbf{ua2}$$

is a solution, since

$$\begin{aligned}
\mathbf{r0} \bullet \mathbf{T}[\mathbf{A}] &= \mathbf{r0} \bullet [t1 \ast \mathbf{ua1} \ast \mathbf{u1} + t2 \ast \mathrm{ua2u2} + t3 \ast \mathbf{ua2} \ast \mathbf{u3}] \\
&= t1 \ast \mathbf{r0} \bullet \mathbf{ua1} \ast \mathbf{u1} \\
&\quad + t2 \ast \mathbf{r0} \bullet \mathbf{ua2} \ast \mathbf{u2} \\
&\quad + t3 \ast \mathbf{r0} \bullet \mathbf{ua2} \ast \mathbf{u3} \\
&= (\mathbf{r1} \bullet \mathbf{u1} / \mathbf{r0} \bullet \mathbf{ua1}) \ast \mathbf{r0} \bullet \mathbf{ua1} \ast \mathbf{u1} \\
&\quad + (\mathbf{r1} \bullet \mathbf{u2} / \mathbf{r0} \bullet \mathbf{ua2}) \ast \mathbf{r0} \bullet \mathrm{ua2u2} \\
&\quad + (\mathbf{r1} \bullet \mathbf{u3} / \mathbf{r0} \bullet \mathbf{ua2}) \ast \mathbf{r0} \bullet \mathbf{ua2} \ast \mathbf{u3} \\
&= \mathbf{r1} \bullet [\mathbf{u1} \ast \mathbf{u1} + \mathbf{u2} \ast \mathbf{u2} + \mathbf{u3u} \ast \mathbf{3}] \\
&= \mathbf{r1}.
\end{aligned} \tag{316}$$

Six similar variations may be formed for any arbitrary choice of any two distinct directions.

Mathematics: Vector Algebra

To find the rank 1 solutions of **r0•T[A]** = **r1** chose any direction **ua1** not orthogonal to **r0**. Next, define

$$t1 \equiv \mathbf{r1 \cdot u1 / r0 \cdot ua1} \quad (317)$$
$$t2 \equiv \mathbf{r1 \cdot u2 / r0 \cdot ua1} \quad (318)$$
$$t3 \equiv \mathbf{r1 \cdot u3 / r0 \cdot ua1}. \quad (319)$$

Then

$$\mathbf{A} = t1 * \mathbf{u1} * \mathbf{ua1} + t2 * \mathbf{u2} * \mathbf{ua1} + t3 * \mathbf{u3} * \mathbf{ua1}$$

is a solution, since

$$\begin{aligned}
\mathbf{r0} \cdot \mathbf{T}[\mathbf{A}] &= \mathbf{r0} \cdot [t1 * \mathbf{ua1} * \mathbf{u1} + t2 * \mathbf{ua1} * \mathbf{u2} + t3 * \mathbf{ua1} * \mathbf{u3}] \\
&= t1 * \mathbf{r0} \cdot \mathbf{ua1} * \mathbf{u1} \\
&\quad + t2 * \mathbf{r0} \cdot \mathbf{ua1} * \mathbf{u2} \\
&\quad + t3 * \mathbf{r0} \cdot \mathbf{ua1} * \mathbf{u3} \\
&= (\mathbf{r1} \cdot \mathbf{u1} / \mathbf{r0} \cdot \mathbf{ua1} *) \mathbf{r0} \cdot \mathbf{ua1} * \mathbf{u1} \\
&\quad + (\mathbf{r1} \cdot \mathbf{u2} / \mathbf{r0} \cdot \mathbf{ua1}) * \mathbf{r0} \cdot \mathbf{ua1} * \mathbf{u2} \\
&\quad + (\mathbf{r1} \cdot \mathbf{u3} / \mathbf{r0} \cdot \mathbf{ua1}) * \mathbf{r0} \cdot \mathbf{ua1} * \mathbf{u3} \\
&= \mathbf{r1} \cdot [\mathbf{u1} * \mathbf{u1} + \mathbf{u2} * \mathbf{u2} + \mathbf{u3} * \mathbf{u3}] \\
&= \mathbf{r1}. \qquad (320)
\end{aligned}$$

For **ua1** = **ur0**

$$\begin{aligned}
\mathbf{A} &= t1 * \mathbf{u1} * \mathbf{ur0} + t2 * \mathbf{u2} * \mathbf{ur0} + t3 * \mathbf{u3} * \mathbf{ur0} \\
&= (\mathbf{r1} \cdot \mathbf{u1}) * \mathbf{u1} * \mathbf{ur0} / r0 \\
&\quad + (\mathbf{r1} \cdot \mathbf{u2}) * \mathbf{u2} * \mathbf{ur0} / r0 \\
&\quad + (\mathbf{r1} \cdot \mathbf{u3}) * \mathbf{u3} * \mathbf{r0} / r0 \\
&= \mathbf{r1} \cdot (\mathbf{u1} * \mathbf{u1} + \mathbf{u2} * \mathbf{u2} + \mathbf{u3} * \mathbf{u3}) * \mathbf{ur0} / r0 \\
&= \mathbf{r1} * \mathbf{qd}(\mathbf{r0}), \qquad (321)
\end{aligned}$$

an outer product.

Furthermore for any **r**

$$\begin{aligned}
\mathbf{r} \cdot \mathbf{T}[\mathbf{A}] &= \mathbf{r} \cdot [t * \mathbf{ua1} * \mathbf{u1} + t2 * \mathbf{ua1} * \mathbf{u2} + t3 * \mathbf{ua1} * \mathbf{u3}] \\
&= t1 * \mathbf{r} \cdot \mathbf{ua1} * \mathbf{u1} + t2 * \mathbf{r} \cdot \mathbf{ua1} * \mathbf{u2} + t3 * \mathbf{r} \cdot \mathbf{ua1} * \mathbf{u3} \\
&= (\mathbf{r1} \cdot \mathbf{u1} / \mathbf{r0} \cdot \mathbf{ua1}) * \mathbf{r} \cdot \mathbf{ua1} * \mathbf{u1} \\
&\quad + (\mathbf{r1} \cdot \mathbf{u2} / \mathbf{r0} \cdot \mathbf{ua1}) * \mathbf{r} \cdot \mathbf{ua1} * \mathbf{u2} \\
&\quad + (\mathbf{r1} \cdot \mathbf{u3} / \mathbf{r0} \cdot \mathbf{ua1}) * \mathbf{r} \cdot \mathbf{ua1} * \mathbf{u3} \\
&= (\mathbf{r} \cdot \mathbf{ua1} / \mathbf{r0} \cdot \mathbf{ua1}) * \mathbf{r1} \cdot [\mathbf{u1} * \mathbf{u1} + u2u2 + \mathbf{u3} * \mathbf{u3}] \\
&= (\mathbf{r} \cdot \mathbf{ua1} / \mathbf{r0} \cdot \mathbf{ua1}) * \mathbf{r1}. \qquad (322)
\end{aligned}$$

Thus the matrix of rank 1 is associated with a unique direction.

Vector Calculus.

In addition to an algebra, the set of vectors, *V3*, like the real number system *R*, is endowed with a topology. Based on its topology *V3* supports limits and thus a calculus.

Functions over *V3* may have limits, may be continuous, may have derivatives in a sense not so much different as analogous to functions over *R*. The discussion of these matters begins with functions which are closely related to real functions over *R.*

Straight Maps in V3

A linear map of the underlying field into the set of vectors produces "straight" figures. There are three such maps:

$$R \longrightarrow V3:\ \mathbf{r}(x1) = \mathbf{r0} + x1 * \mathbf{r1}$$
$$(RxR) \longrightarrow V3:\ \mathbf{r}(x1,x2) = \mathbf{r0} + x1 * \mathbf{r1} + x2 * \mathbf{r2}$$
$$(RxRxR) \longrightarrow V3:\ \mathbf{r}(x1,x2,x3) = \mathbf{r0} + x1 * \mathbf{r1} + x2 * \mathbf{r2} + x3 * \mathbf{r3} \qquad (323)$$

to denote **linear** lines, **linear surfaces** (planes), and **linear regions** in V3.

Here **r0**,**r1**,**r2**,**r3** are fixed, non–parallel vectors and x1,x2,x3 are real variables.

The reader should note the existence of non–linear straight lines, planes and regions:

$$\mathbf{r}(x1) = \mathbf{r0} + f(x1) * \mathbf{r1}$$
$$\mathbf{r}(x1,x2) = \mathbf{r0} + f1(x1,x2) * \mathbf{r1} + f2(x1,x2) * \mathbf{r2}$$
$$\mathbf{r}(x1,x2,x3) = \mathbf{r0} + f1(x1,x2,x3) * \mathbf{r1} + f2(x1,x2,x3) * \mathbf{r2} + f3(x1,x2,x3) * \mathbf{r3}. \qquad (324)$$

Thus "linear" is a subset of "straight".

These concepts allow the same subset of vectors to be considered in different processes.

Mathematics: Vector Calculus

Example:

$$\mathbf{r}(x1) = \mathbf{r0} + x1*x1*\mathbf{r1}$$

is a straight, non–linear line,

$$\mathbf{r}(x1,x2) = \mathbf{r0} + x1*x1*\mathbf{r1} + x2*\mathbf{r2}$$

is a straight, non–linear plane

$$\mathbf{r}(x1,x2,x3) = \mathbf{r0} + x1*x1*\mathbf{r1} + x2*\mathbf{r2} + x3*\mathbf{r3}$$

is a straight, non–linear vicinity.

A straight line need not have derivatives; if it has derivatives they need not be basic.

For a straight line, $\mathbf{r}(x) = \mathbf{r0} + f(x)*\mathbf{r1}$, with a basic derivative

$$D[x](\mathbf{r}(t);dt) = D[x](f(t);dt)*\mathbf{r1}. \tag{325}$$

The unique and constant direction, $\mathbf{u}(\mathbf{r1}) \equiv \mathbf{ur1}$, is called the **straight line's direction**.

A straight surface or **plane** has two principal directions:

$$D[x](\mathbf{r}(t1)|x2;dt1) = D[t](f(t1)|x2;dt1)*\mathbf{r1} \tag{326}$$

and

$$D[t](\mathbf{r}(t2)|x1;dt2) = D[t](f(t2)|x1;dt2)*\mathbf{r2} \tag{327}$$

For fixed x1, from equation (323)

$$\mathbf{r}(x1,x2) - \mathbf{r0} - f(x1)*\mathbf{r1} = f(x2)*\mathbf{r2} \tag{328}$$

defines a set of lines, one for each x1. These are called the **x1 lines of the plane**.

Similarly, for fixed x2,

$$\mathbf{r}(x2,x1) - \mathbf{r0} - f(x2)*\mathbf{r2} = f(x1)*\mathbf{r1} \tag{329}$$

defines a set of lines, one for each x2. These are called the **x2 lines of the plane**.

A vector function **f**(**r**) may be restricted to the line

r(x1) = **r0** + x1∗**r1**

in which case it is written

f(**r**)|**r**(x1)

or more specifically

f(**r**)|(**r0** + x1∗**r1**).

Similarly, functions of *V3* may be restricted to planes or vicinities.

More generally, a restriction may apply to **r**(x1), **r**(x1,x2), or **r**(x1,x2,x3) non–straight subsets of V3. These restrictions are called **curved**. A function may be restricted to a curved line in which case it is also written **f**(**r**)|**r**(x1).

The topology of V3.

Consider the set, *VT*, of subsets of the three dimensional set of vectors which are arbitrary unions of sets

N = {**r**|abs(**r**–**r1**)<er}

for all positive *er* of *R* and all ***r1*** of V3. Then

V3 and the null set are subsets of VT
VT is closed with respect to the intersection
of a finite number of its subsets,
VT is closed with respect to the union
of an arbitrary collection of its members.

VT is thus a **topology** in V3. The members of VT are called the **open** sets of V3.

For a given ***r1*** and a given *er*, the set N(**r1**,er) is called an **open neighborhood** of **r1**.

Definition 11 (restricted topology)
Given
VT, the topology of V3;
U an arbitrary subset of V3;
then

VT|U ≡ {VT ∩ U} is called the **topology of V3 restricted to U**.

end of definition

The sets of VT|U are called the open sets of U. The open sets of U may contains sets which are *not* open in VT.

Let VM be the smallest set of subsets of V3 which includes all its open sets, their complements, and all denumerable unions of the such subsets. The sets of VM are called the **measurable** sets of V3.

Similarly let VM|U be the smallest set of subsets of VT|U which contain all its open sets, their complements in U and all denumerable unions of such subsets. The sets VM|U are called the **measurable** sets of U.

Since

$$0 \leq \text{abs}(\mathbf{r}-\mathbf{r1}) < \infty$$

and $0 = \text{abs}(\mathbf{r}-\mathbf{r1})$ if and only if $\mathbf{r} = \mathbf{r1}$

and $\text{abs}(\mathbf{r}-\mathbf{r1}) = \text{abs}(\mathbf{r1}-\mathbf{r})$

and $\text{abs}(\mathbf{r}-\mathbf{r1}) \leq \text{abs}(\mathbf{r}-\mathbf{r2}) + \text{abs}(\mathbf{r2}-\mathbf{r1})$,

abs may serve as a metric. In VT, both **r** and **r1** may be any vectors of V3; in VT|U, both **r** and **r1** may be any vectors of U.

With this metric in VT any vector ***r1*** is a limit vector of *V3*. A corresponding statement may not be made generally for VT|U.

Curves in V3

A curve, serving as a restricted set, may also form the basis for functional limits.

Curves may be described either with reference to the underlying field *R* or to the set of vectors *V3*. Clearly an unrestricted collection of vectors {**r**} does not describe a curve.

Definition 12 (connected neighborhoods)

Given

er>0;

N(**r1**,er), an open neighborhood of **r1**;

N(**r2**,er), an open neighborhood of **r2**;

if for any er

$$N(\mathbf{r1},\text{er}) \cap N(\mathbf{r2},\text{er}) \text{ is not empty,}$$

then

r1 is connected to r2.

end of definition

A set of vectors, all of whose elements have connected neighborhoods is a **connected set**.

If **r1** and **r2** are connected, the direction, **u**(**r1**– **r2**) is called the **direction of the connection**.

In an arbitrary, connected set of vectors, a single vector may be connected in all directions, in some directions, or only in one direction.

Definition 13 (connected sets: curved lines, surfaces, and vicinities)

A connected set, each of whose vectors is connected in no more than a finite number of directions is a vectorial **curved line**.

A connected set, each of whose vectors is connected in an infinite number of directions each of which can be resolved into two fixed directions is a vectorial **curved surface**.

A connected set, each of whose vectors is connected in all possible directions is a vectorial **curved vicinity**.

end of definition

Connected sets are thus topological ideas defined over a restricted topology of V3. As such, other topological ideas may apply to them.

Definition 14 (limit vectors in lines, surfaces, and vicinities)

Given

S, a subset of V3 consisting only of connected vectors;
r1, a vector in S;
er >0
N(**r1**,er), an open neighborhood of **r1** in S

then

if there exists for any er, another vector **r2** connected to **r1**

r1 is a limit vector of S.

end of definition

So defined, all vectors in lines, surfaces and vicinities are limit vectors in their respective subsets.

Mathematics: Vector Calculus

If S1 and S2 are each connected sets, with S1 ∩ S2 empty, S1 ∪ S2 is said to be **a set composed of two disconnected segments**.

The set of connected vectors between **r1** and **r2** in a curved line is symbolized as C(**r1**,**r2**).

Limits in V3.

In V3, limits of functions are defined in terms of limit vectors in V3 (their domain) and the topologies of their functional ranges. Like those in *R*, limits in V3 are a somewhat fussy subject.

Omni–Directional Limits

Consider now f(**r**) ≡ **r•q**(**r**) = 3. For any **r1** in V3, f(**r**) has a limit which holds in any direction. For this and similar functions, the limit may be specified directly from the topology of V3.

Definition 15 (omni–directional limits in V3)

Given

ef and er, real positive numbers greater than zero;

r1 and its er neighborhood, {**r**|abs(**r**–**r1**)<er};

dr, any vector in the neighborhood;

f(**r**) a given vector function;

if there can be found a vector **f1** such that

for any ef>0 there can always be found an er>0 such that

abs(**f**(**r1**+**dr**) – **f1**)<ef (330)

for *any* **dr** in the neighborhood

then

f1 is the **omni–directional limit of f at r1**.

Similarly, for f(**r**) a given scalar function of V3;

if there can be found a scalar f1 such that

for any ef>0 there can always be found an er>0 such that

abs(f(**r1**+**dr**) – f1)<ef (331)

for *any* **dr** in the neighborhood

then

f1 is the **omni–directional limit of** f **at r1**.

end of definition

Such a limit, if it exists, is written

$$\lim \mathbf{f}(\mathbf{r1+dr}) = \mathbf{f1} \text{ as } \mathbf{dr} \rightarrow \mathbf{0} \quad (332)$$

$$\lim f(\mathbf{r1+dr}) = f1 \text{ as } \mathbf{dr} \rightarrow \mathbf{0} \quad (333)$$

or

$$\mathbf{f}(\mathbf{r1+dr}) \rightarrow \mathbf{f1} \text{ as } \mathbf{dr} \rightarrow \mathbf{0} \quad (334)$$

$$f(\mathbf{r1+dr}) \rightarrow f1 \text{ as } \mathbf{dr} \rightarrow \mathbf{0} \quad (335)$$

A function is said to be **omni–directionally continuous** at **r1** if

$$\mathbf{f}(\mathbf{r1}) = \lim \mathbf{f}(\mathbf{r1+dr}) \text{ as } \mathbf{dr} \rightarrow \mathbf{0} \quad (336)$$

In contrast to real functions, where limits may be defined from only two directions, in V3 limits may be defined for an infinite number of directions.

Consequently, functions can exist which do not have a limit in an omni–directional sense, but which may have limits in a restricted sense. Limits for these functions can be defined from a restricted topology on V3. The restrictions to a straight line or to a curved line are useful. In the case of a straight line the restriction can be characterized by a single direction[30], and so such restricted limits are called **directional limits**.

Accordingly, functions which may not be continuous omni–directionally may nevertheless be continuous directionally.

Mathematics: Vector Calculus

Directional Limits

At **r1** let **ur** ≡ **u**(**r**–**r1**) be a given direction. Then for **f**(**r**) or f(**r**), the quantities **f**(**r1**+dr∗**ur**) or f(**r1**+dr∗**ur**) may have a limit[31] as dr approaches 0.

Definition 16 (directional limit in V3)

Given

ef and dr, real positive numbers greater than zero;
r and **r1**, real vectors in V3;
ur ≡ **u**(**r**–**r1**), a direction;
f(**r**) a given vector function;

if there can be found a vector **f1** such that

for any ef>0 there can always be found a dr>0 whenever

abs(**f**(**r1**+dr∗**ur**) – **f1**)<ef, (337)

then

f1 is the **forward ur directional limit of f at r1**;

if there can be found a vector **f1** such that

abs(**f1** – **f**(**r1**–dr∗**ur**))<ef, (338)

then

f1 is the **backward ur directional limit of f at r1**;

for f(**r**) a given scalar function of V3;

if there can be found a scalar f1 such that

for any ef>0 there can always be found a dr>0 whenever

abs(f(**r1**+dr∗**ur**) – f1)<ef, (339)

then

f1 is the **forward ur directional limit of** f **at r1**;

if there can be found a scalar f1 such that

abs(f1 – f(**r1**–dr∗**ur**))<ef, (340)

then

f1 is the **backward ur directional limit of** f **at r1**.

end of definition

Directional limits are symbolized as

$$\lim_f \mathbf{f}|\mathbf{ur} = \mathbf{f1} \text{ at } \mathbf{r1} \text{ as } dr \longrightarrow 0 \tag{341}$$
$$\lim_f f|\mathbf{ur} = f1 \text{ at } \mathbf{r1} \text{ as } dr \longrightarrow 0 \tag{342}$$
$$\lim_b \mathbf{f}|\mathbf{ur} = \mathbf{f1} \text{ at } \mathbf{r1} \text{ as } dr \longrightarrow 0 \tag{343}$$
$$\lim_b f|\mathbf{ur} = f1 \text{ at } \mathbf{r1} \text{ as } dr \longrightarrow 0 \tag{344}$$

or

$$\mathbf{f}(\mathbf{r1}+\mathbf{dr})|\mathbf{ur}\longrightarrow\mathbf{f1} \text{ as } dr\longrightarrow 0 \tag{345}$$
$$f(\mathbf{r1}+\mathbf{dr})|\mathbf{ur}\longrightarrow f1 \text{ as } dr\longrightarrow 0 \tag{346}$$
$$\mathbf{f}(\mathbf{r1}-\mathbf{dr})|\mathbf{ur}\longrightarrow\mathbf{f1} \text{ as } dr\longrightarrow 0 \tag{347}$$
$$f(\mathbf{r1}-\mathbf{dr})|\mathbf{ur}\longrightarrow f1 \text{ as } dr\longrightarrow 0. \tag{348}$$

Directional limits may be viewed as limits of the restricted function **f**(**r**)|(dr✱**ur**) in the topology restricted to the **ur** direction.

Note that at **r1** the directional limit in one direction **ur1** may differ from one in another direction **ur2**.

These limits need not exist, the function's forward directional limit need not equal its backward directional limit, nor need either equal **f**(**r1**) or f(**r1**).

A function is said to be **continuous at r1 in the ur direction** if

$$\mathbf{f}(\mathbf{r1}) = \lim \mathbf{f}(\mathbf{r1} + dr*\mathbf{ur}) = \lim \mathbf{f}(\mathbf{r1} - dr*\mathbf{ur}) \text{ as } dr \longrightarrow 0. \tag{349}$$

Limits in Quadrant

Are there other classes of functional limits which still preserve a sense of direction, but apply to a larger class of functions than the omni–directional limit? In particular is there a way of grouping directions to obtain a classification of limits intermediate between directional and omni–directional limits?

The arbitrary orientation of V3 provides just such a possibility.

In this classification an arbitrary vector **r**=r✱(c1✱**u1**+c2✱**u2**+c3✱**u3**) is classified by the sign of its directions, c1, c2, and c3. There are 2^3=8 such gross "directions" which are called **quadrants**. The quadrant for which all the ci are non–negative is called the **positive quadrant**.

Mathematics: Vector Calculus

Definition 17 (limits *of* the positive quadrant)
Given
dr1, dr2, dr3 positive real numbers;
f(**r**), a given vectorial function over V3;
if there can be found a vector **f1** such that

lim (**f**(**r1**+dr1∗**u1**)) = **f1** as dr1 ⟶ 0

and

lim (**f**(**r1**+dr2∗**u2**)) = **f1** as dr2 ⟶ 0

and

lim (**f**(**r1**+dr3∗**u3**)) = **f1** as dr3 ⟶ 0 (350)

then

f1 is the **limit of f of the positive quadrant at r1**.

Similarly, for f(**r**) a given scalar function of V3;
if there can be found a scalar f1 such that

lim (f(**r1**+dr1∗**u1**)) = f1 as dr1 ⟶ **0**

and

lim (f(**r1**+dr2∗**u2**)) = f1 as dr2 ⟶ **0**

and

lim (f(**r1**+dr3∗**u3**)) = f1 as dr3 ⟶ **0** (351)

then

f1 is the **limit of** f **of the positive quadrant at r1**;
end of definition

Limits of the positive quadrant may be seen as limits in the topology restricted to the directions **u1**, **u2**, and **u3**.

Limits of functions *of* the positive quadrant are symbolized as:

lim **f(r1+{dr})** = **f1** as **dr+** ⟶ **0** (352)
lim f(**r1+{dr}**) = f1 as **dr+** ⟶ **0** (353)

or

f(r1+{dr})⟶f1 as **dr+⟶0** (354)
f(**r1+{dr}**)⟶f1 as **dr+⟶0**. (355)

Functions with limits of the positive quadrant have the *same* limit in the three directions of the origin. They differ from functions with three different limits in those same directions.

Some functions may also have the same limit for any direction in the positive quadrant.

Definition 18 (limits *in* the positive quadrant)
Given
ur, a direction in the positive quadrant;
f(**r**), a given vectorial function over V3;
if there can be found a vector **f1** for all **ur** in the positive quadrant such that

$$\lim (\mathbf{f}(\mathbf{r1}+dr*\mathbf{ur})) = \mathbf{f1} \text{ as } dr \longrightarrow 0 \qquad (356)$$

then

f1 is the **limit of f in the positive quadrant at r1**.

Similarly, for f(**r**) a given scalar function of V3;
if there can be found a scalar f1 for all **ur** in the positive quadrant such that

$$\lim (f(\mathbf{r1}+dr*\mathbf{ur})) = f1 \text{ as } dr \longrightarrow 0 \qquad (357)$$

then

f1 is the **limit of** f **in the positive quadrant at r1**.

end of definition

Limits of functions *in* the positive quadrant are symbolized as:

$$\lim \mathbf{f}(\mathbf{r1}+\mathbf{dr}) = \mathbf{f1} \text{ as } \mathbf{dr}+ \longrightarrow \mathbf{0} \qquad (358)$$
$$\lim f(\mathbf{r1}+\mathbf{dr}) = f1 \text{ as } \mathbf{dr}+ \longrightarrow \mathbf{0} \qquad (359)$$

or

$$\mathbf{f}(\mathbf{r1}+\mathbf{dr})\longrightarrow\mathbf{f1} \text{ as } \mathbf{dr}+\longrightarrow\mathbf{0} \qquad (360)$$
$$f(\mathbf{r1}+\mathbf{dr})\longrightarrow f1 \text{ as } \mathbf{dr}+\longrightarrow\mathbf{0} \qquad (361)$$

Limits in the positive quadrant may be seen as limits in the topology restricted to the positive quadrant.

If a function has a limit *in* the positive quadrant, it also has the same limit *of* the positive quadrant. The reverse is not generally true.

If **f**(**r1**) = **f1**, the limit in the positive quadrant, ***f*** is called **continuous in the positive quadrant at r1**.

Even if ***f*** is continuous in the positive quadrant at **r1**, it does not follow that it is continuous omni–directionally or even continuous in a direction not in the positive quadrant.

Limits of Section

The idea of the positive quadrant can be broadened to an arbitrary section.

Definition 19 (section)
Given
sri≥0, i=1,2,3, real variables;
uiR1, i=1,2,3, non–parallel directions;
for **r1** as reference
the set

SECT(**r1**,**u1R1**,**u2R1**,**u3R1**)
≡ {r|**r** = **r1** + sr1∗**u1R1** + sr2∗**u2R1** +sr3∗**u3R1**} (362)
for all sri

is called a **section of V3 at r1**.

The subsets
{**r**|**r** = **r1** + sr1∗**u1R1** + sr2∗**u2R1**}
{**r**|**r** = **r1** + sr1∗**u1R1** + sr3∗**u3R1**}
{**r**|**r** = **r1** + sr2∗**u2R1** + sr3∗**u3R1**}
are called the **faces** of the section.
end of definition

A vector **r2** in SECT(**r1**,**u1R1**,**u2R1**,**u3R1**) has two representations[32]:

r2(**u1R1**,**u2R1**,**u3R1**) = **r1** + sr1∗**u1R1**
+ sr2∗**u2R1**
+ sr3∗**u3R1** (363)

and

r2(**u1**,**u2**,**u3**) = **r1** + r1∗**u1** + r2∗**u2** + r3∗**u3** (364)

For vectors represented sectionally the sri are positive, while **uiR1** may have any direction; in other words, vectors represented sectionally always imply the positive–definite convention.

In contrast, in the second representation the ri may be negative, but the **ui** are positive in the given orthogonal orientation, which is the basic convention.

Positive sri imply the vertex of a section excludes reflex angles.

Definition 20 (limits *of* section)

Given

r1 as reference;

SECT(**r1**,**u1R1**,**u2R1**,**u3R1**);

f(**r**), a given vectorial function over V3;

if there can be found a vector **f1** such that

lim (**f**(**r1**+sr1∗**u1R1**)) = **f1** as sr1 ⟶ 0

and

lim (**f**(**r1**+sr2∗**u2R1**)) = **f1** as sr2 ⟶ 0

and

lim (**f**(**r1**+sr3∗**u3R1**)) = **f1** as sr3 ⟶ 0 (365)

then

f1 is the **limit of f of SECT**(**r1**,**u1R1**,**u2R1**,**u3R1**) **at r1**.

Similarly,

for f(**r**) a given scalar function of V3;

if there can be found a scalar f1 such that

lim (f(**r1**+sr1∗**u1R1**)) = f1 as sr1 ⟶ 0

and

lim (f(**r1**+sr2∗**u2R1**)R) = f1 as sr2 ⟶ 0

and

lim (f(**r1**+sr3∗**u3R1**)) = f1 as sr3 ⟶ 0 (366)

then

f1 is the **limit of** f **of SECT**(**r1**,**u1R1**,**u2R1**,**u3R1**) **at r1**.

end of definition

Limits of functions of SECT(**r1**,**u1R1**,**u2R1**,**u3R1**) are symbolized as:

lim **f**(**r1**+{**dR1**}) = **f1** as **dR1** ⟶ **0** (367)

lim f(**r1**+{**dR1**}) = f1 as **dR1** ⟶ **0** (368)

or

f(**r1**+{**dR1**})⟶**f1** as **dR1**⟶**0** (369)

f(**r1**+{**dR1**})⟶f1 as **dR1**⟶**0**. (370)

Limits in Section

Again some functions may have the same limit for every direction in the section.

Definition 21 (limits *in* section)

Given

r1 as reference;

SECT(**r1**,**u1R1**,**u2R1**,**u3R1**);

ur, a direction in SECT(**r1**,**u1R1**,**u2R1**,**u3R1**);

f(**r**), a given vectorial function over V3;

if there can be found a vector **f1** for any **ur** in

SECT(**r1**,**u1R1**,**u2R1**,**u3R1**)

such that

lim (**f**(**r1**+sr∗**ur**)) = **f1** as sr ⟶ 0 (371)

then

f1 is the **limit of f in SECT(r1,u1R1,u2R1,u3R1) at r1**.

Similarly, for f(**r**) a given scalar function of V3;

if there can be found a scalar f1 for any **ur** in

SECT(**r1**,**u1R1**,**u2R1**,**u3R1**)

such that

lim (f(**r1**+sr1∗**u1R1**)) = f1 as sr ⟶ 0 (372)

then

f1 is the **limit of** f **in SECT(r1,u1R1,u2R1,u3R1) at r1**.

end of definition

Limits of functions *in* SECT(**r1**,**u1R1**,**u2R1**,**u3R1**) are symbolized as:

lim **f**(**r1**+**dR1**) = **f1** as **dR1** ⟶ **0** (373)

lim f(**r1**+**dR1**) = f1 as **dR1** ⟶ **0** (374)

or

f(**r1**+**dR1**)⟶**f1** as **dR1**⟶**0** (375)

f(**r1**+**dR1**)⟶f1 as **dR1**⟶**0** (376)

Limits in sections may be seen as limits in the topology restricted to the section.

If a function has a limit *in* a section, it also has the same limit *of* the section.

If **f**(**r1**) = **f1**, the limit in a section, ***f*** is called **sectionally continuous at r1**.

Even if ***f*** is sectionally continuous at **r1**, it does not follow that it is continuous omni–directionally or even continuous in a direction not in the section.

Every function with an omni–directional limit at **r1** also has a sectional limit.

Sectional limits, in turn, are exemplified in quadrant limits on the one hand and, in the limiting case, as directional limits on the other.

Now consider **r**(x) *differentiable* with respect to x. Then since for every x D[x](**r**|**r**(t);dt) is a vector with a connected direction,

CX ≡ {**r**(x)}is a continuous curved line.

Example: **r**(x) = x∗**r1** + sin(x)∗**r2** + cos(x)∗**r3**, for **r1**, **r2**, and **r3** fixed, describes an elliptical helix oriented by location vectors **r1**, **r2** and **r3**.

The linear line

tang(x|x1) = **r**(x1) + x∗D[x1](**r**(t);dt) (377)

is called the tangent line to **r**(x) at x1. The direction of the tangent line is the direction of D[x1](**r**(t);dt).

Arc-Length

The **arc–length** of **r**(x) referenced to **r**(x1) is defined by

s(**r**(x1),**r**(x)) ≡ I[x1,x](abs(D[t](**r**(u);du);dt)) (378)

from a given reference point **r**(x1) to any point **r**(x). Arc–lengths increase monotonically.

Example: For **r**(x)=**r1**+x∗x∗**r2**, a straight, non–linear line,

D[x](**r**(u);du) = 2∗x∗**r2**

the arc–length referenced to **r**(−1),

s(**r**(−1),**r**(1)) = I[−1,1](abs(2∗x∗**r2**);dx)
= abs(**r2**)∗(I[−1,0](abs(2∗x);dx)
+ I[0,1](abs(2∗**x**);dx))
= abs(**r2**)∗(1+1)
= 2∗abs(**r2**),

even though **r**(x) over the image [−1,1] describes a closed path since **r**(1)=**r**(−1).

Example: For **r**(x)=**r0**+x∗**c** , **c** constant

D[x](**r**(u);du) = **c**

and s(**r0**,**r**(x)) = I[0,x](abs(**c**);dt) = x∗abs(**c**).

Example: For **r**(x)=**r0**, **r0** constant

D[x](**r**(u);du) = **0**

and s(**r0**,**r**(x)) = 0.

No matter how the curve may be translated or reoriented, the arc–length of a given section of the curve remains invariable. For this reason it is called an **intrinsic feature** of the curved line in V3.

The arc–length, s(**r**(x1),**r**(x)), is a real function of a real variable.

Monotonic, it may also serve as a compound variable, provided

D[x](**r**(t);dt) ≠ 0.

Consequently,

D[x](**r**(t);dt) = D[s(**r**(x1),**r**(x))](**r**(s);ds)∗D[x](s(**r**(x1),**r**(t));dt). (379)

For D[x](s(**r**(x1),**r**(t));dt) > 0 then

D[s(**r**(x1),**r**(x))](**r**(s);ds) = D[x](**r**(t);dt)/D[x](s(**r**(x1),**r**(t));dt)
= D[x](**r**(t);dt)/abs(D[x](**r**(t);dt)).

Consequently, D[s(**r**(x1),**r**(x))](**r**(s);ds) is a direction[33], so that

D[s(**r**(x1),**r**(x))](**r**(s);ds) = **u**(**r**(x1)) ≡ **ut**(x1) (380)

and

D[x](**r**(t);dt) = **ut**(x)∗D[x](s(**r**(x1),**r**(t));dt). (381)

The direction **ut**(x) is the direction[34] of the tangent line of the curved line at **r**(x).

Process in V3

Functions over the curved line **r**(x) may be either restricted, **f**(x)|**r**(x), or compound, **f**(**r**(x)).

Ideas derived from real functions (such as continuity, step functions, derivatives, and integrals) are thereby extended to the subset {**r**(x)}.

As maps of real numbers into the set of vectors, these restricted functions possess the same properties (and restrictions) as vectors themselves and other additional properties that derive from their status as functions.

Definition 22 (limits of restricted functions along CX)

Given

CX = {**r**(x)}, a curved line in V3;
x1, a real number;
ef and dx, real positive numbers;
r, **r**(x1), real vectors in V3;
f(x)|CX, a given restricted vector function;

if there can be found a vector **f1** such that

for any ef>0 there can always be found a dx>0 whenever

abs(**f**((x1+dx)|CX) − **f1**)<ef, (382)

then

f1 is the forward limit along CX of f at r(x1);

---

if there can be found a vector **f0** such that

$$\mathrm{abs}(\mathbf{f0} - \mathbf{f}(x1-dx)|CX)<ef, \qquad (383)$$

then

f0 is the backward limit along CX of f at r(x1);

for f(**r**) a given scalar function of V3;

if there can be found a scalar f1 such that

for any ef>0 there can always be found a dr>0 whenever

$$\mathrm{abs}(f((x1+dx)|CX) - f1)<ef, \qquad (384)$$

then

f1 is the **forward limit along CX of** f **at r**(x);

if there can be found a scalar f0 such that

$$\mathrm{abs}(f0 - f(x1-dx)|\mathbf{r}(x))<ef, \qquad (385)$$

then

f0 is the **backward limit along CX of** f **at r**(x).

end of definition

Limits of restricted functions along CX are symbolized as

$$\lim_f \mathbf{f}|CX = \mathbf{f1} \text{ at } \mathbf{r}(x1) \text{ as } dx \longrightarrow 0 \qquad (386)$$

$$\lim_f f|CX = f1 \text{ at } \mathbf{r}(x1) \text{ as } dx \longrightarrow 0 \qquad (387)$$

$$\lim_b \mathbf{f}|CX= \mathbf{f0} \text{ at } \mathbf{r}(x1) \text{ as } dx \longrightarrow 0 \qquad (388)$$

$$\lim_b f|CX= f0 \text{ at } \mathbf{r}(x1) \text{ as } dx \longrightarrow 0 \qquad (389)$$

or

$$\mathbf{f}(x1+dx)|CX \longrightarrow \mathbf{f1} \text{ as } dx \longrightarrow 0 \qquad (390)$$

$$f(x1+dx)|CX \longrightarrow f1 \text{ as } dx \longrightarrow 0 \qquad (391)$$

$$\mathbf{f}(x1-dx)|CX \longrightarrow \mathbf{f0} \text{ as } dx \longrightarrow 0 \qquad (392)$$

$$f(x1-dx)|CX \longrightarrow f0 \text{ as } dx \longrightarrow 0. \qquad (393)$$

These limits need not exist, the function's forward limit need not equal its backward limit, nor need either equal **f**(**r**(**x1**)) or f(**r**(**x1**)).

A function is said to be **continuous in** x **at r**(x1) **along CX** if

$$\mathbf{f}(x1)|CX = \lim \mathbf{f}(x1+dx)|CX = \lim \mathbf{f}(x1-dx)|CX \text{ as } dx \longrightarrow 0. \qquad (394)$$

Limits may also be defined for the curved line CR whose domain is defined without reference to *R.* Such limits apply to the domain of the compound function **f**(**r**(x)).

Definition 23 (limits of compound functions along CR)

Given

CR, a curved line in V3;
r1, a vector in CR
ut(**r1**), the direction of CR at **r1**;
ef and dr, real positive numbers;
f(**r**)|CR, a vector function over CR;

for

dr(**r1**) ≡ dr∗**ut**(**r1**), dr>0

if there can be found a vector **f1** such that

for any ef>0 there can always be found a dr>0 whenever

abs(**f**(**r1**+**dr**(**r1**)) – **f1**)<ef, (395)

then

f1 is a limit of f(r) at r1 along CR;

for f(**r**) a given scalar function of V3;

if there can be found a scalar f1 such that

for any ef>0 there can always be found a dr>0 whenever

abs(f(**r1**+**dr**(**r1**)) – f1)<ef, (396)

then

f1 is a **limit of** f(**r**) **at r1 along CR**.

end of definition

Limits of functions along CR are symbolized as

lim **f**(**r1**+**dr**)|CR = **f1** at **r1** as dr ⟶ 0 (397)

or

f(**r1**+**dr**)|CR ⟶**f1** at **r1** as dr⟶0. (398)

Since at **r1** a limit depends on the curve's direction, more than one limit of **f**(**r**) may exist there.

A function is said to be **continuous in r at r1 along CR** if

f(**r1**) = lim **f**(**r1** + **dr**(**r1**)) as dr ⟶ 0 for all possible **dr**(**r1**). (399)

Mathematics: Vector Calculus

A restricted function **f**(x)|CX may have multiple values at **r1**(x) while a compound function **f**(**r**(x)) may have multiple values at **r**(x1). In the former case the multiple functional values are called **stalled** values; in the latter case the multiple functional values are called **branched** values.

It is easily seen that directional limits are a specific instance of limits along CX or CR.

The following table summarizes the symbolism for these various limits.

Type of Limit	Symbol
omni–directional	**f(r1+dr)** ⟶ **f1** as **dr** ⟶ **0**
directional	**f(r1+dr)**|**ur**⟶**f1** as dr⟶0
1st quadrant	
ui only	**f(r1+**{**dr**}**)**⟶**f1** as **dr+**⟶**0**
continuous	**f(r1+dr)**⟶**f1** as **dr+**⟶**0**
sectional	
uRi only	**f(r1+**{**dR1**}**)**⟶**f1** as **dR1**⟶**0**
continuous	**f(r1+dR1)**⟶**f1** as **dR1**⟶**0**
along a curved line	
restricted	**f**(x1+dx)|CX⟶**f1** as dx⟶0
compound	**f(r1+dr(r1)**)|CR⟶**f1 as** dr⟶0

Symbols for Limits in V3

Differentiation in V3

In V3 differentiation may be defined with reference either to the underlying field *R* or to the vector field itself. With respect to *R*, the differentiation maintains order in the usually restricted topology. For this reason it is often referred to as **process**. The unrestricted topology, in contrast, fails to maintain order under its metric. Directional differentiation has aspects of both.

Differentiation by means of scalar increments produces derivatives:

in a given direction and so called **directional**
along CX with respect to x
along CR with respect to r.

Differentiation by means of vector increments arises from three possibilities:

	V3⟶R	called the **divergence**,	symbolized by D•
	V3⟶V3	called the **curl**,	symbolized by **D∧**
and	R⟶V3	called the **gradient**,	symbolized by **D✱**.

Derivatives in *V3*, like those over *R*, belong to functions rather than to relationships and do not necessarily arise from functional values. A smaller class of functions, called **basic**, possess derivatives from which an accurate estimate of a neighboring functional values may be made.

Vector functions and their derivatives in V3, much like real functions in R, may have four possible attributes with regard to continuity.

Now consider the four possible attributes of both functions and their derivatives:

functional continuity at all points of the domain
functional continuity at all points of the domain
except for a finite number of points
functional discontinuity
on a denumerable set of of points of the domain
functional discontinuity
on a non–denumerable set of points of the domain.

For the purposes of Theoretical Physics only the first two attributes will be considered.

Consequently, four categories of functions and their derivatives are considered:

1. continuous functions with continuous derivatives;
2. continuous functions with continuous derivatives except at a finite number of points;
3. continuous functions except at a finite number of points with continuous derivatives;
4. continuous functions except at a finite number of points with continuous derivatives except at a finite number of points.

Directional Derivatives

From directional limits directional derivatives may be defined.

Definition 24 (basic directional derivatives)

Given

r1, a reference vector in V3;

ur ≡ **u**(**r**–**r1**), a direction;

f(**r**) a given vector function;

then if it exists

$\lim_f$((**f**(**r1**+dr✱**ur**)–**f**(**r1**))/dr) as dr✱**ur⟶0** (400)

is called the **basic forward directional derivative of f in the ur direction**;

$\lim_b$((**f**(**r1**)–**f**(**r1**+dr✱**ur**))/dr) as dr✱**ur⟶0** (401)

is called the **basic backward directional derivative of f in the ur direction**;

given f(**r**) a given scalar function of V3;

then if it exists

$\lim_f$((f(**r1**+dr✱**ur**)–f(**r1**))/dr) as dr✱**ur⟶0** (402)

is called the **basic forward directional derivative of** f **in the ur direction**;

$\lim_b$((f(**r1**)–f(**r1**+dr✱**ur**))/dr) as dr✱**ur⟶0** (403)

is called the **basic backward directional derivative of** f **in the ur direction**.

end of definition:

Forward derivatives are symbolized as

$$\mathrm{D}[\mathbf{r1},\mathbf{r1+dr}](\mathbf{f}(\mathbf{r})|\mathbf{ur};\mathrm{dr})$$
$$\mathrm{D}[\mathbf{r1},\mathbf{r1+dr}](\mathrm{f}(\mathbf{r})|\mathbf{ur};\mathrm{dr}). \qquad (404)$$

Backward directional derivatives are symbolized as:

$$\mathrm{D}[\mathbf{r1-dr},\mathbf{r1}](\mathbf{f}(\mathbf{r})|\mathbf{ur};\mathrm{dr})$$
$$\mathrm{D}[\mathbf{r1-dr}\ ,\mathbf{r1}](\mathrm{f}(\mathbf{r})|\mathbf{ur};\mathrm{dr}). \qquad (405)$$

From this basic notation, it is clear the negative derivatives

$$\mathrm{D}[\mathbf{r1+dr},\mathbf{r1}](\mathbf{f}(\mathbf{r})|-\mathbf{ur};\mathrm{dr})$$
$$\mathrm{D}[\mathbf{r1},\mathbf{r1-dr}](\mathbf{f}(\mathbf{r})|-\mathbf{ur};\mathrm{dr}) \qquad (406)$$

may similarly be defined.

If a function has a basic directional derivative at ***r1***, it is called **differentiable** there. If forward directional and backward directional derivatives are equal, the function is said to have a **continuous directional derivative** there. If the function has a continuous directional derivative at each vector along a straight line

$$\mathbf{r1} + \mathrm{r}{*}\mathbf{ur}$$

the function is said to have a continuous directional derivative **everywhere** along the line.

Continuous derivatives along a straight line are written simply as

$$\mathrm{D}[\mathbf{r1},\mathbf{r1+dr}](\mathbf{f}(\mathbf{r})|\mathbf{ur};\mathrm{dr}), \qquad (407)$$

or even

$$\mathrm{D}[\mathbf{r1}](\mathbf{f}(\mathbf{r})|\mathbf{ur};\mathrm{dr}). \qquad (408)$$

For functions with a basic directional derivative

$$\mathbf{f}(\mathbf{r1+dr})|\mathbf{ur} \approx \mathbf{f}(\mathbf{r1}) + \mathrm{dr}{*}\mathrm{D}[\mathbf{r1},\mathbf{r1+dr}](\mathbf{f}(\mathbf{r})|\mathbf{ur};\mathrm{dr}). \qquad (409)$$

Derivatives along CX with respect to x

Derivatives along CX borrow a forward/backward sense from the underlying field.

Definition 25 (basic derivatives along CX with respect to x)

Given

CX = {**r**(x)}, a curve;

f(x)|CX, a restricted function;

then if it exists

lim ((**f**(x1+dx)−**f**(x1))/dx) as dx⟶0 (410)

is called the **basic forward derivative of f|r**(x) at x1;

lim ((**f**(x1)−**f**(x1−dx))/dx) as dx⟶0 (411)

is called the **basic backward derivative of f|r**(x) at x1.

end of definition:

Forward derivatives along CX are written as

D[x,x+dx](**f**|CX;dx) (412)

or

D[x](**f**|CX;d_fx). (413)

Backward derivatives along CX are written as

D[x−dx,x](**f**|CX;dx) (414)

or

D[x1](**f**|CX;d_bx). (415)

From this basic notation, it is clear the negative derivatives

D[x+dx,x](**f**|CX;dx) (416)

and

D[x,x−dx](**f**|CX;dx) (417)

may similarly be defined.

If a function has a basic derivative restricted to CX at **r**(x1), it is called **differentiable** there. If forward and backward derivatives are equal, the function is said to have a **continuous** restricted derivative there. If the function has a continuous restricted derivative at each point of CX the function is said to have a continuous restricted derivative **everywhere** along CX.

Continuous restricted derivatives are written simply as

D[x1](**f**(x)|CX;dx).

For functions with a basic restricted derivative

f(x1+dx) ≈ **f**(x1) + dx∗D[x1,x1+dx](**f**(x)|CX;dx). (418)

Derivatives along CR with respect to r

Definition 26 (basic derivatives along CR with respect to **r**)
Given
CR, a curved line with possibly many branches;
r1, a point in CR;
ut(**r1**), a curve direction at **r1**;
f(**r**), a vector function over CR;
for
dr(**r1**) ≡ dr∗**ut**(**r1**), dr>0
r = **r1** + **dr**(**r1**), another point on a branch
in a neighborhood of **r1**;
then if it exists

lim (**f**(**r1**+**dr**(**r1**))−**f**(**r1**))/dr as dr⟶0 (419)

is called a **basic derivative of f**(r) **at r1 along a branch of CR**.

end of definition:

These derivatives are written as

D[**r1**,**r1**+**dr**](**f**(**r**)|CR;dr) (420)

for each branch **r1**.

If a function has a basic derivative restricted to CR along a branch at ***r1***, it is called **differentiable** in that branch there. If forward and backward derivatives for a branch are equal, the function is said to have a **continuous** restricted derivative along that branch there. If the function has a continuous restricted derivative at each point of CR along the branch, the function is said to have a continuous restricted derivative **everywhere** along the branch. If a function has a continuous restricted derivative along all branches at ***r1*** it is said to have a **continuous** restricted derivative for all branches there. If a function has a continuous restricted derivative along all branches everywhere in CR, it is said to have a **continuous** restricted derivative **everywhere** on CR.

Mathematics: Vector Calculus

When the derivatives for each branch are equal, the limit is written simply as

$$\mathrm{D}[\mathbf{r1}](\mathbf{f}(\mathbf{r})|\mathrm{CR};\mathrm{dr}) \qquad (421)$$

For functions with a basic derivative along CR

$$\mathbf{f}(\mathbf{r1}+\mathbf{dr}(\mathbf{r1})) \approx \mathbf{f}(\mathbf{r1}) + \mathrm{dr}*\mathrm{D}[\mathbf{r1},\mathbf{r1}+\mathbf{dr}(\mathbf{r1})](\mathbf{f}(\mathbf{r})|\mathrm{CR};\mathrm{dr}). \qquad (422)$$

Derivatives of a function as restricted or compound are related according to the following theorem.

Theorem 1 (chain rule)

Given

- C ≡ CR = CX = {**r**(x)}, a curved line;
- **r1** = **r**(x1);
- **ut**(**r1**), the direction of C at **r1**;
- **f**(x)|C, a function with a basic derivative along C;
- **f**(**r**(x)) = **f**(x)|C, a compound vector function;

for

- **r**(x) with a basic derivative at **r1**;
- **dr**(**r1**) ≡dr***ut**(**r1**) ≡ **r** − **r1**, dr>0;

then

$$\mathrm{D}[\mathrm{x1},\mathrm{x1}+\mathrm{dx}](\mathbf{f}(\mathrm{x})|\mathrm{C};\mathrm{dx}) = \mathrm{abs}(\mathrm{D}[\mathrm{x1},\mathrm{x1}+\mathrm{dx}](\mathbf{r}(\mathrm{x})|\mathrm{C};\mathrm{dx}))*\mathrm{D}[\mathbf{r1},\mathbf{r1}+\mathbf{dr}](\mathbf{f}(\mathbf{r})|\mathrm{C};\mathrm{dr}). \qquad (423)$$

Proof:

Since D[x1,x1+dx](**f**(x)|C;dx) has a basic derivative at x1

$$\mathbf{f}(\mathbf{r}(\mathrm{x1}+\mathrm{dx}))\approx \mathbf{f}(\mathbf{r}(\mathrm{x1})) + \mathrm{dx}*\mathrm{D}[\mathrm{x1},\mathrm{x1}+\mathrm{dx}](\mathbf{f}(\mathrm{x})|\mathrm{C};\mathrm{dx}).$$

Further, for **dr**(**r1**) = **r**(x1+dx)−**r**(x1)

$$\mathbf{f}(\mathbf{r}(\mathrm{x1}+\mathrm{dx})) = \mathbf{f}(\mathbf{r1}+\mathbf{dr}).$$

Therefore

$$\mathrm{D}[\mathrm{x1},\mathrm{x1}+\mathrm{dx}](\mathbf{f}(\mathrm{x})|\mathbf{r}(\mathrm{x});\mathrm{dx}) \approx (\mathbf{f}(\mathbf{r}(\mathrm{x1}+\mathrm{dx})) - \mathbf{f}(\mathbf{r}(\mathrm{x1})))/\mathrm{dx} \approx (\mathbf{f}(\mathbf{r1}+\mathbf{dr}(\mathbf{r1})) - \mathbf{f}(\mathbf{r1}))/\mathrm{dr})*(\mathrm{dr}/\mathrm{dx})$$

since dr>0.

Now lim (dr/dx) = abs(D[x1,x1+dx](**r**(x)|C;dx)).

Thus,

$$\mathrm{D}[\mathrm{x1},\mathrm{x1}+\mathrm{dx}](\mathbf{f}(\mathrm{x})|\mathrm{C};\mathrm{dx}) = \mathrm{abs}(\mathrm{D}[\mathrm{x1},\mathrm{x1}+\mathrm{dx}](\mathbf{r}(\mathrm{x})|\mathrm{C};\mathrm{dx})) *\mathrm{D}[\mathbf{r1},\mathbf{r1}+\mathbf{dr}(\mathbf{r1})](\mathbf{f}(\mathbf{r})|\mathrm{C};\mathrm{dr}).$$

qed

Theorem 1 gives the conditions for a valid **chain rule** between process derivatives and local derivatives along a curve.

If the process is stalled, D[x1,x1+dx](**r**(x)|C;dx) = **0**. Then, for a non–null derivative along CX, the derivative along CR must be unbounded.

If the curve branches, then at each branch point

D[x1,x1+dx](**f**(x)|C;dx)
D[x1,x1+dx](**r**(x)|C;dx)
D[**r1**,**r1**+**dr**](**f**(**r**)|C;dr)

may be multi–valued.

The theorem may also be expressed in terms of arc–lengths since

ds(**r1**,**r**(x1+dx)) ≈ abs(**r**(x1+dx) – **r**(x1)). (424)

In terms of arc–lengths the theorem becomes

D[x1,x1+dx](**f**(x)|**r**(x);dx)
= D[x1,x1+dx](s(**r1**,**r**(x1+dx)|C;dx)
∗D[**r1**,**r1**+**dr**(**r1**)](**f**(**r**)|C;ds) (425)

For stalled processes the above result does *not* hold.

Differentiation of a Matrix along Curves

If a matrix **A**=**u1**∗**a1**+**u2**∗**a2**+**u3**∗**a3** is a function of x in all its elements, then the **scalar** derivative of **A** is defined as

D[x1](**A**;dx) ≡ **u1**∗D[x1](**a1**(x);dx)
+ **u2**∗D[x1](**a2**(x);dx)
+ **u3**∗D[x1](**a3**(x);dx). (426)

Then

D[x1](tr[**A**(x)];dx) = tr(D[x1][**A**(x)];dx) (427)
D[x1](**g**[**A**(x)];dx) = **g**[D[x1](**A**(x));dx] (428)
D[x1](**c**[**A**(x)];dx) = **c**[D[x1](**A**(x);dx)] (429)

Mathematics: Vector Calculus

$$\mathrm{D[x1](det[}\mathbf{A}\mathrm{(x)];dx)} = \mathrm{D[x1](}\mathbf{a1}\mathrm{(x);dx)}\bullet(\mathbf{a2}\mathrm{(x1)}\wedge\mathbf{a3}\mathrm{(x1))} + \mathrm{D[x1](}\mathbf{a2}\mathrm{(x);dx)}\bullet(\mathbf{a3}\mathrm{(x1)}\wedge\mathbf{a1}\mathrm{(x1))} + \mathrm{D[x1](}\mathbf{a3}\mathrm{(x);dx)}\bullet(\mathbf{a1}\mathrm{(x1)}\wedge\mathbf{a2}\mathrm{(x1))} \quad (430)$$

$$\mathrm{D[x1](}\mathbf{T}[\mathbf{A}\mathrm{(x)];dx)} = \mathbf{T}[\mathrm{D[x1](}\mathbf{A}\mathrm{(x);dx)]} \quad (431)$$

$$\mathrm{D[x1]([}\mathbf{A}\mathrm{(x)}^{-1}\mathrm{];dx)} = -[\mathbf{A}\mathrm{(x1)}^{-1}]\bullet[\mathrm{D[x1](}\mathbf{A}\mathrm{(x);dx)}]\bullet[\mathbf{A}\mathrm{(x1)}^{-1}] \quad (432)$$

$$\mathrm{D[x1]((}\mathbf{A}\mathrm{(x)+}\mathbf{B}\mathrm{(x));dx)} = \mathrm{D[x1](}\mathbf{A}\mathrm{(x);dx)} + \mathrm{D[x1](}\mathbf{B}\mathrm{(x);dx)} \quad (433)$$

$$\mathrm{D[x1]((}\mathbf{A}\mathrm{(x)}\bullet\mathbf{B}\mathrm{(x));dx)} = \mathbf{A}\mathrm{(x1)}\bullet\mathrm{D[x1](}\mathbf{B}\mathrm{(x);dx)} + \mathrm{D[x1](}\mathbf{A}\mathrm{;dx)}\bullet\mathbf{B}\mathrm{(x1)} \quad (434)$$

$$\mathrm{D[x1]((}\mathbf{r}\mathrm{(x)}\bullet\mathbf{A}\mathrm{(x));dx)} = \mathrm{D[x1](}\mathbf{r}\mathrm{(x);dx)}\bullet\mathbf{A}\mathrm{(x1)} + \mathbf{r}\mathrm{(x1)}\bullet\mathrm{D[x1](}\mathbf{A}\mathrm{(x);dx)} \quad (435)$$

$$\mathrm{D[x1](}\mathbf{r1}\mathrm{(x)}*\mathbf{r2}\mathrm{(x);dx)} = \mathbf{r1}\mathrm{(x1)}*\mathrm{D[x1](}\mathbf{r2}\mathrm{(x);dx)} + \mathrm{D[x1](}\mathbf{r1}\mathrm{(x);dx)}*\mathbf{r2}\mathrm{(x1)} \quad (436)$$

$$\mathrm{tr(D[x1][}\mathbf{r1}\mathrm{(x)}*\mathbf{r2}\mathrm{(x)]dx)} = \mathrm{D[x1](}\mathbf{r1}\mathrm{(x)}\bullet\mathbf{r2}\mathrm{(x);dx)} \quad (437)$$

$$\mathrm{D[x1](}\mathbf{G}(\mathbf{r}\mathrm{(x));dx)} = \mathbf{G}\mathrm{(D[x1](}\mathbf{r}\mathrm{(x);dx))} \quad (438)$$

$$\mathrm{D[x1](}\mathbf{C}(\mathbf{r}\mathrm{(x));dx)} = \mathbf{C}\mathrm{(D[x1](}\mathbf{r}\mathrm{(x);dx)).} \quad (439)$$

If ***A*** is a function of ***r***, a vector variable, in all its elements and

dr ≡ d1∗**u1** + d2∗**u2** + d3∗**u3**
≡ d∗**ur**

then the **directional** derivative of **A** with respect to **dr** referenced to **r1** is defined as

$$\mathrm{D[}\mathbf{r1}\mathrm{](}\mathbf{A};\mathbf{dr}) \equiv \mathbf{u1}*\mathrm{D[}\mathbf{r1}\mathrm{](}\mathbf{a1}\mathrm{;d1)} + \mathbf{u2}*\mathrm{D[}\mathbf{r1}\mathrm{](}\mathbf{a2}\mathrm{;d2)} + \mathbf{u3}*\mathrm{D[}\mathbf{r1}\mathrm{](}\mathbf{a3}\mathrm{;d3).} \quad (440)$$

This derivative is called the **directional** derivative of **A** because

$$\mathrm{d1} = \mathrm{d}*\mathbf{ur}\bullet\mathbf{u1} \quad (441)$$

$$\mathrm{d2} = \mathrm{d}*\mathbf{ur}\bullet\mathbf{u2} \quad (442)$$

$$\mathrm{d3} = \mathrm{d}*\mathbf{ur}\bullet\mathbf{u3}. \quad (443)$$

The **ur•ui** are the directional cosines of **dr**.

We will not pursue further the transformations of V3, a fascinating study in itself, except to state they may be grouped and analyzed according to certain canonical forms[35].

Generic sums and products of functions, provided they apply to the same class of derivatives, are given by:

D[](**f1**+**f2**|u) = D[](**f1**|u) + D[](**f2**|u)
D[](**f1**•**f2**|u) = **f1**•D[](**f2**|u) + D[](**f1**|u)•**f2**
D[]((**f1**∧**f2**)|u) = **f1**∧D[](**f2**|u) + D[](**f1**|u)∧**f2**
D[]((**f1**∗**f2**)|u) = **f1**∗D[](**f2**|u) + D[](**f1**|u)∗**f2**.

(444)

For a constant vector **c** and a vector function **f**(**r**(x)) of the real variable, x, the following results hold:

D[](**c**•**f**|u) = **c**•D[](**f**|u)
D[](**c**∧**f**|u) = **c**∧D[](**f**|u)
D[](**c**∗**f**|u) = **c**∗D[](**f**|u).

(445)

It should be noted that since (**f**|u)•(**f**|u) = abs(**f**|u)∗abs(**f**|u),

f•D[](**f**|u) = abs(**f**)∗D[](abs(**f**)|u) (446)

even when abs(D[](**f**|u)) does *not* equal D[](abs(**f**|u)).

Furthermore, if abs(**f**) is constant, (even though **f** varies)

D[](**f**•**f**|u) = 2∗**f**•D[](**f**|u)
= 0. (447)

In this case **f** and D[](**f**|u) are orthogonal.

These formulas are extended in the usual way to generalized functions.

Example: Let **r**(x) = **r1**∗u(x) + x∗**r2**, for fixed **r1** and **r2**, with u(x) a unit step function. Then D[x](**r**(t);dt)= **r1**∗D[x](u(t);dt) + **r2,** that is, **r**(x) has a change of direction from **ur2** to **ur1** at x=0 which returns to **ur2** with increasing x.

Mathematics: Vector Calculus

Directional Divergences, Curls, and Gradients

The reciprocal vector

$$\mathbf{qd}(\mathbf{r}) \equiv \mathbf{u}(\mathbf{r})/r \qquad (448)$$

defined in equation (151) is called the **directional reciprocal vector** of **r**. It becomes a key factor in the definition of directional gradients.

In contrast to the reciprocal vector defined by equation (156) and elaborated in equation (478)

$$\begin{aligned} &\mathbf{r}\cdot\mathbf{qd}(\mathbf{r}) = 1 \\ &\mathbf{r}\wedge\mathbf{qd}(\mathbf{r}) = \mathbf{0} \\ &\mathbf{r}*\mathbf{qd}(\mathbf{r}) = \mathbf{u}(\mathbf{r})*\mathbf{u}(\mathbf{r}). \end{aligned} \qquad (449)$$

Note that for

$$\mathbf{r1} \equiv r11*\mathbf{u1}+r12*\mathbf{u2}+r13*\mathbf{u3}$$

and

$$\mathbf{r2} \equiv r21*\mathbf{u1}+r22*\mathbf{u2}+r23*\mathbf{u3},$$

qd(**r1+r2**) does not equal **qd**(**r1**)+ **qd**(**r2**), but rather

$$\begin{aligned} \mathbf{qd}(\mathbf{r1}) + \mathbf{qd}(\mathbf{r2}) = {} & (r11*\mathbf{u1} + r12*\mathbf{u2} + r13*\mathbf{u3}) \\ & /(r11^2+r12^2+r13^2) \\ & + (r21*\mathbf{u1} + r22*\mathbf{u2} + r23*\mathbf{u3}) \\ & /(r21^2+r22^2+r23^2) \end{aligned} \qquad (450)$$

while

$$\begin{aligned} \mathbf{qd}(\mathbf{r1+r2}) = {} & ((r11+r21)*\mathbf{u1} + (r12+r22)*\mathbf{u2} + (r13+r23)*\mathbf{u3}) \\ & /((r11+r21)^2+(r11+r22)^2+(r13+r23)^2). \end{aligned} \qquad (451)$$

Definition 27 (basic directional divergences, curls, and gradients)

Given

$\mathbf{ur} \equiv \mathbf{u}(\mathbf{r}-\mathbf{r1})$, a direction;

$\mathbf{dr} \equiv \mathbf{r}-\mathbf{r1}$;

$\mathbf{f}(\mathbf{r})|\mathbf{ur}$, a vector function restricted to the line $\{\mathbf{r} = \mathbf{r1}+\mathbf{dr}\}$;

then

the basic forward directional divergence, curl, and gradient at **r1** in direction **ur** are defined as:

Divergence:

$$D[\mathbf{r1},\mathbf{r1}+\mathbf{dr}]\bullet(\mathbf{f}(\mathbf{r})|\mathbf{ur};\mathbf{dr}) \equiv \lim(\mathbf{f}(\mathbf{r1}+\mathbf{dr})-\mathbf{f}(\mathbf{r1}))\bullet\mathbf{qd}(\mathbf{dr}) \quad (452)$$

Curl:

$$\mathbf{D}[\mathbf{r1},\mathbf{r1}+\mathbf{dr}]\wedge(\mathbf{f}(\mathbf{r})|\mathbf{ur};\mathbf{dr}) \equiv \lim(\mathbf{f}(\mathbf{r1}+\mathbf{dr})-\mathbf{f}(\mathbf{r1}))\wedge\mathbf{qd}(\mathbf{dr}) \quad (453)$$

Gradient:

$$\mathbf{D}[\mathbf{r1},\mathbf{r1}+\mathbf{dr}]*(\mathbf{f}(\mathbf{r})|\mathbf{ur};\mathbf{dr}) \equiv \lim(\mathbf{f}(\mathbf{r1}+\mathbf{dr})-\mathbf{f}(\mathbf{r1}))*\mathbf{qd}(\mathbf{dr}) \quad (454)$$

or

$$\mathbf{D}[\mathbf{r1},\mathbf{r1}+\mathbf{dr}]*(f(\mathbf{r})|\mathbf{ur};\mathbf{dr}) \equiv \lim(f(\mathbf{r1}+\mathbf{dr})-f(\mathbf{r1}))*\mathbf{qd}(\mathbf{dr}). \quad (455)$$

The basic backward directional local divergence, curl and gradient, are defined as:

Divergence:

$$D[\mathbf{r1}-\mathbf{dr},\mathbf{r1}]\bullet(\mathbf{f}(\mathbf{r})|\mathbf{ur};\mathbf{dr}) \equiv \lim(\mathbf{f}(\mathbf{r1})-\mathbf{f}(\mathbf{r1}-\mathbf{dr}))\bullet\mathbf{qd}(\mathbf{dr}) \quad (456)$$

Curl:

$$\mathbf{D}[\mathbf{r1}-\mathbf{dr},\mathbf{r1}]\wedge(\mathbf{f}(\mathbf{r})|\mathbf{ur};\mathbf{dr}) \equiv \lim(\mathbf{f}(\mathbf{r1})-\mathbf{f}(\mathbf{r1}-\mathbf{dr}))\wedge\mathbf{qd}(\mathbf{dr}) \quad (457)$$

Gradient:

$$\mathbf{D}[\mathbf{r1}-\mathbf{dr},\mathbf{r1}]*(\mathbf{f}(\mathbf{r})|\mathbf{ur};\mathbf{dr}) \equiv \lim(\mathbf{f}(\mathbf{r1})-\mathbf{f}(\mathbf{r1}-\mathbf{dr}))*\mathbf{qd}(\mathbf{dr}) \quad (458)$$

or

$$\mathbf{D}[\mathbf{r1}-\mathbf{dr},\mathbf{r1}]*(f(\mathbf{r})|\mathbf{ur};\mathbf{dr}) \equiv \lim(f(\mathbf{r1})-f(\mathbf{r1}-\mathbf{dr}))*\mathbf{qd}(\mathbf{dr}) \quad (459)$$

as $\mathbf{dr}\longrightarrow\mathbf{0}$.

end of definition

Mathematics: Vector Calculus

Definition 28 (basic divergences, curls, and gradients along CR)

Given

CR, a curve;
r1, a point in CR;
ut(**r1**), the direction of the curve at **r1**;
f(**r**)|CR, a vector function over CR;

for

dr(**r1**) ≡ dr∗**ut**(**r1**), dr>0

the basic forward divergence, curl and gradient of **f** along CR are defined at **r1** as:

Divergence:

D[**r1**,**r1+dr**(**r1**)]•(**f**(**r**)|CR;**dr**)
≡ lim(**f**(**r1+dr**(**r1**))–**f**(**r1**))•**qd**(**dr**(**r1**)); (460)

Curl:

D[**r1**,**r1+dr**(**r1**)]∧(**f**(**r**)|CR;**dr**)
≡ lim(**f**(**r1+dr**(**r1**))–**f**(**r1**))∧**qd**(**dr**(**r1**)); (461)

Gradient:

D[**r1**,**r1+dr**(**r1**)]∗(**f**(**r**)|CR;**dr**)
≡ lim(**f**(**r1+dr**(**r1**))–**f**(**r1**))∗**qd**(**dr**(**r1**)); (462)

or

D[**r1**,**r1+dr**(**r1**)](f(**r**)|CR;**dr**)
≡ lim(f(**r1+dr**(**r1**))–**f**(**r1**))∗**qd**(**dr**(**r1**)). (463)

as dr⟶0.

end of definition

The basic divergence of a matrix function **A**(**r**) is defined in the forward direction as:

D[r1,**r1+dr]•(A**(**r**)|u;**dr**) ≡ lim **qd**(**dr**)**•T[A**(**r1+dr**)–**A**(**r1**)] (464)

and in the backward direction as:

D[r1–**dr**,**r1]•(A**(**r**)|u;**dr**) ≡ lim **qd**(**dr**)**•T[A**(**r1**)–**A**(**r1**–**dr**)] (465)

where the restriction *u* refers either to a direction or a curve.

If a function has a basic restricted divergence, curl, and gradient, it is called **differentiable** according to that restriction. If forward and backward divergences, curls, and gradients are equal at some point, the function is said to have a **continuous** restricted derivative according to the restriction there. A function may have a continuous restricted divergence, curl, and gradient **everywhere in the restriction**.

When the forward and backward limits are equal, divergences, curls and gradients are written simply as

$$D[\mathbf{r1},\mathbf{r1}+\mathbf{dr}]\cdot(\mathbf{f}(\mathbf{r})|u;\mathbf{dr}),$$
$$\mathbf{D}[\mathbf{r1},\mathbf{r1}+\mathbf{dr}]\cdot(\mathbf{A}(\mathbf{r})|u;\mathbf{dr}),$$
$$\mathbf{D}[\mathbf{r1},\mathbf{r1}+\mathbf{dr}]\wedge(\mathbf{f}(\mathbf{r})|u;\mathbf{dr}),$$
$$\mathbf{D}[\mathbf{r1},\mathbf{r1}+\mathbf{dr}]\ast(f(\mathbf{r})|u;\mathbf{dr}),$$
or
$$\mathbf{D}[\mathbf{r1},\mathbf{r1}+\mathbf{dr}]\ast(\mathbf{f}(\mathbf{r})|u;\mathbf{dr}).$$

Analogous to derivatives of a real variable, the *everywhere* condition may be expressed as a variable in the reference. Then directional derivatives become functions of **r** and are written $\mathbf{D}[\mathbf{r},\mathbf{r}+\mathbf{dr}]\ast(\mathbf{f}|u;\mathbf{dr})$.

If they exist, basic directional local derivatives are unique.

It should be noted that $\mathbf{D}[\mathbf{r},\mathbf{r}+\mathbf{dr}]\ast(\mathbf{f}|u;\mathbf{dr})$ is a transformation, specifically an outer product whose rank thus equals 1.

Divergences, curls, and gradients along a curve can be expressed in terms directional derivatives along a curve and arc–lengths.

Let **ut** be the direction of the tangent line of **r**(x) at **r**(x1). Then

$$\mathbf{qd}(\mathbf{dr}(x1)) \approx \mathbf{ut}/ds(\mathbf{r}(x1),\mathbf{r}(x1+dx)). \tag{466}$$

Consequently

$$\begin{aligned}
&(\mathbf{f}(\mathbf{r}(x1+dx))-\mathbf{f}(\mathbf{r}(x1)))\ast\mathbf{qd}(\mathbf{dr}(x1)) \\
&\quad\approx (\mathbf{f}(s(\mathbf{r}(x1),\mathbf{r}(x1+dx)))-(s(\mathbf{r}(x1),\mathbf{r}(x1))))\ast\mathbf{ut} \\
&\qquad\qquad /ds(\mathbf{r}(x1),\mathbf{r}(x1+dx)) \\
&\quad\longrightarrow D[0,s(\mathbf{r}(x1),\mathbf{r}(x1+dx))](\mathbf{f}(s)|\mathbf{r}(x);ds)\ast\mathbf{ut}.
\end{aligned}$$

Therefore,

$$\mathbf{D}[\mathbf{r}(x1),\mathbf{r}(x1+dx)]\ast(\mathbf{f}|CX;\mathbf{dr}) = D[0,s(\mathbf{r}(x1),\mathbf{r}(x1+dx))](\mathbf{f}(s)|CX;ds)\ast\mathbf{ut}. \tag{467}$$

Mathematics: Vector Calculus

Similarly,

D[**r**(x1),**r**(x1+dx)]•(**f**|CX;**dr**)
 = D[0,s(**r**(x1),**r**(x1+dx))](**f**(s)|CX;ds)•**ut** (468)
D[**r**(x1),**r**(x1+dx)]∧(**f**|CX;**dr**)
 = D[0,s(**r**(x1),**r**(x1+dx))](**f**(s)|CX;ds)∧**ut** (469)
D[**r**(x1),**r**(x1+dx)]✱(f|CX;**dr**)
 = D[0,s(**r**(x1),**r**(x1+dx))](f(s)|CX;ds)✱**ut** (470)
D[**r**(x1),**r**(x1+dx)]•(**A**|CX;**dr**)
 = **ut**•**T**[D[0,s(**r**(x1),**r**(x1+dx))](**A**(s)|CX;ds)]. (471)

Also

D[0,s(**r**(x1),**r**(x1+dx))](**f**(s)|CX;ds)
 = **ut**•**T**[**D**[**r**(x1),**r**(x1+dx)]✱(**f**|CX;**dr**)]. (472)

This latter equation opens the way for studying derivatives along a curve in terms of the intrinsic functions of **r**(x).

Consider now **f**(**r**(x)) and **r**(x) at **r1**(x1). The triple product

((**f**(**r**(x1+dx)) – **f**(**r**(x1)))/dx)•(**dr**(x)∧**qd**(**dr**(x)))
 = –(**f**(**r**(x1+dx)) – **f**(**r**(x1)))∧**qd**(**dr**(x))•**dr**(x)/dx.

Then as limiting forms,

D[x1,x1+dx](**f**|**r**(x);dx)•**D**[**r**(x1),**r**(x1+dx)]∧(**r**(x);**dr**)
 = –**D**[**r**(x1),**r**(x1+dx)]∧(**f**|**r**(x);**dr**)
 •(D[x1,x1+dx](**r**(x);dx))
 = 0. (473)

Thus **D**[**r**(x1),**r**(x1+dx)]∧(**f**|CX;**dr**) and D[x1,x1+dx](**r**(x)|CX;dx) are orthogonal vectors.

Consider again,

((**f**(**r**(x1+dx)) – **f**(**r**(x1)))/dx)∧(**dr**(x)∧**qd**(**dr**(x)))
 = ((**f**(**r**(x1+dx))–**f**(**r**(x1)))•**qd**(**dr**(x)))✱**dr**(x)/dx
 – (((**f**(**r**(x1+dx))–**f**(**r**(x1)))/dx)•**dr**(x))✱**qd**(**dr**(x)).

Then as a limiting form

D[x1,x1+dx](**f**|**r**(x);dx)**∧D**[**r**(x1),**r**(x1+dx)]**∧**(**r**(x);**dr**)
= **D**[**r**(x1),**r**(x1+dx)]•(**f**|**r**(x);**dr**)∗(D[x1,x1+dx](**r**(x);dx))
– D[x1,x1+dx](**f**|**r**(x);dx)•**ut**(x1)∗**ut**(x1)
= **0**. (474)

Consequently,

D[x1,x1+dx](**f**|CX;dx)•**ut**(x1)∗**ut**(x1)
= **D**[**r**(x1),**r**(x1+dx)]•(**f**|CX;**dr**)∗(D[x1,x1+dx](**r**(x)|CX;dx)). (475)

Consider again

((**f**(**r**(x1+dx))–**f**(**r**(x1)))/dx)∗(**dr**(x)**∧qd**(**dr**(x)))
= – ((**f**(**r**(x1+dx))–**f**(**r**(x1)))∗**qd**(**dr**(x)))•**C**(**dr**(x)/dx).

Then as a limiting form,

D[x1,x1+dx](**f**|CX;dx)∗**D**[**r**(x1),**r**(x1+dx)]**∧**(**r**(x)|CX;**dr**)
= –(**D**[**r**(x1),**r**(x1+dx)]∗(**f**|CX;**dr**))
•**C**(D[x1,x1+dx](**r**(x)|CX;dx))
= [**0**]. (476)

The derivatives considered above need not be basic derivatives. They may apply to functions with the same functional values but with differently valued derivatives.

For functions with *basic* local derivatives:

f(**r**(x1+dx)) ≈ **f**(**r**(x1)) + dx∗D[x1](**f**|CX;dx),
f(**r1**+**dr**) ≈ **f**(**r1**) + dr∗D[**r1**,**r1**+dr∗**ur**](**f**(**r**)|CR;dr),
f(**r1**+**dr**) ≈ f(**r1**) + dr∗D[**r1**,**r1**+dr∗**ur**](f(**r**)|CR;dr),
f(**r1**+**dr**) ≈ **f**(**r1**) + **dr**•D[**r1**,**r1**+**dr**]∗(**f**(**r**)|CR;**dr**),
f(**r1**+**dr**) ≈ f(**r1**) + **dr**•D[**r1**,**r1**+**dr**]∗(f(**r**)|CR;**dr**),
f(**r1**+**dr**) ≈ **f**(**r1**) + **dr**∗D[**r1**,**r1**+**dr**]•(**f**(**r**)|CR;**dr**),
f(**r**(x1+dx)) ≈ **f**(**r**(x1))
+ **dr**(x1)•**D**[**r**(x1),**r**(x1+dx)]∗(**f**(**r**(x))|CR;**dr**),
f(**r**(x1+dx)) ≈ f(**r**(x1))
+ **dr**(x1)•**D**[**r**(x1),**r**(x1+dx)]∗(f(**r**(x))|CR;**dr**),
f(**r**(x1+dx)) ≈ **f**(**r**(x1))
+ **dr**(x1)∗**D**[**r**(x1),**r**(x1+dx)]•(**f**(**r**(x))|CR;**dr**).
(477)

Mathematics: Vector Calculus

Positive Quadrant Divergences, Curls, and Gradients

In contemporary Mathematics the divergence, curl and gradient are defined as functions with basic derivatives in terms of positive quadrant limits.

Consider first for any

$$\begin{aligned}\mathbf{r} &= r*\mathbf{ur}\\ &= r1*\mathbf{u1} + r2*\mathbf{u2} + r3*\mathbf{u3}\\ &= \mathbf{r}*(c1*\mathbf{u1}+c2*\mathbf{u2}+c3*\mathbf{u3})\end{aligned}$$

the reciprocal vector

$$\begin{aligned}\mathbf{q}(\mathbf{r}) &\equiv (\mathbf{u1}/r1 + \mathbf{u2}/r2 + \mathbf{u3}/r3)\\ &= (\mathbf{u1}/c1 + \mathbf{u2}/c2 + \mathbf{u3}/\mathbf{c}3)/r\\ &= (c/r)*(\mathbf{u1}/(c*c1) + \mathbf{u2}/(c*c2) + \mathbf{u3}/(c*c3))\\ &= \mathbf{q}(\mathbf{ur})/r \qquad (478)\end{aligned}$$

for $c \equiv \mathrm{sqrt}(1/c1^2 + 1/c2^2 + 1/c3^2)$, provided that the real divisions are admissible. The quantity c/r is called the **reciprocal length of the vector r**; the vector, **u1**/(c*c1) + **u2**/(c*c2) + **u3**/(c*c3) is called the **reciprocal direction** of **r**.

The following properties hold for these reciprocal vectors:

$$\mathbf{r}\bullet(\mathbf{q}(\mathbf{r})) = 3 \qquad (479)$$

$$\begin{aligned}\mathbf{r}\wedge(\mathbf{q}(\mathbf{r})) = (c2/c3-c3/c2)*\mathbf{u1}&\\ + (c3/c1-c1/c3)*\mathbf{u2}&\\ + (c1/c2-c2/c1)*\mathbf{u3}& \qquad (480)\end{aligned}$$

$$\mathbf{r}*\mathbf{q}(\mathbf{r})=\begin{bmatrix}1 & c1/c2 & c1/c3\\ c2/c1 & 1 & c2/c3\\ c3/c1 & c3/c2 & 1\end{bmatrix} \qquad (481)$$

$$\mathbf{q}(\mathbf{r})*\mathbf{r}=\begin{bmatrix}1 & c2/c1 & c3/c1\\ c1/c2 & 1 & c3/c2\\ c1/c3 & c2/c3 & 1\end{bmatrix} \qquad (482)$$

$$\det(\mathbf{r}*(\mathbf{q}(\mathbf{r}))) = \det((\mathbf{q}(\mathbf{r}))*\mathbf{r}) = 0 \qquad (483)$$

$$D[x](\mathbf{r}(x);dx)\bullet\mathbf{q}(\mathbf{r}(x)) = -\mathbf{r}(x)\bullet(D[x](\mathbf{q}(\mathbf{r}(x));dx)). \qquad (484)$$

Note that for

r1=r11∗**u1**+r12∗**u2**+r13∗**u3** and **r2**=r21∗**u1**+r22∗**u2**+r23∗**u3**,

q(**r1+r2**) does not equal **q**(**r1**)+ **q**(**r2**),

but rather

q(**r1**) + **q**(**r2**)= **q**(**r1+r2**)•**A** (485)

where

$$\mathbf{A}=\begin{bmatrix}(r11+r12)^2/(r11*r12) & 0 & 0\\ 0 & (r21+r22)^2/(r21*r22) & 0\\ 0 & 0 & (r31+r32)^2/(r31*r32)\end{bmatrix}$$ (486)

Definition 29 (basic positive quadrant divergence, curl, gradient)

Given

real numbers dri > 0, i= 1,2,3;

dr ≡ dr1∗**u1** + dr2∗**u2** + dr3∗**u3**;

r1, a reference vector in V3;

f(**r**)=f1(**r**)∗**u1**+f2(**r**)∗**u2**+f3(**r**)∗**u3**;

f(**r**)=f(r1∗**u1**+r2∗**u2**+r3∗**u3**);

for

tr the trace matrix operator;

c the curl matrix operator;

G the diagonal vector operator;

then

lim (**f**(**r1**+dr1∗**u1**)–**f**(**r1**))•**u1**/dr1
+ (**f**(**r1**+dr2∗**u2**)–**f**(**r1**))•**u2**/dr2
+ (**f**(**r1**+dr3∗**u3**)–**f**(**r1**))•**u3**/dr3 (487)
= lim (((**f**(**r1**+dr1∗**u1**)–**f**(**r1**))•**u1**∗**u1**
+ (**f**(**r1**+dr2∗**u2**)–**f**(**r1**))•**u2**∗**u2**
+ (**f**(**r1**+dr3∗**u3**)–f(r1))•**u3**∗**u3**)•**q**(**dr**)) (488)
= lim (tr[(**f**(**r1**+dr1∗**u1**)–**f**(**r1**))∗**u1**
+ (**f**(**r1**+dr2∗**u2**)–**f**(**r1**))∗**u2**
+ (**f**(**r1**+dr3∗**u3**)–**f**(**r1**))∗**u3**]•[**G**(**q**(**dr**))]) (489)
as **dr**+⟶**0**

is called the **positive quadrant divergence of f**(**r**) **at r1**;

---

then also

lim ((**f**(**r1**+dr1∗**u1**)–**f**(**r1**))∧**u1**/dr1
+ (**f**(**r1**+dr2∗**u2**)–**f**(**r1**))∧**u2**/dr2
+ (**f**(**r1**+dr3∗**u3**)–**f**(**r1**))∧**u3**/dr3)
= lim ((**f**(**r1**+dr1∗**u1**)–**f**(**r1**))•(**u3**∗**u2**–**u2**∗**u3**)/dr1
+ (**f**(**r1**+dr2∗**u2**)–**f**(**r1**))•(**u1**∗**u3**–**u3**∗**u1**)/dr2
+ (**f**(**r1**+dr3∗**u3**)–**f**(**r1**))•(**u2**∗**u1**–**u1**∗**u2**)/dr3)
= lim ((((**f**(**r1**+dr3∗**u3**)–**f**(**r1**))/dr3)•**u2**
– ((**f**(**r1**+dr2∗**u2**)–**f**(**r1**))/dr2)•**u3**)∗**u1**
+ (((**f**(**r1**+dr1∗**u1**)–**f**(**r1**))/dr1)•**u3**
– ((**f**(**r1**+dr3∗**u3**)–**f**(**r1**))/dr3)•**u1**)∗**u2**
+ (((**f**(**r1**+dr2∗**u2**)–**f**(**r1**))/dr2)•**u1**
– ((**f**(**r1**+dr1∗**u1**)–**f**(**r1**))/dr1)•**u2**)∗**u3**) (490)
= lim **c**[((**f**(**r1**+dr1∗**u1**)–**f**(**r1**))∗**u1**
+ (**f**(**r1**+dr2∗**u2**)–**f**(**r1**))∗**u2**
+ (**f**(**r1**+dr3∗**u3**)–**f**(**r1**))**u3**]•[**G**(**q**(**dr**))]) (491)
as **dr**+⟶**0**
is called the **positive quadrant curl**[36] of **f**(**r**) at **r1**;

then also

lim ((f(**r1**+dr1∗**u1**)–f(**r1**))∗**u1**/dr1
+ (f(**r1**+dr2∗**u2**)–f(**r1**))∗**u2**/dr2
+ (f(**r1**+dr3∗**u3**)–f(**r1**))∗**u3**/dr3) (492)
= lim([(f(**r1**+dr1∗**u1**)–f(**r1**))∗**u1**
+ (f(**r1**+dr2∗**u2**)–f(**r1**))∗**u2**
+ (f(**r1**+dr3∗**u3**)–f(**r1**))∗**u3**]•[**G**(**q**(**dr**))]) 493)
as **dr**+⟶**0**
is called the **positive quadrant gradient of** f(**r**) **at r1**.
end of definition

The positive quadrant divergence of a vector function is symbolized as
D[**r1**]•(**f**(**r**);**dr**). (494)
The positive quadrant curl of a vector function is symbolized as
D[**r1**]∧(**f**(**r**);**dr**). (495)
The positive quadrant gradient of a scalar function is symbolized as
D[**r1**]∗(f(**r**);**dr**). (496)

These definitions may be expressed in terms of directional derivatives as:

$$\begin{aligned}
D[\mathbf{r1}]\bullet(\mathbf{f}(\mathbf{r});\mathbf{dr}) = {} & D[\mathbf{r1},\mathbf{r1}+dr1\ast\mathbf{u1}](\mathbf{f}(\mathbf{r});\mathbf{dr})\bullet\mathbf{u1} \\
& + D[\mathbf{r1},\mathbf{r1}+dr2\ast\mathbf{u2}](\mathbf{f}(\mathbf{r});\mathbf{dr})\bullet\mathbf{u2} \\
& + D[\mathbf{r1},\mathbf{r1}+dr3\ast\mathbf{u3}](\mathbf{f}(\mathbf{r});\mathbf{dr})\bullet\mathbf{u3} \qquad (497) \\
\mathbf{D}[\mathbf{r1}]\wedge(\mathbf{f}(\mathbf{r});\mathbf{dr}) = {} & D[\mathbf{r1},\mathbf{r1}+dr1\ast\mathbf{u1}](\mathbf{f}(\mathbf{r});\mathbf{dr})\wedge\mathbf{u1} \\
& + D[\mathbf{r1},\mathbf{r1}+dr2\ast\mathbf{u2}](\mathbf{f}(\mathbf{r});\mathbf{dr})\wedge\mathbf{u2} \\
& + D[\mathbf{r1},\mathbf{r1}+dr3\ast\mathbf{u3}](\mathbf{f}(\mathbf{r});\mathbf{dr})\wedge\mathbf{u3} \qquad (498) \\
\mathbf{D}[\mathbf{r1}]\ast(f(\mathbf{r});\mathbf{dr}) = {} & D[\mathbf{r1},\mathbf{r1}+dr1\ast\mathbf{u1}](f(\mathbf{r});\mathbf{dr})\ast\mathbf{u1} \\
& + D[\mathbf{r1},\mathbf{r1}+dr2\ast\mathbf{u2}](f(\mathbf{r});\mathbf{dr})\ast\mathbf{u2} \\
& + D[\mathbf{r1},\mathbf{r1}+dr3\ast\mathbf{u3}](f(\mathbf{r});\mathbf{dr})\ast\mathbf{u3}. \qquad (499)
\end{aligned}$$

Thus these derivatives exist if forward derivatives in the orthogonal directions of the positive quadrant exist.

There is no necessity for $\mathbf{D3}[\mathbf{r1}]\ast(\mathbf{f}(\mathbf{r});\mathbf{dr})$ to equal the negative of $\mathbf{D3}[\mathbf{r1}]\ast(\mathbf{f}(\mathbf{r});-\mathbf{dr})$.

If equal, then the positive quadrant gradient is continuous between the positive quadrant and its corresponding negative quadrant. The condition for this circumstance is simply

$$\begin{aligned}
\lim\,(\mathbf{f}(\mathbf{r1}+dr1\ast\mathbf{u1}) - \mathbf{f}(\mathbf{r1}))/dr1 &= \lim\,(\mathbf{f}(\mathbf{r1}) - f(\mathbf{r1}-dr1\ast\mathbf{u1}))/dr1 \\
\lim\,(\mathbf{f}(\mathbf{r1}+dr2\ast\mathbf{u1}) - \mathbf{f}(\mathbf{r1}))/dr2 &= \lim\,(\mathbf{f}(\mathbf{r1}) - f(\mathbf{r1}-dr2\ast\mathbf{u1}))/dr2 \\
\lim\,(\mathbf{f}(\mathbf{r1}+dr3\ast\mathbf{u1}) - \mathbf{f}(\mathbf{r1}))/dr3 &= \lim\,(\mathbf{f}(\mathbf{r1}) - f(\mathbf{r1}-dr3\ast\mathbf{u1}))/dr3.
\end{aligned}$$

If this condition is met, then the gradient is **continuous in orientation** because, given any quadrant with a gradient, the gradient of the corresponding negative quadrant will be simply the negative of the given quadrant.

However, a gradient continuous in orientation, may not be differentiable in a different orientation. Thus as the attribute of order weakens for ideas of location, a modified definition of continuous differentiability is needed. The modification is called **local differentiability** and is given in Definition 30 below.

If a function has a basic positive quadrant divergence, curl, and gradient at ***r1***, it is called **differentiable** there.

Mathematics: Vector Calculus

The form of a gradient, in common with all local derivatives, depends on the orientation of the origin, but not directly on the value of the function. Likewise, a function's differentiability and continuity in the above directional sense also depend on the given orientation.

The Positive Quadrant Gradient of a Vector Function

Can we define the basic first quadrant gradient of a vector function, **f**(**r**)?

To develop such a definition let us start with a function **f**(**r**)=**c**∗f(r) where ***c*** is a constant vector.

The formal application of the definition of gradient to this function yields the result,

D[**r1**]∗(**c**∗f(**r**);**dr**) = **D**[**r1**]∗(f(**r**)∗**c**;**dr**)
= **c**∗**D**[**r1**]∗(f(**r**);**dr**) (500)

provided the gradient of a constant vector is taken as the zero vector.

The result is an outer product which must be distinguished from
(**D**[**r1**]∗(f(**r**);**dr**))∗**c**.

The positive quadrant gradient of any vector function

f(**r**) = **f**(**r**)•**u1**∗**u1**+**f**(**r**)•**u2**∗**u2**+**f**(**r**)•**u3**∗**u3**
≡ f1(**r**)∗**u1**+f2(**r**)∗**u2**+f3(**r**)∗**u3**

then becomes

D[**r1**]∗(**f**(**r**);**dr**) ≡ **u1**∗**D**[**r1**]∗(f1(**r**);**dr**)
+ **u2**∗**D**[**r1**]∗(f2(**r**);**dr**)
+ **u3**∗**D**[**r1**]∗(f3(**r**);**dr**) (501)
= **u1**∗**u1**•**D**[**r1**]∗(**f**;**dr**)
+ **u2**∗**u2**•**D**[**r1**]∗(**f**;**dr**)
+ **u3**∗**u3**•**D**[**r1**]∗(**f**;**dr**). (502)

Thus

D[**r1**]∗(fi(**r**);**dr**) = **ui**•**D**[**r1**]∗(**f**(**r**);**dr**), i=1,2,3. (503)

Now

ui∗D[r1]∗(fi(r);dr)
= lim(ui∗(ui•(f(r1+dr1∗u1)–f(r1))∗u1
+ ui•(f(r1+dr2∗u2)–f(r1))∗u2
+ ui•(f(r1+dr3∗u3)–f(r1))∗u3)•G(q(dr)))
= ui∗ui•((f(r1+dr1∗u1)–f(r1))∗u1
+ (f(r1+dr2∗u2)–f(r1))∗u2
+ (f(r1+dr3∗u3)–f(r1))∗u3)•G(q(dr))

so that

D[r1]∗(f(r);dr)
= lim (u1∗u1+u2∗u2+u3∗u3)
•((f(r1+dr1∗u1)–f(r1))∗u1
+ (f(r1+dr2∗u2)–f(r1))∗u2
+ (f(r1+dr3∗u3)–f(r1))∗u3)•G(q(dr))
= ((f(r1+dr1∗u1)–f(r1)))∗u1
+ (f(r1+dr2∗u2)–f(r1))∗u2
+ (f(r1+dr3∗u3)–f(r1))∗u3)
•(G(q(dr))   as dr+⟶0   (504)

which accords nicely with the prior definition of a positive quadrant gradient.

We see D[r1]∗(f(r);dr) is a transformation which contemplates nine separate derivatives[37].

As transformations, positive quadrant gradients may have any rank, 0, 1, 2, or 3. Just as matrices of rank 1 need not be outer products, so also positive quadrant gradients of rank 1 need not be directional gradients.

Note that

D[r1]∗(r;dr) = u1∗u1+u2∗u2+u3∗u3 = I   (505)
tr[D[r1]∗(f(r);dr)] = D[r1]•(f(r);dr)   (506)
c[D[r1]∗(f(r);dr)] = D[r1]∧(f(r);dr)   (507)
c[T[D[r1]∗(f(r);dr)]] = –D[r1]∧(f(r);dr)   (508)
C(D[r1]∧(f(r);dr)) = T[D[r1]∗(f(r);dr)]–D[r1]∗(f(r);dr)   (509)
f(r)∧(D[r1]∧(g(r);dr)) = f(r)•T[D[r1]∗(g(r);dr)]
–f(r)•D[r1]∗(g(r);dr).   (510)

Mathematics: Vector Calculus

Note also that

$$\mathbf{D[r1]}\bullet(\mathbf{r};\mathbf{dr}) = \mathbf{dr}\bullet\mathbf{q}(\mathbf{dr}) = 3,\ \text{not } 1 \tag{511}$$

$$\mathbf{D[r1]}\wedge(\mathbf{r};\mathbf{dr}) = \mathbf{0},\ \text{not } \mathbf{dr}\wedge\mathbf{q}(\mathbf{dr}) \tag{512}$$

and

$$\mathbf{D[r1]}*(\mathbf{r};\mathbf{dr}) = \mathbf{I},\ \text{not } \mathbf{dr}*\mathbf{q}(\mathbf{dr}). \tag{513}$$

The positive quadrant gradient of a vector function is a unifying idea which embraces the divergence, curl, and scalar gradients. Consequently, subsequent results will be developed primarily for the gradient of vector functions. Where results for scalar functions, divergences, and curls are obvious they will often be omitted.

A matrix, $\boldsymbol{A}$, may be a local function, $\mathbf{A}(\mathbf{r})$, and so have a positive quadrant divergence defined as:

$$\begin{aligned}\mathbf{D[r1]}\bullet(\mathbf{A}(\mathbf{r});\mathbf{dr}) \equiv \lim\ (&\mathbf{u1}\bullet\mathbf{T}[\mathbf{A}(\mathbf{r1}+\mathrm{dr1}*\mathbf{u1})-\mathbf{A}(\mathbf{r1})]/\mathrm{dr1}\\ &+\mathbf{u2}\bullet\mathbf{T}[\mathbf{A}(\mathbf{r1}+\mathrm{dr2}*\mathbf{u1})-\mathbf{A}(\mathbf{r1})]/\mathrm{dr2}\\ &+\mathbf{u3}\bullet\mathbf{T}[\mathbf{A}(\mathbf{r1}+\mathrm{dr3}*\mathbf{u1})-\mathbf{A}(\mathbf{r1})]/\mathrm{dr3}).\end{aligned} \tag{514}$$

Clearly, both directional local gradients, $\mathbf{D[r1,r1+dr]}*(\mathbf{f}(\mathbf{r});\mathbf{dr})$, and gradients along a curve, $\mathbf{D}[\mathbf{r}(\mathrm{x1}),\mathbf{r}(\mathrm{x1+dx})]*(\mathbf{f}|\mathbf{r}(\mathrm{x});\mathbf{dr})$, differ from the positive quadrant gradient $\mathbf{D[r1]}*(\mathbf{f};\mathbf{dr})$.

Note that in V3 the curls

$\mathbf{D[r1,r1+dr]}\wedge(\mathbf{r};\mathbf{dr})$,

$\mathbf{D}[\mathbf{r}(\mathrm{x1}),\mathbf{r}(\mathrm{x1+dx})]\wedge(\mathbf{r}|\mathbf{r}(\mathrm{x});\mathbf{dr})$,

and $\mathbf{D[r1]}\wedge(\mathbf{r};\mathbf{dr})$

all equal $\mathbf{0}$,

the divergences

$\mathbf{D[r1,r1+dr]}\bullet(\mathbf{r};\mathbf{dr})$

$\mathbf{D}[\mathbf{r}(\mathrm{x1}),\mathbf{r}(\mathrm{x1+dx})]\bullet(\mathbf{r}|\mathbf{r}(\mathrm{x});\mathbf{dr})$

equal 1.

Also

$\mathrm{D}[\mathbf{r1}]\bullet(\mathbf{r};\mathbf{dr}) = 3$

and the matrices

$\mathbf{D[r1,r1+dr]}*(\mathbf{r};\mathbf{dr})$

$\mathbf{D}[\mathbf{r}(\mathrm{x1}),\mathbf{r}(\mathrm{x1+dx})]*(\mathbf{r}|\mathbf{r}(\mathrm{x});\mathbf{dr})$

do not equal $\mathbf{I}$, the identity transformation.

Positive Quadrant Derivatives of Compound Functions.

Local derivatives may also be defined with respect to vector functions other than ***r***.

Let **z**(**r**) = z1∗**u1** + z2∗**u2** + z3∗**u3** be a function defined not necessarily over the positive quadrant but with a image in the positive quadrant. Let the increment in **z** at **z**(**r1**) be the positive quadrant vector

dz ≡ dz1∗**u1**+dz2∗**u2**+dz3∗**u3**.

Then the compound function **f**(**z**(**r**)) is a function of ***z*** as well as ***r***. The positive quadrant compound gradient of ***f*** with respect to ***z*** is defined as

D[**r1**]∗(**f**(**z**);**dz**) ≡ lim ((**f**(**z**(**r1**)+dz1∗**u1**)–**f**(**z**(**r1**)))∗**u1**
+ (**f**(**z**(**r1**)+dz2∗**u2**)–**f**(**z**(**r1**)))∗**u2**
+ (**f**(**z**(**r1**)+dz3∗**u3**)–**f**(**z**(**r1**)))∗**u3**)
•(**G**(**q**(**dz**))) as **+dz⟶0**. (515)

Also

D[**r1**]•(**f**(**z**);**dz**) ≡ tr(**D**[**r1**]∗(**f**(**z**);d**z**)) (516)
D[**r1**]∧(**f**(**z**);**dz**) ≡ **c**(**D**[**r1**]∗(**f**(**z**);d**z**)). (517)

Local derivatives of compound functions require the existence of limits for ***f*** in a section defined by **z**(**r**) just as compound functions of real derivatives require limits not only over the set S over which ***z*** is defined but also limits for ***f*** over the set **z**(S).

In contemporary Mathematics it is customary to write local derivatives without declaring the variable or the reference. An undeclared variable is presumed to refer to **r**; that is, D**:f** ≡ D:(**f**;**dr**) for ":" meaning •, **∧**, or ∗. In Theoretical Physics the variable of differentiation is denoted explicitly for clarity.

Mathematics: Vector Calculus

For constant **r1**, **urr1** the direction of the vector (**r**–**r1**), and **A** a fixed transformation, the following table gives some useful specific results for positive quadrant derivatives.

Function:	D•	**D∧**	**D∗**
urr1=(**r**–**r1**) /abs(**r**–**r1**)	2/abs(**r**–**r1**)	0	(**I**–**urr1**∗**urr1**) /abs(**r**–**r1**)
r–**r1**	3	0	**I** (identity)
(**r**–**r1**)•(**r**–**r1**)			2∗(**r**–**r1**)
abs(**r**–**r1**)			**urr1**
$(\text{abs}(\mathbf{r}-\mathbf{r1}))^{n}$			$(\text{abs}(\mathbf{r}-\mathbf{r1}))^{(n-1)}$ ∗**urr1**
1/abs(**r**–**r1**)			–**urr1** /((**r**–**r1**)•(**r**–**r1**))
urr1 /((**r**–**r1**)•(**r**–**r1**))	0	0	(**I**– 3∗**urr1**)∗**urr1** $/(\text{abs}(\mathbf{r}-\mathbf{r1}))^{3}$
f(**r**)∗(**r**–**r1**)	3∗f +(**r**–**r1**)•**D**∗f	(**r**–**r1**)∧**D**∗f	f∗**I** +(**r**–**r1**)∗**D**∗f
(**r**–**r1**)•**A**	tr[**A**]	**c**[**A**]	**T**[**A**]
r1•**A**•**r**			**r1**•**A**
r•**A**•**r1**			**r1**•**T**[**A**]
(**r**–**r1**)•**A**•(**r**–**r1**)			(**r**–**r1**)•(**A**+**T**[**A**]

Some Specific Positive Quadrant Results

Local Differentiability

The results above apply to functions restricted either to a curve, a direction, or the three orthogonal directions. Expanded results are possible with further functional restrictions.

Functions with the same functional values may have different positive quadrant gradients. Furthermore, even functions with basic directional derivatives in the orthogonal directions of the origin may not relate to gradients in other directions of the positive quadrant.

Definition 30 (continuous local differentiability)

Given

- **u1**,**u2**,**u3** the orthogonal orientation of the origin;
- dr1,dr2,dr3 real numbers;
- **dr** ≡ dr1∗**u1** + dr2∗**u2** + dr3∗**u3**;
- **udr**, the direction of **dr**;
- **f**(**r**), a function over V3;

if for every **udr**

$$
\begin{aligned}
(\mathbf{f}(\mathbf{r1}+\mathbf{dr}) - \mathbf{f}(\mathbf{r1})) = {} & \mathbf{f}(\mathbf{r1}+\mathrm{dr1}*\mathbf{u1}) - \mathbf{f}(\mathbf{r1}) \\
& + \mathbf{f}(\mathbf{r1}+\mathrm{dr2}*\mathbf{u2}) - \mathbf{f}(\mathbf{r1}) \\
& + \mathbf{f}(\mathbf{r1}+\mathrm{dr3}*\mathbf{u3}) - \mathbf{f}(\mathbf{r1}) \quad \text{as } \mathbf{dr}\longrightarrow \mathbf{0}
\end{aligned}
\tag{518}
$$

then

f(**r**) is **continuously locally differentiable at r1**.

end of definition

If ***f*** is locally differentiable for all **r** in a subset S of V3, then **f** is said to be **locally differentiable on S**.

If the function is locally differentiable only for positive dri, the function is said to be **locally differentiable at r1 in the positive quadrant**.

Wherever a function is locally differentiable at **r1** in the positive quadrant, the existence of limits implies it has a positive quadrant gradient there.

The definition of a positive quadrant gradient may be applied to functions which are continuously locally differentiable as:

$$\mathbf{D}[\mathbf{r1}]*(\mathbf{f}(\mathbf{r});\mathbf{dr}) \equiv \lim ((\mathbf{f}(\mathbf{r1}+dr1*\mathbf{u1})-\mathbf{f}(\mathbf{r1}))*\mathbf{u1} + (\mathbf{f}(\mathbf{r1}+dr2*\mathbf{u2})-\mathbf{f}(\mathbf{r1}))*\mathbf{u2} + (\mathbf{f}(\mathbf{r1}+dr3*\mathbf{u3})-\mathbf{f}(\mathbf{r1}))*\mathbf{u3}) \bullet \mathbf{G}(\mathbf{q}(\mathbf{dr})) \quad \text{as } dri \to 0. \qquad (519)$$

Functions with the same functional values may have different positive quadrant gradients even if locally differentiable.

For ***f*** with a *basic* gradient

$$\mathbf{f}(\mathbf{r1}+\mathbf{dr}) - \mathbf{f}(\mathbf{r1}) \approx \mathbf{dr}\bullet(\mathbf{u1}*(\mathbf{f}(\mathbf{r1}+dr1*\mathbf{u1})-\mathbf{f}(\mathbf{r1})))/dr1 + \mathbf{dr}\bullet(\mathbf{u2}*(\mathbf{f}(\mathbf{r1}+dr2*\mathbf{u2})-\mathbf{f}(\mathbf{r1})))/dr2 + \mathbf{dr}\bullet(\mathbf{u3}*(\mathbf{f}(\mathbf{r1}+dr3*\mathbf{u3})-\mathbf{f}(\mathbf{r1})))/dr3 \longrightarrow \mathbf{dr}\bullet\mathbf{T}[\mathbf{D}[\mathbf{r1}]*(\mathbf{f}(\mathbf{r});\mathbf{dr})]. \qquad (520)$$

Consequently for continuously locally differentiable functions with basic gradient at **r1**

$$\mathbf{f}(\mathbf{r1}+\mathbf{dr}) \approx \mathbf{f}(\mathbf{r1}) + \mathbf{dr}\bullet\mathbf{T}[\mathbf{D}[\mathbf{r1}]*(\mathbf{f}(\mathbf{r});\mathbf{dr})]. \qquad (521)$$

A basic positive quadrant gradient of a continuously locally differentiable function at **r1** is called a **basic locally continuous gradient** there. If a function has a basic continuous gradient everywhere in V3, the function is said to have a basic continuous gradient **everywhere**.

The form of the positive quadrant gradient does not apply generally to other quadrants. Whenever the positive quadrant gradient does apply to all other quadrants, the function at **r1** is said to have a **simply continuous gradient** there. If a function has a simply continuous gradient everywhere in V3, the function is said to have a simply continuous gradient **everywhere**.

The gradient of a vectorial function is a matrix. What is its rank?

The answer depends on **f**(**r**). Similar to real functions, a vector function at **r1** may have a gradient unrelated to its neighboring values. Such gradients may have any rank.

If the function has a *basic* gradient, the rank of the gradient depends on how **f**(**r**) varies at **r1**. If **f**(**r**) varies only in one of the designated directions then its gradient has rank 1. Consequently it is possible for a gradient to be [**0**] for a function which does not vary in any of the designated directions, but nevertheless varies in some intermediate direction. If, however, a function is continuous *in* the positive quadrant, then a functional change in an intermediate direction is accompanied by a change in one or more of the designated directions according to the rank of the continuous gradient.

Accordingly, with a changed orientation, a gradient may change not only its form but also its rank.

The promotion of a positive quadrant gradient into a one simply continuous rests on two implications:

the function is continuously differentiable

the differentiability in the positive quadrant extends to the other quadrants.

The second implication is accommodated formally by allowing either one or more of the ***ui*** or the *dri* to be negative.

Now let the curved line CX contain **r1**. Let ***f*** be continuously locally differentiable at **r1**(x1) with a simply continuous gradient.
Let **dr** = **r**(x1+dx) – **r1**(x1). Then

D[**r1**](**f**|CX;dx) = D[x1](**r**(x)|CX;dx)•**T**[**D**[**r1**]∗(**f**(**r**);**dr**)]. (522)

Equation (522) can be viewed as expressing how and under what conditions:

a **chain rule** is valid in V3

an indexed change in a locally differentiable function is related to its change in location.

Mathematics: Vector Calculus

Now let **r**(x) be defined from values over which a scalar *f* with basic derivatives has a constant value. From equation (522) it follows the gradient **D[r1]**∗(f(**r**);**dr**) is orthogonal to the direction of the tangent line of **r**(x) at **r1**. Such traces, **r**(x|f=constant) are called **contour lines**.

For a constant vector function, **f**

D[x1](**r**(x)|CX;dx)•**T**[**D**[**r1**]∗(**f**(**r**);**dr**)] = **0** (523)

that is, the direction of the tangent line of **r**(x) at **r1** falls in the zero subset of the matrix **T**[**D**[**r1**]∗(**f**(**r**);**dr**)].

Now if the scalar function f(**r**) is a constant on a differentiable local surface **r**(x1,x2), then

D[**r**(x1),**r**(x1+dx)]∗(f|CX1;**dr**)

and

D[**r**(x2),**r**(x2+dx)]∗(f|CX2;**dr**)

are both orthogonal to both curves CX1 and CX2 at **r**(x1,x2). it follows that wherever *f* is constant on a differentiable surface, the gradients are orthogonal to that surface at **r**(x1,x2).

The sign of the gradients is to be noted here.

For ***f*** constant on a surface, from equation (473),

D[x1,x1+dx](**r**(x)|CX;dx)•**D**[**r**(x1),**r**(x1+dx)]∧(**f**|CX;**dr**) = 0 (524)

that is, **D[r1]**∧(**f**|CX;**dr**) is orthogonal to the direction of the tangent line of **r**(x1) at **r1**.

Therefore it also follows that for ***f*** constant on a differentiable surface **D**[**r**(x1),**r**(x1+dx)]∧(**f**|CX1;**dr**) and **D**[**r**(x2),**r**(x2+dx)]∧(**f**|CX2;**dr**) are orthogonal to the designated surface at **r**(x1,x2).

Sectional gradients

The idea of a basic positive quadrant gradient can be further generalized by relaxing the requirement for orthogonality.

Definition 31 (basic sectional gradient,divergence, and curl)
Given
SECT(**r1,u1R1**,**u2R1**,**u3R1**) at **r1**;
f(**r**) defined in SECT(**r1,u1R1**,**u2R1**,**u3R1**);
and at **r1**
dR1 = d1R1∗**u1R1**+d2R1∗**u2R1**+d3R1∗**u3R1**, diR1>0;
then the basic **sectional gradient** of a generalized function is defined as

D[**r1**]∗(**f**;**dR1**) ≡ lim (**f**(**r1**+d1R1∗**u1R1**)–**f**(**r1**))∗**u1R1**/d1R1
+ (**f**(**r1**+d2R1∗**u2R1**)–**f**(**r1**))∗**u2R1**/d2R1
+ (**f**(**r1**+d3R1∗**u3R1**)–**f**(**r1**))∗**u3R1**/d3R1
as **dR1**+⟶**0**
(525)

and the related sectional divergence and curl are defined as:

D[**r1**]•(**f**;**dR1**) ≡ lim (**f**(**r1**+d1R1∗**u1R1**)–**f**(**r1**))•**u1R1**/d1R1
+ (**f**(**r1**+d2R1∗**u2R1**)–**f**(**r1**))•**u2R1**/d2R1
+ (**f**(**r1**+d3R1∗**u3R1**)–**f**(**r1**))•**u3R1**/d3R1
as **dR1**+⟶**0**
(526)

D[**r1**]∧(**f**;**dR1**) ≡ lim (**f**(**r1**+d1R1∗**u1R1**)–**f**(**r1**))∧**u1R1**/d1R1
+ (**f**(**r1**+d2R1∗**u2R1**)–**f**(**r1**))∧**u2R1**/d2R1
+ (**f**(**r1**+d3R1∗**u3R1**)–**f**(**r1**))∧**u3R1**/d3R1
as **dR1**+⟶**0**.
(527)

end of definition

Moreover

D[**r1**]•(**f**;**dR1**) = tr[**D3**[**r1**]∗(**f**;**dR1**)] (528)

where tr is the trace matrix operator since

(**f**(**r1**+d1R1∗**u1R1**) – **f**(**r1**))•**u1R1**/d1R1
+ (**f**(**r1**+d2R1∗**u2R1**) – **f**(**r1**))•**u2R1**/d2R1
+(**f**(**r1**+d3R1∗**u3R1**) – **f**(**r1**))•**u3R1**/d3R1

Mathematics: Vector Calculus

$$
\begin{aligned}
&= (\mathbf{f}(\mathbf{r1}+\mathrm{d1R1}\ast\mathbf{u1R1}) - \mathbf{f}(\mathbf{r1}))\cdot\mathbf{I}\cdot\mathbf{u1R1}/\mathrm{d1R1} \\
&\quad + (\mathbf{f}(\mathbf{r1}+\mathrm{d2R1}\ast\mathbf{u2R1}) - \mathbf{f}(\mathbf{r1}))\cdot\mathbf{I}\cdot\mathbf{u2R1}/\mathrm{d2R1} \\
&\quad + (\mathbf{f}(\mathbf{r1}+\mathrm{d3R1}\ast\mathbf{u3R1}) - \mathbf{f}(\mathbf{r1}))\cdot\mathbf{I}\cdot\mathbf{u3R1}/\mathrm{d3R1} \\
&= (\mathbf{f}(\mathbf{r1}+\mathrm{d1R1}\ast\mathbf{u1R1}) - \mathbf{f}(\mathbf{r1})) \\
&\qquad \cdot[\mathbf{u1}\ast\mathbf{u1}+\mathbf{u2}\ast\mathbf{u2}+\mathbf{u3}\ast\mathbf{u3}]\cdot\mathbf{u1R1}/\mathrm{d1R1} \\
&\quad + (\mathbf{f}(\mathbf{r1}+\mathrm{d2R1}\ast\mathbf{u2R1}) - \mathbf{f}(\mathbf{r1})) \\
&\qquad \cdot[\mathbf{u1}\ast\mathbf{u1}+\mathbf{u2}\ast\mathbf{u2}+\mathbf{u3}\ast\mathbf{u3}]\cdot\mathbf{u2R1}/\mathrm{d2R1} \\
&\quad + (\mathbf{f}(\mathbf{r1}+\mathrm{d3R1}\ast\mathbf{u3R1}) - \mathbf{f}(\mathbf{r1})) \\
&\qquad \cdot[\mathbf{u1}\ast\mathbf{u1}+\mathbf{u2}\ast\mathbf{u2}+\mathbf{u3}\ast\mathbf{u3}]\cdot\mathbf{u3R1}/\mathrm{d3R1} \\
&= \mathbf{u1}\cdot(\mathbf{f}(\mathbf{r1}+\mathrm{d1R1}\ast\mathbf{u1R1}) - \mathbf{f}(\mathbf{r1}))\ast\mathbf{u1R1}\cdot\mathbf{u1}/\mathrm{d1R1} \\
&\quad + \mathbf{u2}\cdot(\mathbf{f}(\mathbf{r1}+\mathrm{d1R1}\ast\mathbf{u1R1}) - \mathbf{f}(\mathbf{r1}))\ast\mathbf{u1R1}\cdot\mathbf{u2}/\mathrm{d1R1} \\
&\quad + \mathbf{u3}\cdot(\mathbf{f}(\mathbf{r1}+\mathrm{d1R1}\ast\mathbf{u1R1}) - \mathbf{f}(\mathbf{r1}))\ast\mathbf{u1R1}\cdot\mathbf{u3}/\mathrm{d1R1} \\
&\quad + \mathbf{u1}\cdot(\mathbf{f}(\mathbf{r1}+\mathrm{d2R1}\ast\mathbf{u2R1}) - \mathbf{f}(\mathbf{r1}))\ast\mathbf{u2R1}\cdot\mathbf{u1}/\mathrm{d2R1} \\
&\quad + \mathbf{u2}\cdot(\mathbf{f}(\mathbf{r1}+\mathrm{d2R1}\ast\mathbf{u2R1}) - \mathbf{f}(\mathbf{r1}))\ast\mathbf{u2R1}\cdot\mathbf{u2}/\mathrm{d2R1} \\
&\quad + \mathbf{u3}\cdot(\mathbf{f}(\mathbf{r1}+\mathrm{d2R1}\ast\mathbf{u2R1}) - \mathbf{f}(\mathbf{r1}))\ast\mathbf{u2R1}\cdot\mathbf{u3}/\mathrm{d2R1} \\
&\quad + \mathbf{u1}\cdot(\mathbf{f}(\mathbf{r1}+\mathrm{d3R1}\ast\mathbf{u3R1}) - \mathbf{f}(\mathbf{r1}))\ast\mathbf{u3R1}\cdot\mathbf{u1}/\mathrm{d3R1} \\
&\quad + \mathbf{u2}\cdot(\mathbf{f}(\mathbf{r1}+\mathrm{d3R1}\ast\mathbf{u3R1}) - \mathbf{f}(\mathbf{r1}))\ast\mathbf{u3R1}\cdot\mathbf{u2}/\mathrm{d3R1} \\
&\quad + \mathbf{u3}\cdot(\mathbf{f}(\mathbf{r1}+\mathrm{d3R1}\ast\mathbf{u3R1}) - \mathbf{f}(\mathbf{r1}))\ast\mathbf{u3R1}\cdot\mathbf{u3}/\mathrm{d3R1} \\
&= \mathbf{u1}\cdot(\mathbf{f}(\mathbf{r1}+\mathrm{d1R1}\ast\mathbf{u1R1}) - \mathbf{f}(\mathbf{r1}))\ast\mathbf{u1R1}\cdot\mathbf{u1}/\mathrm{d1R1} \\
&\quad + \mathbf{u1}\cdot(\mathbf{f}(\mathbf{r1}+\mathrm{d2R1}\ast\mathbf{u2R1}) - \mathbf{f}(\mathbf{r1}))\ast\mathbf{u2R1}\cdot\mathbf{u1}/\mathrm{d2R1} \\
&\quad + \mathbf{u1}\cdot(\mathbf{f}(\mathbf{r1}+\mathrm{d3R1}\ast\mathbf{u3R1}) - \mathbf{f}(\mathbf{r1}))\ast\mathbf{u3R1}\cdot\mathbf{u1}/\mathrm{d3R1} \\
&\quad + \mathbf{u2}\cdot(\mathbf{f}(\mathbf{r1}+\mathrm{d1R1}\ast\mathbf{u1R1}) - \mathbf{f}(\mathbf{r1}))\ast\mathbf{u1R1}\cdot\mathbf{u2}/\mathrm{d1R1} \\
&\quad + \mathbf{u2}\cdot(\mathbf{f}(\mathbf{r1}+\mathrm{d2R1}\ast\mathbf{u2R1}) - \mathbf{f}(\mathbf{r1}))\ast\mathbf{u2R1}\cdot\mathbf{u2}/\mathrm{d2R1} \\
&\quad + \mathbf{u2}\cdot(\mathbf{f}(\mathbf{r1}+\mathrm{d3R1}\ast\mathbf{u3R1}) - \mathbf{f}(\mathbf{r1}))\ast\mathbf{u3R1}\cdot\mathbf{u2}/\mathrm{d3R1} \\
&\quad + \mathbf{u3}\cdot(\mathbf{f}(\mathbf{r1}+\mathrm{d1R1}\ast\mathbf{u1R1}) - \mathbf{f}(\mathbf{r1}))\ast\mathbf{u1R1}\cdot\mathbf{u3}/\mathrm{d1R1} \\
&\quad + \mathbf{u3}\cdot(\mathbf{f}(\mathbf{r1}+\mathrm{d2R1}\ast\mathbf{u2R1}) - \mathbf{f}(\mathbf{r1}))\ast\mathbf{u2R1}\cdot\mathbf{u3}/\mathrm{d2R1} \\
&\quad + \mathbf{u3}\cdot(\mathbf{f}(\mathbf{r1}+\mathrm{d3R1}\ast\mathbf{u3R1}) - \mathbf{f}(\mathbf{r1}))\ast\mathbf{u3R1}\cdot\mathbf{u3}/\mathrm{d3R1} \\
&\longrightarrow \mathbf{u1}\cdot\mathbf{D3}[\mathbf{r1}]\ast(\mathbf{f};\mathbf{dR1})\cdot\mathbf{u1} \\
&\qquad + \mathbf{u2}\cdot\mathbf{D3}[\mathbf{r1}]\ast(\mathbf{f};\mathbf{dR1})\cdot\mathbf{u2} \\
&\qquad + \mathbf{u3}\cdot\mathbf{D3}[\mathbf{r1}]\ast(\mathbf{f};\mathbf{dR1})\cdot\mathbf{u3} \\
&= \mathrm{tr}[\mathbf{D3}[\mathbf{r1}]\ast(\mathbf{f};\mathbf{dR1})].
\end{aligned}
$$

Similarly

$$
\mathbf{D3}[\mathbf{r1}]\wedge(\mathbf{f};\mathbf{dR1}) = \mathbf{c}[\mathbf{D3}[\mathbf{r1}]\ast(\mathbf{f};\mathbf{dR1})] \qquad (529)
$$

where **c** is the curl matrix operator since

$$
\begin{aligned}
&\mathbf{c}[\mathbf{D3}[\mathbf{r1}]\ast(\mathbf{f};\mathbf{dR1})] \\
&\quad = \lim \mathbf{c}[(\mathbf{f}(\mathbf{r1}+\mathrm{d1R1}\ast\mathbf{u1R1}) - \mathbf{f}(\mathbf{r1}))\ast\mathbf{u1R1}/\mathrm{d1R1}] \\
&\qquad\quad + \mathbf{c}[(\mathbf{f}(\mathbf{r1}+\mathrm{d2R1}\ast\mathbf{u2R1}) - \mathbf{f}(\mathbf{r1}))\ast\mathbf{u2R1}/\mathrm{d2R1}] \\
&\qquad\quad + \mathbf{c}[(\mathbf{f}(\mathbf{r1}+\mathrm{d3R1}\ast\mathbf{u3R1}) - \mathbf{f}(\mathbf{r1}))\ast\mathbf{u3R1}/\mathrm{d3R1}]
\end{aligned}
$$

$= (\mathbf{f}(\mathbf{r1}+d1R1*\mathbf{u1R1}) - \mathbf{f}(\mathbf{r1},\mathbf{a1}))\wedge\mathbf{u1R1}/d1R1$
$\quad + (\mathbf{f}(\mathbf{r1}+d2R1*\mathbf{u1R1}) - \mathbf{f}(\mathbf{r1}))\wedge\mathbf{u2R1}/d2R1$
$\quad + (\mathbf{f}(\mathbf{r1}+d3R1*\mathbf{u1R1}) - \mathbf{f}(\mathbf{r1}))\wedge\mathbf{u3R1}/d3R1$
$\longrightarrow \mathbf{D3}[\mathbf{r1}]\wedge(\mathbf{f};\mathbf{dR1})$.

The sectional divergence of a matrix function is defined as:

$$\mathbf{D}[\mathbf{r1}]\cdot(\mathbf{A};\mathbf{dR1}) \equiv \lim \mathbf{u1R1}\cdot\mathbf{T}[\mathbf{A}(\mathbf{r1}+d1R1*\mathbf{u1R1})-\mathbf{A}(\mathbf{r1})]/d1R1 + \mathbf{u2R1}\cdot\mathbf{T}[\mathbf{A}(\mathbf{r1}+d2R1*\mathbf{u2R1})-\mathbf{A}(\mathbf{r1})]/d2R1 + \mathbf{u3R1}\cdot\mathbf{T}[\mathbf{A}(\mathbf{r1}+d3R1*\mathbf{u3R1})-\mathbf{A}(\mathbf{r1})]/d3R1 \text{ as } \mathbf{dR1}+\longrightarrow\mathbf{0}. \quad (530)$$

In a given section the above differences are all well–defined and may or may not have limits at **r1**. The existence of sectional gradients implies that a function need only have limits *of* the section. The existence of sectional gradients does not necessarily imply either that sectional vector functions have basic directional derivatives or that they are differentiable.

Conversely, the directional derivative,

$$\mathbf{D}[\mathbf{r1}](\mathbf{f};\mathbf{dR1}) \longleftarrow (\mathbf{f}(\mathbf{r1}+\mathbf{dR1}) - \mathbf{f}(\mathbf{r1}))/dR1 \text{ as } dR1\longrightarrow 0 \quad (531)$$

may exist in the **dR1** direction even though the sectional gradient does not.

Definition 32 (sectional local differentiability)

Given

$\mathbf{dR1} \equiv dR1*\mathbf{udR1} \equiv d1R1*\mathbf{u1R1} + d2R1*\mathbf{u2R1} + d3R1*\mathbf{u3R1}, \quad diR1\geq 0;$

$\mathbf{f}(\mathbf{r})$ defined over SECT(**r1**,**u1R1**,**u2R1**,**u3R1**) at **r1**;

if for every direction **udR1** in **SECT**(**r1**,**u1R1**,**u2R1**,**u3R1**)

$$\mathbf{f}(\mathbf{r1}+\mathbf{dR1}) - \mathbf{f}(\mathbf{r1}) = \mathbf{f}(\mathbf{r1}+d1R1*\mathbf{u1R1}) - \mathbf{f}(\mathbf{r1}) + \mathbf{f}(\mathbf{r1}+d2R1*\mathbf{u2R1}) - \mathbf{f}(\mathbf{r1}) + \mathbf{f}(\mathbf{r1}+d3R1*\mathbf{u3R1}) - \mathbf{f}(\mathbf{r1}) \text{ as } \mathbf{dR1}\longrightarrow\mathbf{0} \quad (532)$$

then

f is **locally differentiable in SECT**(**r1**,**u1R1**,**u2R1**,**u3R1**) **at r1**.

end of definition

Mathematics: Vector Calculus

The gradient of a function locally differentiable in
SECT(**r1**,**u1R1**,**u2R1**,**u3R1**) at **r1**
is called a **continuous gradient in** SECT(**r1**,**u1R1**,**u2R1**,**u3R1**) **at r1**.

For a function locally differentiable with a basic continuous gradient in SECT(**r1**,**u1R1**,**u2R1**,**u3R1**) at **r1**, **f**(**r1**+**dR1**) does *not* approximate
f(**r1**) + **dR1**•**T**[**D**[**r1**]∗(**f**;**dR1**)]
but rather

f(r1+dR1) − f(r1)
≈ f(r1+d1R1∗u1R1) − f(r1)
+ f(r1+d2R1∗u2R1) − f(r1)
+ f(r1+d3R1∗u3R1) − f(r1)
= dR1•[(u2R1∧u3R1)∗(u2R1∧u3R1)
+ (u3R1∧u1R1)∗(u3R1∧u1R1)
+ (u1R1∧u2R1)∗(u1R1∧u2R1)]
•[u1R1∗(f(r1+d1R1∗u1R1) − f(r1))/d1R1
+ u2R1∗(f(r1+d2R1∗u2R1) − f(r1))/d2R1
+ u3R1∗(f(r1+d3R1∗u3R1) − f(r1))/d3R1]
/(u1R1•u2R1∧u3R1)^2
⟶dR1•[(u2R1∧u3R1)∗(u2R1∧u3R1)
+ (u3R1∧u1R1)∗(u3R1∧u1R1)
+ (u1R1∧u2R1)∗(u1R1∧u2R1)]
•T[D[r1]∗(f;dR1)]/(u1R1•u2R1∧u3R1)^2. (533)

A function with basic continuous gradients also has basic continuous sectional gradients.

Relationships between Continuous Gradients, Sectional Gradients, and Directional Gradients.

A transformation relating a section to the positive quadrant is useful. Consider

UR1 ≡ u1∗u1R1 + u2∗u2R1 + u3∗u3R1 (534)

which transforms any vector of the positive quadrant defined by **u1**, **u2**, and **u3** into SECT(**0**,**u1R1**,**u2R1**,**u3R1**).

Since each of the sectional directions may be represented as a vector of directional cosines,

$$\mathbf{uiR1} = \mathrm{ci1}*\mathbf{u1} + \mathrm{ci2}*\mathbf{u2} + \mathrm{ci3}*\mathbf{u3}. \tag{535}$$

The transformation may also be represented as

$$\mathbf{UR1} = \begin{bmatrix} \mathrm{c11} & \mathrm{c12} & \mathrm{c13} \\ \mathrm{c21} & \mathrm{c22} & \mathrm{c23} \\ \mathrm{c33} & \mathrm{c32} & \mathrm{c33} \end{bmatrix} \tag{536}$$

a matrix of directional cosines.

The inverse of **UR1** transforms sectional vectors into positive quadrant vectors.

$$\begin{aligned}[\mathbf{UR1}]^{-1} = [&(\mathbf{u2R1} \wedge \mathbf{u3R1}) * \mathbf{u1} \\ &+ (\mathbf{u3R1} \wedge \mathbf{u1R1}) * \mathbf{u2} \\ &+ (\mathbf{u1R1} \wedge \mathbf{u2R1}) * \mathbf{u3}]/\det[\mathbf{UR1}].\end{aligned}$$

The following sectional relationships mirror their orthogonal counterparts.

From equation (197)

$$\begin{aligned}\det[\mathbf{UR1}] * \mathbf{I} = &(\mathbf{u2R1} \wedge \mathbf{u3R1}) * \mathbf{u1R1} \\ &+ (\mathbf{u3R1} \wedge \mathbf{u1R1}) * \mathbf{u2R1} \\ &+ (\mathbf{u1R1} \wedge \mathbf{u2R1}) * \mathbf{u3R1}\end{aligned} \tag{537}$$

$$\begin{aligned}\mathrm{tr}[\det[\mathbf{UR1}] * \mathbf{I}] &= (\mathbf{u2R1} \wedge \mathbf{u3R1}) \cdot \mathbf{u1R1} \\ &\quad + (\mathbf{u3R1} \wedge \mathbf{u1R1}) \cdot \mathbf{u2R1} \\ &\quad + (\mathbf{u1R1} \wedge \mathbf{u2R1}) \cdot \mathbf{u3R1} \\ &= 3 * \det[\mathbf{UR1}]\end{aligned} \tag{538}$$

$$\begin{aligned}\mathbf{c}(\det[\mathbf{UR1}] * \mathbf{I}) &= (\mathbf{u2R1} \wedge \mathbf{u3R1}) \wedge \mathbf{u1R1} \\ &\quad + (\mathbf{u3R1} \wedge \mathbf{u1R1}) \wedge \mathbf{u2R1} \\ &\quad + (\mathbf{u1R1} \wedge \mathbf{u2R1}) \wedge \mathbf{u3R1} \\ &= \mathbf{0}.\end{aligned} \tag{539}$$

Mathematics: Vector Calculus

Vectors represented sectionally conform to an algebra similar to the algebra for an orthogonal orientation.

Let

$$\begin{aligned}\mathbf{rR1} &\equiv \mathrm{r1R1}*\mathbf{u1R1} + \mathrm{r2R1}*\mathbf{u2R1} + \mathrm{r3R1}*\mathbf{u3R1}\\ &\equiv \mathrm{rR1}*\mathbf{urR1} &\quad(540)\\ \mathbf{r} &\equiv \mathbf{rR1}\cdot\mathbf{UR1}^{-1}\\ &\equiv \mathrm{r1}*\mathbf{u1} + \mathrm{r2}*\mathbf{u2} + \mathrm{r3}*\mathbf{u3}\\ &\equiv \mathrm{r}*\mathbf{ur}\\ &\equiv \mathrm{r}*(\mathrm{c1}*\mathbf{u1} + \mathrm{c2}*\mathbf{u2} + \mathrm{c3}*\mathbf{u3}). &\quad(541)\end{aligned}$$

In general, rR1 does not equal r, not does $\mathbf{urR1} = \mathbf{ur}$.

However,

$$\mathbf{rR1} = \mathbf{r}\cdot\mathbf{uR1} = \mathrm{r}*(\mathrm{c1}*\mathbf{u1R1} + \mathrm{c2}*\mathbf{u2R1} + \mathrm{c3}*\mathbf{u3R1}) \quad (542)$$

so that

$$\begin{aligned}\mathrm{riR1} &= \mathrm{r}*\mathrm{ci}\\ &= \mathrm{ri},\ i=1,2,3 &\quad(543)\\ \mathrm{rR1} &= \mathrm{sqrt}((\mathbf{r}\cdot\mathbf{UR1}\cdot\mathbf{u1})^2 + (\mathbf{r}\cdot\mathbf{UR1}\cdot\mathbf{u2})^2 + (\mathbf{r}\cdot\mathbf{UR1}\cdot\mathbf{u3})^2) &\quad(544)\\ \mathbf{urR1} &= \mathbf{rR1}/\mathrm{rR1}\\ &= (\mathrm{r1R1}/\mathrm{rR1})*\mathbf{u1R1}\\ &\qquad + (\mathrm{r2R1}/\mathrm{rR1})*\mathbf{u2R1}\\ &\qquad + (\mathrm{r3R1}/\mathrm{rR1})*\mathbf{u3R1}. &\quad(545)\end{aligned}$$

Now let $\mathbf{r0R1} = \mathbf{r0}\cdot\mathbf{UR1}$ and $\mathbf{r1R1} = \mathbf{r1}\cdot\mathbf{UR1}$ be two vectors in the same SECT($\mathbf{r1},\mathbf{u1R1},\mathbf{u2R1},\mathbf{u3R1}$).

Then

$$\begin{aligned}\mathbf{r0R1}*\mathbf{r1R1} &= \mathbf{T}[\mathbf{UR1}]\cdot[\mathbf{r0}*\mathbf{r1}]\cdot\mathbf{UR1} &\quad(546)\\ \mathbf{r0R1}\cdot\mathbf{r1R1} &= \mathrm{tr}[\mathbf{T}[\mathbf{UR1}]\cdot[\mathbf{r0}*\mathbf{r1}]\cdot\mathbf{UR1}] &\quad(547)\\ &= \mathbf{r0}\cdot\mathbf{r1}\\ &\qquad + (\mathrm{r01}*\mathrm{r12} + \mathrm{r02}*\mathrm{r11})*\mathbf{u1R1}\cdot\mathbf{u2R1}\\ &\qquad + (\mathrm{r01}*\mathrm{r13} + \mathrm{r03}*\mathrm{r11})*\mathbf{u1R1}\cdot\mathbf{u3R1}\\ &\qquad + (\mathrm{r02}*\mathrm{r13} + \mathrm{r02}*\mathrm{r12})*\mathbf{u2R1}\cdot\mathbf{u3R1} &\quad(548)\\ \mathbf{r0R1}\wedge\mathbf{r1R1} &= \mathbf{c}(\mathbf{T}[\mathbf{UR1}]\cdot[\mathbf{r0}*\mathbf{r1}]\cdot\mathbf{UR1}) &\quad(549)\\ &= (\mathrm{r01}*\mathrm{r12} - \mathrm{r02}*\mathrm{r11})*\mathbf{u1R1}\wedge\mathbf{u2R1}\\ &\qquad + (\mathrm{r03}*\mathrm{r11} - \mathrm{r01}*\mathrm{r13})*\mathbf{u3R1}\wedge\mathbf{u1R1}\\ &\qquad + (\mathrm{r02}*\mathrm{r13} - \mathrm{r02}*\mathrm{r12})*\mathbf{u2R1}\wedge\mathbf{u3R1}. &\quad(550)\end{aligned}$$

Note

D[**r1**]∗(**f**;**dR1**)•[**UR1**]$^{-1}$
≡ lim ((**f**(**r1**+d1R1∗**u1R1**)–**f**(**r1**))∗**u1**/d1R1
+ (**f**(**r1**+d2R1∗**u2R1**)–**f**(**r1**))∗**u2**/d2R1
+ (**f**(**r1**+d3R1∗**u3R1**)–**f**(**r1**))∗**u3**/d3R1)
/det[**UR1**] (551)

and

f(**r1**+diR1∗**uiR1**) – **f**(**r1**)
≈ diR1∗**uiR1**•[**UR1**]$^{-1}$•**T**[**UR1**]$^{-1}$•**T**[**D**[**r1**]∗(**f**;**dR1**)]. (552)

which is a restatement of equation (533).

The relationship between a simply continuous gradient and a sectional gradient is given in the following theorem.

Theorem 2 (relationship between simply continuous and sectional gradients)
Given
f(r) a function with a basic, simply continuous gradient at **r1**;
SECT(**r1,u1R1**,**u2R1**,**u3R1**) a section at **r1**;
then

D[**r1**]∗(**f**;**dR1**)
= [**D**[**r1**]∗(**f**;**dr**)]
•[**u1R1**∗**u1R1** + **u2R1**∗**u2R1** + **u3R1**∗**u3R1**] (553)
= [**D**[**r1**]∗(**f**;**dr**)]•**T**[**UR1**]•**UR1**. (554)

Proof:
[**D**[**r1**]∗(**f**;**dr**)]•[**u1R1**∗**u1R1** + **u2R1**∗**u2R1** + **u3R1**∗**u3R1**]
= d1R1∗**u1R1**•**T**[**D**[**r1**]∗(**f**;**dr**)]∗**u1R1**/d1R1
+ d2R1∗**u2R1**•**T**[**D**[**r1**]∗(**f**;**dr**)]∗**u2R1**/d2R1
+ d3R1∗**u3R1**•**T**[**D**[**r1**]∗(**f**;**dr**)]∗**u3R1**/d3R1
≈ (**f**(**r1**+d1R1∗**u1R1**) – **f**(**r1**))∗**u1R1**/d1R1
+ (**f**(**r1**+d2R1∗**u2R1**) – **f**(**r1**))∗**u2R1**/d2R1
+ (**f**(**r1**+d3R1∗**u3R1**) – **f**(**r1**))∗**u3R1**/d3R1
from equation (521);
⟶ **D**[**r1**]∗(**f**;**dR1**).
Moreover,
T[**UR1**]•**UR1** = [**u1R1**∗**u1R1** + **u2R1**∗**u2R1** + **u3R1**∗**u3R1**].
qed

Mathematics: Vector Calculus

The inverse relationship

D[**r1**]∗(**f**;**dr**)
= **D**[**r1**]∗(**f**;**dR1**)
•[**u1R1**∗**u1R1**+ **u2R1**∗**u2R1**+ **u3R1**∗**u3R1**]$^{-1}$ (555)

also holds for functions with basic, simply continuous gradients at **r1**.

On the other hand, sectional gradients may also be related to directional derivatives.

Theorem 3 (sectional gradients and directional derivatives)
Given
SECT(**r1**,**u1R1**,**u2R1**,**u3R1**), a section at **r1**;
CX a curve into SECT(**r1**,**u1R1**,**u2R1**,**u3R1**) at **r1**≡ **r**(x1);
f(**r**) a function sectionally differentiable
in SECT(**r1**,**u1R1**,**u2R1**,**u3R1**) at **r1**;
dR1 ≡ dR1∗**udR1**
≡ d1R1∗**u1R1** + d2R1∗**u2R1** + d3R1∗**u3R1**;
= **r**(x1+dx) – **r**(x1);
then

D[x1,x1+dx](**f**|CX;dx)
= D[x1,x1+dx](**r**(x)|CX;dx)•[**UR1**$^{-1}$]•**T**[**UR1**$^{-1}$]
•**T**[**D**[**r1**]∗(**f**;**dR1**)]. (556)

Proof:
D[x1,x1+dx](**r**(x);dx)
≈ (**r**(x1+dx) – **r**(x1))/dx
= dR1∗**udR1**/dx
= (d1R1∗**u1R1** + d2R1∗**u2R1** + d3R1∗**u3R1**)/dx.
Then,
D[x1,x1+dx](**r**(x);dx)•[**UR1**$^{-1}$]•**T**[**UR1**$^{-1}$]•**T**[**D**[**r1**]∗(**f**;**dR1**)]
≈ (d1R1∗**u1R1** + d2R1∗**u2R1** + d3R1∗**u3R1**)
•[(**u2R1**∧**u3R1**)∗**u1**
+(**u3R1**∧**u1R1**)∗**u2**
+(**u1R1**∧**u2R1**)∗**u3**]
•[**u1**∗(**u2R1**∧**u3R1**)
+**u2**∗(**u3R1**∧**u1R1**)
+**u3**∗(**u1R1**∧**u2R1**)]
•[**u1R1**∗(**f**(**r1**+**d1R1**∗**u1R1**)–**f**(**r1**))/d1R1
+ **u2R1**∗(**f**(**r1**+**d2R1**∗**u2R1**)–**f**(**r1**))/d2R1
+ **u3R1**∗(**f**(**r1**+**d3R1**∗**u3R1**)–**f**(**r1**))/d3R1]
/((det[**UR1**])^2∗dx)

= (**f**(**r1**+d1R1∗**u1R1**)−**f**(**r1**)
+ **f**(**r1**+d2R1∗**u2R1**)−**f**(**r1**)
+ **f**(**r1**+d3R1∗**u3R1**)−**f**(**r1**))/dx
≈ (**f**(**r1+dR1**(**x1**))−**f**(**r1**))/dx
⟶ D[x1,x1+dx](**f**|**r**(x);dx). qed

Sectional gradients of vector functions are transformations. As transformations they may be used meaningfully only restrictedly.

As transformations

UR1: V3 ⟶ V3
UR1^{-1}:V3 ⟶ V3.

In restriction

UR1: positive quadrant ⟶ SECT(**r1,u1R1**,**u2R1**,**u3R1**)
UR1^{-1}: SECT(**r1,u1R1**,**u2R1**,**u3R1**) ⟶ positive quadrant.
(557)

The use of these transformations restricted to the latter domains is called their **canonical usage**. Only canonical usage is appropriate for sectional gradients. For other than canonical usage, the transformations will generate non-canonical images and so be inappropriate for sectional gradients.

Let F be the image of the gradient of **f**(**r**). Then from equation (552) for a function differentiable in SECT(**r1**,**u1R1**,**u2R1**,**u3R1**)

[**UR1**]$^{-1}$•**T**[**UR1**]$^{-1}$•**T**[**D**[**r1**]∗(**f**;**dR1**)]:
SECT(**r1**,**u1R1**,**u2R1**,**u3R1**) ⟶ F

so that

T[**UR1**]$^{-1}$•**T**[**D**[**r1**]∗(**f**;**dR1**)]: positive quadrant ⟶ F.

If **u1R1** = **u1**, **u2R1** = **u2**, and **u3R1** = **u3**, then **UR1=I** the identity transformation; SECT(**r1,u1R1**,**u2R1**,**u3R1**) is the positive quadrant; and **D**[**r1**]∗(**f**;**dR1**) = **D**[**r1**]∗(**f**;**dr**) for the positive quadrant. Differences in the eight quadrants are thus selected by the sign of **uiR1**. When all eight quadrants have identical sectional gradients, the gradient is then **simply continuous at r1**.

Mathematics: Vector Calculus

On the other hand, when **uiR1⟶ur**, i=1,2,3

$$\mathbf{UR1} = \mathbf{u1}*\mathbf{ur} + \mathbf{u2}*\mathbf{ur} + \mathbf{u3}*\mathbf{ur}$$
$$= (\mathbf{u1}+\mathbf{u2}+\mathbf{u3})*\mathbf{ur},$$

an outer product. Then **UR1** has no inverse.

Then sectional gradient becomes

$$\mathbf{D}[\mathbf{r1}]*(\mathbf{f};\mathbf{dR1}) = \lim\ (\mathbf{f}(\mathbf{r1}+\text{dr1}*\mathbf{ur})-\mathbf{f}(\mathbf{r1}))*\mathbf{ur}/\text{dr1}$$
$$+ (\mathbf{f}(\mathbf{r1}+\text{dr2}*\mathbf{ur})-\mathbf{f}(\mathbf{r1}))*\mathbf{ur}/\text{dr2}$$
$$+ (\mathbf{f}(\mathbf{r1}+\text{dr3}*\mathbf{ur})-\mathbf{f}(\mathbf{r1}))*\mathbf{ur}/\text{dr3}$$
$$= 3*((\mathbf{f}(\mathbf{r1}+\text{dr}*\mathbf{ur})-\mathbf{f}(\mathbf{r1}))/\text{dr})*\mathbf{ur}$$
$$\longrightarrow 3*\mathbf{D}[\mathbf{r1},\mathbf{r1}+\text{dr}*\mathbf{ur}]*(\text{f}(\mathbf{r});\text{dr}*\mathbf{ur}) \qquad (558)$$

or given **r**(x), a non–stalled process curve with

$$\mathbf{r}(\text{x1}) = \mathbf{r1}$$

and

$$\mathbf{r}(\text{x1}+\text{dx}) = \mathbf{r1}+\text{dr}*\mathbf{ur},$$

$$\mathbf{D}[\mathbf{r1}]*(\mathbf{f};\mathbf{dR1})$$
$$= 3*\text{D}[\mathbf{r1},\mathbf{r1}+\mathbf{dr}(\mathbf{r1},\mathbf{r}(\text{x1}+\text{dx}))](\mathbf{f}(\mathbf{r});\text{dr}(\mathbf{r1},\mathbf{r}(\text{x})))*\mathbf{ur} \qquad (559)$$
$$= 3*\text{D}[\text{x1},\text{x1}+\text{dx}](\mathbf{f}|\mathbf{r}(\text{x});\text{dx})/\text{D}[\text{x1},\text{x1}+\text{dx}](\text{dr}(\mathbf{r1},\mathbf{r}(\text{x});\text{dx}))$$
$$*\mathbf{ur}. \qquad (560)$$

Sectional gradients are then reduced to outer products.

Sectional gradients thus occupy an intermediate position between continuous, unrestricted gradients and directional gradients.

Rank of a Transformation

As an idea, *V3* is so defined that **r**, considered no matter how large or how small, or in whatever direction, is never lacking. This property arises from the properties of *R*, the real field underlying *V3*. This sufficiency underlies the conclusion that the positive quadrant gradient **D[r1]**∗(**r**;**dr**) = **I**, the identity matrix.

It is possible, however, to designate specific subsets S of V3 over which the gradient does not equate to the identity matrix. Restricted to surfaces, curves or points, **D[r1]**∗(**r**|S;**dr**) is singular, that is det[**D[r1]**∗(**r**|S;**dr**)] = 0. For example, for constant **r1**,

D[r1]∗(**r1**;**dr**) = [**0**]

for which det[**D[r1]**∗(**r1**;**dr**)] = 0 .

The restriction may also be written as a function **f**:V3⟶S where the restriction is seen as the image of ***f***. The function itself may have a basic gradient restricted to the positive quadrant, a section, or a direction.

The various possibilities correspond to the rank of the gradient as a matrix as shown in the table:

Rank	Determinant	Restriction/Image
0	0	points
1	0	curves
2	0	surfaces
3	non–zero	vicinities

Rank of Matrices in V3

Examples:

If **f=c**, **c** constant, then det[**D[r1]**∗(**f**;**dr**)] = 0, rank **0**.

If **f**=f∗**uc**, **uc**, a fixed direction, det[**D[r1**,**r1**+dr∗**uc]**∗(**f**;**dr**)]= **0**, rank 1.

If **f**=f1∗**uc1**+f2∗**uc2**, two fixed directions, det[**D[r1]**∗(**f**;**dr**)] = **0**, rank 2.

If **f**=f1∗**u1**+f2∗**u2**+f3∗**u3**, then det[**D[r1]**∗(**f**;**dr**)],≠,**0**, rank 3.

Mathematics: Vector Calculus

Local Differentiation of Sums and Products

The following table lists statements true generically.[38] They hold within a class for continuous, positive quadrant, sectional, and directional local derivatives (given appropriate differentiability).

Item:	D• (**divergence**)	**D∧** (**curl**)	**D∗** (**gradient**)
c constant			**0**
c constant	0	**0**	[**0**]
f+g, sum			**D**∗f + **D**∗g
f+g, sum	D•**f** + D•**g**	**D∧f + D∧g**	**D∗f + D∗g**
f∗g, product			f∗**D**∗g + g∗**D**∗f
1/f, reciprocal			–(**D**∗f)/f^2
f∗**g**, product	f∗D•**g** + **g**•**D**∗f	f∗**D∧g** +**g∧**(**D**∗f)	f∗**D∗g** + **g∗D**∗f
f•g, product			**f•D∗g** + **g•D∗f**
f∧g, product	**f•D∧g – g•D∧f**	**f•T[D∗g] – g•T[D∗f] – f∗D•g + g∗D•f**	**C(f)•D∗g – C(g)•D∗f**
f∗g, product	**f∗D•g** +**g•T(D∗f)**		

Gradients of Sums and Products

The table can be used to determine other combinations. For instance for

A = u1∗a1+u2∗a2+u3∗a3

D•(A) = (D•**a1**)∗**u1** + (D•**a2**)∗**u2** + (D•**a3**)∗**u3** (561)

D•(T[A]) = u1•T[D∗a1] + u2•T[D∗a2] + u3•T[D∗a3] (562)

D•(f•A) = (f•u1)∗D•**a1** + **(f•u2)**∗D•**a2** + **(f•u3)**∗D•**a3**
+ **a1•(u1•D∗f) + a2•(u2•D∗f) + a3•(u3•D∗f)** (563)

D•(f•T[A]) = u1•(f•D∗a1 + a1•D∗f)
+ **u2•(f•D∗a2 + a2•D∗f)**
+ **u3•(f•D∗a3 + a3•D∗f)** 564)

D∧(f•A) = (f•u1)∗D∧a1 + (f•u2)∗D∧a2 + (f•u3)∗D∧a3
+ **a1∧(u1•D∗f) + a2∧(u2•D∗f) + a3∧(u3•D∗f)** (565)

D∧(**f**•T[**A**]) = **u1**∧(**f**•**D**∗**a1**) + **u2**∧(**f**•**D**∗**a2**) + **u3**∧(**f**•**D**∗**a3**)
+ (**u1**∧**a1** + **u2**∧**a2** + **u3**∧**a3**)•**D**∗**f** (566)

D∗(**f**•**A**) = (**f**•**u1**)∗**D**∗**a1** + (**f**•**u2**)∗**D**∗**a2** + (**f**•**u3**)∗**D**∗**a3**
+ (**a1**∗**u1** + **a2**∗**u2** + **a3**∗**u3**)•**D**∗**f** (567)

D∗(**f**•T[**A**]) = **u1**∗(**f**•**D**∗**a1**) + **u2**∗(**f**•**D**∗**a2**) + **u3**∗(**f**•**D**∗**a3**)
+ (**u1**∗**a1** + **u2**∗**a2** + **u3**∗**a3**)•**D**∗**f**. (568)

Note[39] that

D•(C(**f**)) = **D**∧**f** (569)
D•T(C(**f**)) = –**D**∧**f** (570)
D∧(**D**∧**f**) = **D**•T[**D**∗**f**] – **D**•**D**∗**f** (571)
D∧(**f**∧**g**) = **D**•T(**f**∗**g**) – **D**•(**f**∗**g**). (572)

Combinations of these local derivatives are also possible. For instance, there often appear

D•(**D**∧**f**) = 0 (573)
D∧(**D**∗**f**) = **0**. (574)

The combination is D•**D**∗ is likewise often useful.

Local Integration

In V3 integration, like differentiation, may be defined with reference either to the underlying field *R* or to the vector field itself.

Integration by means of scalar increments produces three integrals:

in a given direction and so called **directional**
along r(x) **with respect to** x
along r(x) **with respect to r**.

Integration by means of vector increments arises from three possibilities:

V3⟶R called the **invergence**, symbolized by **I•**
V3⟶V3 called the **incurl**, symbolized by **I∧**
and R⟶V3 called the **ingradient**, symbolized by **I∗**

Integration in *V3* arises from the topology of V3 restricted or unrestricted.

Directional Integrals

Definition 33 (directional integrals)
Given
r1, a vector in V3;
r2, a vector in V3;
ur ≡ **u**(**r2**−**r1**), a direction;
f(**r**) a given vector function;
then if it exists

lim dr∗(**f**(**r1**) + **f**(**r1**+dr∗**ur**) + **f**(**r1**+2∗dr∗**ur**) + ... + **f**(**r2**))
as dr⟶0 (575)

is called the
directional integral of f in the ur direction from r1 to r2.
end of definition:

Directional integrals are symbolized as

I[**r1**,**r2**](**f**(**r**)|**ur**;dr). (576)

Mathematics: Vector Integration

For functions with a basic directional derivative in direction **ur** from **r1** to **r2**

I[**r1**,**r2**](D[**r1**,**r1+dr**](**f**(**r**)|**ur**;dr)|**ur**;dr)
        = lim dr∗(D[**r1**,**r1+dr**](**f**(**r**)|**ur**;dr)
                + D[**r1+dr**,**r1+**2∗**dr**](**f**(**r**)|**ur**;dr)
                + ...
                + D[**r2**,**r2+dr**](**f**(**r**)|**ur**;dr))
        = lim dr∗((**f**(**r1+dr**)|**ur**–**f**(**r1**)|**ur**)/dr
                + (**f**(**r1+**2∗**dr**)|**ur**–**f**(**r1+dr**)|**ur**)/dr
                + ...
                + (**f**(**r2+**dr∗**ur**)|**ur**–**f**(**r2**)|**ur**)/dr)
        = lim (**f**(**r2+**dr∗**ur**)|**ur**–**f**(**r1**)|**ur**)

Consequently for functions with basic directional derivatives in direction **ur** from **r1** to **r2**,

lim **f**(**r2+**dr∗**ur**)|**ur**
        = **f**(**r1**) + **I**[**r1**,**r2**](D[**r1**,**r1+dr**](**f**(**r**)|**ur**;dr)|**ur**;dr).   (577)

Equation (577) is the **fundamental theorem of integral calculus for directional integrals**.

Process Integration

The integral of a restricted function **f**|CX is defined from the induced topology. Such integrals are called **integrals along CX with respect to** x.

Definition 34   (integrals along CX with respect to x)

Given
    CX = {**r**}, a curve;
    f|CX, a vector function restricted to CX;
then if it exists

lim dx∗(**f**(**r**(x1)) + **f**(**r**(x1+dx)) + **f**(**r**(x1+2∗dx)) + ... + **f**(**r**(x2)))
                                        as dx⟶0   (578)

    is called the

integral of f|CX **from r**(x1) **to r**(x2).

end of definition:

Integrals of restricted functions along CX with respect to x are symbolized as

$$\mathbf{I}[\mathbf{r}(x1),\mathbf{r}(x2)](\mathbf{f}(x)|CX;dx). \tag{579}$$

For restricted functions with a basic derivative along CX from x1 to x2

$$\begin{aligned}
&\mathbf{I}[\mathbf{r}(x1),\mathbf{r}(x2)](D[\mathbf{r}(x),\mathbf{r}(x+dx)](\mathbf{f}(x)|CX;dx)|CX\ ;dx)\\
&\quad = \lim dx*(D[\mathbf{r}(x1),\mathbf{r}(x1+dx)](\mathbf{f}(x)|CX\ ;dx)\\
&\qquad + D[\mathbf{r}(x1+dx),\mathbf{r}(x1+2*dx)](\mathbf{f}(x)|CX;dx)\\
&\qquad + \ldots\\
&\qquad + D[\mathbf{r}(x2),\mathbf{r}(x2+dx)](\mathbf{f}(x)|CX;dx))\\
&\quad = \lim dx*((\mathbf{f}(x1+dx\ )-\mathbf{f}(x1))/dx\\
&\qquad + (\mathbf{f}(x1+2*dx)-\mathbf{f}(x1+dx))/dx\\
&\qquad + \ldots\\
&\qquad + (\mathbf{f}(x2+dx\ )-\mathbf{f}(x2))/dx)\\
&\quad = \lim (\mathbf{f}(x2+dx\ )-\mathbf{f}(x1)).
\end{aligned}$$

Consequently for functions restricted to CX with basic directional derivatives along the curve from **r**(x1) to **r**(x2),

$$\begin{aligned}
&\lim \mathbf{f}(x2+dx\ )\\
&\qquad = \mathbf{f}(x1) + \mathbf{I}[\mathbf{r}(x1),\mathbf{r}(x2)](D[\mathbf{r}(x),\mathbf{r}(x+dx)](\mathbf{f}(x)|CX;dx)|CX;dx).
\end{aligned} \tag{580}$$

Equation (580) is the **fundamental theorem of integral calculus for integrals of restricted functions over** **r**(x) **with basic directional derivatives** along the curve from **r**(x1) to **r**(x2).

Mathematics: Vector Integration

Definition 35 (integrals along CR with respect to **r**)

Given

CR, a curve in V3;

r1 a location in CR;

ut(**r**), the curve direction at **r** in CR;

for **r** in CR define

$\mathbf{dr}(\mathbf{r}) \equiv dr*\mathbf{ut}(\mathbf{r})$, $dr > 0$;

f(**r**)|CR, a vector function over CR;

then if it exists

$$\begin{aligned}\lim dr*(\mathbf{f}(\mathbf{r1}) &+ \mathbf{f}(\mathbf{r1}+\mathbf{dr}(\mathbf{r1}))\\ &+ \mathbf{f}(\mathbf{r1}+\mathbf{dr}(\mathbf{r1})+\mathbf{dr}(\mathbf{r1}+\mathbf{dr}(\mathbf{r1})))\\ &+ \ldots\\ &+ \mathbf{f}(\mathbf{r}2)) \text{ as } dr \longrightarrow 0\end{aligned} \tag{581}$$

is called the

integral of f(r) from r1 to r2 along CR.

end of definition

Integrals of compound functions along **r**(x) are symbolized as

$$\mathbf{I}[\mathbf{r1},\mathbf{r2}](\mathbf{f}(\mathbf{r})|\mathrm{CR};dr). \tag{582}$$

The end–point of the integration may then be computed as

$$\mathbf{r2} = \mathbf{r1} + \mathbf{I}[\mathbf{r1},\mathbf{r2}](\mathbf{dr}(\mathbf{r})|\mathrm{CR};dr). \tag{583}$$

For compound functions with basic derivatives along CR from **r1**=**r**(x1) to **r2**=**r**(x2)

$$\begin{aligned}&\mathbf{I}[\mathbf{r1},\mathbf{r2}](\mathrm{D}[\mathbf{r},\mathbf{r}+\mathbf{dr}(\mathbf{r})](\mathbf{f}(\mathbf{r})|\mathrm{C};dr)|\mathrm{CR};dr)\\ &\quad = \lim dr*(\mathrm{D}[\mathbf{r1},\mathbf{r1}+\mathbf{dr}(\mathbf{r1})](\mathbf{f}(\mathbf{r})|\mathrm{CR};dr)\\ &\qquad + \mathrm{D}[\mathbf{r1}+\mathbf{dr}(\mathbf{r1}),\ \mathbf{r1}+\mathbf{dr}(\mathbf{r1})+\mathbf{dr}(\mathbf{r1}+\mathbf{dr}(\mathbf{r1}))](\mathbf{f}(\mathbf{r})|\mathrm{CR};dr)\\ &\qquad + \ldots\\ &\qquad + \mathrm{D}[\mathbf{r2},\mathbf{r2}+\mathbf{dr}(\mathbf{r2})](\mathbf{f}(\mathbf{r})|\mathrm{CR};dr))\\ &\quad = \lim \mathbf{f}(\mathbf{r1}+\mathbf{dr}(\mathbf{r1}))|\mathrm{CR}-\mathbf{f}(\mathbf{r1})\\ &\qquad + \mathbf{f}(\mathbf{r1}+\mathbf{dr}(\mathbf{r1})+\mathbf{dr}(\mathbf{r1}+\mathbf{dr}(\mathbf{r1})))|\mathrm{CR}-\mathbf{f}(\mathbf{r1}+\mathbf{dr}(\mathbf{r1}))|\mathrm{CR}\\ &\qquad + \ldots\\ &\qquad + \mathbf{f}(\mathbf{r2}+\mathbf{dr})|\mathrm{CR}-\mathbf{f}(\mathbf{r2})\\ &\quad = \lim \mathbf{f}(\mathbf{r2}+\mathbf{dr}(\mathbf{r2}))|\mathrm{CR}-\mathbf{f}(\mathbf{r1}).\end{aligned}$$

Consequently for compound functions with basic derivative of f(r(x)) along CR from **r1** to **r2**,

lim f(r2+dr(r2))|CR
= f(r1) + I[r1,r2](D[r,r+dr(r)](f(r)|CR;dr)|CR;dr). (584)

Equation (584) is the **fundamental theorem of integral calculus along CR with respect to r**.

Theorem 4 (Integrals over a curve in terms of arc-lengths)

Given

C ≡ CR = CX = {r(x)}, a curve;
r1 = r(x1);
r2 = r(x2);
ut(r), the direction of C at r;
f(r(x)), a compound vector function;

for

r(x) with a basic derivative along C;
dr(r) ≡ dr*ut(r), dr>0;

then

I[r1,r2](f(r)|C);dr) = I[x1,x2](abs((D[r](r(x)|C;dx))*f(x)|C;dx).(585)

Proof:

Let m(xi) ≡ abs(r(xi +dx) − r(xi))/dr be the number of dr intervals in abs(r(xi+dx)−r(xi)).
Let r(xi+dx) ≈ r(xi) + dri = r(xi) + dri1 + dri2 + ... + drim(xi)
where

dri1 ≡ dr*ut(r(x1))
dri2 ≡ dr*ut(r(x1))+dr*ut(r(x1))
....

Then
I[r1,r2](f(r)|C;dr)
= lim dr*(f(r1))
+ f(r1+dr*ut(r1))
+ f(r1+dr*ut(r1)+dr*ut(r1+dr*ut(r1)))
+ ...
+ f(r2)) as dr⟶0

Mathematics: Vector Integration

= lim dr∗(**f**(**r1**)
+ **f**(**r1+dr1**1)
+ **f**(**r1+dr1**1+**dr1**2)
+ ...
+ **f**(**r**(xi+dx)**+dr1**1+**dr1**2**+**...**+dr1**m(x1))
+ ...
+ **f**(x2)) as dr⟶0
= lim (**f**(**r**(x1))+**f**(**r**(x1)**+dr1**1)+**f**(**r**(x1)**+dr1**1+**dr1**2)+ ...)
∗abs(**r**(x1+dx) − **r**(x1))/m(x1)
+ (**f**(**r**(x1+dx) + **f**(**r**(x1+dx)**+dr1**))
+**f**(**r**(x1+dx)**+dr1+dr2**1) + ...)
∗abs(**r**(x1+2∗dx) − **r**(x1+dx))/m(x1+dx)
+ ...
+ (**f**(x2) +**f**(**r**(x2)**+drn**1)+**f**(**r**(x2) **+drn**1+**drn**2) + ...)
∗abs(**r**(x2+dx) − **r**(x2))/m(x2)
= lim dx∗((**f**(**r**(x1))+**f**(**r**(x1)**+dr1**1)+**f**(**r**(x1)**+dr1**1+**dr1**2)+ ...)
/m(x1))
∗abs(**r**(x1+dx) − **r**(x1))/dx
+ ((**f**(**r**(x1+dx)**+dr1**) +**f**(**r**(x1+dx)**+dr1+dr2**1)
+**f**(**r**(x1+dx)**+dr1** **+dr2**1+**dr2**2)+ ...)/m(x1))
∗abs(**r**(x1+2∗dx) − **r**(x1+dx))/dx
+ ...
+ ((**f**(x2)+**f**(**r**(x2)**+drn**1)+**f**(**r**(x2) **+drn**1+**drn**2)
+ ...)/m(x2))
∗abs(**r**(x2+dx) − **r**(x2))/dx).

Now
lim (**f**(**r**(xi)**+dr1**)+**f**(**r**(xi)**+dr1+dr2**1)+**f**(**r**(xi)**+dr1**
+dr21+**dr2**2)+ ...)/m(xi)
= **f**(**r**(xi)) as dr⟶0.

Thus
I[**r1**,**r2**](**f**(**r**)|C);dr)
≈ lim dx∗(**f**(**r**(x1)∗abs(**r**(x1 +dx) − **r**(x1))/dx
+ **f**(**r**(x1+dx))∗abs(**r**(x1+2∗dx) − **r**(x1))/dx
+ ...
+ **f**(**r**(x2))∗abs(**r**(x2+dx) − **r**(x2))/dx)
as dx⟶0
= **I**[x1,x2](abs(D[**r**](**r**(x)|C;dx))∗**f**(x)|C;dx).
qed

For a stalled process,
D[**r**](**r**(x)|C;dx) = **0**
r(x2) = **r**(x1)
I[**r1**,**r2**](**f**(**r**(x))|C;dr) = **0**
even though **I**[x1,x2](**f**(x)|C;dx) may not be zero.

If the curve branches, integrals may be assigned to each separate branch.

Generic formulas for integrating sums and products of functions, provided they apply to the same class of integrals, are:

$$
\begin{aligned}
&\mathbf{I}[\,]((\mathbf{f1}+\mathbf{f2})|\mathrm{u}) = \mathbf{I}[\,](\mathbf{f1}|\mathrm{u}) + \mathbf{I}[\,](\mathbf{f2}|\mathrm{u})\\
&\mathbf{I}[\,](\mathrm{D}[\,](\mathbf{f1}\bullet\mathbf{f2}|\mathrm{u})|\mathrm{u}) = \mathbf{I}[\,](\mathbf{f1}\bullet\mathrm{D}[\,](\mathbf{f2}|\mathrm{u})|\mathrm{u}) + \mathbf{I}[\,](\mathrm{D}[\,](\mathbf{f1}|\mathrm{u})\bullet\mathbf{f2}|\mathrm{u})\\
&\mathbf{I}[\,](\mathrm{D}[\,](\mathbf{f1}\wedge\mathbf{f2}|\mathrm{u})|\mathrm{u}) = \mathbf{I}[\,](\mathbf{f1}\wedge\mathrm{D}[\,](\mathbf{f2}|\mathrm{u})|\mathrm{u}) + \mathbf{I}[\,](\mathrm{D}[\,](\mathbf{f1}|\mathrm{u})\wedge\mathbf{f2}|\mathrm{u})\\
&\mathbf{I}[\,](\mathrm{D}[\,](\mathrm{f1}\ast\mathrm{f2}|\mathrm{u})|\mathrm{u}) = \mathbf{I}[\,](\mathrm{f1}\ast\mathrm{D}[\,](\mathrm{f2}|\mathrm{u})|\mathrm{u}) + \mathbf{I}[\,](\mathrm{D}[\,](\mathrm{f1}|\mathrm{u})\ast\mathrm{f2}|\mathrm{u}).
\end{aligned}
\tag{586}
$$

For a constant vector $\boldsymbol{c}$,

$$
\begin{aligned}
&\mathbf{I}[\,](\mathbf{c}\bullet\mathbf{f}|\mathrm{u}) = \mathbf{c}\bullet\mathbf{I}[\,](\mathbf{f}|\mathrm{u}),\\
&\mathbf{I}[\,](\mathbf{c}\wedge\mathbf{f}|\mathrm{u}) = \mathbf{c}\wedge\mathbf{I}[\,](\mathbf{f}|\mathrm{u}),\\
&\mathbf{I}[\,](\mathbf{c}\ast\mathbf{f}|\mathrm{u}) = \mathbf{c}\ast\mathbf{II}[\,](\mathbf{f}|\mathrm{u}).
\end{aligned}
\tag{587}
$$

Also,

$$
\mathrm{abs}(\mathbf{I}[\,](\mathbf{f}|\mathrm{u})) \le \mathbf{I}[\,](\mathrm{abs}(\mathbf{f})|\mathrm{u}).
\tag{588}
$$

Mathematics: Vector Integration

Directional Invergences, Incurls, and Ingradients

Definition 36 (directional integrals)

Given

r1, a vector in V3;

r2, another vector in V3;

ur ≡ **u**(**r2**−**r1**), a direction;

dr ≡ dr∗**ur** ;

then

the forward directional invergence, incurl and ingradient in direction **ur** from **r1** to **r2** are defined as:

Invergence:

lim (**f**(**r1**) + **f**(**r1**+**dr**)|**ur** + **f**(**r1**+2∗**dr**)|**ur** + ... + **f**(**r2**))•**dr**, as dr⟶0 (589)

Incurl:

lim (**f**(**r1**) + **f**(**r1**+**dr**)|**ur** + **f**(**r1**+2∗**dr**)|**ur** + ... + **f**(**r2**))∧**dr**, as dr⟶0 (590)

Ingradient:

lim (**f**(**r1**) + **f**(**r1**+**dr**)|**ur** + **f**(**r1**+2∗**dr**)|**ur** + ... + **f**(**r2**))∗**dr**, as dr⟶0 (591)

or

lim (f(**r1**) + f(**r1**+**dr**)|**ur** + f(**r1**+2∗**dr**)|**ur** + ... + f(**r2**))∗**dr**, as dr⟶0. (592)

end of definition

Directional invergences, incurls and ingradients are symbolized as

I[**r1**,**r2**]•(**f**(**r**)|**ur**;**dr**)

I[**r1**,**r2**]∧(**f**(**r**)|**ur**;**dr**)

I[**r1**,**r2**]∗(**f**(**r**)|**ur**;**dr**). (593)

Invergences, Incurls, and Ingradients along curves

Integration aimed at inverting divergences, curls and gradients along curves may also be defined.

Definition 37 (invergences, incurls and ingradients along CX)

Given

CX = {**r**(x)}, a curve;

dx >0;

dr(x) ≡ **r**(x+dx) − **r**(x);

then the directional invergence, incurl and ingradient along CX from **r**(x1) to **r**(x2) are defined as:

Invergence:

lim (**f**(**r**(x1))•**dr**(x1)
+ **f**(**r**(x1+dx))•**dr**(x1+dx)
+ **f**(**r**(x1+2∗dx))•**dr**(x1+2∗dx)
+ ...
+ **f**(**r**(x2)))•**dr**(x2), as dx⟶0 (594)

Incurl:

lim (**f**(**r**(x1))∧**dr**(x1)
+ **f**(**r**(x1+dx))∧**dr**(x1+dx)
+ **f**(**r**(x1+2∗dx))∧**dr**(x1+2∗dx)
+ ...
+ **f**(**r**(x2)))∧**dr**(x2), as dx⟶0 (595)

Ingradient:

lim (**f**(**r**(x1))∗**dr**(x1)
+ **f**(**r**(x1+dx))∗**dr**(x1+dx)
+ **f**(**r**(x1+2∗dx))∗**dr**(x1+2∗dx)
+ ...
+ **f**(**r**(x2)))∗**dr**(x2), as dx⟶0 (596)

or

lim (f(**r**(x1))∗**dr**(x1)
+ f(**r**(x1+dx))∗**dr**(x1+dx)
+ f(**r**(x1+2∗dx))∗**dr**(x1+2∗dx)
+ ...
+ f(**r**(x2)))∗**dr**(x2), as dx⟶0. (597)

end of definition

Mathematics: Vector Integration

Invergences, incurls and ingradients along CX are symbolized as

I[**r**(x1),**r**(x2)]•(**f**(**r**)|CX;**dr**)
I[**r**(x1),**r**(x2)]∧(**f**(**r**)|CX;**dr**)
I[**r**(x1),**r**(x2)]∗(**f**(**r**)|CX;**dr**).

(598)

Definition 38 (invergences, incurls and ingradients along CR)

Given

CR = a continuous curve;
dr >0;
dr(**r**) ≡ dr∗**ut**(**r**), **r** in CR;

then

the directional invergence, incurl, and ingradient along CR from **r1** to **r2** are defined as:

Invergence:
lim **f**(**r1**)•**dr**(**r1**)
+ **f**(**r1**+**dr**(**r1**))•**dr**(**r1**+**dr**(**r1**))
+ ...
+ **f**(**r2**)•**dr**(**r2**), as dr⟶0 (599)

Incurl:
lim **f**(**r1**)∧**dr**(**r1**)
+ **f**(**r1**+**dr**(**r1**))∧**dr**(**r1**+**dr**(**r1**))
+ ...
+ **f**(**r2**)∧**dr**(**r2**), as dr⟶0 (600)

Ingradient:
lim **f**(**r1**)∗**dr**(**r1**)
+ **f**(**r1**+**dr**(**r1**))∗**dr**(**r1**+**dr**(**r1**))
+ ...
+ **f**(**r2**)∗**dr**(**r2**), as dr⟶0 (601)

or
lim f(**r1**)∗**dr**(**r1**) + f(**r1**
+**dr**(**r1**))∗**dr**(**r1**+**dr**(**r1**))
+ ...
+ f(**r2**)∗**dr**(**r2**), as dr⟶0. (602)

end of definition

Invergences, incurls and ingradients along CR are symbolized as

I[**r1**,**r2**]•(**f**(**r**)|CR;**dr**)
I[**r1**,**r2**]∧(**f**(**r**)|CR;**dr**)
I[**r1**,**r2**]∗(**f**(**r**)|CR;**dr**). (603)

Since abs(**dr**(**r**)) = dr, we may define

I[**r1**,**r2**]∗(**f**(**r**)|CR;abs(**dr**))
≡ lim **f**(**r1**)∗abs(**dr**(**r1**))
+ **f**(**r1**+**dr**(**r1**))∗abs(**dr**(**r1**+**dr**(**r1**)))
+ ...
+ **f**(**r2**)∗abs(**dr**(**r2**)) as dr⟶0. 604)

Equation (604) restates equation (581).

Directional integration of transformations, **A**(**r**(x)) or **A**(**r**), are defined similarly as

I[**r**(x1),**r**(x2)]•(**A**(**r**(x))|CX;**dr**)
≡ lim (**A**(**r**(x1))∗**dr**(x1)
+ **A**(**r**(x1+dx))∗**dr**(x1+dx)
+ **A**(**r**(x1+2∗dx))∗**dr**(x1+2∗dx)
+ ...
+ **A**(**r**(x2)))∗**dr**(x2) as dx⟶0 (605)

I[**r1**,**r2**]•(**A**(**r**)|CR;**dr**),
≡ lim **A**(**r1**)∗**dr**(**r1**)
+ **A**(**r1**+**dr**(**r1**))∗**dr**(**r1**+**dr**(**r1**))
+ ...
+ **A**(**r2**)∗**dr**(**r2**) as dr⟶0. (606)

Mathematics: Vector Integration

For constant c1, **c1** and c2, **c2** with f1, **f1** and f2, **f2** defined on CX or CR corresponding to the condition u over C

$$\mathbf{I}[C]\bullet((c1\ast\mathbf{f1}+c2\ast\mathbf{f2})|u;\mathbf{dr}) = c1\ast\mathbf{I}[C]\bullet(\mathbf{f1}|u;\mathbf{dr}) + c2\ast\mathbf{I}[C]\bullet(\mathbf{f2}|u;\mathbf{dr}) \tag{607}$$

$$\mathbf{I}[C]\bullet((\mathbf{c1}\ast f1+\mathbf{c2}\ast f2)|u;\mathbf{dr}) = \mathbf{c1}\bullet\mathbf{I}[C]\ast(f1|u;\mathbf{dr}) + \mathbf{c2}\bullet\mathbf{I}[C]\ast(f2|u;\mathbf{dr}) \tag{608}$$

$$\mathbf{I}[C]\bullet((\mathbf{c1}\wedge\mathbf{f1}+\mathbf{c2}\wedge\mathbf{f2})|u;\mathbf{dr}) = \mathbf{c1}\bullet\mathbf{I}[C]\wedge(\mathbf{f1}|u;\mathbf{dr}) + \mathbf{c2}\bullet\mathbf{I}[C]\wedge(\mathbf{f2}|u;\mathbf{dr}) \tag{609}$$

$$\mathbf{I}[C]\bullet((\mathbf{c1}\ast\mathbf{f1}+\mathbf{c2}\ast\mathbf{f2})|u;\mathbf{dr}) = \mathbf{c1}\ast\mathbf{I}[C]\bullet(\mathbf{f1}|u;\mathbf{dr}) + \mathbf{c2}\ast\mathbf{I}[C]\bullet(\mathbf{f2}|u;\mathbf{dr}) \tag{610}$$

$$\mathbf{I}[C]\wedge((c1\ast\mathbf{f1}+c2\ast\mathbf{f2})|u;\mathbf{dr}) = c1\ast\mathbf{I}[C]\wedge(\mathbf{f1}|u;\mathbf{dr}) + c2\ast\mathbf{I}[C]\wedge(\mathbf{f2}|u;\mathbf{dr}) \tag{611}$$

$$\mathbf{I}[C]\wedge((\mathbf{c1}\ast f1+\mathbf{c2}\ast f2)|\mathbf{m1};\mathbf{dr}) = \mathbf{c1}\wedge\mathbf{I}[C]\ast(f1|\mathbf{m1};\mathbf{dr}) + \mathbf{c2}\wedge\mathbf{I}[C]\ast(f2|\mathbf{m1};\mathbf{dr}) \tag{612}$$

$$\begin{aligned}\mathbf{I}[C]\wedge((\mathbf{c1}\wedge\mathbf{f1}+\mathbf{c2}\wedge\mathbf{f2})|u;\mathbf{dr}) &= \mathbf{c1}\bullet\mathbf{T}[\mathbf{I}[C]\ast(\mathbf{f1}|u;\mathbf{dr}) - \mathbf{c1}\bullet\mathbf{I}[C]\ast(\mathbf{f1}|u;\mathbf{dr}) \\ &\quad + \mathbf{c2}\bullet\mathbf{T}[\mathbf{I}[C]\ast(\mathbf{f2}|u;\mathbf{dr}) - \mathbf{c2}\bullet\mathbf{I}[C]\ast(\mathbf{f1}|u;\mathbf{dr})\end{aligned} \tag{613}$$

$$\mathbf{I}[C]\ast((c1\ast f1+c2\ast f2)|u;\mathbf{dr}) = c1\ast\mathbf{I}[C]\ast(f1|u;\mathbf{dr}) + c2\ast\mathbf{I}[C]\ast(f2|u;\mathbf{dr}) \tag{614}$$

$$\mathbf{I}[C]\ast((\mathbf{c1}\bullet\mathbf{f1}+\mathbf{c2}\bullet\mathbf{f2})|u;\mathbf{dr}) = \mathbf{c1}\bullet\mathbf{I}[C]\ast(\mathbf{f1}|u;\mathbf{dr}) + \mathbf{c2}\bullet\mathbf{I}[C]\ast(\mathbf{f2}|u;\mathbf{dr}) \tag{615}$$

$$\mathbf{I}[C]\ast((c1\ast\mathbf{f1}+c2\ast\mathbf{f2})|u;\mathbf{dr}) = c1\ast\mathbf{I}[C]\ast(\mathbf{f1}|u;\mathbf{dr}) + c2\ast\mathbf{I}[C]\ast(\mathbf{f2}|u;\mathbf{dr}) \tag{616}$$

$$\mathbf{I}[C]\ast((\mathbf{c1}\ast f1+\mathbf{c2}\ast f2)|u;\mathbf{dr}) = \mathbf{c1}\ast\mathbf{I}[C]\ast(f1|u;\mathbf{dr}) + \mathbf{c2}\ast\mathbf{I}[C]\ast(f2|u;\mathbf{dr}) \tag{617}$$

$$\mathbf{I}[C]\ast((\mathbf{c1}\wedge\mathbf{f1}+\mathbf{c2}\wedge\mathbf{f2})|u;\mathbf{dr}) = \mathbf{C}(\mathbf{c1})\bullet\mathbf{I}[C]\ast(\mathbf{f1}|u;\mathbf{dr}) + \mathbf{C}(\mathbf{c2})\bullet\mathbf{I}[C]\ast(\mathbf{f2}|u;\mathbf{dr}) \tag{618}$$

where **C** is the curl matrix operator.

For a subclass of restricted functions the following theorem applies.

Theorem 5 (interior cancellation over curves)
Given:
CX, a curve;
f(**r**(x)), a function with have a basic gradient along CX;
dr(xi) ≡ **r**(xi+dx) – **r**(xi) for **r**(xi) in CX;
then

I[**r**(x1),**r**(xm)]•(**D**[**r**(x),**r**(x+dx)]∗(**f**(**r**)|CX;**dr**)|CX;**dr**)
= lim **f**(**r**(xm)+**dr**(xm)) – **f**(**r**(x1)) as dx⟶0. (619)

Proof:
I[**r**(x1),**r**(xm)]•(**D**[**r**(xi)]∗(**f**(**r**(x))|CX;**dr**)|CX;**dr**)
= lim **I**[C(**r**(x1),**r**(xm))]•((**f**(**r**(xi)+**dr**(xi))–**f**(**r**(xi)))
∗**qd**(**dr**(xi))|CX;**dr**)
= lim (**f**(**r**(x1)+**dr**(x1))–**f**(**r**(x1)))∗**qd**(**dr**(x1))•**dr**(x1)
+ (**f**(**r**(x1+**dr**(x1)+**dr**(x1+dx))–**f**(**r**(x1+**dr**(x1)))
∗**qd**(**dr**(x1+dx))•**dr**(x1+dx)
+ ...
+ (**f**(**r**(xm)+**dr**(xm))–**f**(**r**(xm)))∗**qd**(**dr**(xm))•**dr**(xm)
= lim **f**(**r**(x1)+**dr**(x1))–**f**(**r**(x1))
+ **f**(**r**(x1+dx)+**dr**(x1+dx))–**f**(**r**(x1+**dr**(x1)))
+ ...
+ **f**(**r**(xm)+**dr**(xm))–**f**(**r**(xm))
= lim (**f**(**r**(xm)+**dr**(xm)) – **f**(**r**(x1))) as dx⟶0. qed

Corollary
Given:
CR, a curve;
f(**r**), a function with have a basic gradient along CR;
dr(**r**) ≡ dr∗**ut**(**r**), **r** in CR;
then

I[**r1**,**rm**]•(**D**[**r**,**r**+**dr**(**r**)]∗(**f**(**r**)|CR;**dr**)|CR;**dr**)
= lim (**f**(**rm**)+**dr**(**rm**)) – **f**(**r1**), as dr⟶0. (620)

Equations (619) and (620) are the **fundamental theorem of integral vector calculus extended to curves**.

Mathematics: Vector Integration

Step functions over curves.

Since they are real functions, the unit step functions defined in equations 31 and 32 may be multiplied as scalar functions with vector functions.

Now functions
restricted to a curve CX with reference to x
restricted to a direction **ur** with reference to r
or restricted to a connected curve CR with reference to **r**
preserve some order.

The previous definitions for unit step functions are expanded accordingly.

Definition 39 (step functions along curves)

Given

x, a real variable;

for

r1, a constant vector;
ur, a given direction;
r = **r1** + x∗**ur**, a line of vectors;

$$u(x - xi)|\mathbf{ur} \equiv 1, \quad x - xi > 0$$
$$\equiv 0, \quad x - xi \leq 0 \qquad (621)$$
$$v(x - xi)|\mathbf{ur} \equiv 1, \quad x - xi \geq 0$$
$$\equiv 0, \quad x - xi < 0 \qquad (622)$$

for

r(x) a curve over CX

$$u(\mathbf{r}(x) - \mathbf{r}(xi))|CX \equiv 1, \quad x - xi > 0$$
$$\equiv 0, \quad x - xi \leq 0 \qquad (623)$$
$$v(\mathbf{r}(x) - \mathbf{r}(xi))|CX \equiv 1, \quad x - xi \geq 0$$
$$\equiv 0, \quad x - xi < 0 \qquad (624)$$

for

r in CR, a curve from **r1** to **r2**

$$u(\mathbf{r} - \mathbf{ri})|CR \equiv 1, \quad \mathbf{r} \text{ in } (\mathbf{ri},\mathbf{r2}]$$
$$\equiv 0, \quad \mathbf{r} \text{ in } [\mathbf{r1},\mathbf{ri}] \qquad (625)$$
$$v(\mathbf{r} - \mathbf{ri})|CR \equiv 1, \quad \mathbf{r} \text{ in } [\mathbf{ri},\mathbf{r2}]$$
$$\equiv 0, \quad \mathbf{r} \text{ in } [\mathbf{r1},\mathbf{ri}). \qquad (626)$$

end of definition

Results similar to previous results for real functions hold.

Given

the generic condition c as **ur** or CX or CR;
x generically either the directional variable of the one in CX or CR;

D[x](u(t–xi)|c;d_ft) = 0, x≠xi
= lim (1/dx) x = xi (627)

D[x](v(t–xi)|c;d_bt) = 0, x≠xi
= lim (1/dx) x=xi; (628)

for ***f***(x)|c a generalized function

(**f**(x)|c)∗D[x](u(t–xi);d_ft)
= **f**(xi)∗D[xi](u(t–xi);d_ft), x=xi
= **0** otherwise; (629)

u(x–xi)∗D[x](**f**(t)|c;d_fx)
= **0** x≤xi
= D[x](**f**(t);d_fx) x>xi (630)

D[x](**f**(t)∗u(t–xi)|c;d_ft)
= **0** x<xi
= lim **f**(xi+dx)|c/dx x=xi
= D[x](**f**(t);d_fx) x>xi
(631)

I[](**f**(x)∗D[x](u(t–xi)|c;d_ft)|c;dx)
= **f**(xi) (632)

I[](**f**(x)∗D[x](v(t–xi)|;d_bt)|c;dx)
= **f**(xi) (633)

for ***f***(x)|c a continuous, bounded, and asymptotic function

I[]((u(t–xi))∗D[x](**f**(t)|c;d_ft)|c;dx)
= –**f1**, limit of **f** from above as x ⟶xi (634)

I[]((v(t–xi))∗D[x](**f**(t)|c;d_bt)|c;dx)
= –**f0**, limit of **f** from below as x ⟶xi (635)

I[](D[x]((u(t–xi))∗**f**(t)|c;d_ft)|c;dx)
= **0** (636)

I[](D[x]((v(t–xi))∗**f**(t)|c;d_bt)|c;dx)
= **0**. (637)

Mathematics: Vector Integration

If further ***f*** is continuous at *xi*

I[](**f**(x)∗D[t](u(t–xi)|;d_ft)|c;dx)
= –**I**[](D[x](**f**(t)|;d_ft)∗u(x–xi)|c;dx) (638)
I[](**f**(x)∗D[t](v(t–xi)|;d_bt)|;dx)
= –**I**[](D[x](**f**(t)|;d_bt)∗v(x–xi)|;dx). (639)

Over the curve from **r**(x0) to **r**(x2)

I[**r**(x0),**r**(xm)](D[**r**(x),**r**(x+dx)](f(**r**(x))∗u(**r**(x)–**r**(xi))|CX;dx)|CX;dx)
= lim **f**(**r**(xm+dx))|CR – **f**(**r**(x0)) (640)
I[**r1**(x0),**r**(xm)](D[**r**(x–dx),**r**(x)](f(**r**(x))∗v(**r**(x)–**r**(xi))|CX;dx)|CX;dx)
= **f**(**r**(xm))–lim **f**(r(x0 – dx))|CR . (641)
I[**r1**,**rm**](D[**r**,**r+dr**,](**f**(**r**)∗u(**r**–**ri**)|CR;dr)|CR;dr)
= lim **f**(**rm**+**dr**(**rm**))|CR –**f** (**r1**). (642)
I[**r1**,**rm**](D[**r**–**dr**,**r**](**f**(**r**)∗v(**r**–**ri**)|CR;dr)|CR;dr)
= **f**(**rm**)–lim **f**(**r1**–**dr**(**rm**))|CR. (643)

More generally, if
f1(x∗**ur**) is a continuous function in r in direction **ur;**
g(x), the anti–derivative of **f1**,
that is, D[x](**g**(t∗**ur**)|**ur**;dt) = **f1**(x∗**ur**);
x0<xi<xm;
d constant;
then the function
f(x∗**ur**) = **f1**(x∗**ur**) + **d**∗u(x–xi)
is integrated as
I[x0∗**ur**,xm∗**ur**]((**f**(x∗**ur**) + **d**∗u(x–xi))|**ur**;dx)
= **g**(xm∗**ur**) – **g**(x0∗**ur**) + **d**∗(xm–xi). (644)

If
f1(**r**(x)) is a continuous function along CX;
g(**r**(x)), the anti–derivative of **f1;**
that is, D[**r**(x),**r**(x+dx)](**g**(**r**(t))|CX;dt) = **f1**(**r**(x));
x0<xi<xm;
d constant;

then the function

f(**r**(x)) = **f1**(**r**(x)) + **d**∗u(x−xi)|CX

is integrated as

I[**r**(x0),**r**(xm)]((**f**(**r**(x))|CX + **d**∗u(x−xi)|CX)|CX;dx)
= **g**(**r**(xm)) − **g**(**r**(x0)) + **d**∗(xm−xi). (645)

If

f1(**r**) is a continuous function along CR;
g(**r**), the anti−derivative of **f1**, that is,
D[**r+dr**](**g**(**r**)|CR;dr) = **f1**(**r**);
r1 in [**r0**,**rm**];
d constant;

then the function

f(**r**) = **f1**(**r**) + **d**∗u(**r**−**ri**)|CR

is integrated as

I[**r0**,**rm**]((**f**(**r**)|CR + **d**∗**u**(**r**−**ri**)|CR)|CR;dr)
= **g**(**rm**) − **g**(**r0**) + **d**∗**I**[**ri**,**rm**](1|CR;dr). (646)

Equations (644), (645), and (646) are the **fundamental theorem of integral calculus of functions over curves in V3 extended to functions with a finite step discontinuity**.

In addition to derivatives, step functions may also possess gradients. Functions with similar functional values may have different gradients.

Step functions with basic directional gradients or basic gradients along a curve are defined as follows.

Definition 40 (basic gradients of step functions)

Given

x, a real variable;

for

r1, a constant vector;
ur, a given direction;
r = **r1** + x∗**ur,** a line of vectors;
dr ≡ dx∗**ur**;

Mathematics: Vector Integration

then

the basic **directional gradient of a directional step function in direction ur** is defined as

D[**r**,**r**+**dr**]∗(u(x – xi)|**ur**;**dr**)
≡ lim(u(x+dx – xi)|**ur** – u(x – xi)|**ur**)∗**qd**(**dr**)
= **0**, x ≠ xi
= **ur**/dx, x = xi
D[**r**,**r**+**dr**]∗(v(x – xi)|**ur**;**dr**)
≡ lim(v(x+dx – xi)|**ur** – v(x – xi)|**ur**)∗**qd**(**dr**)
= **0**, for all x
D[**r** – **dr**,**dr**]∗(u(x – xi)|**ur**;**dr**)
≡ lim(u(x – xi)|**ur** – u(x – dx – xi)|**ur**)∗**qd**(**dr**)
= **0**, for all x
D[**r** – **dr**,**dr**]∗(v(x – xi)|**ur**;**dr**)
≡ lim(v(x – xi)|**ur** – v(x – dx – xi)|**ur**)∗**qd**(**dr**)
= **0**, x ≠ xi
= **ur**/dx, x = xi
as dx⟶0 (647)

for

r(x) a curve over CX;

then

the basic gradient of a step function along CX is defined as
D[**r**(x),**r**(x)+**dr**(x)]∗(u(**r**(x) – **r**(xi))|CX;**dr**(x))
≡ lim(u(**r**(x)+**dr**(x)– **r**(xi))|CX –u(**r**(x) – **r**(xi))|CX)
∗**qd**(**dr**(**r**(xi)))
= **0**, **r** ≠ **r**(xi) in CX
= **qd**(**dr**(**r**(xi))), **r**=**r**(xi)
D[**r**(x),**r**(x)+**dr**(x)]∗(v(**r**(x) – **r**(xi))|CX;**dr**(x))
≡ lim(v(**r**(x)+**dr**(x)– **r**(xi))|CX –v(**r**(x) – **r**(xi))|CX)
∗**qd**(**dr**(**r**(xi)))
= **0**, for all **r**(x) in CX
D[**r**(x)–**dr**(x),**r**(x)]∗(u(**r**(x) – **r**(xi))|CX;**dr**(x))
≡ lim(u(**r**(xi)– **r**(x)–**dr**(x))|CX –u(**r**(x) – **r**(xi))|CX)
∗**qd**(**dr**(**r**(xi)))
= **0**, for all **r**(x) in CX

D[**r**(x)–**dr**(x),**r**(x)]∗(v(**r**(x) – **r**(xi))|CX;**dr**(x))
≡ lim(v(**r**(xi)– **r**(x)–**dr**(x))|CX –v(**r**(x) – **r**(xi))|CX)
∗**qd**(**dr**(**r**(xi)))
= **0**, **r**(x)≠**r**(xi) in CX
= **qd**(**dr**(**ri**)), **r**(x)=**r**(xi)
as **dr**(**r**(x))⟶**0** (648)

for
r in CR a curve from **r1** to **r2**;
then
the basic gradient of a step function along CR is defined as
D[**r**,**r**+**dr**(**r**)]∗(u(**r** – **ri**)|CR;**dr**)
≡ lim(u(**r**+**dr**(**r**)– **ri**)|CR –u(**r** – **ri**)|CR)∗**qd**(**dr**(**r**))
= **0**, **r** ≠ **ri** in CR
= **qd**(**dr**(**ri**)), **r**=**ri**
D[**r**,**r**+**dr**(**r**)]∗(v(**r** – **ri**)|CR;**dr**)
≡ lim(v(**r**+**dr**(**r**) – **ri**)|CR – v(**r** – **ri**)|CR)∗**qd**(**dr**(**r**))
= **0**, for all **r** in CR
D[**r** – **dr**(**r**),**r**]∗(u(**r** – **ri**)|CR;**dr**)
≡ lim(u(**ri** – **r** – –**dr**(**r**))|CR – u(**r** – **ri**)|CR)∗**qd**(**dr**(**r**))
= **0**, for all **r** in CR
D[**r** – **dr**(**r**),**r**]∗(v(**r** – **ri**)|CR;**dr**)
≡ lim(v(**ri** – **r**– **dr**(**r**))|CR v(**r** – **ri**)|CR)∗**qd**(**dr**(**r**))
= **0**, **r**≠**ri** in CR
= **qd**(**dr**(**ri**)), **r**=**ri**
as **dr**(**r**)⟶**0**. (649)
end of definition

Thus, like real impulse functions, these gradients are impulsive in amplitude but in addition have a definite direction.

Now for ***f*** a generalized function over CR, consider

(**f**|CR)∗(**D**[**r**,**r**+**dr**]∗(u(**r**–**ri**)|CR);**dr**)
= **f**(**ri**)∗**qd**(**dr**(**ri**)), **r**=**ri**
= [**0**], otherwise; (650)
(u(**r**–**ri**)|CR)∗**D**[**r**,**r**+**dr**]∗(**f**|CR;**dr**)
= [**0**], **r** in [**r0**,**ri**]
= **D**[**r**,**r**+**dr**]∗(**f**|CR;**dr**), otherwise; (651)

Mathematics: Vector Integration

and

$$\begin{aligned}
&\mathbf{D}[\mathbf{r},\mathbf{r+dr}]*((\mathbf{f}|\mathrm{CR})*u(\mathbf{r}-\mathbf{ri})|\mathrm{CR};\mathbf{dr})\\
&\qquad = [\mathbf{0}] && \mathbf{r\ in}\ \text{in}\ [\mathbf{r0},\mathbf{ri})\\
&\qquad = \lim \mathbf{f}(\mathbf{ri}+\mathbf{dr}(\mathbf{ri}))*\mathbf{qd}(\mathbf{dr}(\mathbf{ri})), && \mathbf{r}=\mathbf{ri}\\
&\qquad = \mathbf{D}[\mathbf{r},\mathbf{r+dr}]*(\mathbf{f}|\mathrm{CR};\mathbf{dr}), && \mathbf{r}\ \text{in}\ (\mathbf{ri},\mathbf{r2}].
\end{aligned} \tag{652}$$

If f is continuous at **ri** these considerations are sufficiently written as

$$\begin{aligned}
&\mathbf{D}[\mathbf{r},\mathbf{r+dr}]*((\mathbf{f}|\mathrm{CR})*u(\mathbf{r}-\mathbf{ri})|\mathrm{CR};\mathbf{dr})\\
&\qquad = \mathbf{f}(\mathbf{ri})*\mathbf{D}[\mathbf{r},\mathbf{r+dr}](u(\mathbf{r}-\mathbf{ri})|\mathrm{CR};\mathbf{dr})\\
&\qquad\qquad + v(\mathbf{r}-\mathbf{ri})|\mathrm{CR}*\mathbf{D}[\mathbf{r},\mathbf{r+dr}]*(\mathbf{f}|\mathrm{CR};\mathbf{dr}).
\end{aligned} \tag{653}$$

Similarly,

$$\begin{aligned}
&\mathbf{D}[\mathbf{r}-\mathbf{dr},\mathbf{r}\ ]*(\mathbf{f}|\mathrm{CR})*v(\mathbf{r}-\mathbf{ri})|\mathrm{CR});\mathbf{dr})\\
&\qquad = \mathbf{f}(\mathbf{ri})*\mathbf{D}[\mathbf{r}-\mathbf{dr},\mathbf{r}](v(\mathbf{r}-\mathbf{ri})|\mathrm{CR};\mathbf{dr})\\
&\qquad\qquad + v(\mathbf{r}-\mathbf{ri})|\mathrm{CR}*\mathbf{D}[\mathbf{r}-\mathbf{dr},\mathbf{r}]*(\mathbf{f}|\mathrm{CR};\mathbf{dr})
\end{aligned} \tag{654}$$

$$\begin{aligned}
&\mathbf{D}[\mathbf{r},\mathbf{r+dr}]*((\mathbf{f}|\mathbf{ur})*u(\mathbf{r}-\mathbf{ri})|\mathbf{ur};\mathbf{dr})\\
&\qquad = \mathbf{f}(\mathbf{ri})*\mathbf{D}[\mathbf{r},\mathbf{r+dr}]*(u(\mathbf{r}-\mathbf{ri})|\mathbf{ur};\mathbf{dr})\\
&\qquad\qquad + (v(\mathbf{r}-\mathbf{ri})|\mathbf{ur})*\mathbf{D}[\mathbf{r},\mathbf{r+dr}]*(\mathbf{f}|\mathbf{ur};\mathbf{dr})
\end{aligned} \tag{655}$$

$$\begin{aligned}
&\mathbf{D}[\mathbf{r}-\mathbf{dr},\mathbf{r}]*((\mathbf{f}|\mathbf{ur})*v(\mathbf{r}-\mathbf{ri})|\mathbf{ur};\mathbf{dr})\\
&\qquad = \mathbf{f}(\mathbf{ri})*\mathbf{D}[\mathbf{r}-\mathbf{dr},\mathbf{r}]*(v(\mathbf{r}-\mathbf{ri})|\mathbf{ur};\mathbf{dr})\\
&\qquad\qquad + (v(\mathbf{r}-\mathbf{ri})|\mathbf{ur})*\mathbf{D}[\mathbf{r}-\mathbf{dr},\mathbf{r}]*(\mathbf{f}|\mathbf{ur};\mathbf{dr}).
\end{aligned} \tag{656}$$

Now for ***f*** a generalized function over a curve,

$$\begin{aligned}
&\mathbf{I}[\mathrm{CR}]\bullet((\mathbf{f}|\mathrm{CR})*\mathbf{D}[\mathbf{r},\mathbf{r+dr}]*(u(\mathbf{r}-\mathbf{ri})|\mathrm{CR};\mathbf{dr})|\mathrm{CR};\mathbf{dr})\\
&\qquad = \lim \mathbf{f}(\mathbf{ri})*\mathbf{qd}(\mathbf{dr}(\mathbf{ri}))\bullet\mathbf{dr}(\mathbf{ri})\\
&\qquad = \mathbf{f}(\mathbf{ri}).
\end{aligned} \tag{657}$$

In addition,

$$\begin{aligned}
&\mathbf{I}[\mathrm{CR}]\bullet((\mathbf{f}|\mathrm{CR})\wedge\mathbf{D}[\mathbf{r},\mathbf{r+dr}]*(u(\mathbf{r}-\mathbf{ri})|\mathrm{CR};\mathbf{dr})|\mathrm{CR};\mathbf{dr})\\
&\qquad = \mathbf{f}(\mathbf{ri})\wedge\mathbf{qd}(\mathbf{dr}(\mathbf{ri}))\bullet\mathbf{dr}(\mathbf{ri})\\
&\qquad = \mathbf{f}(\mathbf{ri})\bullet\mathbf{qd}(\mathbf{dr}(\mathbf{ri}))\wedge\mathbf{dr}(\mathbf{ri})\\
&\qquad = 0
\end{aligned} \tag{658}$$

$$\begin{aligned}
&\mathbf{I}[\mathrm{CR}]\wedge((\mathbf{f}|\mathrm{CR})*\mathbf{D}[\mathbf{r},\mathbf{r+dr}]*(u(\mathbf{r}-\mathbf{ri})|\mathrm{CR};\mathbf{dr})|\mathrm{CR};\mathbf{dr})\\
&\qquad = \mathbf{0}
\end{aligned} \tag{659}$$

$$\begin{aligned}
&\mathbf{I}[\mathrm{CR}]\wedge((\mathbf{f}|\mathrm{CR})\wedge\mathbf{D}[\mathbf{r},\mathbf{r+dr}]*(u(\mathbf{r}-\mathbf{ri})|\mathrm{CR};\mathbf{dr})|\mathrm{CR};\mathbf{dr})\\
&\qquad = \lim \mathbf{f}(\mathbf{ri})\bullet\mathbf{dr}(\mathbf{ri})*\mathbf{qd}(\mathbf{dr}(\mathbf{ri})) - \mathbf{f}(\mathbf{ri})
\end{aligned} \tag{660}$$

I[CR]*((f|CR)*D[r,r+dr]*(u(r−ri)|CR;dr)|CR;dr)
= lim f(ri)*qd(dr(ri))*dr(ri) (661)
I[CR]*((f|CR)*D[r,r+dr]∧(u(r−ri)|CR;dr)|CR;dr)
= C(f(ri))•(qd(dr(ri))*dr(ri)) (662)
I[CR]*((f|CR)•D[r,r+dr]∧(u(r−ri)|CR;dr)|CR;dr)
= f(ri)•qd(dr(ri))*dr(ri). (663)

For v(r−ri),

I[CR]•((f|CR)*D[r−dr,r]*(v(r−ri)|CR;dr)|CR;dr)
= f(ri) (664)
I[CR]•((f|CR)∧D[r−dr,r]*(v(r−ri)|CR;dr)|CR;dr)
= 0 (665)
I[CR]∧((f|CR)*D[r−dr,r]*(v(r−ri)|CR;dr)|CR;dr)
= 0 (666)
I[CR]∧((f|CR)∧D[r−dr,r]*(v(r−ri)|CR;dr)|CR;dr)
= f(ri)•dr(ri)*qd(dr(ri)) − f(ri) (667)
I[CR]*((f|CR)*D[r−dr,r]*(v(r−ri)|CR;dr)|CR;dr)
= f(ri)*qd(dri)*dri (668)
I[CR]*((f|CR)*D[r−dr,r]∧(v(r−ri)|CR;dr)|CR;dr)
= C(f(ri))•(qd(dr(ri))*dr(ri)) (669)
I[CR]*((f|CR)•D[r−dr,r]∧(v(r−ri)|CR;dr)|CR;dr)
= f(ri)•qd(dr(ri))*dr(ri). (670)

Similar results hold for directional functions.

Now consider for *f* bounded and continuous except possibly at ri interior to (r0,rm) where

lim f(r(i+1))|CR ≡ f1
lim f(r(i−1))|CR ≡ f0

I[r0,rm]•((u(r−ri)|CR)*D[r,r+dr(r)]*(f(r)|CR;dr)|CR;dr)
= lim D[r(i+1),r(i+1)+dr(r(i+1))]*(f(r)|CR;dr)•dr(r(i+1))
+ ...
+ D[rm,r+dr(rm)]*(f(r)|CR;dr)*dr(rm)

= lim (f(r(i+2))|CR – f(r(i+1))|CR)
∗qd(dr(r(i+1)))•dr(r(i+1))
+ ...
+(f(r(m+1))|CR – f(rm)|CR)∗qd(dr(rm))•dr(rm)
= lim – f(r(i+1))|CR) + (f(r(m+1))|CR
= f(rm) – f1.

Thus,

I[r0,rm]•((u(r–ri)|CR)∗D[r,r+dr(r)]∗(f(r)|CR;dr)|CR;dr)
= f(rm) – f1 (671)

I[r0,rm]•((u(r–ri)|CR)∗D[r,r+dr(r)]∧(f(r)|CR;dr)|CR;dr)
= 0 (672)

I[r0,rm]∧((u(r–ri)|CR)∗D[r,r+dr(r)]∧(f(r)|CR;dr)|CR;dr)
= f(rm)•[dr(rm)∗qd(dr(rm)) – I]
– f1•[dr(r(i+1))∗qd(dr(r(i+1))) – I] (673)

I[r0,rm]∗((u(r–ri)|CR)∗D[r,r+dr(r)]∧(f(r)|CR;dr)|CR;dr)
= C(f(rm))•[qd(dr(rm))∗dr(rm)]
– C(f1)•[qd(dr(r(i+1)))∗dr(r(i+1))] (674)

I[r0,rm]∗((u(r–ri)|CR)∗D[r,r+dr(r)]•(f(r)|CR;dr)|CR;dr)
= f(rm)•[qd(dr(rm))∗dr(rm)]
– f1•[qd(dr(r(i+1)))∗dr(r(i+1))]. (675)

Similarly,

I[r0,rm]•((v(r–ri)|CR)∗D[r–dr(r),r]∗(f(r)|CR;dr)|CR;dr)
= f(rm) – f0 (676)

I[r0,rm]•((v(r–ri)|CR)∗D[r–dr(r),r]∧(f(r)|CR;dr)|CR;dr)
= 0 (677)

I[r0,rm]∧((v(r–ri)|CR)∗D[r–dr(r),r]∧(f(r)|CR;dr)|CR;dr)
= f(rm)•[dr(rm)∗qd(dr(rm)) – I]
– f1•[dr(r(i–1))∗qd(dr(r(i–1))) – I] (678)

I[r0,rm]∗((v(r–ri)|CR)∗D[r–dr(r),r]∧(f(r)|CR;dr)|CR;dr)
= C(f(rm))•[qd(dr(rm))∗dr(rm)]
– C(f1)•[qd(dr(r(i–1)))∗dr(r(i–1))] (679)

I[r0,rm]∗((v(r–ri)|CR)∗D[r–dr(r),r]•(f(r)|CR;dr)|CR;dr)
= f(rm)•[qd(dr(rm))∗dr(rm)]
– f1•[qd(dr(r(i–1)))∗dr(r(i–1))]. (680)

Similar results hold for directional functions.

Next consider

I[**r0**,**rm**]•(**D**[**r**,**r**+**dr**(**r**)]∗(u(**r**−**ri**)|CR)∗(**f**(**r**)|CR;**dr**)|CR;**dr**)
= lim **f**(**r**(i+1))|CR∗**qd**(**dr**(**ri**))•**dr**(**ri**)
+ (**f**(**r**(i+2))|CR − **f**(**r**(i+1))|CR)
∗**qd**(**dr**(**r**(i+1)))•**dr**(**r**(i+1))
+ ...
+(**f**(**r**(m+1))|CR − **f**(**rm**)|CR)
∗**qd**(**dr**(**rm**))•**dr**(**rm**)
= lim **f**(**r**(m+1))|CR
= **f**(**rm**).

Thus,

I[**r0**,**rm**]•(**D**[**r**,**r**+**dr**(**r**)]∗(u(**r**−**ri**)|CR)∗(**f**(**r**)|CR;**dr**)|CR;**dr**)
= **f**(**rm**) (681)

I[**r0**,**rm**]•(**D**[**r**,**r**+**dr**(**r**)]∗(u(**r**−**ri**)|CR)∧(**f**(**r**)|CR;**dr**)|CR;**dr**)
= 0 (682)

I[**r0**,**rm**]∧(**D**[**r**,**r**+**dr**(**r**)]∗(u(**r**−**ri**)|CR)∧(**f**(**r**)|CR;**dr**)|CR;**dr**)
= **f**(**rm**)•[**dr**(**rm**)∗**qd**(**dr**(**rm**)) − **I**] (683)

I[**r0**,**rm**]∗(**D**[**r**,**r**+**dr**(**r**)]∗(u(**r**−**ri**)|CR)∧(**f**(**r**)|CR;**dr**)|CR;**dr**)
= **C**(**f**(**rm**))•[**qd**(**dr**(**rm**))∗**dr**(**rm**)] (684)

I[**r0**,**rm**]∗(**D**[**r**,**r**+**dr**(**r**)]∗(u(**r**−**ri**)|CR)•(**f**(**r**)|CR;**dr**)|CR;**dr**)
= **f**(**rm**)•[**qd**(**dr**(**rm**))∗**dr**(**rm**)]. (685)

Similarly,

I[**r0**,**rm**]•(**D**[**r**−**dr**(**r**),**r**]∗(v(**r**−**ri**)|CR)∗(**f**(**r**)|CR;**dr**)|CR;**dr**)
= **f**(**rm**) (686)

I[**r0**,**rm**]•(**D**[**r**−**dr**(**r**),**r**]∗(v(**r**−**ri**)|CR)∧(**f**(**r**)|CR;**dr**)|CR;**dr**)
= 0 (687)

I[**r0**,**rm**]∧(**D**[**r**−**dr**(**r**),**r**]∗(v(**r**−**ri**)|CR)∧(**f**(**r**)|CR;**dr**)|CR;**dr**)
= **f**(**rm**)•[**dr**(**rm**)∗**qd**(**dr**(**rm**)) − **I**] (688)

I[**r0**,**rm**]∗(**D**[**r**−**dr**(**r**),**r**]∗(v(**r**−**ri**)|CR)∧(**f**(**r**)|CR;**dr**)|CR;**dr**)
= **C**(**f**(**rm**))•[**qd**(**dr**(**rm**))∗**dr**(**rm**)] (689)

I[**r0**,**rm**]∗(**D**[**r**−**dr**(**r**),**r**]∗(v(**r**−**ri**)|CR)•(**f**(**r**)|CR;**dr**)|CR;**dr**)
= **f**(**rm**)•[**qd**(**dr**(**rm**))∗**dr**(**rm**)]. (690)

Mathematics: Vector Integration

Consequently for $\boldsymbol{f}$ continuous at $\boldsymbol{ri}$,

$$\mathbf{I}[\mathbf{r0},\mathbf{rm}]\cdot(\mathbf{D}[\mathbf{r},\mathbf{r}+\mathbf{dr}(\mathbf{r})]*(u(\mathbf{r}-\mathbf{ri})|CR)*(\mathbf{f}(\mathbf{r})|CR;\mathbf{dr})|CR;\mathbf{dr})$$
$$= \mathbf{I}[\mathbf{r0},\mathbf{rm}]\cdot((u(\mathbf{r}-\mathbf{ri})|CR)*\mathbf{D}[\mathbf{r},\mathbf{r}+\mathbf{dr}(\mathbf{r})]*(\mathbf{f}(\mathbf{r})|CR;\mathbf{dr})|CR;\mathbf{dr})$$
$$+ \mathbf{I}[\mathbf{r0},\mathbf{rm}]\cdot((\mathbf{f}|CR)*\mathbf{D}[\mathbf{r},\mathbf{r}+\mathbf{dr}]*(u(\mathbf{r}-\mathbf{ri})|CR;\mathbf{dr});\mathbf{dr}). \quad (691)$$

Thus, the integral product rule holds only selectively for these many invergences, incurls, and ingradients over curved lines.

Now let C1 = [**r0**,**ri**) and C2 = [**ri**,**rm**]. Then for f=1

$$\mathbf{I}[C1]\cdot(\mathbf{D}[\mathbf{r},\mathbf{r}+\mathbf{dr}(\mathbf{r})]*(u(\mathbf{r}-\mathbf{ri})|CR;\mathbf{dr})|CR;\mathbf{dr}) = 0 \quad (692)$$

while

$$\mathbf{I}[C2]\cdot(\mathbf{D}[\mathbf{r},\mathbf{r}+\mathbf{dr}(\mathbf{r})]*(u(\mathbf{r}-\mathbf{ri})|CR;\mathbf{dr})|CR;\mathbf{dr}) = 1. \quad (693)$$

Similarly, for C1 = [**r0**,**ri**) and C2 = [**ri**,**rm**]

$$\mathbf{I}[C1]\cdot(\mathbf{D}[\mathbf{r},\mathbf{r}+\mathbf{dr}(\mathbf{r})]*(v(\mathbf{r}-\mathbf{ri})|CR;\mathbf{dr})|CR;\mathbf{dr}) = 0 \quad (694)$$

while

$$\mathbf{I}[C2]\cdot(\mathbf{D}[\mathbf{r},\mathbf{r}+\mathbf{dr}(\mathbf{r})]*(v(\mathbf{r}-\mathbf{ri})|CR;\mathbf{dr})|CR;\mathbf{dr}) = 1. \quad (695)$$

Now let

$$C1 \equiv [\mathbf{r1},\mathbf{r2}] \subset CR \equiv [\mathbf{r0},\mathbf{rm2}]. \quad (696)$$

Then

$$\mathbf{I}[C1]*(\mathbf{f}|CR;\mathbf{dr}) = \mathbf{I}[CR]*(v(\mathbf{r}-\mathbf{r1})-u(\mathbf{r}-\mathbf{r2}))*(\mathbf{f}(|CR);\mathbf{dr}). \quad (697)$$

In this way even though $\boldsymbol{f}$ may be unbounded on some portions of a curve, it may be still be integrated over portions over which it is bounded.

Now for ***d*** a vector constant let

$$\mathbf{f}(\mathbf{r})|CR = \mathbf{f1}(\mathbf{r})|CR + \mathbf{d}*u(\mathbf{r}-\mathbf{ri}) \qquad (698)$$

where ***f1*** is bounded and continuous over C such that .

$$\mathbf{I}[CR]*(\mathbf{f1}(\mathbf{r})|CR;\mathbf{dr}) = \mathbf{G}(\mathbf{rm}) - \mathbf{G}(\mathbf{r0}). \qquad (699)$$

Then

$$\mathbf{I}[CR]*(\mathbf{f}|CR;\mathbf{dr}) = \mathbf{I}[CR]*(\mathbf{f1}|CR;\mathbf{dr}) + \mathbf{d}*\mathbf{I}[\mathbf{ri},\mathbf{rm}]*(1|CR;\mathbf{dr}) \qquad (700)$$

$$= \mathbf{G}(\mathbf{rm}) - \mathbf{G}(\mathbf{r0}) + \mathbf{d}*\mathbf{I}[\mathbf{ri},\mathbf{rm}]*(1|CR;\mathbf{dr}). \qquad (701)$$

Similarly for

$$\mathbf{f}(\mathbf{r})|CR = \mathbf{f1}(\mathbf{r})|CR + \mathbf{d}*v(\mathbf{r}-\mathbf{ri}) \qquad (702)$$

$$\mathbf{I}[CR]*(\mathbf{f}|CR;\mathbf{dr}) = \mathbf{G}(\mathbf{rm}) - \mathbf{G}(\mathbf{r0}) + \mathbf{d}*\mathbf{I}[\mathbf{ri},\mathbf{rm}]*(1|CR;\mathbf{dr}). \qquad (703)$$

Equation (703) is also an **extension to the fundamental theorem of integral calculus**.

Analogous conclusions apply to invergences and incurls. These results further expand the fundamental theorem of integral calculus to functions with a finite number of finite discontinuities in a given direction or over curves.

Mathematics: Vector Integration

The following development establishes a theory of local integration[40] which inverts sectional gradients.

Vector measures in V3

Similar to integration over *R*, integration over *V3* involves partitioning and measuring.

Definition 41 (partitions of V3)

Given

p1, p2, p3, positive real numbers;

u1R1, **u2R1**, **u3R1**, non–parallel directions;

for

n>0 a given integer;

i,j,k= – $(n^2–1)$,...,–,1,0,1,2,3,...n^2;

(i–1)/n≤t1<i/n for abs(t1)≤n;

(j–1)/n≤t2<j/n for abs(t2)≤n;

(k–1)/n≤t3<k/n for abs(t3)≤n;

Pn(i,j,k) ≡ {**r**=t1∗p1∗**u1R1** + t2∗p2∗**u2R1** + t3∗p3∗**u3R1**}
and Pn∞ otherwise. (704)

end of definition

Any **r** of V3 may be placed into one and only one of the subsets of Pn. Pn is thus a partition of V3 into $(2*n^2)^3 + 1$ disjoint subsets Pn(i,j,k) called **cells**. The density of cells is n^3/(p1∗p2∗p3).

The location of each cell is defined as a specific location within the cell.

Definition 42 (location of a cell in a partition)

Given

Pn(i,j,k), a cell in partition Pn;

then

r(Pn(i,j,k)) ≡ (i – ½)∗p1∗**u1R1**/n
+ (j – ½)∗p2∗**u2R1**/n
+ (k – ½)∗p3∗**u3R1**/n (705)

r(Pn∞)≡ 0.

end of definition

For each Pn(i,j,k) of Pn let[41]

$$\mathbf{mn}(Pn(i,j,k)) \equiv ((\mathbf{r}(Pn(i+1,j,k)) - \mathbf{r}(Pn(i,j,k)))*\mathbf{u1R1} + (\mathbf{r}(Pn(i,j+1,k)) - \mathbf{r}(Pn(i,j,k)))*\mathbf{u2R1} + (\mathbf{r}(Pn(i,j,k+1)) - \mathbf{r}(Pn(i,j,k)))*\mathbf{u3R1})/n^2 = (p1*\mathbf{u1R1}+p2*\mathbf{u2R1}+p3*\mathbf{u3R1})/n^3$$

with $\mathbf{mn}(Pn\infty) \equiv \mathbf{0}$. (706)

Then each cell, with the exception of Pn∞, has the same vectorial value.

For a measurable set M of V3 in a given partition Pn, let *bn(M∩Pn)* be the number of cells for which M∩Pn(i,j,k) is non–empty[42].

Then define[43] the **measure of M with respect to the partition Pn** as

$$\mathbf{mn}(M\cap Pn) \equiv bn(M\cap Pn)*(\mathbf{mn}(Pn(i,j,k))) = (p1*\mathbf{u1R1}+p2*\mathbf{u2R1}+p3*\mathbf{u3R1})*bn(M\cap Pn)/n^3. \quad (707)$$

The measure of *V3* with respect to Pn is thus

$$\mathbf{mn}(V3\cap Pn) = (8*n^3 + 1/n^3)*(p1*\mathbf{u1R1}+p2*\mathbf{u2R1}+p3*\mathbf{u3R1}).$$

708)

Definition 43 (measure of a measurable set M)

Given

$\mathbf{pR1} = (p1*\mathbf{u1R1}+p2*\mathbf{u2R1}+p3*\mathbf{u3R1})$;

M a measurable set of V3;

Pn, a sequenced set of partitions of V3 based on **pR1**;

mn(M∩Pn), the measure of M with respect to the partition Pn;

then

$$\mathbf{m}(M) \equiv \lim(\mathbf{mn}(M\cap Pn)) \text{ as } n\longrightarrow\infty \quad (709)$$

$$\equiv m(M)*\mathbf{pR1}. \quad (710)$$

end of definition

The measure of M is thus seen to consist of a scalar

$$m(M) = \lim\ (1/n^3)*bn(M\cap Pn) \text{ as } n\longrightarrow\infty \quad (711)$$

and a vector,

$$\mathbf{pR1} \equiv p1*\mathbf{u1R1}+p2*\mathbf{u2R1}+p3*\mathbf{u3R1}, \quad (712)$$

called the **partition vector**.

Mathematics: Vector Integration

The measure of a given measurable set in V3 depends on the partition vector. Where the dependence needs to be made explicit, the measure of M is written as a condition:

$$\mathbf{m}(\mathrm{M}|\mathbf{pR1}) = \mathrm{m}(\mathrm{M}|\mathbf{pR1})*\mathbf{pR1}. \qquad (713)$$

It is instructive to examine this measure in terms of associated volumes.

Theorem 6 (Measure and volumes)

Given

M, a measurable set of V3;

vol(M), volume of M;

pR1 ≡ p1∗**u1R1**+p 2∗**u2R1**+p3∗**u3R1**, a partition vector;

for

p ≡ p1∗**u1** + p2∗**u2** +p3∗**u3**;

UR1 ≡ u1∗u1R1 + u2∗u2R1 +u3∗3R1;

then

$$\mathrm{vol}(\mathrm{Pn}(i,j,k)) = \det[\mathbf{G}(\mathbf{p})]*\det[\mathbf{UR1}]/n^3 \qquad (714)$$

$$= \mathrm{vol}(\mathrm{P1}(i,j,k))/n^3 \qquad (715)$$

$$\mathbf{mn}(\mathrm{Pn}(i,j,k)) = \mathrm{vol}(\mathrm{Pn}(i,j,k))*\mathbf{pR1}/(\det[\mathbf{G}(\mathbf{p})]*\det[\mathbf{UR1}]) \qquad (716)$$

$$= \mathrm{vol}(\mathrm{Pn}(i,j,k))*\mathbf{pR1}/\mathrm{vol}(\mathrm{P1}(i,j,k)) \qquad (717)$$

$$\mathbf{mn}(\mathrm{M}|\mathbf{pR1}) = \mathrm{bn}(\mathrm{M}\cap\mathrm{Pn})*\mathrm{vol}(\mathrm{Pn}(i,j,k))*\mathbf{pR1} /(\det[\mathbf{G}(\mathbf{p})]*\det[\mathbf{UR1}]) \qquad (718)$$

$$= \mathrm{bn}(\mathrm{M}\cap\mathrm{Pn})*\mathrm{vol}(\mathrm{Pn}(i,j,k))*\mathbf{pR1}/\mathrm{vol}(\mathrm{P1}(i,j,k)) \qquad (719)$$

$$\mathbf{m}(\mathrm{M}|\mathbf{pR1}) = \mathrm{vol}(\mathrm{M})*\mathbf{pR1}/\mathrm{vol}(\mathrm{P1}(i,j,k)). \qquad (720)$$

Proof:

$$\mathrm{vol}(\mathrm{P1}(i,j,k)) = (\mathrm{p1}*\mathbf{u1R1})\cdot(\mathrm{p2}*\mathbf{u2R1})\wedge(\mathrm{p3}*\mathbf{u3R1})$$
$$= \det[\mathbf{G}(\mathbf{p})]*\det[\mathbf{UR1}].$$

Consequently,

$$\mathrm{vol}(\mathrm{Pn}(i,j,k)) = \det[\mathbf{G}(\mathbf{p})]*\det[\mathbf{UR1}]/n^3$$

Again,

$$\mathrm{mn}(\mathrm{Pn}(i,j,k)) = 1/n^3$$
$$= \mathrm{vol}(\mathrm{Pn}(i,j,k))/(\det[\mathbf{G}(\mathbf{p})]*\det[\mathbf{UR1}])$$
$$= \mathrm{vol}(\mathrm{Pn}(i,j,k))/\mathrm{vol}(\mathrm{P1}(i,j,k))$$
$$\mathrm{mn}(\mathrm{M}\cap\mathrm{Pn}) = \mathrm{bn}(\mathrm{M}\cap\mathrm{Pn})/n^3$$
$$= \mathrm{bn}(\mathrm{M}\cap\mathrm{Pn})*\mathrm{vol}(\mathrm{Pn}(i,j,k))/\mathrm{vol}(\mathrm{P1}(i,j,k))$$

Finally for a measurable set

$$\lim \mathrm{bn}(\mathrm{M}\cap\mathrm{Pn})*\mathrm{vol}(\mathrm{Pn}(i,j,k)) = \mathrm{vol}(\mathrm{M}).$$

qed

For M1 and M2 measurable and disjoint,

$$\mathbf{m}(M1 \cup M2) = \mathbf{m}(M1) + \mathbf{m}(M2) = (m(M1) + m(M2)) * \mathbf{pR1} \qquad (721)$$

Moreover,

$$\mathbf{m}(Pn(i,j,k)) = \mathbf{mn}(Pn(i,j,k)) \qquad (722)$$

since for

$$\mathbf{mn}(Pn(i,j,k)) = (p1 * \mathbf{u1R1} + p2 * \mathbf{u2R1} + p3 * \mathbf{u3R1})/n^3$$

$$\mathbf{m2n}(Pn(i,j,k)) = 8 * (p1 * \mathbf{u1R1} + p2 * \mathbf{u2R1} + p3 * \mathbf{u3R1}) / (2 * n)^3 = \mathbf{mn}(Pn(i,j,k)).$$

Consequently the limit, which exists for measurable sets, is established by the subsequence **mn**,**m2n**,**m4n**,**m8n**,...

Definition 44 (Surface of M)

Given

$\mathbf{pR1} = (p1 * \mathbf{u1R1} + p2 * \mathbf{u2R1} + p3 * \mathbf{u3R1})$;

M, a measurable set of V3;

Pn, a sequenced set of partitions of V3 based on **pR1**;

mn($M \cap Pn$), the measure of M with respect to the partition Pn;

then

if $Pn(i,j,k) \cap M = Pn(i,j,k)$ (723)

Pn(i,j,k) is called an **cell interior to** M

if $Pn(i,j,k) \cap M = \text{null}$ (724)

Pn(i,j,k) is called an **cell exterior to** M

if $Pn(i,j,k) \cap M \subset Pn(i,j,k)$ properly (725)

Pn(i,j,k) is called a **surface cell of** M.

end of definition

The set

$$SM \equiv \{\lim \mathbf{r}(Pn(i,j,k)) \mid Pn(i,j,k) \text{ is a surface cell}\} \qquad (726)$$

is called the **surface of** M.

Mathematics: Vector Integration

The surface of M may be further distinguished:
for **pR1** = pR1∗**pR1** for any pR1>0

SMPE|**pR1** ≡ {**r** in SM |(**r+pR1**/n^3 is in V3 – (M∪SM) (727)
is called the **exterior positive surface of** M

SMPI|**pR1** ≡ {**r** in SM |(**r+pR1**/n^3 is in (M – SM) (728)
is called the **interior positive surface of** M

SMP0|**pR1** ≡ {**r** in SM |(**r+pR1**/n^3 is in SM (729)
is called the **zero positive surface of** M

SMNE|**pR1** ≡ {**r** in SM |(**r–pR1**/n^3 is in V3 – (M∪SM) (730)
is called the **exterior negative surface of** M

SMNI|**pR1** ≡ {**r** in SM |(**r–pR1**/n^3 is in (M – SM) (731)
is called the **interior negative surface of** M

SMN0|**pR1** ≡ {**r** in SM |(**r–pR1**/n^3 is in SM (732)
is called the **zero negative surface of** M.

The sets SMPE|**pR1**, SMPI|**pR1**, and SMP0|**pR1**, partition SM as do the sets SMNE|**pR1**, SMNI|**pR1**, and SMN0|**pR1**.

Theorem 7 (Measure of a translated set)

Given

a = a1∗**u1** + a2∗**u2** + a3∗**u3**, a translation vector;
M, a set with measure
m(M) = m(M)∗**pR1**;

for

MA = {**a** + **r**|**r** in M}

then

m(MA) = **m**(M). (733)

Proof:

If M is measurable, then MA is also measurable. If {Pn} is a nested sequence of partitions of V3, then so also is {**a**+Pn}. Moreover, **m**(M) is unchanged in {**a**+Pn}. Now **m**(MA) = **m**(M) in {**a**+Pn}. So also then in {Pn}. qed

The following theorem relates measures to different partitions, enabling partitions to be chosen conveniently.

Theorem 8 (Measure of a dilated set)
Given

pR1 ≡ p1R1∗**u1R1**+p2R1∗**u2R1**+p3R1∗**u3R1**, a partition vector;

M, a measurable set;

m(M|**pR1**) ≡ m(M|**pR1**)∗**pR1**;

pR2 ≡ p1R2∗**u1R2**+p2R2∗**u2R2**+p3R2∗**u3R2**,
a second partition vector;

p1 ≡ p1R1∗**u1**+p2R1∗**u2**+p3R1∗**u3**;

p2 ≡ p1R2∗**u1**+p2R2∗**u2**+p3R2∗**u3**;

UR1 = **u1**∗**u1R1** + **u2**∗**u2R1** + **u3**∗**u3R1**;

UR2 = **u1**∗**u1R2** + **u2**∗**u2R2** + **u3**∗**u3R2**;

then

m(M|**pR2**) = m(M|**pR1**)
∗det[**UR1**]∗det[**G**(**p1**)]/(det[**UR2**]∗det[**G**(**p2**)]). (734)

Proof:

m(M|**pR1**) = vol(M)/(det[**UR1**]∗det[**G**(**p1**)])
m(M|**pR2**) = vol(M)/(det[**UR2**]∗det[**G**(**p2**)]).
Consequently,
m(M|**pR1**)∗det[**UR1**]∗det[**G**(**p1**)]
= m(M|**pR2**)∗det[**UR2**]∗det[**G**(**p2**)]
= vol(M). qed

Mathematics: Vector Integration

Definition 45 (Invergences, incurls, and ingradients over V3)

Given

f, a measurable function over V3;

M, a measurable set of V3;

pR1 = pR1∗**uR1**

= p1∗**u1R1**+p2∗**u2R1**+p3∗**u3R1**, a partition vector;

for

S[bn(M∩Pn)], the sum over which M∩Pn is non–empty

I[M]•(**f**(**r**)|**m**;**dR1**)

≡ lim S[bn(M∩Pn)](**f**(**r**(Pn(i,j,k)))•**m**(Pn(i,j,k)))

= lim S[bn(M∩Pn)](**f**(**r**(Pn(i,j,k)))•**pR1**/n^3) (735)

I[M]∧(**f**(**r**)|**m**;**dR1**)

≡ lim S[bn(M∩Pn)](**f**(**r**(Pn(i,j,k)))∧**m**(Pn(i,j,k)))

= lim S[bn(M∩Pn)](**f**(**r**(Pn(i,j,k)))∧**pR1**/n^3) (736)

I[M]∗(**f**(**r**)|**m**;**dR1**)

≡ lim S[bn(M∩Pn)](**f**(**r**(Pn(i,j,k)))∗**m**(Pn(i,j,k)))

= lim S[bn(M∩Pn)](**f**(**r**(Pn(i,j,k)))∗**pR1**/n^3) (737)

as n⟶∞.

end of definition

For simplicity define

dR1 ≡ **pR1**/n^3 (738)

The measure of M can thus be written as

m(M) = m(M)∗**pR1**

= lim bn(M∩Pn)∗(**pR1**/n^3)

= lim bn(M∩Pn)∗**dR1**. (739)

The invergence, incurl, and ingradient may be also written:

I[M]•(**f**(**r**)|**m**;**dR1**) ≡ lim S[bn(M∩Pn)](**f**(**r**(Pn(i,j,k)))•**dR1**) (740)

I[M]∧(**f**(**r**)|**m**;**dR1**) ≡ lim S[bn(M∩Pn)](**f**(**r**(Pn(i,j,k)))∧**dR1**) (741)

I[M]∗(**f**(**r**)|**m**;**dR1**) ≡ lim S[bn(M∩Pn)](**f**(**r**(Pn(i,j,k)))∗**dR1**) (742)

as n⟶∞.

For[44] abs(**m**(Pn(i,j,k))):V3⟶R :

I[M](**f**(**r**)|m;dR1)
≡ lim (S[bn(M∩Pn)](**f**(**r**(Pn(i,j,k)))∗abs(**m**(Pn(i,j,k))))) (743)
= lim S[bn(M∩Pn)](**f**(**r**(Pn(i,j,k)))/n^3)
∗sqrt((**pR1•u1**)^2 +(**pR1•u2**)^2 +(**pR1•u3**)^2). (744)

Similarly, the integration of transformation, **A**(**r**) is defined as

I[M]•(**A**(**r**)|**m**;**dR1**)
≡ lim S[bn(M∩Pn)](**m**(Pn(i,j,k))•**T**[**A**(**r**(Pn(i,j,k)))]). (745)

Since

lim S[bn(M∩Pn)](**f**(**r**(Pn(i,j,k)))•**pR1**/n^3)
= lim S[bn(M∩Pn)](**f**(**r**(Pn(i,j,k))/n^3))•**pR1** (746)

invergences, incurls and ingradients may be also written as:

I[M]•(**f**(**r**)|**m**;**dR1**) ≡ **I**[M](**f**(**r**)|**m**)•**pR1** (747)
I[M]∧(**f**(**r**)|**m**;**dR1**) ≡ **I**[M](**f**(**r**)|**m**)∧**pR1** (748)
I[M]∗(**f**(**r**)|**m**;**dR1**) ≡ **I**[M](**f**(**r**)|**m**)∗**pR1** (749)

where
I[M](**f**(**r**)|**m**) ≡ lim S[bn(M∩Pn)](**f**(**r**(Pn(i,j,k))/n^3)). (750)

Integrals may also be defined in terms of functions of the partition vector **pR1** as in

I[M]•(**f**(**r**)|a;**C**(**dR1**))
≡ lim S[bn(M∩Pn)](**f**(**r**(Pn(i,j,k)))•**C**(**pR1**)/n^3) (751)
= lim S[bn(M∩Pn)](**f**(**r**(Pn(i,j,k)))•**C**(**dR1**)) (752)
where **C** is the curl vector operator, or, for another example, as
I3[M]•(**f**(**r**)|a;**dR1•UR1**$^{-1}$)
≡ lim S[bn(M∩Pn)](**f**(**r**(Pn(i,j,k)))•(**pR1•UR1**$^{-1}$)/n^3) (753)
= lim S[bn(M∩Pn)](**f**(**r**(Pn(i,j,k)))•(**dR1•UR1**$^{-1}$)). (754)

Mathematics: Vector Integration

Just as continuous gradients may be expressed as sectional gradients (but not necessarily vice–versa) so may integrals of functions with continuous gradients be expressed in terms of a sectional partition.

Theorem 9 (partition)

Given

M a measurable set in V3;

f a continuous function over M;

$\mathbf{pR1} = \mathrm{p1R1} \ast \mathbf{u1R1} + \mathrm{p2R1} \ast \mathbf{u2R1} + \mathrm{p3R1} \ast \mathbf{u3R1}$, a partition vector;

$\mathbf{I}[\mathrm{M}] \ast (\mathbf{f};\mathbf{dR1}) \equiv \mathbf{I}[\mathrm{M}](\mathbf{f}(\mathbf{r})|\mathbf{m1}) \ast \mathbf{pR1}$;

for

$\mathbf{pR2} = \mathrm{p1R2} \ast \mathbf{u1R2} + \mathrm{p2R2} \ast \mathbf{u2R2} + \mathrm{p3R2} \ast \mathbf{u3R2}$,
a second partition vector;

$\mathbf{p1} \equiv \mathrm{p1R1} \ast \mathbf{u1} + \mathrm{p2R1} \ast \mathbf{u2} + \mathrm{p3R1} \ast \mathbf{u3}$;

$\mathbf{p2} \equiv \mathrm{p1R2} \ast \mathbf{u1} + \mathrm{p2R2} \ast \mathbf{u2} + \mathrm{p3R2} \ast \mathbf{u3}$;

$\mathbf{UR1} \equiv \mathbf{u1} \ast \mathbf{u1R1} + \mathbf{u2} \ast \mathbf{u2R1} + \mathbf{u3} \ast \mathbf{u3R1}$;

$\mathbf{UR2} \equiv \mathbf{u1} \ast \mathbf{u1R2} + \mathbf{u2} \ast \mathbf{u2R2} + \mathbf{u3} \ast \mathbf{u3R2}$;

then

$$\mathrm{I}[\mathrm{M}]\cdot(\mathbf{f};\mathbf{dR2}) = \mathrm{I}[\mathrm{M}](\mathbf{f}(\mathbf{r})|\mathbf{m1})\cdot\mathbf{pR2} \ast \det[\mathrm{G}(\mathbf{p1})] \ast \det[\mathbf{UR1}] /(\det[\mathrm{G}(\mathbf{p2}) \ast \det[\mathbf{UR2}]) \quad (755)$$

$$\mathbf{I}[\mathrm{M}]\wedge(\mathbf{f};\mathbf{dR2}) = \mathbf{I}[\mathrm{M}](\mathbf{f}(\mathbf{r})|\mathbf{m1})\wedge\mathbf{pR2} \ast \det[\mathrm{G}(\mathbf{p1})] \ast \det[\mathbf{UR1}] /(\det[\mathrm{G}(\mathbf{p2})] \ast \det[\mathbf{UR2}]) \quad (756)$$

$$\mathbf{I}[\mathrm{M}] \ast (\mathbf{f};\mathbf{dR2}) = \mathbf{I}[\mathrm{M}](\mathbf{f}(\mathbf{r})|\mathbf{m1}) \ast \mathbf{pR2} \ast \det[\mathrm{G}(\mathbf{p1})] \ast \det[\mathbf{UR1}] /(\det[\mathrm{G}(\mathbf{p2})] \ast \det[\mathbf{UR2}]). \quad (757)$$

Proof:

Designate the n1 partition under **pR1** as PR1n1;
designate the n2 partition under **pR2** as PR2n2.

$\mathrm{mn1}(\mathrm{M} \cap \mathrm{Pn1}|\mathbf{pR1}) = \mathrm{bn1}(\mathrm{M} \cap \mathrm{Pn1}|\mathbf{pR1})/\mathrm{n1}^3$

$\mathrm{mn2}(\mathrm{M} \cap \mathrm{Pn2}|\mathbf{pR2}) = \mathrm{bn2}(\mathrm{M} \cap \mathrm{Pn2}|\mathbf{pR2})/\mathrm{n2}^3$

By the dilation theorem

$\mathrm{m}(\mathrm{PRn1(i,j,k)}|\mathbf{pR2}) = \mathrm{m}(\mathrm{PR1n(i,j,k)}|\mathbf{pR1}) \ast \det[\mathbf{UR1}] \ast \det[\mathbf{G}(\mathbf{p1})] /(\det[\mathbf{UR2}] \ast \det[\mathbf{G}(\mathbf{p2})])$

For n1=n2

$\mathrm{bn2}(\mathrm{PRn1(i,j,k)} \cap \mathrm{Pn2}) = \det[\mathbf{UR1}] \ast \det[\mathbf{G}(\mathbf{p1})]/(\det[\mathbf{UR2}] \ast \det[\mathbf{G}(\mathbf{p2})])$

Thus one cell under the PR1n1 partition is occupied by
$\det[\mathbf{UR1}] \ast \det[\mathbf{G}(\mathbf{p1})]/(\det[\mathbf{UR2}] \ast \det[\mathbf{G}(\mathbf{p2})])$
cells[45] under the PR2n1 partition.

Now adjust the ration n2/n1 such that

bn2(PRn1(i,j,k)∩Pn2) ≈ 1.

For such a ratio and number of cells in the PR2n2 partition equals the number of cells in the PR1n1 partition. Thus there exists a 1–1 correspondence between **r**(Pn1(i,,j,k)) and **r**(Pn2(i',j',k')).

Further

$(n1/n2)^3$ = det[**UR1**]∗det[**G**(**p1**)]/(det[**UR2**]∗det[**G**(**p2**)])

bn1(M∩Pn1)] ≈ bn2(M∩Pn2)].

Moreover,with the ratio n2/n1 maintained

r(Pn2(i',j',k')) ⟶ **r**(Pn1(i,j,k)) as n1 ⟶ ∞

and since **f** is continuous,

f(**r**(PR2n2(i',j',k'))) ⟶ **f**(**r**(PR1n1(i,j,k))).

Consequently,

lim S[bn1(M∩Pn1)](**f**(**r**(Pn(i,j,k)))/$n1^3$)

= lim S[bn2(M∩Pn2)](**f**(**r**(Pn(i',j',k')))/$n2^3$)

∗det[**UR2**]∗det[**G**(**p2**)]/(det[**UR1**]∗det[**G**(**p1**)]).

Consequently,

I[M]•(**f**;**dR2**) ≡ **I**[M](**f**(**r**)|**m1**)•**pR2**

∗det[G(**p1**)]∗det[**UR1**]/(det[G(**p2**)]∗det[**UR2**])

I[M]∧(**f**;**dR2**) ≡ **I**[M](**f**(**r**)|**m1**)∧**pR2**}

∗det[G(**p1**)]∗det[**UR1**]/(det[G(**p2**)]∗det[**UR2**])

I[M]∗(**f**;**dR2**) ≡ **I**[M](**f**(**r**)|**m1**)∗**pR2**}

∗det[G(**p1**)]∗det[**UR1**]/(det[G(**p2**)]∗det[**UR2**]).

qed

Although a specific partition vector must be referenced for a proper definition, the following generic properties of these integrals are true for any given partition vector, **pR1**.

If M is the intersection of all open sets containing **r1** and **f** is bounded at **r1**

I[**r1**]•(**f**|**m**;**dR1**) = lim (S[bn(**r1**∩Pn)](**f**•**m**(Pn(i,j,k))))

= 0

I[**r1**]∧(**f**|**m**;**dR1**) = lim (S[bn(**r1**∩Pn)](**f**∧**m**(Pn(i,j,k))))

= **0**

I[**r1**]∗(**f**|**m**;**dR1**) = lim (S[bn(**r1**∩Pn)](**f**∗**m**(Pn(i,j,k))))

= [**0**].

(758)

Mathematics: Vector Integration

If **c** is a constant vector, for any measurable set M

$$\begin{aligned}
\mathrm{I}[M]\cdot(\mathbf{c}|\mathbf{m};\mathbf{dR1}) &= \lim\,(S[bn(M\cap Pn)](\mathbf{c}\cdot\mathbf{m}(Pn(i,j,k))))\\
&= \mathbf{c}\cdot\mathbf{m}(M)\\
&= m(M)\ast\mathbf{c}\cdot\mathbf{pR1}
\end{aligned}\tag{759}$$

$$\begin{aligned}
\mathbf{I}\wedge[M]\wedge(\mathbf{c}|\mathbf{m};\mathbf{dR1}) &= \lim\,(S[bn(M\cap Pn)](\mathbf{c}\wedge\mathbf{m}(Pn(i,j,k))))\\
&= \mathbf{c}\wedge\mathbf{m}(M)\\
&= m(M)\ast\mathbf{c}\wedge\mathbf{pR1}
\end{aligned}\tag{760}$$

$$\begin{aligned}
\mathbf{I}[M]\ast(\mathbf{c}|\mathbf{m};\mathbf{dR1}) &= \lim\,(S[bn(M\cap Pn)](\mathbf{c}\ast\mathbf{m}(Pn(i,j,k))))\\
&= \mathbf{c}\ast\mathbf{m}(M)\\
&= m(M)\ast\mathbf{c}\ast\mathbf{pR1}.
\end{aligned}\tag{761}$$

In particular

$$\mathbf{I}[M]\ast(1|\mathbf{m};\mathbf{dR1}) = \mathbf{m}(M).\tag{762}$$

For constant c1, **c1** and c2, **c2** with f1, **f1** and f2, **f2** bounded on M,

$$\begin{aligned}
&\mathrm{I}[M]\cdot(c1\ast\mathbf{f1}+c2\ast\mathbf{f2}|\mathbf{m};\mathbf{dR1})\\
&\qquad= c1\ast\mathrm{I}[M]\cdot(\mathbf{f1}|\mathbf{m};\mathbf{dR1}) + c2\ast\mathrm{I}[M]\cdot(\mathbf{f2}|\mathbf{m};\mathbf{dR1})
\end{aligned}\tag{763}$$

$$\begin{aligned}
&\mathbf{I}[M]\cdot(\mathbf{c1}\ast f1+\mathbf{c2}\ast f2|\mathbf{m};\mathbf{dR1})\\
&\qquad= \mathbf{c1}\cdot\mathbf{I}[M](f1|\mathbf{m};\mathbf{dR1}) + \mathbf{c2}\cdot\mathbf{I}[M]\ast(f2|\mathbf{m};\mathbf{dR1})
\end{aligned}\tag{764}$$

$$\begin{aligned}
&\mathrm{I}[M]\cdot(\mathbf{c1}\wedge\mathbf{f1}+\mathbf{c2}\wedge\mathbf{f2}|\mathbf{m};\mathbf{dR1})\\
&\qquad= \mathbf{c1}\cdot\mathrm{I}[M]\wedge(\mathbf{f1}|\mathbf{m};\mathbf{dR1}) + \mathbf{c2}\cdot\mathrm{I}[M]\wedge(\mathbf{f2}|\mathbf{m};\mathbf{dR1})
\end{aligned}\tag{765}$$

$$\begin{aligned}
&\mathbf{I}[M]\cdot(\mathbf{c1}\ast\mathbf{f1}+\mathbf{c2}\ast\mathbf{f2}|\mathbf{m};\mathbf{dR1})\\
&\qquad= \mathbf{c1}\ast\mathrm{I}[M]\cdot(\mathbf{f1}|\mathbf{m};\mathbf{dR1}) + \mathbf{c2}\ast\mathrm{I}[M]\cdot(\mathbf{f2}|\mathbf{m};\mathbf{dR1})
\end{aligned}\tag{766}$$

$$\begin{aligned}
&\mathbf{I}[M]\wedge(c1\ast\mathbf{f1}+c2\ast\mathbf{f2}|\mathbf{m};\mathbf{dR1})\\
&\qquad= c1\ast\mathbf{I}[M]\wedge(\mathbf{f1}|\mathbf{m};\mathbf{dR1}) + c2\ast\mathbf{I}[M]\wedge(\mathbf{f2}|\mathbf{m};\mathbf{dR1})
\end{aligned}\tag{767}$$

$$\begin{aligned}
&\mathbf{I}[M]\wedge(\mathbf{c1}\ast f1+\mathbf{c2}\ast f2|\mathbf{m};\mathbf{dR1})\\
&\qquad= \mathbf{c1}\wedge\mathbf{I}[M](f1|\mathbf{m};\mathbf{dR1}) + \mathbf{c2}\wedge\mathbf{I}[M]\ast(f2|\mathbf{m};\mathbf{dR1})
\end{aligned}\tag{768}$$

$$\begin{aligned}
&\mathbf{I}[M]\wedge(\mathbf{c1}\wedge\mathbf{f1}+\mathbf{c2}\wedge\mathbf{f2}|\mathbf{m};\mathbf{dR1})\\
&\qquad= \mathbf{c1}\cdot\mathbf{T}(\mathbf{I}[M]\ast(\mathbf{f1}|\mathbf{m};\mathbf{dR1})) - \mathbf{c1}\cdot\mathbf{I}[M]\ast(\mathbf{f1}|\mathbf{m};\mathbf{dR1})\\
&\qquad\quad + \mathbf{c2}\cdot\mathbf{T}(\mathbf{I}[M]\ast(\mathbf{f2}|\mathbf{m};\mathbf{dR1}))\\
&\qquad\quad - \mathbf{c2}\cdot\mathbf{I}[M]\ast(\mathbf{f2}|\mathbf{m};\mathbf{dR1})
\end{aligned}\tag{769}$$

$$\mathbf{I}[M]*(c1*f1+c2*f2|\mathbf{m};\mathbf{dR1}) = c1*\mathbf{I}[M]*(f1|\mathbf{m};\mathbf{dR1}) + c2*\mathbf{I}[M]*(f2|\mathbf{m};\mathbf{dR1}) \tag{770}$$

$$\mathbf{I}[M]*(\mathbf{c1}\cdot\mathbf{f1}+\mathbf{c2}\cdot\mathbf{f2}|\mathbf{m};\mathbf{dR1}) = \mathbf{c1}\cdot\mathbf{I}[M]*(\mathbf{f1}|\mathbf{m};\mathbf{dR1}) + \mathbf{c2}\cdot\mathbf{I}[M]*(\mathbf{f2}|\mathbf{m};\mathbf{dR1}) \tag{771}$$

$$\mathbf{I}[M]*(c1*\mathbf{f1}+c2*\mathbf{f2}|\mathbf{m};\mathbf{dR1}) = c1*\mathbf{I}[M]*(\mathbf{f1}|\mathbf{m};\mathbf{dR1}) + c2*\mathbf{I}[M]*(\mathbf{f2}|\mathbf{m};\mathbf{dR1}) \tag{772}$$

$$\mathbf{I}[M]*(\mathbf{c1}*f1+\mathbf{c2}*f2|\mathbf{m};\mathbf{dR1}) = \mathbf{c1}*\mathbf{I}[M]*(f1|\mathbf{m};\mathbf{dR1}) + \mathbf{c2}*\mathbf{I}[M]*(f2|\mathbf{m};\mathbf{dR1}) \tag{773}$$

$$\mathbf{I}[M]*(\mathbf{c1}\wedge\mathbf{f1}+\mathbf{c2}\wedge\mathbf{f2}|\mathbf{m};\mathbf{dR1}) = \mathbf{C}(\mathbf{c1})\cdot\mathbf{I}[M]*(\mathbf{f1}|\mathbf{m};\mathbf{dR1})+ \mathbf{C}(\mathbf{c2})\cdot\mathbf{I}[M]*(\mathbf{f2}|\mathbf{m};\mathbf{dR1}). \tag{774}$$

Since the partition vector is specified in the notation, the reference to the measure in the symbol for integration is usually suppressed.

For a subset of functions with basic derivatives and integrals, the following theorem applies.

Theorem 10 (interior cancellation)

Given

$\mathbf{f}$, a function with a basic, continuous gradient $\mathbf{D}[\mathbf{r}]*(\mathbf{f};\mathbf{dr})$ in V3;

$\mathbf{pR1} = p1*\mathbf{u1R1} + p2*\mathbf{u2R1} + p3*\mathbf{u3R1}$, a partition vector;

$\mathbf{I}[V3]\cdot(\mathbf{f};\mathbf{dR1}) = g$ bounded;

for

$\mathbf{dR1} = dR1*\mathbf{uR1};$

$= d1R1*\mathbf{u1R1} + d2R1*\mathbf{u2R1} + d3R1*\mathbf{u3R1}$

$= \mathbf{pR1}/n^3$

$\mathbf{UR1} = \mathbf{u1}*\mathbf{u1R1} + \mathbf{u2}*\mathbf{u2R1} + \mathbf{u3}*\mathbf{u3R1};$

then

$$\mathbf{D}[\mathbf{r}]*(\mathbf{I}[V3]\cdot(\mathbf{f};\mathbf{dR1});\mathbf{dr}) = \mathbf{I}[V3]\cdot(\mathbf{D}[\mathbf{r}]*(\mathbf{f};\mathbf{dr});\mathbf{dR1}) = \mathbf{0}. \tag{775}$$

Proof:

Since g is constant, $\mathbf{D}[\mathbf{r1}]*(g;\mathbf{dr}) = \mathbf{0}$.

Again, the bounded integral implies $\mathbf{f}(r*\mathbf{ur})\longrightarrow\mathbf{0}$ as $r\longrightarrow\infty$ for any $\mathbf{ur}$.

Moreover,

I[V3]•(**D**[**r1**]∗(**f**;**dr**);**dR1**)
= lim S[bn(V3∩Pn)](**pR1**•**T**[**D**[**r**(Pn(i,j,k))]∗(**f**;**dr**)])/n^3
= lim S[bn(V3∩Pn)](**pR1**•[**UR1**$^{-1}$]•**T**[**UR1**]$^{-1}$
•**T**[**D**[**r**(Pn(i,j,k))]∗(**f**;**dR1**)])/n^3
= lim S[bn(V3∩Pn)]
(d1R1∗(**f**(**r**(Pn(i,j,k)+d1R1∗**u1R1**)) − **f**(**r**(Pn(i,j,k))))/d1R1
+ d2R1∗(**f**(**r**(Pn(i,j,k)+d2R1∗**u2R1**))
− **f**(**r**(Pn(i,j,k))))/d2R1
+ d3R1∗(**f**(**r**(Pn(i,j,k)+d3R1∗**u3R1**))
− **f**(**r**(Pn(i,j,k))))/d3R1)
= lim S[bn(V3∩Pn)](**f**(**r**(Pn(i+1,j,k))) − **f**(**r**(Pn(i,j,k)))
+ **f**(**r**(Pn(i,j+1,k))) − **f**(**r**(Pn(i,j,k)))
+ **f**(**r**(Pn(i,j,k+1))) − **f**(**r**(Pn(i,j,k))))

observing

r(Pn(i,j,k)+d1R1∗**u1R1**) = **r**(Pn(i,j,k)+p1R1∗**u1R1**/n^3)
= **r**(Pn(i+1,j,k)).

The sum over the index i, for illustration, becomes

(...+ **f**(**r**(Pn(i+2,j,k))) − **f**(**r**(Pn(i+1,j,k)))
+ **f**(**r**(Pn(i+1,j,k))) − **f**(**r**(Pn(i,j,k)))
+ **f**(**r**(Pn(i,j,k)))
− **f**(**r**(Pn(i−1,j,k)))
+...)
⟶ (**f**(r∗**u1R1**) + **f**(r∗(−**u1R1**))) as r⟶∞
= **0**.

The same is true for indices j and k.

The same is true for any **ur** = **u**(**r**(Pn(i,j,k))) since **ur** equals some weighted combination of the indices.

Therefore, **I**[V3]•(**D**[**r1**]∗(**f**;**dr**);**dR1**) = **0**. qed

The rule n^3∗dR1 = pR1 is called the **matched integration rule**[46].

Corollary

Given:

f(**r**), a function with a basic sectional gradient everywhere in V3 with bounded invergence;

then

I[V3]•(**D**[**r**]∗(**f**;**dR1**)•[**UR1**]$^{-1}$•**T**[**UR1**]$^{-1}$;**dR1**) = **0**. (776)

Suppose for a given partition based on **pR1**, a gradient **D**[**r**]∗(**f**;**dR2**) is constant over M. Then

I[M]•(**D**[**r**]∗(**f**;**dR2**);**dR1**)
=lim S[bn(M∩Pn)](**m**(Pn(i,j,k))•**T**[**D**[**r**(Pn(i,j,k))]∗(**f**;**dR2**)])
=lim S[bn(M∩Pn)](**m**(Pn(i,j,k)))•**T**[**D**[**r**]∗(**f**;**dR2**)]
=**m**(M|**pR1**)•**T**[**D**[**r**]∗(**f**;**dR2**)] (777)
= vol(M)∗**pR1**•**T**[**D**[**r**]∗(**f**;**dR2**)]/vol(P1(i,j,k)). (778)

Moreover, for **f**(**r**) a function with a basic, simply continuous *constant* gradient

I[M]•(**D**[**r**]∗(**f**;**dR2**)•[**UR2**]$^{-1}$•**T**[**UR2**]$^{-1}$;**dR1**)
= vol(M)∗**pR1**•[**UR2**]$^{-1}$•**T**[**UR2**]$^{-1}$•**T**[**D**[**r**]∗(**f**;**dR2**)]
/vol(P1(i,j,k))
= vol(M)∗**pR1**•**T**[**D**[**r**]∗(**f**;**dr**)]/vol(P1(i,j,k)). (779)

More generally, for **f** continuously differentiable

I[M]•(**D**[**r**]∗(**f**;**dR1**)•[**UR1**]$^{-1}$•**T**[**UR1**]$^{-1}$;**dR1**)
=lim S[bn(M∩Pn)](**m**(Pn(i,j,k))
•[**UR1**]$^{-1}$•**T**[**UR1**]$^{-1}$•**T**[**D**[**r**(Pn(i,j,k))]∗(**f**;**dR1**)])
=lim S[bn(M∩Pn)]((p1R1∗**u1** + p2R1∗**u2** + p3R1∗**u3**)
/n^{3})
•(**u1**∗(**f**(**r**(Pn(i,j,k))+d1R1∗**u1R1**) − **f**(**r**(Pn(i,j,k))))/d1R1
+ **u2**∗(**f**(**r**(Pn(i,j,k))+d2R1∗**u2R1**) − **f**(**r**(Pn(i,j,k))))/d2R1
+ **u3**∗(**f**(**r**(Pn(i,j,k))+d3R1∗**u3R1**) −**f**(**r**(Pn(i,j,k))))/d3R1)
=lim S[bn(M∩Pn)](**f**(**r**(Pn(i,j,k))+d1R1∗**u1R1**)
− **f**(**r**(Pn(i,j,k)))
+**f**(**r**(Pn(i,j,k))+d2R1∗**u2R1**) − **f**(**r**(Pn(i,j,k)))
+**f**(**r**(Pn(i,j,k))+d3R1∗**u3R1**) − **f**(**r**(Pn(i,j,k))))
=lim S[bn(M∩Pn)](**f**(**r**(Pn(i,j,k)+**dR1**)) − **f**(**r**(Pn(i,j,k))))
=lim (S[bn(SMPE|**pR1**∩Pn)](**f**(**r**(Pn(i,j,k))+**dR1**))
− S[bn(SMPI|**pR1**∩Pn)](**f**(**r**(Pn(i,j,k)))) (780)
≡ I[SMPI|**pR1**,SMPE|**pR1**](**f**(**sr**)). (781)

Equation (780) is the **fundamental theorem of vector calculus for sectional gradients over a set M in V3**.

Mathematics: Vector Integration

If further ***f*** is asymptotic

$$\begin{aligned}
&\mathbf{I}[\mathrm{V3}]\cdot(\mathbf{D}[\mathbf{r}]\ast(\mathbf{f};\mathbf{dR1})\cdot[\mathbf{UR1}]^{-1}\bullet\mathbf{T}[\mathbf{UR1}]^{-1};\mathbf{dR1})\\
&\qquad = \mathbf{I}[\mathrm{M}\cup\mathrm{SM}]\bullet(\mathbf{D}[\mathbf{r}]\ast(\mathbf{f};\mathbf{dR1})\bullet[\mathbf{UR1}]^{-1}\bullet\mathbf{T}[\mathbf{UR1}]^{-1};\mathbf{dR1})\\
&\qquad\quad + \mathbf{I}[\mathrm{V3}-(\mathrm{M}\cup\mathrm{SM})]\bullet(\mathbf{D}[\mathbf{r}]\ast(\mathbf{f};\mathbf{dR1})\bullet[\mathbf{UR1}]^{-1}\bullet\mathbf{T}[\mathbf{UR1}]^{-1};\mathbf{dR1})\\
&\qquad = \mathbf{0}. \qquad (782)
\end{aligned}$$

Thus for this case

$$\begin{aligned}
&\mathbf{I}[\mathrm{V3}-(\mathrm{M}\cup\mathrm{SM})]\bullet(\mathbf{D}[\mathbf{r}]\ast(\mathbf{f};\mathbf{dR1})\bullet[\mathbf{UR1}]^{-1}\bullet\mathbf{T}[\mathbf{UR1}]^{-1};\mathbf{dR1})\\
&\qquad = -\,\mathbf{I}[\mathrm{SMPI}|\mathbf{pR1},\mathrm{SMPE}|\mathbf{pR1}](\mathbf{f}(\mathbf{sr})). \qquad (783)
\end{aligned}$$

For functions for which cancellation operates as above, sectional divergences and curls are expressed similarly.

$$\begin{aligned}
&\mathbf{I}[\mathrm{M}]\cdot(\mathbf{D}[\mathbf{r}]\wedge(\mathbf{f};\mathbf{dR1});\mathbf{dR1})\\
&\qquad = \mathbf{I}[\mathrm{SMPI}|\mathbf{pR1},\mathrm{SMPE}|\mathbf{pR1}](\mathbf{D}[\mathbf{sr}]\wedge(\mathbf{f};\mathbf{dR1})\bullet\mathbf{dR1}) \qquad (784)\\
&\mathbf{I}[\mathrm{M}]\wedge(\mathbf{D}[\mathbf{r}]\wedge(\mathbf{f};\mathbf{dR1});\mathbf{dR1})\\
&\qquad = \mathbf{I}[\mathrm{SMPI}|\mathbf{pR1},\mathrm{SMPE}|\mathbf{pR1}](\mathbf{D}[\mathbf{sr}]\wedge(\mathbf{f};\mathbf{dR1})\wedge\mathbf{dR1}) \qquad (785)\\
&\mathbf{I}[\mathrm{M}]\ast(\mathbf{D}[\mathbf{r}]\bullet(\mathbf{f};\mathbf{dR1});\mathbf{dR1})\\
&\qquad = \mathbf{I}[\mathrm{SMPI}|\mathbf{pR1},\mathrm{SMPE}|\mathbf{pR1}](\mathbf{D}[\mathbf{sr}]\bullet(\mathbf{f};\mathbf{dR1})\ast\mathbf{dR1}) \qquad (786)\\
&\mathbf{I}[\mathrm{M}]\ast(\mathbf{D}[\mathbf{r}]\wedge(\mathbf{f};\mathbf{dR1});\mathbf{dR1})\\
&\qquad = \mathbf{I}[\mathrm{SMPI}|\mathbf{pR1},\mathrm{SMPE}|\mathbf{pR1}](\mathbf{D}[\mathbf{sr}]\wedge(\mathbf{f};\mathbf{dR1})\ast\mathbf{dR1}) \qquad (787)
\end{aligned}$$

and for scalar functions

$$\begin{aligned}
&\mathbf{I}[\mathrm{M}]\bullet(\mathbf{D}[\mathbf{r}]\ast(\mathrm{f};\mathbf{dR1})\bullet[\mathbf{UR1}]^{-1}\bullet\mathbf{T}[\mathbf{UR1}]^{-1};\mathbf{dR1})\\
&\qquad = \mathbf{I}[\mathrm{SMPI}|\mathbf{pR1},\mathrm{SMPE}|\mathbf{pR1}](\mathrm{f}(\mathbf{sr})) \qquad (788)\\
&\mathbf{I}[\mathrm{M}]\wedge(\mathbf{D}[\mathbf{r}]\ast(\mathrm{f};\mathbf{dR1});\mathbf{dR1})\\
&\qquad = \mathbf{I}[\mathrm{SMPI}|\mathbf{pR1},\mathrm{SMPE}|\mathbf{pR1}](\mathbf{D}[\mathbf{sr}]\ast(\mathrm{f};\mathbf{dR1})\wedge\mathbf{dR1}) \qquad (789)\\
&\mathbf{I}[\mathrm{M}]\ast(\mathbf{D}[\mathbf{r}]\ast(\mathrm{f};\mathbf{dR1});\mathbf{dR1})\\
&\qquad = \mathbf{I}[\mathrm{SMPI}|\mathbf{pR1},\mathrm{SMPE}|\mathbf{pR1}]((\mathbf{D}[\mathbf{sr}]\ast(\mathrm{f};\mathbf{dR1})\ast\mathbf{dR1}). \qquad (790)
\end{aligned}$$

These equations are extensions of the f**undamental theorem of the calculus for V3**.

Now let {Mi} be a properly nested sequence of subsets of M.

For **f**(**r**) be continuous over M, let

$$\mathbf{G}(\mathrm{Mi}) \equiv \mathbf{I}[\mathrm{Mi}]*(\mathbf{f};\mathbf{dR1}). \quad (791)$$

Then for Mi ⊂ Mj

$$\mathbf{I}[\mathrm{Mj} - \mathrm{Mi}]*(\mathbf{f};\mathbf{dR1}) = \mathbf{I}[\mathrm{Mj}]*(\mathbf{f};\mathbf{dR1}) - \mathbf{I}[\mathrm{Mi}]*(\mathbf{f};\mathbf{dR1})$$
$$= \mathbf{G}(\mathrm{Mj}) - \mathbf{G}(\mathrm{Mi}). \quad (792)$$

STEP FUNCTIONS IN V3

In V3 the idea of order is incompatible with the usual topology. As substitute consider the interior, exterior, and surface cells of a given measurable set M in a given partition. Then *V3* can be separated into parts over whose surface *SM* the discontinuity in step functions may be defined.

It is useful to name the following sets

EMPEF|**pR1** ≡ {**r**|**r**=**res**+dR1∗**uR1**, **res** in SMPE|**pR1**; dR1>0} (793)
EMPEB|**pR1** ≡ {**r**|**r**=**res**+dR1∗**uR1**, **res** in SMPE|**pR1**; dR1<0} (794)
EMPIF|**pR1** ≡ {**r**|**r**=**ris**+dR1∗**uR1**, **ris** in SMPI|**pR1**; dR1>0} (795)
EMPIB|**pR1** ≡ {**r**|**r**=**ris**+dR1∗**uR1**, **ris** in SMPI|**pR1**; dR1<0} (796)
EMNEF|**pR1** ≡ {**r**|**r**=**res**+dR1∗**uR1**, **res** in SMNE|**pR1**; dR1>0} (797)
EMNEB|**pR1** ≡ {**r**|**r**=**res**+dR1∗**uR1**, **res** in SMNE|**pR1**; dR1<0} (798)
EMNIF|**pR1** ≡ {**r**|**r**=**ris**+dR1∗**uR1**, **ris** in SMNI|**pR1**; dR1>0} (799)
EMNIB|**pR1** ≡ {**r**|**r**=**ris**+dR1∗**uR1**, **ris** in SMNI|**pR1**; dR1<0}.(800)

Mathematics: Vector Integration

Definition 46 (point step functions in V3)

Given

$\mathbf{pR1} \equiv pR1*\mathbf{uR1}$, a partition vector;

Pn, a set of nested partitions based on **pR1**;

M, a measurable set;

SM, the surface of M;

then

for a given **ser** in SMPE|**pR1**

$$u(\mathbf{r}-\mathbf{ser}|M) \equiv 0, \quad \mathbf{r} = \mathbf{ser}+dR1*\mathbf{uR1},\ dR1 \le 0$$
$$\equiv 1, \quad \text{otherwise} \qquad (801)$$
$$v(\mathbf{r}-\mathbf{ser}|M) \equiv 0, \quad \mathbf{r} = \mathbf{ser}+dR1*\mathbf{uR1},\ dR1 < 0$$
$$\equiv 1, \quad \text{otherwise} \qquad (802)$$

for a given **sir** in SMPI|**pR1**

$$u(\mathbf{r}-\mathbf{sir}|M) \equiv 0, \quad \mathbf{r} = \mathbf{sir}+dR1*\mathbf{uR1},\ dR1 \le 0$$
$$\equiv 1, \quad \text{otherwise} \qquad (803)$$
$$v(\mathbf{r}-\mathbf{sir}|M) \equiv 0, \quad \mathbf{r} = \mathbf{sir}+dR1*\mathbf{uR1},\ dR1 < 0$$
$$\equiv 1, \quad \text{otherwise} \qquad (804)$$

are **unit point step functions in V3**.

end of definition

Now let

$$\mathbf{dR1} \equiv dR1*\mathbf{uR1} \qquad (805)$$

For a given **ser** in SMPE|**pR1**

$$\mathbf{D}[\mathbf{r},\mathbf{r}+\mathbf{dR1}]*(u(\mathbf{r}-\mathbf{ser})|M;\mathbf{dR1}) = \mathbf{0} \qquad \mathbf{r} \neq \mathbf{ser} \qquad (806)$$
$$\mathbf{D}[\mathbf{r},\mathbf{r}-\mathbf{dR1}]*(u(\mathbf{r}-\mathbf{ser})|M;\mathbf{dR1}) = \mathbf{0} \qquad \mathbf{r} \neq \mathbf{ser} \qquad (807)$$
$$\mathbf{D}[\mathbf{r},\mathbf{r}+\mathbf{dR1}]*(u(\mathbf{r}-\mathbf{ser})|M;\mathbf{dR1}) = \lim \mathbf{uR1}/dR1 \qquad \mathbf{r} = \mathbf{ser} \qquad (808)$$
$$\mathbf{D}[\mathbf{r},\mathbf{r}-\mathbf{dR1}]*(u(\mathbf{r}-\mathbf{ser})|M;\mathbf{dR1}) = \mathbf{0} \qquad \mathbf{r} = \mathbf{ser}. \qquad (809)$$

Similarly,

$$\mathbf{D}[\mathbf{r},\mathbf{r}+\mathbf{dR1}]*(v(\mathbf{r}-\mathbf{ser})|M;\mathbf{dR1}) = \mathbf{0} \qquad \mathbf{r} \neq \mathbf{ser} \qquad (810)$$
$$\mathbf{D}[\mathbf{r},\mathbf{r}-\mathbf{dR1}]*(v(\mathbf{r}-\mathbf{ser})|M;\mathbf{dR1}) = \mathbf{0} \qquad \mathbf{r} \neq \mathbf{ser} \qquad (811)$$
$$\mathbf{D}[\mathbf{r},\mathbf{r}+\mathbf{dR1}]*(v(\mathbf{r}-\mathbf{ser})|M;\mathbf{dR1}) = \mathbf{0} \qquad \mathbf{r} = \mathbf{ser} \qquad (812)$$
$$\mathbf{D}[\mathbf{r},\mathbf{r}-\mathbf{dR1}]*(v(\mathbf{r}-\mathbf{ser})|M;\mathbf{dR1}) = -\lim \mathbf{uR1}/dR1 \qquad \mathbf{r} = \mathbf{ser}. \qquad (813)$$

A similar result holds for any **ser** in SMPE|**pR1**

Now let ***f*** be a generalized function over V3.

For **ser** in SMPE|**pR1**

D[**r**,**r+dR1**]∗(**f**(**r**)∗u(**r**−**ser**)|M;**dR1**)
  = **D**[**r**,**r+dR1**]∗(**f**;**dR1**) **r** ≠ **ser**+dR1∗**uR1**
  = **D**[**r**,**r+dR1**]∗(**f**;**dR1**) **r** = **ser**+dR1∗**uR1** dR1 > 0
  = [**0**] **r** = **ser**+dR1∗**uR1** dR1 < 0
  = lim **f**(**ser+dR1**)∗**uR1**/dR1 **r** = **ser** (814)

D[**r**,**r**−**dR1**]∗(**f**(**r**)∗u(**r**−**ser**)|M;**dR1**)
  = **D**[**r**,**r**−**dR1**]∗(**f**;**dR1**) **r** ≠ **ser**+dR1∗**uR1**
  = **D**[**r**,**r**−**dR1**]∗(**f**;**dR1**) **r** = **ser**+dR1∗**uR1** dR1 > 0
  = [**0**] **r** = **ser**+dR1∗**uR1** dR1 ≤ 0 (815)

f(**r**)∗**D**[**r**,**r+dR1**]∗(u(**r**−**ser**)|M;**dR1**)
  = [**0**] **r** ≠ **ser**
  = lim **f**(**ser**)∗**uR1**/dR1 **r** = **ser** (816)

f(**r**)∗**D**[**r**,**r**−**dR1**]∗(u(**r**−**ser**)|M;**dR1**)
  = [**0**] (817)

(u(**r**−**ser**)|M)∗**D**[**r**,**r+dR1**]∗(**f**;**dR1**)
  = **D**[**r**,**r+dR1**]∗(**f**;**dR1**) **r** ≠ **ser**+dR1∗**uR1**
  = **D**[**r**,**r+dR1**]∗(**f**;**dR1**) **r** = **ser**+dR1∗**uR1** dR1>0
  = [**0**] **r** = **ser**+dR1∗**uR1** dR1≤ 0 (818)

(u(**r**−**ser**)|M)∗**D**[**r**,**r**−**dR1**]∗(**f**;**dR1**)
  = **D**[**r**,**r**−**dR1**]∗(**f**;**dR1**), **r** ≠ **ser**+dR1∗**uR1**
  = **D**[**r**,**r**−**dR1**]∗(**f**;**dR1**), **r** = **ser**+dR1∗**uR1** , dR1 > 0
  = [**0**], **r** = **ser**+dR1 ∗**uR1**, dR1≤ 0.
  (819)

Likewise for v(**r**−**ser**)|M

D[**r**,**r+dR1**]∗(**f**(**r**)∗v(**r**−**ser**)|M;**dR1**)
  = **D**[**r**,**r+dR1**]∗(**f**;**dR1**) **r** ≠ **ser**+dR1∗**uR1**
  = **D**[**r**,**r+dR1**]∗(**f**;**dR1**) **r** = **ser**+dR1∗**uR1** dR1 ≥ 0
  = **0**, **r** = **ser**+dR1∗**uR1** dR1 < 0 (820)

D[**r**,**r**−**dR1**]∗(**f**(**r**)∗v(**r**−**ser**)|M;**dR1**)
  = **D**[**r**,**r**−**dR1**]∗(**f**;**dR1**) **r** ≠ **ser**+dR1∗**uR1**
  = **D**[**r**,**r**−**dR1**]∗(**f**;**dR1**) **r** = **ser**+dR1∗**uR1** dR1 > 0
  = [**0**] **r** = **ser**+dR1∗**uR1** dR1 < 0
  = lim −**f**(**ser**)∗**uR1**/dR1 **r** = **ser** (821)

Mathematics: Vector Integration

f(**r**)∗**D**[**r**,**r**+**dR1**]∗(v(**r**−**ser**)|M;**dR1**)
  = [**0**] (822)

f(**r**)∗**D**[**r**,**r**−**dR1**]∗(v(**r**−**ser**)|M;**dR1**)
  = [**0**] **r** ≠ **ser**
  = lim −**f**(**ser**)∗**uR1**/dR1 **r** = **ser** (823)

(v(**r**−**ser**)|M)∗**D**[**r**,**r**+**dR1**]∗(**f**;**dR1**)
  = **D**[**r**,**r**+**dR1**]∗(**f**;**dR1**) **r** ≠ **ser**+dR1∗**uR1**
  = **D**[**r**,**r**+**dR1**]∗(**f**;**dR1**) **r** = **ser**+dR1∗**uR1** dR1 ≥ 0
  = [**0**] **r** = **ser**+dR1∗**uR1** dR1 < 0 (824)

(v(**r**−**ser**)|M)∗**D**[**r**,**r**−**dR1**]∗(**f**;**dR1**)
  = **D**[**r**,**r**−**dR1**]∗(**f**;**dR1**) **r** ≠ **ser**+dR1∗**uR1**
  = **D**[**r**,**r**−**dR1**]∗(**f**;**dR1**) **r** = **ser**+dR1∗**uR1** dR1 ≥ 0
  = [**0**] **r** = **ser**+dR1∗**uR1** dR1 < 0. (825)

Similar results hold for **sir** in SMPI|**pR1**.

For ***f*** continuous at ***ser*** or ***sir***

D[**r**,**r**+**dR1**]∗(**f**(**r**)∗u(**r**−**ser**)|M;**dR1**)
  = **f**(**r**)∗**D**[**r**,**r**+**dR1**]∗(u(**r**−**ser**)|M;**dR1**)
    + v(**r**−**ser**)|M∗**D**[**r**,**r**+**dR1**]∗(**f**;**dR1**) (826)

D[**r**,**r**+**dR1**]∗(**f**(**r**)∗v(**r**−**ser**)|M;**dR1**)
  = **f**(**r**)∗**D**[**r**,**r**+**dR1**]∗(v(**r**−**ser**)|M;**dR1**)
    + v(**r**−**ser**)|M∗**D**[**r**,**r**+**dR1**]∗(**f**;**dR1**) (827)

D[**r**,**r**−**dR1**]∗(**f**(**r**)∗u(**r**−**ser**)|M;**dR1**)
  = **f**(**r**)∗**D**[**r**,**r**−**dR1**]∗(u(**r**−**ser**)|M;**dR1**)
    + u(**r**−**ser**)|M∗**D**[**r**,**r**−**dR1**]∗(**f**;**dR1**) (828)

D[**r**,**r**−**dR1**]∗(**f**(**r**)∗v(**r**−**ser**)|M;**dR1**)
  = **f**(**r**)∗**D**[**r**,**r**−**dR1**]∗(v(**r**−**ser**)|M;**dR1**)
    + v(**r**−**ser**)|M∗**D**[**r**,**r**−**dR1**]∗(**f**;**dR1**). (829)

Corresponding statements may be made for ***sir*** in SMPI.

Now let ***f*** be a generalized function over V3 and let

pR1 ≡ pR1∗**uR1** (830)

be a partition vector with corresponding set of partitions Pn.

Then

lim(S[bn(V3∩Pn)](**f**(**r**(Pn(i,j,k)))∗**D**[**r**,**r+dR1**]∗(u(**r**−**ser**)|M;**dR1**)
•**m**(Pn(i,j,k))
= lim lim **f(ser)**∗**pR1**/(n^3 ∗**dR1**), n^3∗**dR1** = **pR1**.

With the limits so linked,

I[V3]•(**f**(**r**)∗**D**[**r**,**r+dR1**]∗(u(**r**−**ser**)|M;**dR1**);**dR1**)
= **f**(**ser**) **ser** in SMPE|**pR1**. (831)

Likewise,

I[V3]•(**f**(**r**)∗**D**[**r**,**r+dR1**]∗(u(**r**−**sir**)|M;**dR1**);**dR1**)
= **f**(**sir**) **sir** in SMPI|**pR1** (832)
I[V3]•(**f**(**r**)∗**D**[**r**,**r**−**dR1**]∗(v(**r**−**ser**)|M;**dR1**);**dR1**)
= −**f**(**ser**) **ser** in SMPE|**pR1** (833)
I[V3]•(**f**(**r**)∗**D**[**r**,**r**−**dR1**]∗(v(**r**−**sir**)|M;**dR1**);**dR1**)
= −**f**(**sir**) **sir** in SMPI|**pR1**. (834)

These equations may be taken as functions over subsets of of SM.

Now let ***f*** be continuous over V3.

I[V3]•(u(**r**−**ser**)∗**D**[**r**,**r+dR1**]∗(**f**(**r**);**dR1**)|M;**dR1**)
= lim S[bn(V3∩Pn)]((u(**r**(Pn(i,j,k))−**ser**)|M)
∗**D**[**r**(Pn(i,j,k)),**r**(Pn(i,j,k));**dR1**]∗(**f**(**r**);**dR1**)•**m**(Pn(i,j,k)))
= lim lim −**f**(**ser+dR1**)∗**pR1**/(n^3 ∗**dR1**), n^3∗dR1 = pR1
= −**f**(**ser**). (835)

Likewise,
I[V3]•(u(**r**−**sir**)∗**D**[**r**,**r+dR1**]∗(**f**(**r**)|M;**dR1**);**dR1**)
= −**f**(**sir**) (836)
I[V3]•(v(**r**−**ser**)∗**D**[**r**,**r**−**dR1**]∗(**f**(**r**)|M;**dR1**);**dR1**)
= **f**(**ser**) (837)
I[V3]•(v(**r**−**sir**)∗**D**[**r**,**r**−**dR1**]∗(**f**(**r**)|M;**dR1**);**dR1**)
= **f**(**sir**). (838)

Mathematics: Vector Integration

Further still for I[V3]•(**f**;**dR1**) bounded,

I[V3]•(**D**[**r**,**r+dR1**]∗(((u(**r**–**ser**)|M)∗**f**);**dR1**);**dR1**)
= lim lim S[bn(V3**∩**Pn)]
(**f**(**r**(Pn(i+1,j+1,k+1)))
∗(u(**r**(Pn(i+1,j+1,k+1))–**ser**)|M)
–**f**(**r**(Pn(i,j,k)))
∗(u(**r**(Pn(i,j,k))–**ser**)|M))
= **f**(**r**(Pn(i1+1,j1+1,k1+1)))
∗(u(**r**(Pn(i1+1,j1+1,k1+1))–**ser**)|M)
–f(**r**(Pn(i1,j1,k1)))
∗(u(**r**(Pn(i1,j1,k1))–**ser**)|M)
+ f(**r**(Pn(i1+2,j1+2,k1+2)))
∗(u(**r**(Pn(i1+2,j1+2,k1+2))–**ser**)|M)
–f(**r**(Pn(i1+1,j1+1,k1+1)))
∗(u(**r**(Pn(i1+1,j1+1,k1+1))–**ser**)|M)
+ ...
= **0**. (839)

Likewise,

I[V3]•(**D**[**r**,**r+dR1**]∗((u(**r**–**sir**)|M)∗**f**;**dR1**);**dR1**) = **0** (840)
I[V3]•(**D**[**r**,**r**–**dR1**]∗((v(**r**–**ser**)|M)∗**f**;**dR1**);**dR1**) = **0** (841)
I[V3]•(**D**[**r**,**r**–**dR1**]∗((v(**r**–**sir**)|M)∗**f**;**dR1**);**dR1**) = **0**. (842)

Other integral operations are possible.

I[V3]•(**f**(**r**)∧**D**[**r**,**r+dR1**]∗(u(**r–ser**)|M;**dR1**);**dR1**)
= lim (**f**(**ser**)∧**D**[**r**]∗(u(**r–ser**)|M);**dR1**)•**m**(Pn(i,j,k))
= **f**(**ser**)•(**D**[**r**,**r+dR1**]∗(u(**r–ser**)|M;**dR1**)∧**m**(Pn(i,j,k)))
= 0 (843)
I[V3]∧(f(**r**)∗**D**[**r**,**r+dR1**]∗(u(**r–ser**)|M;**dR1**);**dR1**)
= **0** (844)
I[V3]∧(**f**(**r**)∧**D**[**r**,**r+dR1**]∗(u(**r–ser**)|M;**dR1**);**dR1**)
= **f**(**ser**)•((**uR1**∗**uR1**) – **I**) (845)
I[V3]∗(f(**r**)∗**D**[**r**,**r+dR1**]∗(u(**r–ser**)|M;**dR1**);**dR1**)
= f(**ser**)∗(**uR1**∗**uR1**) (846)
I[V3]∗(f(**r**)∧**D**[**r**,**r+dR1**]∗(u(**r–ser**)|M;**dR1**);**dR1**)
= **C**(**f**(**ser**))•(**uR1**)∗(**uR1**) (847)

I[V3]∗(**f**(**r**)•**D**[**r**,**r+dR1**]∗(u(**r**−**ser**)|M;**dR1**);**dR1**)
= **f**(**ser**)•(**uR1**∗**uR1**). (848)

Similar results[47] hold for v(**r**−**ser**) and ***sir***.

The above development may be repeated for sectional gradients with additional complications.[48]

Definition 47 (local step functions in V3)
Given
pR1 ≡ pR1∗**uR1,** a partition vector;
Pn, a set of nested partitions based on **pR1**;
M, a measurable set;
SM, the surface of M;
then

u(**r**)|M ≡ 0, **r** in M∪SM
≡ 1, otherwise (849)
v(**r**)|M ≡ 0, **r** in M−SM
≡ 1, otherwise (850)

are **local unit step functions in V3**
end of definition

Now let
dR1 ≡ d1R1∗**u1R1** + d2R1∗**u2R1** + d3R1∗**u3R1** (851)

Then
D[**r**]∗(u(**r**|M);**dR1**) = **0** **r** in M−SM (852)
D[**r**]∗(u(**r**|M);**dR1**) = **0** **r** in V3−(M∪SM) (853)
D[**r**]∗(u(**r**|M);**dR1**) = lim u(**r**+d1R1∗**u1R1**|M)∗**u1R1**/d1R1
+ u(**r**+d2R1∗**u2R1**|M)∗**u2R1**/d2R1
+ u(**r**+d3R1∗**u3R1**|M)∗**u3R1**/d3R1
r in SM. (854)

For **r** in SM then, **D**[**r**]∗(u(**r**|M);**dR1**) may equal **0**, or may be impulsive in one, two, or all three sectional directions.

Mathematics: Vector Integration

Similarly,

D[**r**]∗(v(**r**|M);**dR1**) = **0**   **r** in M–SM   (855)
D[**r**]∗(v(**r**|M);**dR1**) = **0**   **r** in V3–(M∪SM)   (856)
D[**r**]∗(v(**r**|M);**dR1**) = lim ((v(**r**+d1R1∗**u1R1**|M)–)∗**u1R1**)/d1R1
+ ((v(**r**+d2R1∗**u2R1**|M)–1)∗**u2R1**)/d2R1
+ ((v(**r**+d3R1∗**u3R1**|M)–1)∗**u3R1**)/d3R1,
r in SM.   (857)

For **r** in SM then, **D**[**r**]∗(v(**r**|M);**dR1**) may equal **0**, or may be impulsive in one, two, or all three sectional directions.

Let ***sr*** be an element of SM and **r** = **sr**+diR1∗**uiR1**. The following table gives the relationship between the differences in u(**r**|M) and v(**r**|M).

Location of r	u(**r**|M)–u(**sr**|M)	v(**r**|M)–v(**sr**|M)
V3–(M∪SM)	1	0
SM	0	0
M–SM	0	–1

Local Step Functions at Surfaces

Thus where the gradient of u(**sr**|M) is impulsive, it is always positively impulsive, whereas where the gradient of v(**sr**|M) is impulsive, it is always negatively impulsive. Moreover in those directions where the gradient of u(**sr**|M) is impulsive, the gradient of v(**sr**|M) is zero, and vice–versa.

Clearly, **D**[**r**]∗(u(**r**|M);**dR1**) and **D**[**r**]∗(v(**r**|M);**dR1**) are *not* continuously differentiable.

Now let ***f*** be a generalized function over V3.

D[**r**]∗(**f**(**r**)∗u(**r**)|M;**dR1**)
= **D**[**r**]∗(**f**;**dR1**)   **r** in V3–(M∪SM)
= lim **f**(**r**+d1R1∗**u1R1**)∗u(**r**+d1R1∗**u1R1**)∗**u1R1**/d1R1
+ **f**(**r**+d2R1∗**u2R1**)∗u(**r**+d2R1∗**u2R1**)∗**u2R1**/d2R1
+**f**(**r**+d3R1∗**u3R1**)∗u(**r**+d3R1∗**u3R1**)∗**u3R1**/d3R1
r in SM
= [**0**]   **r** in M–SM   (858)

f(r)*D[r]*(u(r)|M;dR1)
= [0] r not in SM
= f(r)* (lim u(r+d1R1*u1R1)*u1R1/d1R1
+ u(r+d2R1*u2R1)*u2R1/d2R1
+ u(r+d3R1*u3R1)*u3R1/d3R1)
r in SM (859)

(u(r)|M)*D[r]*(f;dR1)
= D[r]*(f;dR1) r in V3–(MuSM)
= [0] r in SM
= [0] r in M–SM (860)

D[r]*(f(r)*v(r)|M;dR1)
= D[r]*(f;dR1) r in V3–(MuSM)
= lim (f(r+d1R1*u1R1)*v(r+d1R1*u1R1)*u1R1–f(r))
/d1R1
+(f(r+d2R1*u2R1)*v(r+d2R1*u2R1)*u2R1–f(r))
/d2R1
+(f(r+d3R1*u3R1)*v(r+d3R1*u3R1)*u3R1–f(r))
/d3R1
r in SM
= [0] r in M–SM (861)

f(r)*D[r]*(v(r)|M;dR1)
= [0] r not in SM
= f(r)* (lim (v(r+d1R1*u1R1)–1)*u1R1/d1R1
+ (v(r+d2R1*u2R1)–1)*u2R1/d2R1
+(v(r+d3R1*u3R1)–1)*u3R1/d3R1)
r in SM (862)

(v(r)|M)*D[r]*(f;dR1)
= D[r]*(f;dR1) r in V3–(MuSM)
= D[r]*(f;dR1) r in SM
= [0] r in M–SM. (863)

Mathematics: Vector Integration

Now let

pR1 ≡ p1R1∗**u1R1** + p21R1∗**u2R1** +p3R1∗**u3R1** (864)

be a partition vector with corresponding set of partitions Pn and

dR1 ≡ **u1R1**/d1R1 + **u2R1**/d2R1 +**u3R1**/d3R1 (865)
p1 ≡ p1R1∗**u1** + p2R1∗**u2** +p3R1∗**u3** (866)
d1 ≡ **u1**/d1R1 + **u2**/d1R1 +**u3**/d1R1. (867)

Then

lim S[bn(V3∩Pn)](**f**(**r**(Pn(i,j,k)))
∗(**D**[**r**(Pn(i,j,k)))]∗(u(**r**)|M;**dR1**)
+ **D**[**r**(Pn(i,j,k))]∗(v(**r**)|M;**dR1**))
•[**UR1**]$^{-1}$•**T**[**UR1**]$^{-1}$
•**m**(Pn(i,j,k))
= lim S[bn(SM∩Pn)](**f**(**sr**(Pn(i,j,k)))
∗(**D**[**sr**(Pn(i,j,k))]∗(u(**r**)|M;**dR1**)
+ **D**[**sr**(Pn(i,j,k))]∗(v(**r**)|M;**dR1**))
•**m**(Pn(i,j,k)))
= lim S[bn(SM∩Pn)](**f**(**sr**(Pn(i,j,k)))
∗(((u(**sr**+d1R1∗**u1R1**)|M
+ v(**sr**+d1R1∗**u1R1**)|M−1)∗**u1R1**/d1R1)
+ ((u(**sr**+d2R1∗**u2R1**)|M
+ v(**sr**+d2R1∗**u2R1**)|M−1)∗**u2R1**/d2R1)
+ ((u(**sr**+d3R1∗**u3R1**)|M
+ v(**sr**+d3R1∗**u3R1**)|M−1)∗**u3R1**/d3R1))
•[**UR1**]$^{-1}$•**T**[**UR1**]$^{-1}$
•(p1R1∗**u1R1** + p2R1∗**u2R1** +p3R1∗**u3R1**)/n^3
= lim S[bn(SM∩Pn)](**f**(**sr**(Pn(i,j,k)))
∗((u(**sr**+d1R1∗**u1R1**)|M
+ v(**sr**+d1R1∗**u1R1**)|M−1)
∗(p1R1/d1R1)/n^3
+ (u(**sr**+d2R1∗**u2R1**)|M
+ v(**sr**+d2R1∗**u2R1**)|M−1)
∗(p2R1/d2R1)/n^3
+ (u(**sr**+d3R1∗**u3R1**)|M
+ v(**sr**+d3R1∗**u3R1**)|M−1)
∗(p3R1/d3R1)/n^3)).

Let the limits be taken with

diR1 = piR1/n^3.

With the limits so linked,

I[V3]•(**f**(**r**)∗(**D**[**r**]∗(u(**r**)|M;**dR1**) + **D**[**r**]∗(v(**r**)|M;**dR1**))
•[**UR1**]$^{-1}$•**T**[**UR1**]$^{-1}$;**dR1**)
= lim S[bn(SM∩Pn)](**f**(**sr**(Pn(i,j,k)))
∗(u(**sr**+d1R1∗**u1R1**)|M
+ v(**sr**+d1R1∗**u1R1**)|M−1
+ u(**sr**+d2R1∗**u2R1**)|M
+ v(**sr**+d2R1∗**u2R1**)|M−1
+ u(**sr**+d3R1∗**u3R1**)|M
+ v(**sr**+d3R1∗**u3R1**)|M−1)) (868)
≡ **I**[SM](**f**(**sr**)∗(u(**sr**+d1R1∗**u1R1**)|M
+ v(**sr**+d1R1∗**u1R1**)|M−1
+ u(**sr**+d2R1∗**u2R1**)|M
+ v(**sr**+d2R1∗**u2R1**)|M−1
+ u(**sr**+d3R1∗**u3R1**)|M
+ v(**sr**+d3R1∗**u3R1**)|M−1)). (869)

If
sr+diR1∗**uiR1** is in V3−(M∪SM)
then u(**sr**+diR1∗**uiR1**|M)+ v(**sr**+diR1∗**uiR1**|M)−1 = 1.
If
sr+diR1∗**uiR1** is in M−SM
then u(**sr**+diR1∗**uiR1**|M)+ v(**sr**+diR1∗**uiR1**|M)−1 = −1.
If
sr+diR1∗**uiR1** is in SM
then u(**sr**+diR1∗**uiR1**|M)+ v(**sr**+diR1∗**uiR1**|M)−1 = 0.

Now let ***f*** be continuous, bounded and asymptotic over V3

I[V3]•(v(**r**)|M −u(**r**)|M))∗(**D**[**r**]∗(**f**(**r**);**dR1**))
•[**UR1**]$^{-1}$•**T**[**UR1**]$^{-1}$;**dR1**)
= **I**[SM]•(**D**[**r**]∗(**f**(**r**);**dR1**)•[**UR1**]$^{-1}$•**T**[**UR1**]$^{-1}$;**dR1**) (870)
= lim S[bn(SM∩Pn)]
(**f**(**sr**(Pn(i,j,k))+d1R1∗**u1R1**)−**f**(**sr**(Pn(i,j,k)))
+ **f**(**sr**(Pn(i,j,k))+d2R1∗**u2R1**)−**f**(**sr**(Pn(i,j,k)))
+ **f**(**sr**(Pn(i,j,k))+d3R1∗**u3R1**)−**f**(**sr**(Pn(i,j,k))))
= **0** (871)

since ***f*** is continuous.

Mathematics: Vector Integration

Also,

I[V3]•((**D**[**r**]∗(**f**(**r**)∗u(**r**)|M;**dR1**)+**D**[**r**]∗(**f**(**r**)∗v(**r**)|M;**dR1**))
•[**UR1**]$^{-1}$•**T**[**UR1**]$^{-1}$;**dR1**)
= lim S[bn(V3∩Pn)]((**D**[**r**(Pn(i,j,k))]∗(**f**(**r**)∗u(**r**)|M;**dR1**)
+ **D**[**r**(Pn(i,j,k))]∗(**f**(**r**)∗v(**r**)|M;**dR1**))
•[**UR1**]$^{-1}$•**T**[**UR1**]$^{-1}$•**m**(Pn(i,j,k)))
= lim(S[bn(V3∩Pn)](((**f**(**r**(Pn(i,j,k))+d1R1∗**u1R1**)
∗u(**r**+d1R1∗**u1R1**)|M
–**f**(**r**(Pn(i,j,k)))∗u(**r**(Pn(i,j,k)))|M
+ **f**(**r**(Pn(i,j,k))+d1R1∗**u1R1**)
∗v(**r**+d1R1∗**u1R1**)|M
–**f**(**r**(Pn(i,j,k)))∗v(**r**(Pn(i,j,k)))|M)
∗**u1R1**/d1R1
+ (**f**(**r**(Pn(i,j,k))+d2R1∗**u2R1**)
∗u(**r**+d2R1∗**u2R1**)|M
–**f**(**r**(Pn(i,j,k)))∗u(**r**(Pn(i,j,k)))|M
+ **f**(**r**(Pn(i,j,k))+d2R1∗**u2R1**)
∗v(**r**+d2R1∗**u2R1**)|M
–**f**(**r**(Pn(i,j,k)))∗v(**r**(Pn(i,j,k)))|M)
∗**u2R1**/d2R1
+ (**f**(**r**(Pn(i,j,k))+d3R1∗**u3R1**)
∗u(**r**+d3R1∗**u3R1**)|M
–**f**(**r**(Pn(i,j,k)))∗u(**r**(Pn(i,j,k)))|M
+ **f**(**r**(Pn(i,j,k))+d3R1∗**u3R1**)
∗v(**r**+d3R1∗**u3R1**)|M
–**f**(**r**(Pn(i,j,k)))∗v(**r**(Pn(i,j,k)))|M)
∗**u3R1**/d3R1)
•[**UR1**]$^{-1}$•**T**[**UR1**]$^{-1}$
•(p1R1∗**u1R1** + p2R1∗**u2R1** +p3R1∗**u3R1**)/n^{3}
= lim S[bn(SM∩Pn)](**f**(**sr**(Pn(i,j,k))+d1R1∗**u1R1**)
∗(u(**sr**+d1R1∗**u1R1**)|M
+ v(**sr**+d1R1∗**u1R1**)|M)
– **f**(**sr**(Pn(i,j,k)))
+ **f**(**sr**(Pn(i,j,k))+d2R1∗**u2R1**)
∗(u(**sr**+d2R1∗**u2R1**)|M
+ v(**sr**+d2R1∗**u2R1**)|M)
–**f**(**sr**(Pn(i,j,k)))

+ **f**(**r**(Pn(i,j,k))+d3R1∗**u3R1**)
∗u(**r**+d3R1∗**u3R1**)|M
+ v(**sr**+d3R1∗**u3R1**)|M)
−**f**(**sr**(Pn(i,j,k)))
≡ **I**[SM](**f**(**sr**(Pn(i,j,k))+d1R1∗**u1R1**)
∗(u(**sr**+d1R1∗**u1R1**)|M
+ v(**sr**+d1R1∗**u1R1**)|M)
− **f**(**sr**(Pn(i,j,k)))
+ **f**(**sr**(Pn(i,j,k))+d2R1∗**u2R1**)
∗(u(**sr**+d2R1∗**u2R1**)|M
+ v(**sr**+d2R1∗**u2R1**)|M)
−**f**(**sr**(Pn(i,j,k)))
+ **f**(**r**(Pn(i,j,k))+d3R1∗**u3R1**)
∗u((**r**+d3R1∗**u3R1**)|M
+ v(**sr**+d3R1∗**u3R1**)|M)
−**f**(**sr**(Pn(i,j,k)))). (872)

This is the **fundamental theorem of integral calculus over measurable sets in V3.**

As with directional gradients, other integral operations are possible.

I[V3]•(f(**r**)∗(**D**[**r**]∗(u(**r**)|M;**dR1**) + **D**[**r**]∗(v(**r**)|M;**dR1**))
•[**UR1**]$^{-1}$•**T**[**UR1**]$^{-1}$;**dR1**)
= **I**[SM](f(**sr**)
∗(u(**sr**+d1R1∗**u1R1**|M) + v(**sr**+d1R1∗**u1R1**)|M−1
+ u(**sr**+d2R1∗**u2R1**)|M + v(**sr**+d2R1∗**u2R1**)|M−1
+ u(**sr**+d3R1∗**u3R1**)|M + v(**sr**+d3R1∗**u3R1**)|M−1))
(873)

I[V3]•(**f**(**r**)∧(**D**[**r**]∗(u(**r**)|M;**dR1**) + **D**[**r**]∗(v(**r**)|M;**dR1**))
•[**UR1**]$^{-1}$•**T**[**UR1**]$^{-1}$;**dR1**)
= lim S[bn(SM∩Pn)](**f**(**sr**(Pn(i,j,k)))
•((u(**sr**+d1R1∗**u1R1**)|M
+ v(**sr**+d1R1∗**u1R1**)|M−1)∗**u1R1**/d1R1
+ (u(**sr**+d2R1∗**u2R1**)|M
+ v(**sr**+d2R1∗**u2R1**)|M−1)∗**u2R1**/d2R1
+ (u(**sr**+d3R1∗**u3R1**)|M
+ v(**sr**+d3R1∗**u3R1**)|M−1)∗**u3R1**/d3R1)

Mathematics: Vector Integration

•(p1R1*u2R1∧u3R1
+ p2R1*u3R1∧u1R1
+ p3R1*u1R1∧u2R1))
= lim S[bn(SM∩Pn)](f(sr(Pn(i,j,k)))
•((u(sr+d1R1*u1R1)|M
+ v(sr+d1R1*u1R1)|M−1)
*u1R1∧(u2R1∧u3R1)
+ (u(sr+d2R1*u2R1)|M
+ v(sr+d2R1*u2R1)|M−1)
*u2R1∧(u3R1∧u1R1)
+ (u(sr+d3R1*u3R1)|M
+ v(sr+d3R1*u3R1)|M−1)
*u3R1∧(u1R1∧u2R1))
= I[SM](f(sr)
•((u(sr+d1R1*u1R1)|M
+ v(sr+d1R1*u1R1)|M−1)
*u1R1∧(u2R1∧u3R1)
+ (u(sr+d2R1*u2R1)|M
+ v(sr+d2R1*u2R1)|M−1)
*u2R1∧(u3R1∧u1R1)
+ (u(sr+d3R1*u3R1)|M
+ v(sr+d3R1*u3R1)|M−1)
*u3R1∧(u1R1∧u2R1))) (874)

I[V3]•(f(r)∧(D[r]*(u(r)|M;dR1) + D[r]*(v(r)|M;dR1));dR1)
= I[SM](f(sr)
•C((u(sr+d1R1*u1R1)|M
+ v(sr+d1R1*u1R1)|M−1)*u1R1/d1R1
+ (u(sr+d2R1*u2R1)|M
+ v(sr+d2R1*u2R1)|M−1)*u2R1/d2R1
+ (u(sr+d3R1*u3R1)|M
+ v(sr+d3R1*u3R1)|M−1)*u3R1/d3R1)
•pR1) (875)

I[V3]∧(f(r)∧(D[r]*(u(r)|M;dR1) + D[r]*(v(r)|M;dR1));dR1)
= I[SM](f(sr)
•C((u(sr+d1R1*u1R1)|M
+ v(sr+d1R1*u1R1)|M−1)*u1R1/d1R1
+ (u(sr+d2R1*u2R1)|M
+ v(sr+d2R1*u2R1)|M−1)*u2R1/d2R1

+ (u(sr+d3R1*u3R1)|M
+ v(sr+d3R1*u3R1)|M−1)*u3R1/d3R1)
•C(pR1)) (876)

I[V3]∧(f(r)*(D[r]*(u(r)|M;dR1) + D[r]*(v(r)|M;dR1));dR1)
= I[SM](f(sr)
*((u(sr+d1R1*u1R1)|M
+ v(sr+d1R1*u1R1)|M−1)*u1R1/d1R1
+ (u(sr+d2R1*u2R1)|M
+ v(sr+d2R1*u2R1)|M−1)*u2R1/d2R1
+ (u(sr+d3R1*u3R1)|M
+ v(sr+d3R1*u3R1)|M−1)*u3R1/d3R1)
•C(dR1)) (877)

I[V3]*(f(r)*(D[r]*(u(r)|M;dR1) + D[r]*(v(r)|M;dR1));dR1)
= I[SM](f(sr)
*((u(sr+d1R1*u1R1)|M
+ v(sr+d1R1*u1R1)|M−1)*u1R1/d1R1
+ (u(sr+d2R1*u2R1)|M
+ v(sr+d2R1*u2R1)|M−1)*u2R1/d2R1
+ (u(sr+d3R1*u3R1)|M
+ v(sr+d3R1*u3R1)|M−1)*u3R1/d3R1)
*dR1) (878)

I[V3]*(f(r)•(D[r]*(u(r)|M;dR1) + D[r]*(v(r)|M;dR1));dR1)
= I[SM](f(sr)
•((u(sr+d1R1*u1R1)|M
+ v(sr+d1R1*u1R1)|M−1)*u1R1/d1R1
+ (u(sr+d2R1*u2R1)|M
+ v(sr+d2R1*u2R1)|M−1)*u2R1/d2R1
+ (u(sr+d3R1*u3R1)|M
+ v(sr+d3R1*u3R1)|M−1)*u3R1/d3R1)
*dR1) (879)

I[V3]*(f(r)∧(D[r]*(u(r)|M;dR1) + D[r]*(v(r)|M;dR1));dR1)
= I[SM](f(sr)
∧((u(sr+d1R1*u1R1)|M
+ v(sr+d1R1*u1R1)|M−1)*u1R1/d1R1
+ (u(sr+d2R1*u2R1)|M
+ v(sr+d2R1*u2R1)|M−1)*u2R1/d2R1

$$\begin{aligned}&+(\mathrm{u}(\mathbf{sr}+\mathrm{d3R1}\ast\mathbf{u3R1})|\mathrm{M}\\&\qquad+\mathrm{v}(\mathbf{sr}+\mathrm{d3R1}\ast\mathbf{u3R1})|\mathrm{M}-1)\ast\mathbf{u3R1}/\mathrm{d3R1})\\&\qquad\qquad\ast\mathbf{dR1})\end{aligned}\tag{880}$$

$$\begin{aligned}&\mathbf{I}[\mathrm{V3}]\ast(\mathrm{f}(\mathbf{r})\ast(\mathbf{D}[\mathbf{r}]\ast(\mathrm{u}(\mathbf{r})|\mathrm{M};\mathbf{dR1})+\mathbf{D}[\mathbf{r}]\ast(\mathrm{v}(\mathbf{r})|\mathrm{M};\mathbf{dR1}));\mathbf{dR1})\\&\quad=\mathbf{I}[\mathrm{SM}](\mathbf{f}(\mathbf{sr})\\&\qquad\ast((\mathrm{u}(\mathbf{sr}+\mathrm{d1R1}\ast\mathbf{u1R1}|\mathrm{M})\\&\qquad\qquad+\mathrm{v}(\mathbf{sr}+\mathrm{d1R1}\ast\mathbf{u1R1}|\mathrm{M})-1)\ast\mathbf{u1R1}/\mathrm{d1R1}\\&\qquad+(\mathrm{u}(\mathbf{sr}+\mathrm{d2R1}\ast\mathbf{u2R1}|\mathrm{M})\\&\qquad\qquad+\mathrm{v}(\mathbf{sr}+\mathrm{d2R1}\ast\mathbf{u2R1}|\mathrm{M})-1)\ast\mathbf{u2R1}/\mathrm{d2R1}\\&\qquad+(\mathrm{u}(\mathbf{sr}+\mathrm{d3R1}\ast\mathbf{u3R1}|\mathrm{M})\\&\qquad\qquad+\mathrm{v}(\mathbf{sr}+\mathrm{d3R1}\ast\mathbf{u3R1}|\mathrm{M})-1)\ast\mathbf{u3R1}/\mathrm{d3R1})\\&\qquad\qquad\qquad\ast\mathbf{dR1}).\end{aligned}\tag{881}$$

For ***f*** or *f* continuous, bounded and asymptotic over V3

$$\begin{aligned}&\mathbf{I}[\mathrm{V3}]\bullet((\mathrm{v}(\mathbf{r})|\mathrm{M}-\mathrm{u}(\mathbf{r})|\mathrm{M})\ast(\mathbf{D}[\mathbf{r}]\ast(\mathbf{f}(\mathbf{r});\mathbf{dR1}))\bullet[\mathbf{UR1}]^{-1}\bullet\mathbf{T}[\mathbf{UR1}]^{-1};\mathbf{dR1})\\&\quad=\mathbf{I}[\mathrm{SM}]\bullet(\mathbf{D}[\mathbf{rs}]\ast(\mathbf{f}(\mathbf{r});\mathbf{dR1})\bullet[\mathbf{UR1}]^{-1}\bullet\mathbf{T}[\mathbf{UR1}]^{-1};\mathbf{dR1})\\&\quad=\lim\,(\mathbf{f}(\mathbf{sr}+\mathrm{d1R1}\ast\mathbf{u1R1})-\mathbf{f}(\mathbf{sr})\\&\qquad+\mathbf{f}(\mathbf{sr}+\mathrm{d2R1}\ast\mathbf{u2R1})-\mathbf{f}(\mathbf{sr})\\&\qquad+\mathbf{f}(\mathbf{sr}+\mathrm{d3R1}\ast\mathbf{u3R1})-\mathbf{f}(\mathbf{sr}))\\&\quad=\mathbf{0}\end{aligned}\tag{882}$$

$$\begin{aligned}&\mathbf{I}[\mathrm{V3}]\bullet((\mathrm{v}(\mathbf{r})|\mathrm{M}-\mathrm{u}(\mathbf{r})|\mathrm{M})\ast(\mathbf{D}[\mathbf{r}]\ast\mathrm{f}(\mathbf{r});\mathbf{dR1}))\\&\qquad\qquad\bullet[\mathbf{UR1}]^{-1}\bullet\mathbf{T}[\mathbf{UR1}]^{-1};\mathbf{dR1})\\&\quad=0\end{aligned}\tag{883}$$

$$\begin{aligned}&\mathbf{I}[\mathrm{V3}]\bullet((\mathrm{v}(\mathbf{r})|\mathrm{M}-\mathrm{u}(\mathbf{r})|\mathrm{M})\ast(\mathbf{D}[\mathbf{r}]\wedge(\mathbf{f}(\mathbf{r});\mathbf{dR1}));\mathbf{dR1})\\&\quad=\mathbf{I}[\mathrm{SM}](\mathbf{D}[\mathrm{sr}]\wedge(\mathbf{f}(\mathbf{r});\mathbf{dR1})\bullet\mathbf{dR1})\end{aligned}\tag{884}$$

$$\begin{aligned}&\mathbf{I}[\mathrm{V3}]\wedge((\mathrm{v}(\mathbf{r})|\mathrm{M}-\mathrm{u}(\mathbf{r})|\mathrm{M})\ast(\mathbf{D}[\mathbf{r}]\wedge(\mathbf{f}(\mathbf{r});\mathbf{dR1}));\mathbf{dR1})\\&\quad=\mathbf{I}[\mathrm{SM}](\mathbf{D}[\mathbf{r}]\wedge(\mathbf{f}(\mathbf{r});\mathbf{dR1})\wedge\mathbf{dR1})\end{aligned}\tag{885}$$

$$\begin{aligned}&\mathbf{I}[\mathrm{V3}]\wedge((\mathrm{v}(\mathbf{r})|\mathrm{M}-\mathrm{u}(\mathbf{r})|\mathrm{M})\ast(\mathbf{D}[\mathbf{r}]\ast(\mathrm{f}(\mathbf{r});\mathbf{dR1}));\mathbf{dR1})\\&\quad=\mathbf{I}[\mathrm{SM}](\mathbf{D}[\mathbf{r}]\ast(\mathrm{f}(\mathbf{r});\mathbf{dR1})\wedge\mathbf{dR1})\end{aligned}\tag{886}$$

$$\begin{aligned}&\mathbf{I}[\mathrm{V3}]\ast((\mathrm{v}(\mathbf{r})|\mathrm{M}-\mathrm{u}(\mathbf{r})|\mathrm{M})\ast(\mathbf{D}[\mathbf{r}]\bullet(\mathbf{f}(\mathbf{r});\mathbf{dR1}));\mathbf{dR1})\\&\quad=\mathbf{I}[\mathrm{SM}](\mathbf{D}[\mathbf{r}]\bullet(\mathbf{f}(\mathbf{r});\mathbf{dR1})\ast\mathbf{dR1})\end{aligned}\tag{887}$$

I[V3]∗((v(**r**)|M−u(**r**)|M)∗(**D**[**r**]∧(**f**(**r**);**dR1**));**dR1**)
  = **I**[SM](**D**[**r**]∧(**f**(**r**);**dR1**)∗**dR1**) (888)

I[V3]∗((v(**r**)|M−u(**r**)|M)∗(**D**[**r**]∗(f(**r**);**dR1**));**dR1**)
  = **I**[SM](**D**[**r**]∗(f(**r**);**dR1**)∗**dR1**). (889)

If the surface of a set M is differentiable at **sr1**, there exists at **sr1** three well–defined orthogonal vectors[49]: two tangential unit vectors **ut1**(**sr1**),and **ut2**(**sr1**), and a orthogonal vector **un**(**sr1**). For **un** constructed to be the "outward" orthogonal into V3–(M∪SM) this local orientation may be used to define both directional and sectional gradients.

Differentiable surfaces at **sr1** may be classified as following:

sr1+d1∗**ut1** in (M–SM) or in V3 – (M∪SM)
sr1+d2∗**ut2** in (M–SM) or in V3 – (M∪SM)
sr1–d1∗**ut1** in (M–SM) or in V3 – (M∪SM)
sr1–d2∗**ut2** in (M–SM) or in V3 – (M∪SM)
sr1+d3∗**un** in (M–SM) or in V3 – (M∪SM)

Directional and sectional gradients may then be defined according a local system of quadrant gradients. There are 32 classifications of such differentiable surfaces. For each classification eight local quadrant gradients **D**[**sr1**]∗(u(**r**|M);**dRn**) may be defined.

Example. Let **rs1**+d1∗**ut1**, **rs1**–d1∗**ut1**, and **rs1**–d2∗**ut2** all lie in SM for d1 and d2 positive and sufficiently small. Let **rs1**+d2∗**ut2** and **rs1**+d3∗**un** lie in V3–(M∪SM) for d2 and d3 positive and sufficiently small. For this classification,

D[**sr1**]∗(u(**sr1**|M);**dR1n**) ≡ u(d1∗**ut1**)∗**ut1**/d1
  + u(d2∗**ut2**)∗**ut2**/d2
  + u(d3∗**un**)∗**un**/d3
  = **qd**(d2∗**ut2**) + **qd**(d3∗**un**)

D[**sr1**]∗(u(**sr1**|M);**dR2n**) ≡ (u(–d1∗**ut1**))∗**ut1**/d1
  + (u(+d2∗**ut2**))∗**ut2**/d2
  + (u(+d3∗**un**))∗**un**/d3
  = **qd**(d2∗**ut2**) + **qd**(d3∗**un**)

Mathematics: Vector Integration

D[**sr1**]∗(u(**sr1**|M);**dR3n**) ≡ (u(+d1∗**ut1**))∗**ut1**/d1
+ (u(−d2∗**ut2**))∗**ut2**/d2
+ (u(+d3∗**un**))∗**un**/d3
= **qd**(d3∗**un**)

D[**sr1**]∗(u(**sr1**|M);**dR4n**) ≡ (u(−d1∗**ut1**))∗**ut1**/d1
+ (u(−d2∗**ut2**))∗**ut2**/d2
+ (u(+d3∗**un**))∗**un**/d3
= **qd**(d3∗**un**)

D[**sr1**]∗(u(**sr1**|M);**dR5n**) ≡ (u(+d1∗**ut1**))∗**ut1**/d1
+ (u(+d2∗**ut2**))∗**ut2**/d2
+ (u(−d3∗**un**))∗**un**/d3
= **qd**(d2∗**ut2**)

D[**sr1**]∗(u(**sr1**|M);**dR6n**) ≡ (u(−d1∗**ut1**))∗**ut1**/d1
+ (u(+d2∗**ut2**))∗**ut2**/d2
+ (u(−d3∗**un**))∗**un**/d3
= **qd**(d2∗**ut2**)

D[**sr1**]∗(u(**0**|**sr1**);**dR7n**) ≡ (u(+d1∗**ut1**))∗**ut1**/d1
+ (u(−d2∗**ut2**))∗**ut2**/d2
+ (u(−d3∗**un**))∗**un**/d3
= **0**

D[**sr1**]∗(u(**sr1**|M);**dR8n**) ≡ (u(−d1∗**ut1**))∗**ut1**/d1
+ (u(−d2∗**ut2**))∗**ut2**/d2
+ (u(−d3∗**un**))∗**un**/d3
= **0**.

Definition 48 (directional derivative and gradients orthogonal to a surface)

Given

M, a measurable set with a differentiable surface SM;
sr, an element of SM;
un(**sr**), the outward unit orthogonal vector of M at **sr**;

for

f, a function with a basic derivative at **sr** in direction **un**(**rs**);

lim(f(**sr**+d∗**un**(**sr**))–f(**sr**))/d   as d⟶0   (890)

is called the **basic directional derivative orthogonal to** M **at rs**

lim(f(**sr**+d∗**un**(**sr**))–f(**sr**))∗**qd**(d∗**un**(**sr**))   as d⟶0   (891)

is called the **basic directional gradient orthogonal to M at rs**.

end of definition

Orthogonal directional derivatives are written

D[**sr**,**sr**+**dun**(**sr**)](**f**(**r**)|M;dr).   (892)

Orthogonal directional gradients are written

D[**sr**,**sr**+**dun**(**sr**)]∗(**f**(**r**)|M;**dun**(**sr**)).   (893)

The orthogonal unit point step functions are written

u(**r**–**sr**|**un**(**sr**))   (894)

and

v(**r**–**sr**|**un**(**sr**))   (895)

with accompanying gradients written as

D[**sr**,**sr**+**dun**(**sr**)]∗(u(**r**–**sr**)|M;**dun**(**sr**))   (896)

and

D[**sr**,**sr**+**dun**(**sr**)]∗(v(**r**–**sr**)|M;**dun**(**sr**)).   (897)

Now for ***f*** continuous, bounded, and asymptotic

I[V3](u(**r**–**sr**|**un**(**sr**))∗(**D**[**r**,**r**+**dun**(**sr**)]•(**f**(**r**)|**un**(**sr**);**dr**))
;abs(**dun**(**sr**)))
= lim (**f**(**sr**+d∗**un**(**sr**))•**un**(**sr**)/d)∗abs(d∗**un**(**sr**))
= **f**(**sr**)•**un**(**sr**)   (898)

and

I[V3]((1–u(**r**–**sr**|**un**(**sr**)))∗**D**[**r**,**r**+**dun**(**sr**)]•(**f**(**r**)|**un**(**sr**);**dr**)
; abs(**dun**(**sr**)))
= –**f**(**sr**)•**un**(**sr**). (899)

A similar integration may be performed for each element of SM.

An analogous development is made for local step functions.

Definition 49 (sectional gradients orthogonal to a surface)
Given
M, a measurable set with a differentiable surface SM;
sr, an element of SM;
ut1(**sr**), **ut2**(**sr**), **un**(**sr**),
orthogonal and tangent vectors of M at **sr**;
for[50]
us1(**sr**) = (**un**(**sr**) + **ut1**(**sr**)+ **ut2**(**sr**))/√3
us2(**sr**) = (2∗**un**(**sr**) – (√3 + 1)∗**ut1**(**sr**)
+ (√3 – 1)∗**ut2**(**sr**))/(2∗√3)
us3(**sr**) = (2∗**un**(**sr**) + (√3 – 1)∗**ut1**(**sr**)
– (√3 + 1)∗**ut2**(**sr**))/(2∗√3)
f a function with a basic sectional gradient at **sr**

lim (f(**sr**+d1∗**us1**(**sr**))–f(**sr**))∗**us1**(**sr**)/d1
+ (f(**sr**+d2∗**us2**(**sr**))–f(**sr**))∗**us2**(**sr**)/d2
+ (f(**sr**+d3∗**us3**(**sr**))–f(**sr**))∗**us3**(**sr**)/d3
sr in SM (900)

is called the **basic orthogonal gradient to** M **at sr**.
end of definition

Orthogonal gradients are written for section
SECT(**sr**,**us1**(**sr**),**us2**(**sr**),**us**(**sr**))
as
D[**sr**]∗(**f**(**r**)|M;**dS**(**sr**)). (901)

As illustration, suppose the local surface planar at **sr1** with **sr1+**d1∗**ut1** and **sr1+**d2∗**ut2** lie in a planar surface. Then

D[**sr1**]∗(u(**r**|M);**dRn**) = lim **un**/d3 (902)
= **D**[**sr1**,**sr1+dRn**]∗(u(**r**–**ser**)|M;**dRn**).

Contrariwise,
D[**sr1**]∗(v(**r**|M);**dRn**) = **0**. (903)

In the "inward" direction –**un** into M–SM
D[**sr1**]∗(u(**r**|M);**dRn**$^-$) = **0** (904)

whereas,

D[**sr1**]∗(v(**r**|M);**dRn**$^-$)
= lim **un**/d3 (905)
=–**D**[**sr1**,**sr1**–**dRn**$^+$]∗(v(**r**–**ser**)|M;**dRn**)
D[**sr**]∗(u(**r**|M)|M;**dS**(**sr**))
= lim (**us1**/d1 + **us2**/d2 + **us3**/d3) (906)
D[**sr**]∗(v(**r**|M)|M;**dS**(**sr**)) = **0**.
(907)

Now let **u1+u2+u3** be a partition vector and ***f*** continuously locally differentiable with a basic gradient and bounded over a measurable set M with differentiable surface SM at each point **sr** of which are defined a set of orthogonal directions **ut1**(**sr**), **ut2**(**sr**), and **un**(**sr**).

Then because of interior cancellation, **I**[M](D[**r**)]•(**f**(**r**);**dr**)∗abs(**dr**) has values only at the surface. Each cell in Pn at the surface may lie partly in M and partly outside of M. The interior part of the cell undergoes interior cancellation, while the exterior remains to be summed. The fraction of the cell corresponding to

f(**sr**(Pn(i,j,k))+d∗**ui**)

is

((**ut1**∧**ut2**)•**ui** /n^3)/n^3 = (**ut1**∧**ut2**)•**ui**

and thus invariant with n.

Mathematics: Vector Integration

Applying these weights to the cells at the surface

I[M]•(D[r]∗(f(r);dr);dr)
= lim S[bn(SM∩Pn)](f(sr(Pn(i,j,k))+d∗u1)
∗(ut1(sr)∧ut2(sr))•u1
+ f(sr(Pn(i,j,k))+d∗u2)
∗(ut1(sr)∧ut2(sr))•u2
+ f(sr(Pn(i,j,k))+d∗u3)
∗(ut1(sr)∧ut2(sr))•u3)
= lim S[bn(SM∩Pn)](f(sr(Pn(i,j,k))+d∗u1)∗u1•un(sr)
+ f(sr(Pn(i,j,k))+d∗u2)∗u2•un(sr)
+ f(sr(Pn(i,j,k))+d∗u3)∗u3•un(sr))
= I[SM](f(sr)∗(u1+u2+u3)•un(sr)) (908)

I[M]•(D[r]∧(f(r);dr);dr) = 0 (909)

I[M]∧(D[r]∧(f(r);dr);dr)
= I[SM]([u1∗u1 – I]•f(sr)∗u1
+[u2∗u2 – I]•f(sr)∗u2
+[u3∗u3 – I]•f(sr)∗u3)•un(sr) (910)

I[M]∗(D[r]•(f(r);dr);dr)
= I[SM](un(sr)•([u1∗f(sr)]•[u1∗u1]
+ [u2∗f(sr)]•[u2∗u2]
+ [u3∗f(sr)]•[u3∗u3])) 911)

I[M]∗(D[r]]∧(f(r);dr);dr)
= I[SM](C(f(sr))•([u1∗u1]∗(u1•un)
+ [u2∗u2]∗(u2•un)
+ [u3∗u3]∗(u3•un))). (912)

A variation of these results forms the foundation of the divergence theorem[51].

I[M](D[r)]•(f(r);dr)∗abs(dr)
= lim S[bn(SM∩Pn)](f(sr(Pn(i,j,k))+d∗u1)•u1
∗(ut1(sr)∧ut2(sr))•u1
+ f(sr(Pn(i,j,k))+d∗u2)•u2
∗(ut1(sr)∧ut2(sr))•u2
+ f(sr(Pn(i,j,k))+d∗u3)•u3)
∗(ut1(sr)∧ut2(sr))•u3)

= lim S[bn(SM∩Pn)]
(**f**(**sr**(Pn(i,j,k))+d∗**u1**)•**u1**∗**u1**•**un**(**sr**)
+ **f**(**sr**(Pn(i,j,k))+d∗**u2**)•**u2**∗**u2**•**un**(**sr**)
+ **f**(**sr**(Pn(i,j,k))+d∗**u3**)•**u3**∗**u3**•**un**(**sr**))
=**I**[SM](**f**(**sr**)•**un**(**sr**)). (913)

An extension of equation (913) underlies Green's theorem[52].

Let f(**r**) and g(**r**) be locally differentiable with basic gradients and bounded over M. Then

I[M](D[**r**]•(f(**r**)∗**D**[**r**]∗(g(**r**);**dr**)
−g(**r**)∗**D**[**r**]∗(f(**r**);**dr**);**dr**);abs(**dr**))
= **I**[M]((f(**r**)∗D[**r**]•(**D**[**r**]∗(g(**r**);**dr**);**dr**)
+ D[**r**]∗(f(**r**);**dr**)•**D**[**r**]∗(g(**r**);**dr**)
−g(**r**)∗D[**r**]•(**D**[**r**]∗(f(**r**);**dr**);**dr**)
−D[**r**]∗(f(**r**);**dr**)•**D**[**r**]∗(g(**r**);**dr**));abs(**dr**))
= **I**[M]((f(**r**)∗D[**r**]•(**D**[**r**]∗(g(**r**);**dr**);**dr**)
−g(**r**)∗D[**r**]•(**D**[**r**]∗(f(**r**);**dr**);**dr**));abs(**dr**))
= **I**[SM](f(**sr**)∗**D**[**sr**]∗(g(**r**);**dr**)
−g(**r**)∗**D**[**sr**]∗(f(**r**);**dr**))•**un**(**sr**). (914)

Variations may likewise be formulated for curls, gradients, invergences, incurls and ingradients.

For vector functions **f**(**r**) and **g**(**r**) under the same conditions

I[M]((**D**[**r**]∧(**g**(**r**);**dr**)•**T**[(**D**[**r**]∗(**f**(**r**);**dr**);**dr**));abs(**dr**))
= **I**[SM](**f**(**sr**)∗**D**[**sr**]∧(**g**(**r**);**dr**))•**un**(**sr**). (915)

But again the surface may not be differentiable[53] at **sr**. Such cases may still be analyzed by sectional gradients.

Analysis by sectional gradients embraces quadrant gradients. Sectional gradients of u((**r**|M)) may be used generally not only for differentiable surfaces[54], but also those with angles, corners, etc. Where the surface becomes an inner cusp, the sectional gradient may even become a directional gradient.

Mathematics: Vector Integration

In the sequel we assume surfaces with no more than a denumerable number of such irregularities; results are stated in terms of sectional gradients.

Example. Let **f=r−sr0** and SM be a differentiable surface containing **sr0** which supports an orthogonal orientation.

D[r]∗(f;dr) = I
D[r]∧(f;dr) = 0
D[r]•(f;dr) = 3.

Sectionally for **dR1**

D[r]∗(f;dR1) = u1R1∗u1R1 +u2R1∗u2R1 +u3R1∗u3R1
D[r]∧(f;dR1) = 0
D[r]•(f;dR1) = 3.

Directionally for **ur**

D[r,r+dr]∗(f|ur;dr) = ur∗ur
D[r,r+dr]∧(f|ur;dr) = 0
D[r,r+dr]•(f|ur;dr) = 1
I[r1,r2]∗(f|ur;dr) = (r2∗r2−r1∗r1 +r1∗r2−r2∗r1)/2
    −sr0∗(r2−r1)
I[r1,r2]∧(f|ur;dr) = (r1−sr0)∧(r2−r1)
I[r1,r2]•(f|ur;dr) = (r2•r2−r1•r1)/2 −sr0•(r2−r1).

For a given measurable set M

I[M]•(D[r]∗(f;dr);dr) = m(M) = vol(M)∗(u1+u2+u3)
I[M]∗(D[r]•(f;dr);dr) = 3∗m(M)
I[V3]•(f(r)∗D[r,r+dR1]∗(u(r−ser)|M;dR1);dR1)
    = ser −sr0
I[V3]•(f(r)∗D[r,r+dR1]∗(v(r−ser)|M;dR1);dR1)
    = −ser +sr0
I[V3]∗(u(r−ser)∗D[r,r+dR1]•(f(r);dR1)|M;dR1)
    = I[V3]∗(u(r−ser)|M;dR1), unbounded.

I[V3]•(f(r)∗(D[r]∗(u(r)|M;dR1)
    + D[r]∗(v(r)|M;dR1))•[UR1]$^{-1}$•T[UR1]$^{-1}$;dR1)
    = I[SM](r −sr0) ∗(u(sr+d1R1∗u1R1)|M
        + v(sr+d1R1∗u1R1)|M−1
        + u(sr+d2R1∗u2R1)|M
        + v(sr+d2R1∗u2R1)|M−1
        + u(sr+d3R1∗u3R1)|M
        + v(sr+d3R1∗u3R1)|M−1).

I[V3]∗((v(r)|M −u(r)|M)∗D[r]•(f(r);dr);dr)
    = 3∗I[SM]∗(1;dr)
    = 0.

I[M]•(D[r]∗(f(r);dr);dr)
    = I[M]•(I;dr)
    = I[SM](sr−sr0)∗(u1+u2+u3)•un(sr).

Mathematics: Vector Integration

Now let **a** = a1∗**u1** + a2∗**u2** + a3∗**u3** be a constant vector with **G**(**a**) as its diagonal matrix. Then **r**•**G**(**a**) = **a**•**G**(**r**) is a **dilation** of V3.

Consider now the dilation of the measurable set M symbolized as M•**G**(**a**) and the step function u(**r**|M•**G**(**a**)). Then

u(**r**)|M = u(**r**•**G**(**a**)|M•**G**(**a**)). (916)

Moreover,

D[**r**•**G**(**a**)]∗(u(**r**•**G**(**a**)|M•**G**(**a**));**dR1**•**G**(**a**))
= lim (u((**r**+d1R1∗**u1R1**)•**G**(**a**))|**M**•**G**(**a**)
−u(**r**•**G**(**a**)|M•**G**(**a**))∗**u1R1**•**G**(**a**)
/ abs(d1R1∗**u1R1**•**G**(**a**)))
+ (u((**r**+d2R1∗**u2R1**)•**G**(**a**))|**M**•**G**(**a**)
−u(**r**•**G**(**a**)|M•**G**(**a**))∗**u2R1**•**G**(**a**)
/ abs(d2R1∗**u2R1**•**G**(**a**)))
+ (u((**r**+d3R1∗**u3R1**)•**G**(**a**))|**M**•**G**(**a**)
−u(**r**•**G**(**a**)|M•**G**(**a**))∗**u3R1**•**G**(**a**)
/ abs(d3R1∗**u3R1**•**G**(**a**)))
= lim ((u(**r**+d1R1∗**u1R1**)|M−u(**r**)|M)∗**u1R1**
/ abs(d1R1∗**u1R1**•**G**(**a**))
+ (u(**r**+d2R1∗**u2R1**)|M−u(**r**)|M)∗**u2R1**
/ abs(d2R1∗**u2R1**•**G**(**a**))
+ (u(**r**+d3R1∗**u3R1**)|M−u(**r**)|M)∗**u3R1**
/ abs(d3R1∗**u3R1**•**G**(**a**)))•**G**(**a**)
= lim ((u(**r**+d1R1∗**u1R1**)|M−u(**r**)|M)∗**u1R1**/d1R1
+ (u(**r**+d2R1∗**u2R1**)|M−u(**r**)|M)∗**u2R1**/d2R1
+ (u(**r**+d3R1∗**u3R1**)|M−u(**r**|)M)∗**u3R1**/d3R1)
•[**UR1**]$^{-1}$
•(**u1**∗**u1**/abs(**u1R1**•**G**(**a**))
+ **u2**∗**u2**/abs(**u2R1**•**G**(**a**))
+ **u3**∗**u3**/abs(**u3R1**•**G**(**a**)))
•[**UR1**]•**G**(**a**).

Thus

D[**r**•**G**(**a**)]∗(u(**r**•**G**(**a**))|M•**G**(**a**);**dR1**•**G**(**a**))
= **D**[**r**]∗(u(**r**|M);**dR1**)•[**UR1**]$^{-1}$
•(**u1**∗**u1**/abs(**u1R1**•**G**(**a**))
+ **u2**∗**u2**/abs(**u2R1**•**G**(**a**))
+ **u3**∗**u3**/abs(**u3R1**•**G**(**a**)))
•[**UR1**]•**G**(**a**). (917)

Likewise,

D[**r**•**G**(**a**)]∗(v(**r**•**G**(**a**))|M•**G**(**a**);**dR1**•**G**(**a**))
= **D**[**r**]∗(v(**r**|M);**dR1**)•[**UR1**]$^{-1}$
•(**u1**∗**u1**/abs(**u1R1**•**G**(**a**))
+ **u2**∗**u2**/abs(**u2R1**•**G**(**a**))
+ **u3**∗**u3**/abs(**u3R1**•**G**(**a**)))
•[**UR1**]•**G**(**a**). (918)

For[55] **a** = a

D[a∗**r**]∗(u(a∗**r**)|**a**∗M;a∗**dR1**)
= (a/abs(a))∗**D**[**r**]∗(u(**r**|)M;**dR1**) (919)
D[a∗**r**]∗(v(a∗**r**)|a∗M;a∗**dR1**)
= (a/abs(a))∗**D**[**r**]∗(v(**r**|M);**dR1**). (920)

Local functions with finite step discontinuities at points or surfaces may be decomposed as:

f(**r**) = **f0**(**r**) + S[n](**fn**) + S[m](**fm**) (921)

where

f0 has no discontinuities;
fn = **a1n**∗u(**r**–**rn**) in section R1n
= **a2n**∗u(**r**–**rn**) in section R2n
= ...
fm = **b1m**∗u(**r**–**srm**|RSm) in section RS1m
= **b2m**∗u(**r**–**srm**|RSm) in section RS1m
= ...

Mathematics: Vector Integration

The above results may be further elaborated in terms of specific coordinate systems, not necessarily orthogonal, in the set of vectors. Well chosen coordinates can often make the solution to a particular problem more apparent; but poorly chosen coordinates can often obscure solutions. In this text, results are expressed generically in terms of the arbitrary orientation of the origin and impose no specific system of coordinates.

Summary of local integration.

Given bounded functions with measure **m** over a measurable set M of either a curve C or V3, the following statements hold for local integrals whether directional or sectional:

Item:	I[E]•(**invergence**)	**I**[E]∧(**incurl**)	**I**[E]∗(**ingradient**)
c constant			c∗**m**(E)
c constant	**c**•**m**(E)	**c**∧**m**(E)	**c**∗**m**(E)
c∗f			c∗I∗f
c∗**f**	c∗I•**f**	c∗**I**∧**f**	c∗**I**∗**f**
c∗f	**c**•I∗f	**c**∧**I**∗f	**c**∗**I**∗f
c•**f**			c•I∗f
c∧f	c•I∧f	**c**•(**T**[**I**∗**f**]−**I**∗**f**)	**C**(**c**)•**I**∗**f**
c∗f	**c**∗I•**f**		
f+g			**I**∗f + **I**∗g
f+g	I•f + I•g	**I**∧f + **I**∧g	**I**∗f + **I**∗g
D∗(f∗g)	I•(f∗**D**∗g) +I•(g∗**D**∗f)	**I**∧(f∗**D**∗g) +**I**∧(g∗**D**∗f)	**I**∗(f∗**D**∗g) +**I**∗(g∗**D**∗f)
(**D**∗f)/f²	−I•(**D**∗(1/f))	−**I**∧(**D**∗(1/f))	−**I**∗(**D**∗(1/f))
D•(f∗**g**)			I∗(f∗D•**g**) +I∗(**g**•**D**∗f)
D∧(f∗**g**)	I•(f∗**D**∧**g**) +I•(**g**∧**D**∗f)	I∧(f∗**D**∧**g**) +I∧(**g**∧**D**∗f)	I∗(f∗**D**∧**g**) +I∗(**g**∧**D**∗f)
D∗(f∗**g**)	I•(f∗**D**∗**g**) +I•(**g**∗**D**∗f)	I∧(f∗**D**∗**g**) +I∧(**g**∗**D**∗f)	I∗(f∗**D**∗**g**) +I∗(**g**∗**D**∗f)

Mathematics: Vector Integration

Item:	I[E]•(invergence)	I[E]∧(incurl)	I[E]*(ingradient)
D*(f•g)	I•(f•D*g) +I•(g•D*f)	I∧(f•D*g) +I∧(g•D*f)	I*(f•D*g) +I*(g•D*f)
D•(f∧g)			I*(g•D∧f) –I*(f•D∧g)
D∧(f∧g)	I•(f•T[D*g] –I•(g•T[D*f]) –I•(f*D•g) +I•(g*D•f)	I∧(f•T[D*g] –I∧(g•T[D*f]) –I∧(f*D•g) +I∧(g*D•f)	I*(f•T[D*g]))) –I*(g•T[D*f]) –I*(f*D•g) +I*(g*D•f)
D*(f∧g)	I•(C(f)•D*g) –I•(C(g)•D*f)	I∧(C(f)•D*g) –I∧(C(g)•D*f)	
D•(f*g)	I•(f*D•g) + I•(g•T[D*f])	I∧(f*D•g) + I∧(g•T[D*f])	I*(f*D•g) + I*(g•T[D*f])

Integrals of Sums and Products

As with the table of local derivatives, this table can be used to elaborate other combinations.

Mathematics: Vector Integration

The following two tables reference the symbolism used to distinguish the types of local derivatives and integrals. The symbols for gradients and ingradients are also extended to divergences, curls, invergences and incurls.

Type	Derivative	Gradient
Directional	D[**r1**,**r1+dr**](**f**(**r**)|**ur**;dx)	**D**[**r1**,**r1+dr**]∗(**f**(**r**)|**ur**;**dr**)
Along **r**(x)		
with respect to x	D[x,x+dx](**f**(x)|CX;dx)	
with respect to **r**	D[**r1**,**r1+dr**](**f**(**r**)|CR;dr)	**D**[**r1**,**r1+dr**]∗(**f**(**r**)|CR;**dr**)
positive quadrant		**D**[**r1**]∗(**f**(**r**);**dr**)
Sectional		**D**[**r1**]∗(**f**;**dR1**)

Symbols for local Gradients

Type	Integral	Ingradient
Directional	I[**r1**,**r2**](**f**(**r**)|**ur**;dr)	I[**r1**,**r2**]∗(**f**(**r**)|**ur**;**dr**)
Along **r**(x)		
with respect to x	I[x1,x2](**f**(x)|CX;dx)	I[**r**(x1),**r**(x2)]∗(**f**(x)|CX;**dr**)
with respect to **r**	I[**r1**,**r2**](**f**(**r**)|CR;dr)	I[**r1**,**r2**]∗(**f**(**r**)|CR;**dr**)
positive quadrant		I[M]∗(**f**(**r**);**dr**)
Sectional		I[M]∗(**f**;**dR1**)

Symbols for local Ingradients

DIFFERENCE BETWEEN MATHEMATICS AND THEORETICAL PHYSICS.

The wonderful mathematical world of understandable ideas for which we now have a symbology and a vocabulary (and a few results), is not Theoretical Physics. Physics is the study of reality observable as extended, moving, or forcing. How are these mathematical ideas transformed into ideas useful for the study of Physics, that is, into Theoretical Physics?

Is there a problem? Aren't all mathematical ideas relevant for Physics? Inasmuch as ideas arise from sensible perceptions, a natural concordance reigns between ideas and reality. We easily associate our perception of an object with a location vector in V3, and the extension of such an object with a set of such vectors.

Nevertheless, identifying mathematical ideas with Theoretical Physics raises serious obstacles. For example, if **r0**, **r1**, and **r2** are three location vectors, the valid mathematical expression (**r0**∧**r1**)–**r2** may be computed and evaluated numerically, but what can possibly be its physical meaning?

Units

To start the solution of the first problem, let us attach to the three physical attributes--extension, motion and force--separate labels. The labels are called **physical units**. These physical units can be thought of as an elementary type of measurement[56]. The following table gives our usage.

Object observable as:	**Physical unit**:
extended	L
moving	V
forcing	F

Labels for Physical Units

Theoretical Physics

Note that with this assignment the border between observation and ideas has already been crossed. When the motion of an object is labeled in terms of V units, more is being asserted than mere observation. So beyond being labels, physical units also point to ideas: the idea of extension, the idea of motion, the idea of force.

Since physical units symbolize ideas, combinations may be created to suit our purpose. Such combinations are indicated by the common symbols of multiplication or division. For example, we might create ideas of Theoretical Physics to which we could assign units like L∗L, F∗L, F∗V, V∗L, F∗V/L, etc. None of these created ideas is directly observable; they are said to be observable only from observing the primary observables of Physics.

The symbols of multiplication and division used for units are not to be confused with mathematical multiplication or division. Numbers may be multiplied or divided; physical units are not numbers. The reason for using the mathematical symbols appears below.

In the example above for (**r0∧r1**)–**r2**, if **r2** is a location vector, it would be appropriate to associate it with the physical idea of extension. We should thus attach the physical unit, L, to **r2**. But what do we make of **r0∧r1**? What physical unit does it have?

Geometrical intuition can help here. When **r0** and **r1** are location vectors (physical unit L), cross multiplication in the vector field would refer to a parallelogram formed by the two vectors.

In a similar way, the triple product would refer to a parallelepiped. Thus if **r0**, **r1**, and **r2** are three location vectors, the combination

$$\mathbf{r0}\bullet(\mathbf{r1}\wedge\mathbf{r2}) = (\mathbf{r0}\wedge\mathbf{r1})\bullet\mathbf{r2}$$

is a scalar quantity which specifically equates to the of the parallelepiped described by the three vectors.

Obviously, the volume of a physical object is not directly observed. We observe only its extension, from which we can calculate its volume. This idea of volume, derived from the primary idea of extension, what physical unit should it have?

The example suggests the physical unit of the derivative ideas of Theoretical Physics should reflect the mathematical operations used in their calculation. Since volumes are calculated by two multiplications of three variables of extension, the physical unit of volume is well designated L✱L✱L.

So we begin our transformation of mathematical ideas into Theoretical Physics by restricting mathematical ideas by the following **unit rules**:

1. Mathematical entities are transformed into Theoretical Physics by associating with each separate mathematical entity a primary or derived physical unit. A neutral physical unit is allowable.

2. In Theoretical Physics addition and subtraction are restricted to ideas with identical physical units and produce a result having the same physical unit.

   Multiplication or division as operations are applied to ideas of Theoretical Physics with either the same or different physical units. These operations create derivative ideas of Theoretical Physics with physical units corresponding to the mathematical operations used in their calculation.

3. The equations (statements) of Theoretical Physics must be consistent in physical units.

The derived ideas of Theoretical Physics are said to be "observable" only by reference to the three primary physical observables combined as ideas according to the mathematical expression governing their units.

In the example above, the physical unit of **r0∧r1**, (L✱L), does not match the physical unit of **r2**, (L). Consequently, even though (**r0∧r1**)–**r2** is a valid mathematical expression, it cannot express a valid idea of Theoretical Physics.

Theoretical Physics

What unit should V3 have?

Consider vectorial addition, **+**:V3xV3⟶V3. We see that **+** relates three sets of vectors all whose elements have the same physical units. Consider next vectorial multiplication, **∧**:V3xV3⟶V3. Here none of the set of vectors need have the same unit, but the units of the range must be the product of the units of the two domains. Likewise scalar multiplication, **∗**:RxV3⟶V3 shows Theoretical Physics deals with sets of vectors of different dimensions as well as different units.

In effect, Theoretical Physics expands one mathematical entity like V3 into a class of objects infinite in number distinguished and constrained by their physical units.

Example. A plane is sometimes described as a set of vectors, **r**, such that

$$(\mathbf{r}-\mathbf{r0})\bullet\mathbf{r3} = 0$$

for fixed **r0** and **r3**. The plane is orthogonal to **r3** and passes through the point defined by **r0**.

This description, however, can be deceiving. The prior description of the linear plane in equation (323)

$$\mathbf{r}(t1,t2) - \mathbf{r0} = t1 \ast \mathbf{r1} + t2 \ast \mathbf{r2}$$

is a description of each point in the plane. That is, if **r** is a location vector, the physical unit of each vector **r**, **r0**, **r1**, and **r2** is L, not L∗L.

Let us try to reconcile the two descriptions. Construct

$$(\mathbf{r}-\mathbf{r0})\bullet(\mathbf{r1}\wedge\mathbf{r2}) = (t1 \ast \mathbf{r1} + t2 \ast \mathbf{r2})\bullet(\mathbf{r1}\wedge\mathbf{r2}) = 0.$$

The vector, **r1∧r2**, has physical units L∗L which corresponds to **r3** in the description of the plane above. It follows that **r** and **r3** have different physical units, with **r3** having physical units of L∗L. The above description of the plane as (**r**–**r0**)•**r3** = 0 describes not so much a plane as a degenerate volume, that is one with zero–volume. Here the "0" is not a mere number, but carries the physical unit, L∗L∗L.

This example shows how easily a correct mathematical formulation can lead to physical misinterpretation.

Using the above unit rules, the following mathematical ideas are transformed into Theoretical Physics by attaching to each the corresponding physical units.[57]

Idea	Physical Unit
pure number	neutral
location vector	L
area	L∗L
volume	L∗L∗L
energy	F∗L
power	F∗V
time	L/V
exp(x)	neutral
i(x)	1/x, (impulse function)
u(x)	neutral, (step function)
D[](;dx)	1/x
I[](;dx)	x
abs(**r**)	L, (the length of **r**)
ur	neutral, (the direction of **r**)
s	L, (arc length)
q(**r**)	1/L, (reciprocal vector)
D[]∗(;**dr**)	1/L
I[]•(;**dr**)	L

Some Physical Units

Although the ideas of Theoretical Physics are generated from mathematical ideas, as ideas they are different and distinct.

The distinction is evident because legitimate mathematical operations on mathematical entities may not be legitimate operations on the corresponding ideas of Theoretical Physics. In effect, Theoretical Physics constrains mathematical ideas in their operations while expanding them by adjoining physical units.

Theoretical Physics

The assignment of physical units to mathematical ideas is only a first, necessary constraint for creating the ideas of Theoretical Physics. To understand what more is needed, let us now look at ideas more particularly suited to the observation of extended objects.

THE IDEA OF A PARTICLE

Reality observed as extended implies the observed object consists of parts. Because it has parts, the observed object can be measured. Nevertheless in the literature of modern Physics, it is common to assume an indivisible, extension–less particle (atom) having a given property like velocity or mass or charge.

From a strictly atomic point of view, the idea of mass or energy, to cite instances, is absurd. The absurdity lies in giving physical significance to the mathematical idea of a point. All our observations show the primacy of extension to motion or force.

A mathematical point is never observed. A mere collection of points do not necessarily make up either a line, or a plane, or a volume. In these latter mathematical structures we recognize relationships between constituent parts; indeed the study of these relationships is usually far more rewarding than studying a mere assemblage of points. Physics likewise should be furnished with a fundamental concept that allows it to study a wide range of relationships.

This fundamental concept in Theoretical Physics is called a **particle**. To further the transformation of Mathematics into Physics, a clear idea of a particle is needed.

Definition 50 (particle)

Given

X, the set of vectors describing the location of an extended object referenced to the location of an observer taken as the origin;

x, a vector in X;

VT(**x**), the set of all open sets of X containing the vector **x**;

then

A **particle**, (**x**), is the intersection of all the subsets of VT(**x**).

end of definition

The correspondence made in Theoretical Physics between the physical location of an observable object and V3 allows a particle to refer to the smallest observable part of the object. In this sense (**x**) is a set function mapping observable objects into V3.

$$(\mathbf{x})\text{: \{Extended objects\}} \longrightarrow \text{V3}. \tag{922}$$

The definition of a particle implies that as a mathematical limiting process (**x**) = **x**. However, by its definition the particle (**x**) is not a point; rather, it is a topological entity––a limiting process corresponding to a narrowing focus of observation matched to intersections of open sets in X. This distinction has consequences for Theoretical Physics in two critical ways:

- the particle relates the ideas of Theoretical Physics to extended objects,
- the particle incorporates the limitations of physical observation.

It is sufficient to choose any denumerable properly nested subsets of VT(**x**) whose intersection is **x**. Symbolize these denumerable subsets as {VTn}; then {VTn} is ordered by inclusion.

An example of {VTn} is {**r**(n)|abs(**r**–**x**)<1/n} for n=1,2,....

If there exists an m, such that for every VTn included in VTm an idea of Theoretical Physics can be attached, then the idea is attributed to **x**. The terminology, "the particle (**x**) has such and such a **property**" is used to denote this condition. The idea of a particle is thus an operational concept, much like the notion of continuity in Mathematics.

The physicist has only limited means of observation.[58] However, it is possible, often with instruments, to make increasingly refined observations. This process of increasingly refined physical observation corresponds to the operational definition of a physical particle.

Theoretical Physics

A crucial difference between Mathematics and Theoretical Physics comes from this ability to make increasingly refined observations. The mathematician asserts that *all* included sets in VTm possess the given property. The physicist asserts only that the property is true for *all observable* sets included in VTm. Beyond observation the physicist is reduced to inference or conjecture. The physicist retains the possibility that future more refined observations may reveal physical structures not suggested by the limiting process used in Mathematics. In effect, the assertions of Physics are made contingent on a level of resolution[59].

> Reflection: The theologian, Thomas Aquinas, argues: "Now, two things must be considered in the case of any provident agent––namely premeditation of the order and the establishment of the premeditated order––in things that are subject to providence. The first of these pertains to the cognitive power while the second belongs to the operative. Between the two there is this difference: in the act of premeditating the order, the more perfect that providence is, the more can the order of providence be extended to the smallest details. The fact that we are not able to think out, ahead of time, the order of all particular events in regard to matters to be arranged by us stems from the deficiency of our knowledge, which cannot embrace all singular things. However, the more a person is able to think ahead about a plurality of singular things, the more adroit does he become in his foresight. But, in regard to imposing the premeditated order on things, the providence of a governing agent is more noble and perfect the more universal it is and the more it accomplished his premeditated plan by means of a plurality of ministers because this controlling of ministers occupies an important place in the order that pertains for foresight. Moreover, divine providence must consist in the highest perfection, since He is absolutely and universally perfect." Summa Contra Gentiles 94:10.

The human history of Physics is the laborious enlargement of knowledge by understanding and unifying a widening scope of physical observations. The boundary of this knowledge is an unknown mystery which mankind painstakingly, but not necessarily monotonically, elucidates into science.

MATERIAL AND LOCAL REFERENCES

To further the transformation of mathematical ideas into Theoretical Physics, a way is needed to identify change in the observed object.

So Theoretical Physics requires a starting condition as a reference for the change.

Reference first a hypothetical, material **universe at rest**––a universe in which nothing is moving[60]. Next, associate this hypothetical universe of physical matter with V3.

There are two ways of making this association.

In the first way, associate every particle (**x**) of the material universe at rest with a location, **x**, of V3, denoting its distance and direction referred to the observer. Of course, there may be locations **x** of V3 for which nothing material exists to be observed. Assign to those locations the idea of **null matter,** that is, a location in V3 for which no matter exists in the universe of rest**.** Then each location of V3 is associated in the universe at rest with a material object, either observable or null. Thus associated, V3 is properly labeled V3(**x**)

In this way the universe at rest serves as a reference for both the primary physical observables and also the derived ideas of Theoretical Physics.

Accordingly the phrase, "the material particle, **x**," means the observable physical matter denoted by (**x**) which was at location **x** in the universe at rest.

As a set function, **x** identifies extended physical matter by associating it uniquely with a set of location vectors in the hypothetical universe.

Let S(**x1**) be an *open* neighborhood of ***x1*** corresponding to physical matter. Suppose, subsequently, this physical matter identified by S(**x1**) undergoes change. Let the physical observation be such that the observed matter before the change is identical with the observed matter after the change. Then any changes of physical matter may be referenced to S(**x1**).

For example, call *size0* the initial volume and *size1* the changed volume of the physical material identified by S(**x1**). Since S is open, both *size0* and *size1* are positive real numbers. The ratio size1/size0 can therefore be constructed meaningfully. Moreover, since this ratio can be constructed for any open S(**x1**), the observational limit of this ratio is *properly* attached to the particle ***x1*** as a **property**.

With reference to this hypothetical universe, directly observable primary changes in matter can be described as functions of **x** symbolized as follows:

Observable	At Rest	Subsequently	Idea
extension	**x**	**r**(**x**)	location
motion	**0**	**v**(**x**)	velocity

Material References

Consider the function, **r**:V3(**x**)⟶V3(**r**), denoted by **r**(**x**) which describes the change, if any, in the location of an observable particle, (**x**). For instance, **r**(**x**) = **x** indicates that a given particle **x** is at its reference location.

With the assumption of "null" matter for logical completion, **r**(**x**) not only describes the location of an individual particle **x**, but is also a bijective map of V3 into and onto itself. Considered as a bijective transformation, **r**(**x**) will then generate an inverse transformation, **x**(**r**).

We may interpret **x**(**r**) specifically. Fixing our attention on a particular location **r1**, **x**(**r1**) describes the different observable particles which occupy the location **r1**.

Thus the second way of associating the material universe with V3 becomes apparent. Let S(**r1**) be an open neighborhood of V3(**r**). Then S(**r1**) can serve as a reference for change in matter occupying S(**r1**).

For example, call *sprd0* the amount of matter occupying S(**r1**) in the universe at rest and *sprd1* the amount of matter occupying S(**r1**) subsequently. Again, S(**r1**) being open, both sprd0 and sprd1 are positive real numbers. The ratio sprd1/sprd0 can thus be constructed meaningfully for any S(**r1**) and so the observational limit of this ratio is properly attached to **x**(**r1**) as a property.

In terms of primary functions, then, Theoretical Physics proposes this schema:

$$\mathbf{x}:\{\text{extended objects + null matter}\} \longleftrightarrow V3(\mathbf{x}) \qquad (923)$$
$$\mathbf{r}:\{\text{physical location}\} \longleftrightarrow V3(\mathbf{r}) \qquad (924)$$
$$\mathbf{r}:V3(\mathbf{x}) \longrightarrow V3(\mathbf{r}) \qquad (925)$$
$$\mathbf{x}:V3(\mathbf{r}) \longrightarrow V3(\mathbf{x}). \qquad (926)$$

The identifying set functions, **x** or **r**, with their domains in observable universe and their ranges in V3, continue their role as identifiers through the functions **r**(**x**) and **x**(**r**) reflecting changes in the observable universe as transformations of V3.

Reflection: Another way to express this schema: the physical universe is its matter. By His act of creation God materialized the idea of space. In so doing He gives space a new material existence without in the least compromising space as an idea.

God's purpose is revealed by his Messiah, Jesus, who comes to spiritualize matter, that is, to give matter a new spiritual existence. Jesus does this by revealing Himself as God's incarnation, that is God becoming part of the material universe without in the least compromising His divinity.

The two transformations, **r**(**x**) and **x**(**r**) correspond to two different ways of observing physical reality. We may either fix our attention on an object as it moves from one location to another, or alternatively we may fix our attention on only one location and observe what passes through it.

The first way of observation, **r**(**x**), is called the **material** reference; the second of way of observation, **x**(**r**), is called the **fixed–local**, or often simply the **local** reference.

The two references of Theoretical Physics, as we shall see, lead to a duality of descriptions and analyses of physical reality. Any statement made in the material reference corresponds to a dual statement in the local reference, that is f(**r**(**x**)) = f(**x**(**r**)). For clarity, important dual propositions will be demonstrated explicitly.

Theoretical Physics

Reflection. The duality in Theoretical Physics is a reflection in the physical world of the centrality of Jesus Christ, who revealed Himself by his life, death, and resurrection as the unique divine/human duality. As such, He is the origin, the one who continues, and the destiny of the material universe. It is in Him that the "groanings" of material creation find resolution. While the mystery of Jesus is valued especially for enlightenment of the human condition, it is not without light for the physical universe too.

To distinguish the global function from a particular reference the following symbology is used:

	Global	Local
location	**r**(**x**) or **r**|**x**	**r**(**x1**) or **r**|**x1**
matter	**x**(**r**) or **x**|**r**	**x**(**r1**) or **x**|**r1**

Symbols for Global or Local References

Since **x**(**r1**) means **x** is at **r1**, it follows that

$$\mathbf{r}(\mathbf{x}(\mathbf{r1})) = \mathbf{r1}. \tag{927}$$

Dually,

$$\mathbf{x}(\mathbf{r}(\mathbf{x1})) = \mathbf{x1}. \tag{928}$$

Observations

Physical observations are ordered. Theoretical Physics proposes numbering the ordered observations as a real variable, symbolized as "*a*".

To appreciate the topological construction in Theoretical Physics from the perspective of the universe at rest, let **rx** be a particle in the universe at rest whose location is **xr**. Then **rx** maps into an element of V3(**x**) while **xr** maps into an element of V3(**r**). In the university at rest V3(**x**) and V3(**r**) are related by an identity mapping, for which we put **rx**=**xr** meaning not that a location is a particle, but that **xr** as an element of V3(**r**) has the same location as **rx** as an element of V3(**x**).

Now let RX be a connected open set of physical matter containing **rx**. In the universe at rest RX is identified with XR, a congruent open set.

At observation *a1* let **r1** = **r**(**rx**,a1); let the image of RX under **r**(**x**,a1) be **r**(RX,a1) ≡ Ra1 which consequently must contain **r1**. If **r** is continuous over **x** at *a1*, Ra1 will also be open. The material occupying Ra1 is appropriately described by **x**(Ra1,a1) ≡ RX. The sets XR and Ra1 are generally different subsets of V3(**r**). They are mapped, however, from the same subset RX of V3(**x**).

Likewise, for the same observation a1, let **x1** = **x**(**xr**,a1); let the image of XR under **x**(**r**,a1) be **x**(XR,a1) ≡ Xa1 which consequently must contain **x1**. If **x** is continuous over **r** at a1, Xa1 will also be open. The set of locations occupied by Xa1 is appropriately described by **r**(Xa1,a1) ≡ XR. The sets RX and Xa1 are generally different subsets of V3(**x**). They are mapped from the same subset XR of V3(**r**). Each observation therefore corresponds to a mapping between V3(**x**) and V3(**r**) with the universe at rest corresponding to the identity mapping.

We also put **r**(XR,a1) = XR and **x**(RX,a1) = RX denoting the observer's continuing ability to identify location and matter.

While the perspective from the universe at rest exhibits duality nicely, we are usually concerned with a local perspective at observation *a1*. For reference take the particle **x1** whose location at observation *a1* is **r1**(**x1**,a1) and the local or material topologies referenced to **r1** and **x1**. This referencing focuses attention on **r1**(**x1**,a1), but obscures the overall duality, which, however, is not totally lost. To avoid confusion the reader should be clear about the mappings of neighborhoods of **x1** and **r1** to the universe at rest.

Theoretical Physics

We may now express a fundamental principle distinguishing both Physics and Theoretical Physics from Mathematics.

The Principle of Non–Collocation

For the same observation,
two different particles cannot occupy the same position.

Symbolically,

$$\{\mathbf{r}(\mathbf{x2},a1) = \mathbf{r}(\mathbf{x1},a1)\} \longrightarrow \mathbf{x2}=\mathbf{x1} \tag{929}$$

and

$$\{\mathbf{x}(\mathbf{r2},a1) = \mathbf{x}(\mathbf{r1},a1)\} \longrightarrow \mathbf{r2}=\mathbf{r1}. \tag{930}$$

As a first consequence of the principle of non–collocation, notice that in Theoretical Physics $\mathbf{r}(\mathbf{x1},a1)$ cannot be decomposed into

$$r1(\mathbf{x1},a1)*\mathbf{u1} + r2(\mathbf{x1},a1)*\mathbf{u2} + r3(\mathbf{x1},a1)*\mathbf{u3}$$

but rather

$$\mathbf{r}(\mathbf{x1},a1) = r1(\mathbf{y1},a1)*\mathbf{u1} + r2(\mathbf{y2},a1)*\mathbf{u2} + r3(\mathbf{y3},a1)*\mathbf{u3} \tag{931}$$

for some **y1**, **y2**, and **y3**, because **x1** at observation *a1* occupies location $\mathbf{r}(\mathbf{x1},a1)$, not $r1*\mathbf{u1}$.

The two related functions, $\mathbf{r}(\mathbf{x},a)$ and $\mathbf{x}(\mathbf{r},a)$ are called **primary functions** of Theoretical Physics.

DERIVATIVES OF THEORETICAL PHYSICS.

Since the variable *a* is a real variable, **r**(**x**,a) symbolizes an infinite[61], ordered set of observations. Accordingly **r**(**x**,a) is an infinite, ordered set of transformations V3(**x**)⟶V3(**r**) indexed by a real variable *a*. This same set of observations is differently designated by **x**(**r**,a).

Let **f**(**r**(**x**,a)) be a companion and compatible (that is, having the same physical units) set of generalized functions, vector or scalar, of these observations. Changes in ***f*** may then be referred to changes in **r**, **x**, or a. What can be meant by derivatives of ***f*** with respect to these variables?

Logically one may attempt to hold none, one, two or all three variables fixed.

Holding none or all of the variables fixed repudiates the assumption of the given relationship **r**(**x**,a).

Holding any two of the three variables fixed, makes the two fixed variables independent of the third. This implies, for instance, holding one fixed location for one specific material particle while the indexed observation varies. Consequently all derivatives of the fixed variables for this condition are trivially zero.

The only meaningful condition upon which to develop the idea of a derivative is to hold only one of the three variables fixed. The only three ways of so doing are shown in the following table.

Condition	Derivatives
x fixed	**D∗(f|x;dr)**
	D(**f**|**x**;da)
r fixed	**D∗(f|r;dx)**
	D(**f**|**r**;da)
a fixed	**D∗(f**|a;**dr)**
	D∗(f|a;**dx)**

Derivatives of Theoretical Physics

Theoretical Physics: Derivatives

For **x** fixed, ***f*** is defined over local **traces** **r**(a|**x**); for **r** fixed, ***f*** is defined over material **tracks**, **x**(a|**r**); for *a* fixed, ***f*** is defined over **r**(**x**|a) for the given observation, *a*.

The symbols in the table are open to more than one interpretation; so further clarification is needed. The first way interprets the derivative as restricted to a given local trace or material track (**r**(**x1**,a) or **x**(**r1**,a) is referenced); a second way, as unrestricted in V3, (**r**(**x**,a1) or **x**(**r**,a1) is referenced). Whenever the interpretation is doubtful, the intended interpretation will indicated by writing the derivative as **D1∗** or D1 for the first way and **D3∗** or D3 for the latter[62].

Example: For **r**(**x**,a)−**r1**=**x∗**exp(−a), **r1** constant, a>0,

	D1∗(**r**|a;**dr**) = **ut**(**x**)**∗q**(**ut**(**x**))
while	**D3∗**(**r**|a;**dr**) = **I**;
also	**D1∗**(**r**|a;**dx**) = **ut**(**x**)**∗q**(**ut**(**x**))**∗**exp(−a)
while	**D3∗**(**r**|a;**dx**) = **I∗**exp(−a).
also	D1(**r**|**x1**;da) = −**x1∗**exp(−a)
while	D3(**r**|**x**;da) = −**x∗**exp(−a)

where **ut**(**x**) is the direction of **r**(**x**,a) − **r1**.

The example shows that usually it is not necessary to distinguish D1 from D3.

As in Mathematics, the derivatives of Theoretical Physics comprehend
- indexed derivatives
- directional derivatives
- directional gradients
- sectional gradients

and may vary from location to location, particle to particle and observation to observation.

The derivatives of a generalized function in Theoretical Physics need not exist, need not be basic, need not be continuous.

The idea of rank is given significance for Theoretical Physics by applying it to material concepts.

If **f** is any vectorial function then the rank of **D[r1]∗(f;dr)** in Mathematics specifies whether **f** is applied to points only, curves only, surfaces only or generally in V3. Likewise, the rank of **D[x1]∗(f;dx)** in Theoretical Physics specifies an action on corresponding subsets of physical matter.

However, in contrast to V3(**r**) or V3(**x**) where there is always continuity and sufficiency, the mappings of **f** may be discontinuous or directional. We are thus forced to deal carefully with the derivatives which may not be continuous. In contrast to the primary functions **r**(**x**,a) and **x**(**r**,a) which are governed by the principle of non–collocation, the generalized functions **f**(**r**(**x**,a)) of Theoretical Physics may have multiple values at **f**(**r1**(**x1**,a1)), one value for each section.

Derivatives of the primary functions.

How do the local derivatives of Mathematics translate into those of Theoretical Physics?

To come to an answer to this question consider first the mathematical idea of arc–length. Since the variable **r**(a|**x1**) describes a curve in V3(**r**), previous results based on arc–lengths, s(**r**(a)), apply immediately. In particular

$$\begin{aligned} D[a1](\mathbf{r}|\mathbf{x1};da) &= D[a1](s(\mathbf{r}(a|\mathbf{x1}));da)*\mathbf{utr}(\mathbf{x1},a1) && (932)\\ &= D[a1](sr(\mathbf{x1});da)*\mathbf{utr}(\mathbf{x1},a1) && (933)\end{aligned}$$

where **utr**(**x1**,a1) is the unit tangent of the curve. Call s(**r**(a|**x1**)) ≡ sr(**x1**) the **materially referenced arc–length** and **utr**(**x1**,a1) the **materially referenced unit tangent**.

Similarly, locally referenced arc–lengths and locally referenced unit tangents for **x**(a|**r1**) may also be defined, that is,

$$\begin{aligned} D[a1](\mathbf{x}|\mathbf{r1};da) &= D[a1](s(\mathbf{x}(a|\mathbf{r1}));da)*\mathbf{utx}(\mathbf{r1},a1) && (934)\\ &= D[a1](sx(\mathbf{r1});da)*\mathbf{utx}(\mathbf{r1},a1). && (935)\end{aligned}$$

Clearly sr(**x1**) , referring to a curve of locations occupied by **x1**, differs from sx(**r1**) ≡ s(**x**(a|**r1**)), referring to the material track of particles passing through **r1**. Call sx(**r1**) the **locally referenced arc–length** and **utx**(**r1**,a1) the **locally referenced unit tangent**.

Theoretical Physics: Derivatives

Furthermore, in addition to the arc–lengths, sr(**x1**) and sx(**r1**), yet other arc–lengths may be defined. Given the curve **r**(**x1**,a) referenced to **r1**(**x1**,a1), then

$$\mathbf{r}(\mathbf{x1},a1+da) = \mathbf{r2}(\mathbf{x2},a1)$$

for some **x2** and a1+da.

It follows

$$\mathbf{r}(\mathbf{x1},a1+da) - \mathbf{r}(\mathbf{x1},a1) = \mathbf{r2}(\mathbf{x2},a1) - \mathbf{r}(\mathbf{x1},a1),$$

and as a limit

$$\begin{aligned} D[a1](sr(\mathbf{x1},a1);da) &= D[a1](s(\mathbf{r}(a|\mathbf{x1}));da) \\ &= D[a1](s(\mathbf{r}(\mathbf{x}|a1));da). \end{aligned} \tag{936}$$

The materially referenced arc–length, s(**r**(a|**x1**)) differs from the materially referenced arc–length, s(**r**(**x**|a1)), inasmuch as they refer to different material particles. Yet they derive from the same local trace.

Dually,

$$\begin{aligned} D[a1](sx(\mathbf{r1},a1);da) &= D[a1](s(\mathbf{x}(a|\mathbf{r1}));da) \\ &= D[a1](s(\mathbf{x}(\mathbf{r}|a1));da) \end{aligned} \tag{937}$$

where **x**(**r1**,a1+da) – **x**(**r1**,a1) = **y2**(**s2**,a1) – **x1**(**r1**,a1) for some particle **y2**.

Again the locally referenced arc–length, s(**x**(a|**r1**)) differs from the locally referenced arc–length, s(**x**(**r**|a1)), even though they derive from the same material track.

The subtlety of symbolism for these new arc–lengths can be alleviated by writing them explicitly in the basic convention,

$$D[\mathbf{r2},\mathbf{r1}](s(\mathbf{r}(\mathbf{x}|a1));da) \equiv D[a1](s(\mathbf{r}(\mathbf{x}|a1));da) \tag{938}$$

and

$$D[\mathbf{y2},\mathbf{x1}](s(\mathbf{x}(\mathbf{r}|a1));da) \equiv D[a1](s(\mathbf{x}(\mathbf{r}|a1));da). \tag{939}$$

Similarly, for directional gradients,

D1[**r1**]∗(sr|**x1**;**dr**(**x1**)) = **D1**[**r2**,**r1**]∗(s(**r**(**x**|a1));**dr**(a1))
= **utr**(**x1**,a1). (940)

and

D1[**x1**]∗(sx|**r1**;**dx**(**r1**)) = **D1**[**y2**,**x1**]∗(s(**x**(**r**|a1));**dx**(a1))
= **utx**(**r1**,a1). (941)

Each of these derivatives, just as in Mathematics, may be defined in a forward or backward sense.

We see again the single mathematical idea of arc−length transformed into a panoply of related ideas for Theoretical Physics.

In Mathematics **dr** is defined first as a vectorial increment in the sense of positive quadrant and secondly as restricted to a curve, **dr**(x).

Theoretical Physics has a richer set of increments, many interrelated. A convention is needed to describe the relationship between the particle, its locations, and the observation. The following table illustrates our usage for describing the changes of **r**(**x**,a) from observation *a1* to a subsequent observation *a1+da* referenced to **r1**(**x1**,a1):

index	**s2**	**r1**	**r2**
a1	**x2**	**x1**	**y2**
a1+da	–	**x2**	**x1**

index	**y2**	**x1**	**x2**
a1	**r2**	**r1**	**s2**
a1+da	–	**r2**	**r1**

Designations for the Forward Convention

Thus[63] we symbolize) **r2** ≡ **r**(**x1**,a1+da) from which **y2** ≡ **x**(**r2**,a1; and **x2** ≡ **x**(**r1**,a1+da) from which **s2** ≡ **r**(**x2**,a1). These definitions imply neither the existence of **s2**(**x1**,a) nor the existence of **y2**(**r1**,a) for any observation, *a*. The particle **x1** may never be observed at **s2**. This sense of the definitions, which reference a1<a1+da is called a **forward** sense because a forward sense of the observational index is referenced.

Theoretical Physics: Derivatives

A backward sense which references a1–da<a1 may also be defined as follows:

index	r0	r1	s0
a1–da	**x1**	**x0**	–
a1	**y0**	**x1**	**x0**

index	x0	x1	y0
a1–da	**r1**	**r0**	–
a1	**s0**	**r1**	**r0**

Designations for the Backward Convention

In the backward sense, **r0** ≡ **r**(**x1**,a1–da) from which **y0** ≡ **x**(**r0**,a1); and **x0** ≡ **x**(**r1**,a1–da) from which **s0** ≡ **r**(**x0**,a1). These definitions imply neither the existence of **s0**(**x1**,a) nor of **y0**(**r1**,a) for any observation, *a*.

This symbolism puts **r0**, **r1** and **r2** on the trace of **x1** as **r0**(**x1**,a1–da), **r1**(**x1**,a1), and **r2**(**x1**,a1+da) and **x2**, **x1**, **x0** on the material track at **r1** as **x0**(**r1**,a1–da), **x1**(**r1**,a1), and **x2**(**r1**,a1+da). This usage defines other particles **y0**, **y2** and other locations **s0**, **s2**.

The panoramic convention is shown in the following compounded table:

Index	s0	r0	r1	r2	s2		y0	x0	x1	x2	y2	Index
a1–da		**x1**	**x0**					**r1**	**r0**			a1–da
a1	**x0**	**y0**	**x1**	**y2**	**x2**		**r0**	**s0**	**r1**	**s2**	**r2**	a1
a1+da			**x2**	**x1**					**r2**	**r1**		a1+da

Panoramic Reference Designations

Using the positive definite convention[64] the forward increments are defined as:

$$\mathbf{d}_f(\mathbf{r}|\mathbf{x1}) \equiv \mathbf{r2}(\mathbf{x1},\mathrm{a1+da}) - \mathbf{r1}(\mathbf{x1},\mathrm{a1}) \qquad (942)$$
$$\mathbf{d}_f(\mathbf{r}|\mathrm{a1+da}) \equiv \mathbf{r2}(\mathbf{x1},\mathrm{a1+da}) - \mathbf{r1}(\mathbf{x2},\mathrm{a1+da}) \qquad (943)$$
$$\mathbf{d}_f(\mathbf{r}|\mathrm{a1}) \equiv \mathbf{r2}(\mathbf{y2},\mathrm{a1}) - \mathbf{r1}(\mathbf{x1},\mathrm{a1}). \qquad (944)$$

Dual forward increments are defined as:

$$\mathbf{d}_f(\mathbf{x}|\mathbf{r1}) \equiv \mathbf{x2}(\mathbf{r1},a1+da) - \mathbf{x1}(\mathbf{r1},a1) \tag{945}$$
$$\mathbf{d}_f(\mathbf{x}|a1+da) \equiv \mathbf{x2}(\mathbf{r1},a1+da) - \mathbf{x1}(\mathbf{r2},a1+da) \tag{946}$$
$$\mathbf{d}_f(\mathbf{x}|a1) \equiv \mathbf{x2}(\mathbf{s2},a1) - \mathbf{x1}(\mathbf{r1},a1). \tag{947}$$

For the backward sense[65]

$$\mathbf{d}_b(\mathbf{r}|\mathbf{x1}) \equiv \mathbf{r1}(\mathbf{x1},a1) - \mathbf{r0}(\mathbf{x1},a1-da) \tag{948}$$
$$\mathbf{d}_b(\mathbf{r}|a1-da) \equiv \mathbf{r1}(\mathbf{x0},a1-da) - \mathbf{r0}(\mathbf{x1},a1-da) \tag{949}$$
$$\mathbf{d}_b(\mathbf{r}|a1) \equiv \mathbf{r1}(\mathbf{x1},a1) - \mathbf{r0}(\mathbf{y0},a1). \tag{950}$$

and dually:

$$\mathbf{d}_b(\mathbf{x}|\mathbf{r1}) \equiv \mathbf{x1}(\mathbf{r1},a1) - \mathbf{x0}(\mathbf{r1},a1-da) \tag{951}$$
$$\mathbf{d}_b(\mathbf{x}|a1-da) \equiv \mathbf{x1}(\mathbf{r0},a1-da) - \mathbf{x0}(\mathbf{r1},a1-da) \tag{952}$$
$$\mathbf{d}_b(\mathbf{x}|a1) \equiv \mathbf{x1}(\mathbf{r1},a1) - \mathbf{x0}(\mathbf{s0},a1). \tag{953}$$

From these definitions[66] it follows that

$$\mathbf{d}_f(\mathbf{r}|\mathbf{x1}) = \mathbf{d}_f(\mathbf{r}|a1+da) = \mathbf{d}_f(\mathbf{r}|a1) \tag{954}$$
$$\mathbf{d}_b(\mathbf{r}|\mathbf{x1}) = \mathbf{d}_b(\mathbf{r}|a1-da) = \mathbf{d}_b(\mathbf{r}|a1) \tag{955}$$
$$\mathbf{d}_f(\mathbf{x}|\mathbf{r1}) = \mathbf{d}_f(\mathbf{x}|a1+da) = \mathbf{d}_f(\mathbf{x}|a1) \tag{956}$$
$$\mathbf{d}_b(\mathbf{x}|\mathbf{r1}) = \mathbf{d}_b(\mathbf{x}|a1-da) = \mathbf{d}_b(\mathbf{x}|a1). \tag{957}$$

Consequently, with regard to the primary local and material designators, results may be stated in terms of four increments:

$\mathbf{d}_f(\mathbf{r}|\mathbf{x1})$, $\mathbf{d}_b(\mathbf{r}|\mathbf{x1})$, $\mathbf{d}_f(\mathbf{x}|\mathbf{r1})$, and $\mathbf{d}_b(\mathbf{x}|\mathbf{r1})$

despite their having arisen from other increments.

Limits taken with respect to any of the three variables become directional limits.

For a given limit, any one of the conditions

$\mathbf{r2} \longrightarrow \mathbf{r1}$	or	$\mathbf{r0} \longrightarrow \mathbf{r1}$
$\mathbf{x2} \longrightarrow \mathbf{x1}$	or	$\mathbf{x0} \longrightarrow \mathbf{x1}$
$a1+da \longrightarrow a1.$	or	$a1-da \longrightarrow a1$

implies the other two.

Furthermore, any increment $\mathbf{d} \equiv d1*\mathbf{u1}+d2*\mathbf{u2}+d3*\mathbf{u3}$ implies that any di is a function of the other two d's through $\mathbf{r}(\mathbf{x1},a)$ or $\mathbf{x}(\mathbf{r1},a)$.

Theoretical Physics: Derivatives

Indexed derivatives for the primary functions, then, may be defined meaningfully as

$$D[a1](\mathbf{r}|\mathbf{x1};d_f a) \longleftarrow \mathbf{d_f}(\mathbf{r}|\mathbf{x1})/da \quad (958)$$
$$D[a1](\mathbf{x}|\mathbf{r1};d_f a) \longleftarrow \mathbf{d_f}(\mathbf{x}|\mathbf{r1})/da \quad (959)$$
$$D[a1](\mathbf{r}|\mathbf{x1};d_b a) \longleftarrow \mathbf{d_b}(\mathbf{r}|\mathbf{x1})/da \quad (960)$$
$$D[a1](\mathbf{x}|\mathbf{r1};d_b a) \longleftarrow \mathbf{d_b}(\mathbf{x}|\mathbf{r1})/da. \quad (961)$$

as a1+da⟶a1 or a1–da⟶a1. These derivatives are generalizations of one–sided derivatives in R.

The symbolism above is stated in the positive definite convention. The same ideas may be stated in the basic convention which may be presented as a fully elaborated symbology or, at the cost of some ambiguity, a less elaborated one. Thus further:

$$\mathbf{d_f}(\mathbf{r}|a1+da)/da$$
$$\longrightarrow D[\mathbf{r1}(\mathbf{x},a1+da),\mathbf{r2}(\mathbf{x1},a1+da)](\mathbf{r}|a1+da;da)$$
(fully elaborate)
$$\equiv D[\mathbf{r1},\mathbf{r2}](\mathbf{r}|a1+da;da)$$
(less elaborate)
$$= D[a1](\mathbf{r}|\mathbf{x1};d_f a) \quad (962)$$

$$\mathbf{d_f}(\mathbf{x}|a1+da)/da$$
$$\longrightarrow D[\mathbf{x1}(\mathbf{r},a1+da),\mathbf{x2}(\mathbf{r1},a1+da)](\mathbf{x}|a1+da;da)$$
(fully elaborate)
$$\equiv D[\mathbf{x1},\mathbf{x2}](\mathbf{x}|a1+da;da)$$
(less elaborate)
$$= D[a1](\mathbf{x}|\mathbf{r1};d_f a) \quad (963)$$

$$\mathbf{d_b}(\mathbf{r}|a1-da)/da$$
$$\longrightarrow D[\mathbf{r}(\mathbf{x1},a1-da),\mathbf{r1}(\mathbf{x},a1-da)](\mathbf{r}|a1-da;da)$$
(fully elaborate)
$$\equiv D[\mathbf{r0},\mathbf{r1}](\mathbf{r}|a1-da;da)$$
(less elaborate)
$$= D[a1](\mathbf{r}|\mathbf{x1};d_b a) \quad (964)$$

$$\mathbf{d_b}(\mathbf{x}|a1-da)/da$$
$$\longrightarrow D[\mathbf{x}(\mathbf{r1},a1-da),\mathbf{x1}(\mathbf{r},a1-da)](\mathbf{x}|a1-da;da)$$
(fully elaborate)
$$\equiv D[\mathbf{x0},\mathbf{x1}](\mathbf{r}|a1-da;da)$$
(less elaborate)
$$= D[a1](\mathbf{x}|\mathbf{r1};d_b a) \quad (965)$$

and

$$
\begin{aligned}
&\mathbf{d}_f(\mathbf{r}|a1)/da \\
&\qquad \longrightarrow D[\mathbf{r1}(\mathbf{x1},a1),\mathbf{r2}(\mathbf{y2},a1)](\mathbf{r}|a1;da) \\
&\qquad\qquad\qquad \text{(fully elaborate)} \\
&\qquad \equiv D[\mathbf{r1},\mathbf{r2}](\mathbf{r}|a1;da) \\
&\qquad\qquad\qquad \text{(less elaborate)} \\
&\qquad = D[a1](\mathbf{r}|\mathbf{x1};d_f a) \qquad (966) \\
&\mathbf{d}_f(\mathbf{x}|a1)/da \\
&\qquad \longrightarrow D[\mathbf{x1}(\mathbf{r1},a1),\mathbf{x2}(\mathbf{s2},a1)](\mathbf{r}|a1;da) \\
&\qquad\qquad\qquad \text{(fully elaborate)} \\
&\qquad \equiv D[\mathbf{x1},\mathbf{x2}](\mathbf{x}|a1;da) \\
&\qquad\qquad\qquad \text{(less elaborate)} \\
&\qquad = D[a1](\mathbf{x}|\mathbf{r1};d_f a) \qquad (967) \\
&\mathbf{d}_b(\mathbf{r}|a1)/da \\
&\qquad \longrightarrow D[\mathbf{r0}(\mathbf{y0},a1),\mathbf{r1}(\mathbf{x1},a1)](\mathbf{r}|a1;da) \\
&\qquad\qquad\qquad \text{(fully elaborate)} \\
&\qquad \equiv D[\mathbf{r0},\mathbf{r1}](\mathbf{r}|a1;da) \\
&\qquad\qquad\qquad \text{(less elaborate)} \\
&\qquad = D[a1](\mathbf{r}|\mathbf{x1};d_b a) \qquad (968) \\
&\mathbf{d}_b(\mathbf{x}|a1)/da \\
&\qquad \longrightarrow D[\mathbf{x0}(\mathbf{s0},a1),\mathbf{x1}(\mathbf{r1},a1)](\mathbf{x}|a1;da) \\
&\qquad\qquad\qquad \text{(fully elaborate)} \\
&\qquad \equiv D[\mathbf{x0},\mathbf{x1}](\mathbf{x}|a1;da) \\
&\qquad\qquad\qquad \text{(less elaborate)} \\
&\qquad = D[a1](\mathbf{x}|\mathbf{r1};d_b a). \qquad (969)
\end{aligned}
$$

In possible contrast to other functions, in Theoretical Physics the derivatives of *primary* functions always exist as basic derivatives. They need not, however, be continuous.

The condition

$$D[a](\mathbf{r}|\mathbf{x1};da) = \mathbf{0} \qquad (970)$$

is called **materially stalled**.

The joint conditions

$$
\begin{aligned}
D[a1](\mathbf{r}|\mathbf{x1};d_f a) &= \mathbf{0}, \\
D[a1](\mathbf{r}|\mathbf{x1};d_b a) &\neq \mathbf{0}
\end{aligned}
\qquad (971)
$$

are called **materially stopped** at observation *a1*.

Theoretical Physics: Derivatives

The joint conditions

$$\mathrm{D[a1]}(\mathbf{r}|\mathbf{x1};\mathrm{d_f a}) \neq \mathbf{0},$$
$$\mathrm{D[a1]}(\mathbf{r}|\mathbf{x1};\mathrm{d_b a}) = \mathbf{0} \qquad (972)$$

are called **materially started** at observation *a1*.

Clearly

$$\{\mathrm{D[a1]}(\mathbf{r}|\mathbf{x1};\mathrm{da}) = \mathbf{0}\} \leftrightarrow \{\mathrm{D[a1]}(\mathbf{x}|\mathbf{r1};\mathrm{da}) = \mathbf{0}\}$$
$$\leftrightarrow \{\mathrm{D}[\mathbf{r1},\mathbf{r2}](\mathbf{r}|\mathrm{a1+da;da}) = \mathbf{0}\}$$
$$\leftrightarrow \{\mathrm{D}[\mathbf{r1},\mathbf{r2}](\mathbf{r}|\mathrm{a1+da;da}) = \mathbf{0}\}$$
$$\leftrightarrow \{\mathrm{D}[\mathbf{r0},\mathbf{r1}](\mathbf{r}|\mathrm{a1-da;da}) = \mathbf{0}\}$$
$$\leftrightarrow \{\mathrm{D}[\mathbf{x0},\mathbf{x1}](\mathbf{x}|\mathrm{a1-da;da}) = \mathbf{0}\}.$$

Thus a materially stalled condition implies a locally stalled condition.

However stalled conditions do not imply

$$\mathrm{D}[\mathbf{x0},\mathbf{x1}](\mathbf{x}|\mathrm{a1;da}) = \mathbf{0}$$

nor

$$\mathrm{D}[\mathbf{r0},\mathbf{r1}](\mathbf{r}|\mathrm{a1;da}) = \mathbf{0}.$$

The increments of Theoretical Physics open the possibility of constructing entities such as $\mathbf{d_f}(\mathbf{r}|\mathbf{x1})*\mathbf{qd}(\mathrm{d_b}(\mathbf{x}|\mathbf{r1}))$. For the primary functions **r** and **x** there are 4*4=16 such distinct possibilities which Theoretical Physics contemplates. With reference **r1**(**x1**,a1):

$$\mathbf{d_f(r|x1)*qd(d_f(r|x1)) \rightarrow D1[r1,r2]*(r|x1;d_f r)}$$
$$= \mathbf{utr_f(x1},\mathrm{a1})*\mathbf{utr_f(x1},\mathrm{a1}) \qquad (973)$$
$$\mathbf{d_f(r|x1)*qd(d_f(x|r1)) \rightarrow D1[r1,r2]*(r|x1;d_f x)} \qquad (974)$$
$$\mathbf{d_f(r|x1)*qd(d_b(r|x1)) \rightarrow D1[r1,r2]*(r|x1;d_b r)} \qquad (975)$$
$$\mathbf{d_f(r|x1)*qd(d_b(x|r1)) \rightarrow D1[r1,r2]*(r|x1;d_b x)} \qquad (976)$$
$$\mathbf{d_b(r|x1)*qd(d_f(r|x1)) \rightarrow D1[r0,r1]*(r|x1;d_f r)} \qquad (977)$$
$$\mathbf{d_b(r|x1)*qd(d_f(x|r1)) \rightarrow D1[r0,r1]*(r|x1;d_f x)} \qquad (978)$$
$$\mathbf{d_b(r|x1)*qd(d_b(r|x1)) \rightarrow D1[r0,r1]*(r|x1;d_b r)}$$
$$= \mathbf{utr_b(x1},\mathrm{a1})*\mathbf{utr_b(x1},\mathrm{a1}) \qquad (979)$$
$$\mathbf{d_b(r|x1)*qd(d_b(x|r1)) \rightarrow D1[r0,r1]*(r|x1;d_b x)} \qquad (980)$$
$$\mathbf{d_f(x|r1)*qd(d_f(r|x1)) \rightarrow D1[x1,x2]*(x|r1;d_f r)} \qquad (981)$$
$$\mathbf{d_f(x|r1)*qd(d_f(x|r1)) \rightarrow D1[x1,x2]*(x|r1;d_f x)}$$
$$= \mathbf{utx_f(r1},\mathrm{a1})*\mathbf{utx_f(r1},\mathrm{a1}) \qquad (982)$$

$$\mathbf{d_f(x|r1)*qd(d_b(r|x1)) \longrightarrow D1[x1,x2]*(x|r1;d_b r)} \tag{983}$$
$$\mathbf{d_f(x|r1)*qd(d_b(x|r1)) \longrightarrow D1[x1,x2]*(x|r1;d_b x)} \tag{984}$$
$$\mathbf{d_b(x|r1)*qd(d_f(r|x1)) \longrightarrow D1[x0,x1]*(x|r1;d_f r)} \tag{985}$$
$$\mathbf{d_b(x|r1)*qd(d_f(x|r1)) \longrightarrow D1[x0,x1]*(x|r1;d_f x)} \tag{986}$$
$$\mathbf{d_b(x|r1)*qd(d_b(r|x1)) \longrightarrow D1[x0,x1]*(x|r1;d_b r)} \tag{987}$$
$$\mathbf{d_b(x|r1)*qd(d_b(x|r1)) \longrightarrow D1[x0,x1]*(x|r1;d_b x)} = \mathbf{utx_b(r1},a1)*\mathbf{utx_b(r1},a1) \quad \text{as } \mathbf{d \longrightarrow 0}. \tag{988}$$

It also follows that

$$\mathbf{D1[r1,r2]*(r|x1;d_b x)\cdot D1[x0,x1]*(x|r1;d_f r)} = \mathbf{D1[r1,r2]*(r|x1;d_f r)} \tag{989}$$

as well as similar operations.

Under stalled conditions, these gradients become indeterminate.

The trace of **x1** includes **r0**, **r1**, and **r2**; the track of particles through **r1** includes **x0**, **x1**, and **x2**. With reference to **r1**(**x1**,a1) an intrinsic representation can be made in both a forward and backward sense for both the local trace and the material track. Thus

$$D[a1)](\mathbf{r}|\mathbf{x1};d_f a) \equiv D[a1](srf(\mathbf{x1};da)*\mathbf{utrf}(\mathbf{x1},a1) \tag{990}$$
$$D[a1)](\mathbf{r}|\mathbf{x1};d_b a) \equiv D[a1](srb(\mathbf{x1};da)*\mathbf{utrb}(\mathbf{x1},a1) \tag{991}$$
$$D[a1)](\mathbf{x}|\mathbf{r1};d_f a) \equiv D[a1](sxf(\mathbf{r1};da)*\mathbf{utxf}(\mathbf{r1},a1) \tag{992}$$
$$D[a1)](\mathbf{x}|\mathbf{r1};d_b a) \equiv D[a1](sxb(\mathbf{r1};da)*\mathbf{utxb}(\mathbf{r1},a1) \tag{993}$$

from which 16 combinations may be defined of the form

$$\mathbf{D1[r1(x1},a1)]*(srf(\mathbf{x1});\mathbf{d_f r}) \equiv D[a1](srf(\mathbf{x1});da)*\mathbf{utrf}(\mathbf{x1},a1) \tag{994}$$
$$= D(\mathbf{r}|\mathbf{x1};d_f a). \tag{995}$$

These relationships allow 16 further intrinsic representations of the form:

$$\mathbf{D1[r1(x1},a1)]*(\mathbf{r|x1;d_f x}) = D(srf(\mathbf{x1});d(sxf(\mathbf{r1})))*\mathbf{utrf}(\mathbf{x1},a1)*\mathbf{utxf}(\mathbf{r1},a1). \tag{996}$$

Again we see the directional derivatives of Theoretical Physics related to mathematical directional derivatives as both restricted by the local/material reference and yet also expanded in number and descriptive possibilities.

Theoretical Physics: Derivatives

The transformation **r**(**x**,a1) also has gradients. Again let **r1**(**x1**,a1) be the reference of interest. Then, in contrast to the directional case, the gradient **D3** does not depend on an interval of the observational index, but can be defined for one observation alone.

In Mathematics the simply continuous gradient is a linear operator, which is to say that as a transformation it transforms lines into lines. For Theoretical Physics a stronger concept of gradient is needed based on the relationship between physical matter and its location.

Although even if

r1(**x1**,a1) = r1(**y1**,a1)∗**u1** + r2(**y2**,a1)∗**u2** + r3(**y3**,a1)∗**u3**

for some **y1**,**y2**,**y3**, this condition does not mean the gradients of Theoretical Physics can be similarly decomposed. Indeed even

r1(**x**,a1+da)∗**u1**

does not necessarily equal

r(**y1**,a1+da).

At the reference **r1**(**x1**,a1), consider two arbitrary sections, one material

SECT(**x1**,**u1X1**,**u2X1**,**u3X1**)

and the other local,

SECT(**r1**,**u1R1**,**u2R1**,**u3R1**).

The matrix

UX1 ≡ u1∗u1X1 + u2∗u2X1 + u3∗u3X1 (997)

transforms any first quadrant vector **d** ≡ d1∗**u1**+d2∗**u2**+d3∗**u3**, di≥0 into a sectional vector in the material section SECT(**0**,**u1X1**,**u2X1**,**u3X1**) as

d•**UX1** = (d1∗**u1** + d2∗**u2** + d3∗**u3**)•**UX1**
= d1∗**u1X1**+d2∗**u2X1**+d3∗**u3X1**
≡ **dX1** (998)

so that[67]

dX1•[**UX1**$^{-1}$] = d1∗**u1** + d2∗**u2** + d3∗**u3**
= **d**. (999)

Likewise the matrix

$$\mathbf{UR1} \equiv \mathbf{u1}*\mathbf{u1R1} + \mathbf{u2}*\mathbf{u2R1} + \mathbf{u3}*\mathbf{u3R1} \qquad (1000)$$

transforms **d** into the local sectional vector

$$\mathbf{dR1} \equiv d1*\mathbf{u1R1}+d2*\mathbf{u2R1}+d3*\mathbf{u3R1}$$

so that

$$\mathbf{dR1}\cdot[\mathbf{UR1}^{-1}] = \mathbf{d} = \mathbf{dX1}\cdot[\mathbf{UX1}^{-1}]. \qquad (1001)$$

Now let **r** be in SECT(**r1**,**u1R1**,**u2R1**,**u3R1**).

Then for

$$\mathbf{r}-\mathbf{r1} = \mathbf{dR1} \qquad (1002)$$

$$(\mathbf{r}-\mathbf{r1})\cdot[\mathbf{UR1}^{-1}] = \mathbf{dX1}\cdot[\mathbf{UX1}^{-1}] \qquad (1003)$$

$$(\mathbf{r}-\mathbf{r1})\cdot[\mathbf{UR1}^{-1}]\cdot\mathbf{UX1} = \mathbf{dX1} \qquad (1004)$$

$$= \mathbf{x}-\mathbf{x1}. \qquad (1005)$$

Thus offsets in any arbitrary local section
SECT(**r1**,**u1R1**,**u2R1**,**u3R1**)
are linked to offsets in some material section
SECT(**x1**,**u1X1**,**u2X1**,**u3X1**)
by:

$$[\mathbf{UR1}^{-1}]\cdot\mathbf{UX1} : \mathbf{SECT}(\mathbf{r1},\mathbf{u1R1},\mathbf{u2R1},\mathbf{u3R1}) \longrightarrow \mathbf{SECT}(\mathbf{x1},\mathbf{u1X1},\mathbf{u2X1},\mathbf{u3X1}) \qquad (1006)$$

and

$$[\mathbf{UX1}^{-1}]\cdot\mathbf{UR1} : \mathbf{SECT}(\mathbf{x1},\mathbf{u1X1},\mathbf{u2X1},\mathbf{u3X1}) \longrightarrow \mathbf{SECT}(\mathbf{r1},\mathbf{u1R1},\mathbf{u2R1},\mathbf{u3R1}). \qquad (1007)$$

Theoretical Physics: Derivatives

Moreover

$$[\mathrm{T}[\mathbf{UR1}]]\cdot[\mathbf{UX1}] = \mathbf{u1R1}\ast\mathbf{u1X1} + \mathbf{u2R1}\ast\mathbf{u2X1} + \mathbf{u3R1}\ast\mathbf{u3X1} \tag{1008}$$

$$[\mathrm{T}[\mathbf{UX1}]]\cdot[\mathbf{UR1}] = \mathbf{u1X1}\ast\mathbf{u1R1} + \mathbf{u2X1}\ast\mathbf{u2R1} + \mathbf{u3X1}\ast\mathbf{u3R1} \tag{1009}$$

$$\begin{aligned}
[\mathbf{UR1}^{-1}]\cdot[\mathrm{T}[\mathbf{UX1}^{-1}]] = {} & ((\mathbf{u2R1}\wedge\mathbf{u3R1})\ast(\mathbf{u2X1}\wedge\mathbf{u3X1}) \\
& + (\mathbf{u3R1}\wedge\mathbf{u1R1})\ast(\mathbf{u3X1}\wedge\mathbf{u1X1}) \\
& + (\mathbf{u1R1}\wedge\mathbf{u2R1})\ast(\mathbf{u1X1}\wedge\mathbf{u2X1})) \\
& /(\det[\mathbf{UR1}]\ast\det[\mathbf{UX1}]) && (1010) \\
= {} & [[\mathrm{T}[\mathbf{UX1}]]\cdot[\mathbf{UR1}]]^{-1} && (1011)
\end{aligned}$$

$$\begin{aligned}
[\mathbf{UX1}^{-1}]\cdot[\mathrm{T}[\mathbf{UR1}^{-1}] = {} & ((\mathbf{u2X1}\wedge\mathbf{u3X1})\ast(\mathbf{u2R1}\wedge\mathbf{u3R1}) \\
& + (\mathbf{u3X1}\wedge\mathbf{u1X1})\ast(\mathbf{u3R1}\wedge\mathbf{u1R1}) \\
& + (\mathbf{u1X1}\wedge\mathbf{u2X1})\ast(\mathbf{u1R1}\wedge\mathbf{u2R1})) \\
& /(\det[\mathbf{UR1}]\ast\det[\mathbf{UX1}]) && (1012) \\
= {} & [\mathrm{T}[\mathbf{UR1}]\cdot[\mathbf{UX1}]]^{-1} && (1013) \\
= {} & \mathrm{T}[\mathbf{UR1}^{-1}]\cdot\mathrm{T}[\mathbf{UX1}^{-1}]. && (1014)
\end{aligned}$$

The transformations **UR1** and **UX1** relate two sections, but the relationship is not in general one governed by **r**(**x**,a1). Although

$$\mathbf{x}(\mathbf{r1},\mathrm{a1}) = \mathbf{x1} \text{ and } \mathbf{r}(\mathbf{x1},\mathrm{a1}) = \mathbf{r1},$$

it does not necessarily follow that

$$\mathbf{x}(\mathbf{r1}+\mathbf{dR1},\mathrm{a1}) = \mathbf{x1}+\mathbf{dX1}.$$

Under **r**(**x**,a1) however, the particles

$$\mathbf{x1}+\mathbf{dX1}$$
$$\mathbf{x1}+\mathrm{d1X1}\ast\mathbf{u1X1}$$
$$\mathbf{x1}+\mathrm{d2X1}\ast\mathbf{u2X1}$$
$$\mathbf{x1}+\mathrm{d3X1}\ast\mathbf{u3X1}$$

all have specific locations which may not lie in

$$\mathrm{SECT}(\mathbf{r1},\mathbf{u1R1},\mathbf{u2R1},\mathbf{u3R1}).$$

Generally, the local image of material sections under **r**(**x**,a1) are not sections in a mathematical sense at all, since the image of **r**(**x**,a1) need not be a linear function of its domain. In general

$$\begin{aligned}
&\mathbf{r}(\mathbf{x1}+\mathrm{d1X1}\ast\mathbf{u1X1}) \\
&\quad = \mathbf{r1}+\mathrm{d1R1}(\mathrm{d1X1}\ast\mathbf{u1X1})\ast\mathbf{u1R1}(\mathrm{d1X1}\ast\mathbf{u1X1}). && (1015)
\end{aligned}$$

Definition 51 (linearity)

Given

d, a real number;

{**x1** + d∗**ux**} a set of particles on a line;

if there exist

t1, a real constant;

ur, a direction;

such that

$$\mathbf{r}(\mathbf{x1} + d*\mathbf{ux}, a1) = \mathbf{r}(\mathbf{x1}, a1) + t1*d*\mathbf{ur} \tag{1016}$$

the particles in {**x1** + d∗**ux**} are said to be **linearly related locally at observation** a1.

Given

{**r1** + d∗**ur**} a set of locations on a line;

if there exist

t1, a real constant;

ux, a direction;

such that

$$\mathbf{x}(\mathbf{r1} + d*\mathbf{ur}, a1) = \mathbf{x}(\mathbf{r1}, a1) + t1*d*\mathbf{ux} \tag{1017}$$

the locations in {**r1** + d∗**ur**} are said to be **linearly related materially at observation** a1.

end of definition

If particles in {**x1** + d∗**ux**} are linearly related to locations {**r1** + d∗**ur**}, then locations {**r1** + d∗**ur**} are linearly related to particles {**x1** + d∗**ux**}

If the linear relationship extends only to a line segment, the particles and their locations are said to be linearly related on the corresponding line segments, material or local.

Theoretical Physics: Derivatives

Suppose now all the particles in SECT(**x1**,**u1X1**,**u2X1**,**u3X1**) at observation *a1* are linearly related to their locations. Then

$$\mathbf{r}(\mathbf{x1} + \text{diX1} * \mathbf{uiX1}) = \mathbf{r1} + \text{pi} * \text{diX1} * \mathbf{uiR1},\ \text{i}=1,2,3 \quad (1018)$$

and consequently

$$\mathbf{x}(\mathbf{r1} + \text{diR1} * \mathbf{uiR1}) = \mathbf{x1} + \text{diR1} * \mathbf{uiX1}/\text{pi},\ \text{i}=1,2,3. \quad (1019)$$

Thus

$$\text{pi} \equiv \text{diR1}/\text{diX1}\ \text{i}=1,2,3. \quad (1020)$$

In regions of linearity under **r**(**x**,a1), the image of a section is likewise a section. Such sections are called **mutually linearly related.**

Although a material section may be mutually linearly related to a local section, the primary functions may not be locally differentiable there.

Definition 52 (continuously mutually and locally differentiable sections)

Given at observation a1

SECT(**x1**,**u1X1**,**u2X1**,**u3X1**)

and SECT(**r1**,**u1R1**,**u2R1**,**u3R1**),

related linearly and mutually;

if

for every offset

$$\mathbf{dX1} \equiv \text{d1X1} * \mathbf{u1X1} + \text{d2X1} * \mathbf{u2X1} + \text{d3X1} * \mathbf{u3X1}$$

in SECT(**x1**,**u1X1**,**u2X1**,**u3X1**)

$$\begin{aligned}\mathbf{r}(\mathbf{x1} + \mathbf{dX1}) - \mathbf{r}(\mathbf{x1}) &= \mathbf{r}(\mathbf{x1} + \text{d1X1} * \mathbf{u1X1}) - \mathbf{r1} \\ &\quad + \mathbf{r}(\mathbf{x1} + \text{d2X1} * \mathbf{u2X1}) - \mathbf{r1} \\ &\quad + \mathbf{r}(\mathbf{x1} + \text{d3X1} * \mathbf{u3X1}) - \mathbf{r1}\end{aligned} \quad (1021)$$

and for every offset

$$\mathbf{dR1} \equiv \text{d1R1} * \mathbf{u1R1} + \text{d2R1} * \mathbf{u2R1} + \text{d3R1} * \mathbf{u3R1}$$

in SECT(**r1**,**u1R1**,**u2R1**,**u3R1**)

$$\begin{aligned}\mathbf{x}(\mathbf{r1} + \mathbf{dR1}) - \mathbf{x}(\mathbf{r1}) &= \mathbf{x}(\mathbf{r1} + \text{d1R1} * \mathbf{u1R1}) - \mathbf{x1} \\ &\quad + \mathbf{x}(\mathbf{r1} + \text{d2R1} * \mathbf{u2R1}) - \mathbf{x1} \\ &\quad + \mathbf{x}(\mathbf{r1} + \text{d3R1} * \mathbf{u3R1}) - \mathbf{x1}\end{aligned} \quad (1022)$$

then

the sections are said to be **continuously mutually and locally differentiable**.

end of definition

Theoretical Physics:Derivatives

In Theoretical Physics, sections are constrained more particularly.[68]

- *Sections at **r1(x1**,a1) are restricted to intersections with open neighborhoods of **r1(x1,a1**) on which **r(x**,a1) may be linearized*[69]
- *The sections are continuously mutually and locally differentiable*[70].

So constrained, just as **r1** is called the location of the particle **x1** at observation a1, SECT(**r1**,**u1R1**,**u2R1**,**u3R1**) is called the local section of the material section SECT(**x1**,**u1X1**,**u2X1**,**u3X1**) at **r1**(**x1**,a1).

To emphasize this relationship of

SECT(**r1**,**u1R1**,**u2R1**,**u3R1**)

with

SECT(**x1**,**u1X1**,**u2X1**,**u3X1**)

we will sometimes designate the local section of

SECT(**x1**,**u1X1**,**u2X1**,**u3X1**)

as

SECT(**r1**,**u1R**(**X1**),**u2R**(**X1**),**u3R**(**X1**)).

Conversely, to emphasize this relationship SECT(**x1**,**u1X1**,**u2X1**,**u3X1**) with SECT(**r1**,**u1R1**,**u2R1**,**u3R1**) we will sometimes designate the material section of

SECT(**r1**,**u1R1**,**u2R1**,**u3R1**)

as

SECT(**x1**,**u1X**(**R1**),**u2X**(**R1**),**u3X**(**R1**)).

Now let SECT(**x1**,**u1X1**,**u2X1**,**u3X1**) and SECT(**r1**,**u1R1**,**u2R1**,**u3R1**) be such mutually related sections over which **r**(**x**,a1) and **x**(**r**,a1) are mutually and locally differentiable at **r1**(**x1**,a1). Consequently the sectional gradients

D3[**x1**]∗(**r**(**x**,a)|a1;**dX1**)
 = lim (**r**(**x1**+d1X1∗**u1X1**) – **r**(**x1**))
  ∗**u1X1**/d1X1
 + (**r**(**x1**+d2X1∗**u2X1**) – **r**(**x1**))
  ∗**u2X1**/d2X1
 + (**r**(**x1**+d3X1∗**u3X1**) – **r**(**x1**))
  ∗**u3X1**/d3X1
 = p1∗**u1R1**∗**u1X1**
 + p2∗**u2R1**∗**u2X1**
 + p3∗**u3R1**∗**u3X1**, pi >0. (1023)

Theoretical Physics: Derivatives

D3[**r1**]∗(**x**(**r**,a)|a1;**dR1**)
= lim (**x**(**r1**+d1R1∗**u1R1**) – **x**(**r1**))
∗**u1R1**/d1R1
+ (**x**(**r1**+d2R1∗**u2R1**) – **x**(**r1**))
∗**u2R1**/d2R1
+ (**x**(**r1**+d3R1∗**u3R1**) – **x**(**r1**))
∗**u3R1**/d3R1
= **u1X1**∗**u1R1**/p1
+ **u2X1**∗**u2R1**/p2
+ **u3X1**∗**u3R1**/p3, pi >0. (1024)

In Theoretical Physics the gradients of primary functions always exist. By the principle of non–collocation the gradients of primary functions always have rank 3.

Theorem 11 (gradients of primary functions)
Given
r(**x**,a1) and **x**(**r**,a1), an observation of particles
and their locations at observation a1;
mutually related sections
SECT(**r1**,**u1R1**,**u2R1**,**u3R1**) and
SECT(**x1**,**u1X1**,**u2X1**,**u3X1**);
dX1 = d1X1∗**u1X1**+d2X1∗**u2X1**+d3X1∗**u3X1**;
for
UR1 = **u1**∗**u1R1** + **u2**∗**u2R1** +**u3**∗**u3R1**;
UX1 = **u1**∗**u1X1** + **u2**∗**u2X1** +**u3**∗**u3X1**;
r(**x1** + d1X1∗**u1X1**,a1)–**r1**= d1R1∗**u1R1**;
r(**x1** + d2X1∗**u2X1**,a1)–**r1**= d2R1∗**u2R1**;
r(**x1** + d3X1∗**u3X1**,a1)–**r1**= d3R1∗**u3R1**;
dR1 = d1R1∗**u1R1**+d2R1∗**u2R1**+d3R1∗**u3R1**;
p = lim (d1R1/d1X1)∗**u1**
+ (d2R1/d2X1)∗**u2**
+ (d3R1/d3X1)∗**u3**;
then

D3[**x1**]∗(**r**(**x**,a1);**dX1**) = **T**[**UR1**]•**G**(**p**)•**UX1** (1025)

and

D3[**r1**]∗(**x**(**r**,a1);**dR1**) = **T**[**UX1**]•[**G**(**p**)]$^{-1}$•**UR1**. (1026)

Proof:

D3[**r1**(**x1**,a1)]∗(**r**(**x**,a1);**dX1**)
= lim (**r**(**x1** + d1X1∗**u1X1**,a1)−**r1**)∗**u1X1**/d1X1
+ (**r**(**x1** + d2X1∗**u2X1**,a1)−**r1**)∗**u2X1**/d2X1
+ (**r**(**x1** + d3X1∗**u3X1**,a1)−**r1**)∗**u3X1**/d3X1
= lim d1R1∗**u1R1**∗**u1X1**/d1X1
+ d2R1∗**u2R1**∗**u2X1**/d2X1
+ d3R1∗**u3R1**∗**u3X1**/d3X1
= lim d1R1∗**u1R1**∗**u1**•**u1**∗**u1X1**/d1X1
+ d2R1∗**u2R1**∗**u2**•**u2**∗**u2X1**/d2X1
+ d3R1∗**u3R1**∗**u3**•**u3**∗**u3X1**/d3X1
= lim **u1R1**∗**u1**∗(d1R1/d1X1)•**u1**∗**u1X1**
+ **u2R1**∗**u2**•(d2R1/d2X1)∗**u2**∗**u2X1**
+**u3R1**∗**u3**•(d3R1/d3X1)∗**u3**∗**u3X1**
= lim **u1R1**∗**u1**∗(d1R1/d1X1)•**u1**∗**u1**•**u1**∗**u1X1**
+ **u2R1**∗**u2**•(d2R1/d2X1)•**u1**∗**u1**∗**u2**∗**u2X1**
+**u3R1**∗**u3**•(d3R1/d3X1)•**u1**∗**u1**∗**u3**∗**u3X1**
= lim **T**[**UR1**]
•**G**((d1R1/d1X1)∗**u1**
+(d2R1/d2X1)∗**u2**
+(d3R1/d3X1)∗**u3**)
•**UX1**
= **T**[**UR1**•**G**(**p**)•**UX1**.

This proves the first proposition.
Now since **r**(**x1** + diX1∗**uiX1,a1**)−**r1**= diR1∗**uiR1**
x(**r1** + diR1∗**uiR1,**a1)= **x1** + diX1∗**uiX1**.
Moreover,
lim **G**((d1X1/d1R1)∗**u1**+(d2X1/d2R1)∗**u2**+(d3X1/d3R1)∗**u3**)
= $\mathbf{G}^{-1}$(**p**).
Dually then
D3[**x1**(**r1**,a1)]∗(**x**(**r**,a1);**dR1**)
= **T**[**UX1**]•$\mathbf{G}^{-1}$(**p**)•**UR1**.

qed.

Corollary

For continuous gradients in mutually related sectors

dR1 = **dX1**•[$\mathbf{UX1}^{-1}$]•**T**[$\mathbf{UX1}^{-1}$]•**T**[**D3**[**x1**]∗(**r**(**x**,a1);**dX1**)]
= **dX1**•[$\mathbf{UX1}^{-1}$]•**G**(**p**)•**UR1** (1027)
dX1 = **dR1**•[$\mathbf{UR1}^{-1}$]•**T**[$\mathbf{UR1}^{-1}$]•**T**[**D3**[**r1**]∗(**x**(**r**,a1;**dR1**)}
= **dR1**•[$\mathbf{UR1}^{-1}$]•$\mathbf{G}^{-1}$(**p**)•**UX1**. (1028)

Proof:

$\mathbf{dX1}\bullet[\mathbf{UX1}^{-1}]\bullet\mathbf{G(p)}\bullet\mathbf{UR1}$

$=(\mathrm{d1X1}*\mathbf{u1}+\mathrm{d2X1}*\mathbf{u2}+\mathrm{d3X1}*\mathbf{u3})\bullet\mathbf{G(p)}\bullet\mathbf{UR1}$

$=(\mathrm{d1R1}*\mathbf{u1}+\mathrm{d2R1}*\mathbf{u2}+\mathrm{d3R1}*\mathbf{u3})\bullet\mathbf{UR1}$

$=\mathrm{d1R1}*\mathbf{u1R1}+\mathrm{d2R1}*\mathbf{u2R1}+\mathrm{d3R1}*\mathbf{u3R1})$.

From the theorem

$\mathbf{dX1}\bullet[\mathbf{UX1}^{-1}]\bullet\mathbf{G}^{-1}(\mathbf{p})\bullet\mathbf{UR1}$

$= \mathbf{dX1}\bullet[\mathbf{UX1}^{-1}]\bullet\mathbf{T}[\mathbf{UX1}^{-1}]\bullet\mathbf{D3}[\mathbf{x1}]*(\mathbf{r}(\mathbf{x},\mathrm{a1});\mathbf{dX1})$.

The dual is proven similarly. qed

The corollary gives the relationship between a particle

$\mathbf{x1} + \mathbf{dX1} \equiv \mathbf{x1} + \mathrm{d1X1}*\mathbf{u1X1} + \mathrm{d2X1}*\mathbf{u2X1} + \mathrm{d3X1}*\mathbf{u3X1}$

in section SECT(**x1**,**u1X1**,**u2X1**,**u3X1**)

and its location at

$\mathbf{r}(\mathbf{x1} + \mathbf{dX1}) = \mathbf{r1} + \mathbf{dR1}$

$\equiv \mathbf{r1} + \mathrm{d1R1}*\mathbf{u1R1} + \mathrm{d2R1}*\mathbf{u2R1} + \mathrm{d3R1}*\mathbf{u3R1}$

in the mutually related section SECT(**r1**,**u1R1**,**u2R1**,**u3R1**).

Corollary

For **r**(**x**,a1) with simply continuous gradients at **r1**(**x1**,a1)

$$\mathbf{dr} = \mathbf{dx}\bullet\mathbf{T}[\mathbf{D3}[\mathbf{x1}]*(\mathbf{r}(\mathbf{x},\mathrm{a1});\mathbf{dx})] = \mathbf{dx}\bullet\mathbf{G(p)} \quad (1029)$$

$$\mathbf{dx} = \mathbf{dr}\bullet\mathbf{T}[\mathbf{D3}[\mathbf{r1}]*(\mathbf{x}(\mathbf{r},\mathrm{a1});\mathbf{dr})] = \mathbf{dr}\bullet\mathbf{G}^{-1}(\mathbf{p}). \quad (1030)$$

Corollary

$$\mathbf{T}[\mathbf{UR1}]^{-1}\bullet\mathbf{D3}[\mathbf{x1}]*(\mathbf{r}(\mathbf{x},\mathrm{a1});\mathbf{dX1})\bullet\mathbf{UX1}^{-1}\bullet\mathbf{T}[\mathbf{UX1}]^{-1}\bullet\mathbf{D3}[\mathbf{r1}]*(\mathbf{x}(\mathbf{r},\mathrm{a1});\mathbf{dR1})\bullet\mathbf{UR1}^{-1}$$
$$= \mathbf{T}[\mathbf{UX1}]^{-1}\bullet\mathbf{D3}[\mathbf{r1}]*(\mathbf{x}(\mathbf{r},\mathrm{a1});\mathbf{dR1})\bullet\mathbf{UR1}^{-1}\bullet\mathbf{T}[\mathbf{UR1}]^{-1}\bullet\mathbf{D3}[\mathbf{x1}]*(\mathbf{r}(\mathbf{x},\mathrm{a1});\mathbf{dX1})\bullet\mathbf{UX1}^{-1}$$
$$= \mathbf{I}. \quad (1031)$$

Proof:

$\mathbf{T}[\mathbf{UR1}]^{-1}\bullet\mathbf{D3}[\mathbf{x1}]*(\mathbf{r}(\mathbf{x},\mathrm{a1});\mathbf{dX1})\bullet\mathbf{UX1}^{-1} = \mathbf{G(p)}$

$\mathbf{T}[\mathbf{UX1}]^{-1}\bullet\mathbf{D3}[\mathbf{r1}]*(\mathbf{x}(\mathbf{r},\mathrm{a1});\mathbf{dR1})\bullet\mathbf{UR1}^{-1} = \mathbf{G}^{-1}(\mathbf{p})$.

qed

Corollary

D3[**x1**]∗(**r**(**x**,a)|a1;**dX1**)•[**UX1**$^{-1}$]
•[**T**[**UX1**]$^{-1}$]•**D3**[**r1**]∗(**x**(**r**,a)|a1;**dR1**)
= **u1R1**∗**u1R1** + **u2R1**∗**u2R1** + **u3R1**∗**u3R1** (1032)

D3[**r1**]∗(**x**(**r**,a)|a1;**dR1**)•[**UR1**$^{-1}$]
•[**T**[**UR1**]$^{-1}$]•**D3**[**x1**]∗(**r**(**x**,a)|a1;**dX1**)
= **u1X1**∗**u1X1** + **u2X1**∗**u2X1** + **u3X1**∗**u3X1** (1033)

D3[**x1**]∗(**r**(**x**,a)|a1;**dX1**)•[**UX1**$^{-1}$]
•[**T**[**UR1**]$^{-1}$]•**D3**[**x1**]∗(**r**(**x**,a)|a1;**dX1**)
= p1^2∗**u1R1**∗**u1X1**
+ p2^2∗**u2R1**∗**u2X1**
+ p3^2∗**u3R1**∗**u3X1** (1034)

D3[**r1**]∗(**x**(**r**,a)|a1;**dR1**)•[**UR1**$^{-1}$]
•[**T**[**UX1**]$^{-1}$]•**D3**[**r1**]∗(**x**(**r**,a)|a1;**dR1**)
= **u1X1**∗**u1R1**/p1^2
+**u2X1**∗**u2R1**/p2^2
+ **u3X1**∗**u3R1**/p3^2. (1035)

Gradients of vectors are likewise transformations, but in Theoretical Physics they may not be used unrestrictedly. Suppose, for instance

d ≡ d1∗**u1**+d2∗**u2**+d3∗**u3**, di≥0,

a first quadrant vector. Then

d•**T**[**D3**[**r1**]∗(**r**(**x**,a)|a1;**dX1**)]
= (d1∗p1∗**u1**•**u1X1**
+ d2∗p1∗**u2**•**u1X1**
+ d3∗p1∗**u3**•**u1X1**)∗**u1R1**
+ (d1∗p2∗**u1**•**u2X1**
+ d2∗p2∗**u2**•**u2X1**
+ d3∗p2∗**u3**•**u2X1**)∗**u2R1**
+ (d1∗p3∗**u1**•**u3X1**
+ d2∗p3∗**u2**•**u3X1**
+ d3∗p3∗**u3**•**u3X1**)∗**u3R1**

which lies in SECT(**r1**,**u1R1**,**u2R1**,**u3R1**) provided linearity holds.

Theoretical Physics: Derivatives

Likewise

d•T[D3[r1]∗(**r**(**x**,a)|a1;**dX1**)]•[**UX1**$^{-1}$]]
= (d1∗**u1**+d2∗**u2**+d3∗**u3**)
•(**u1∗u2X1∧u3X1**
+ u2∗u231∧u1X1
+ u3∗u1X1∧u2X1)
•(p1∗**u1X1∗u1R1**
+ p2∗**u2X1∗u1R2**
+ p3∗ **u3X1∗u1R31**)/det[**UX1**]
= d1∗p1∗**u1R1** + d2∗p2∗**u2R1** + d3∗p3∗**u3R1** (1036)

which lies in SECT(**r1**,**u1R1**,**u2R1**,**u3R1**) provided linearity holds.

Likewise[71], for **dX1** = d1X1∗**u1X1 +** d2X1∗**u2X1 +** d3X1∗**u3X1** in SECT(**x1**,**u1X1**,**u2X1**,**u3X1**)

dX1•[**UX1**$^{-1}$]•**T[D3[r1]**∗(**r**(**x**,a)|a1;**dX1**)•[**UX1**$^{-1}$]]
= (d1X1∗**u1X1**+d2X1∗**u2X1**+d3X1∗**u3X1**)
•(**u2X1∧u3X1∗u1**
+ u231∧u1X1∗u2
+ u1X1∧u2X1∗u3)
•(**u1∗u2X1∧u3X1**
+ u2∗u231∧u1X1
+ u3∗u1X1∧u2X1)
•(p1∗**u1X1∗u1R1**
+ p2∗**u2X1∗u1R2**
+ p3∗ **u3X1∗u1R31**)/det[**UX1**]
= d1X1∗p1∗**u1R1**
+ d2X1∗p2∗**u2R1**
+ d3X1∗p3∗**u3R1**. (1037)

Dually,

dR1•[**UR1**$^{-1}$]•**T[D3[r1]**∗(**r**(**x**,a)|a1;**dR1**)•[**UR1**$^{-1}$]]
= d1R1∗**u1R1**/p1 + d2R1∗**u2R1**/p2 + d3R1∗**u3R1**/p3.
(1038)

Consequently, *provided* the sections are mutually related, the gradients of primary functions act as transformations

D3[**r1**]∗(**r**(**x**,a)|a1;**dX1**):
SECT(**x1**,**u1X1**,**u2X1**,**u3X1**) ⟶SECT(**r1**,**u1R1**,**u2R1**,**u3R1**)
D3[x1]∗(**x**(**r**,a)|a1;**dR1**):
SECT(**r1**,**u1R1**,**u2R1**,**u3R1**) ⟶SECT(**x1**,**u1X1**,**u2X1**,**u3X1**).

They act in conjunction with the transformations

UR1: SECT(0,u1,u2,u3) ⟶**SECT(0,u1R1,u2R1,u3R1)**
UX1: SECT(0,u1,u2,u3) ⟶**SECT(0,u1X1,u2X1,u3X1)**
[**UR1**]$^{-1}$:**SECT(0,u1R1,u2R1,u3R1)** ⟶**SECT(0,u1,u2,u3)**
[**UX1**]$^{-1}$:**SECT(0,u1X1,u2X1,u3X1)** ⟶**SECT(0,u1,u2,u3)**.

Using these rules other allowable transformations may be formulated.

Since the gradients of the primary functions have rank 3, they have inverses. Their forms may be obtained from equation (1031).

[**D3**[**r1**]∗(**r**(**x**,a)|a1;**dX1**)]$^{-1}$
= [**UX1**$^{-1}$]•**T**[**UX1**]$^{-1}$•[**D3**[**r1**]∗(**x**(**r**,a)|a1;**dR1**)]
•[**UR1**$^{-1}$]•**T**[**UR1**]$^{-1}$ (1039)
[**D3**[**x1**]∗(**x**(**r**,a)|a1;**dR1**)]$^{-1}$
= [**UR1**$^{-1}$]•**T**[**UR1**]$^{-1}$•[**D3**[**x1**]∗(**r**(**x**,a)|a1;**dX1**)]
•[**UX1**$^{-1}$]•**T**[**UX1**]$^{-1}$. (1040)

In Theoretical Physics, these transformations have specific domains which must be respected. Domain and ranges are not to be concatenated haphazardly. It is usually preferable to modulate the concatenation through the first quadrant.

Clearly the sectional gradient of Theoretical Physics, as an idea, differs from the sectional gradient of Mathematics.

In summary:
at **r1**(**x1**,a1),
in all cases from the existence of the observation **r**(**x**,a1)

{**r**(**x1** + **dX1**) = **r1** + **dR1**} ⟷ {**x**(**r1** + **dR1**) = **x1** + **dX1**};

the relationship, however, may not be linear;
for mutual and linear material and local sections

D3[**r1**]∗(**r**(**x**,a)|a1;**dX1**)

and **D3**[**x1**]∗(**x**(**r**,a)|a1;**dR1**) exist

dX1•[**UX1**$^{-1}$]•T[**UX1**$^{-1}$]•T[**D3**[**r1**]∗(**r**(**x**,a)|a1;**dX1**)]
= d1R1∗**u1R1** + d2R1∗**u2R1** + d3R1∗**u3R1**
dR1•[**UR1**$^{-1}$]•T[**UR1**$^{-1}$]•T[**D3**[**x1**]∗(**x**(**r**,a)|a1;**dR1**)]
= d1X1∗**u1X1** + d2X1∗**u2X1** + d3X1∗**u3X1**;

for mutual and linear material and local sections which are also mutually continuously differentiable,

r(**x1** + **dX1**) – **r1** = **r**(**x1** + d1X1∗**u1X1**) – **r1**
+ **r**(**x1** + d2X1∗**u2X1**) – **r1**
+ **r**(**x1** + d3X1∗**u3X1**) – **r1**
x(**r1** + **dR1**) – **x1** = **x**(**r1** + d1R1∗**u1R1**) – **x1**
+ **x**(**r1** + d2R1∗**u2R1**) –**x1**
+ **x**(**r1** + d3R1∗**u3R1**) – **x1**
dX1•[**UX1**$^{-1}$]•T[**UX1**$^{-1}$]•T[**D3**[**r1**]∗(**r**(**x**,a)|a1;**dX1**)] = **dR1**
dR1•[**UR1**$^{-1}$]•T[**UR1**$^{-1}$]•T[**D3**[**x1**]∗(**x**(**r**,a)|a1;**dR1**)] = **dX1**.

Relabeling sections

What is the effect of relabeling the sectional vectors?

For illustration, consider SECT(**x1**,**u1X1**,**u2X1**,**u3X1**) with the labels for **u1X1** and **u2X1** interchanged. Clearly as sets

SECT(**x1**,**u1X1**,**u2X1**,**u3X1**) equals SECT(**x1**,**u2X1**,**u1X1**,**u3X1**).

But,

UX11 ≡ u1∗u1X1 + u2∗u2X1 + u3∗u3X1

clearly differs from

UX12 ≡ u1∗u2X1 + u2∗u1X1 + u3∗u3X1.

However
D3[**x1**]∗(**r**(**x**,a)|a1;**dX11**)
= lim (**r**(**x1**+d1∗**u1X1**)–**r**(**x1**))∗**u1X1**/d1
+ (**r**(**x1**+d2∗**u2X1**)–**r**(**x1**))∗**u2X1**/d2
+ (**r**(**x1**+d3∗**u3X1**)–**r**(**x1**))∗**u3X1**/d3

and
D3[**x1**]∗(**r**(**x**,a)|a1;**dX12**)
= lim(**r**(**x1**+d1∗**u2X1**)−**r**(**x1**))∗**u2X1**/d1
+ (**r**(**x1**+d2∗**u1X1**)−**r**(**x1**))∗**u1X1**/d2
+ (**r**(**x1**+d3∗**u3X1**)−**r**(**x1**))∗**u3X1**/d3
remain unchanged.

Furthermore,

dX1 = d1X1∗**u1X11** + d2X1∗**u2X11** + d3X1∗**u3X11**
= d2X1∗**u1X12** + d1X1∗**u2X12** + d3X1∗**u3X12**.

Thus,

dX1•[**UX11**]$^{-1}$•**T**[**UX11**$^{-1}$]•**T**[**D3**[**r1**]∗(**r**(**x**,a)|a1;**dX11**)]
= d1R1∗**u1R1** + d2R1∗**u2R1** + d3R1∗**u3R1**
dX1•[**UX12**]$^{-1}$•**T**[**UX12**$^{-1}$]•**T**[**D3**[**r1**]∗(**r**(**x**,a)|a1;**dX12**)]
= d2R1∗**u2R1** + d1R1∗**u1R1** + d3R1∗**u3R1**.

Consequently, relabeling has no effects on the transformations.

The matrix,
D3[**r1**]∗(**r**(**x**,a)|a1;**dX1**)•[**UX1**$^{-1}$]
and its transpose,
T[**D3**[**r1**]∗(**r**(**x**,a)|a1;**dX1**)•[**UX1**$^{-1}$]]
are invariant with respect to the order of the directions defining
SECT(**x1**,**u1X1**,**u2X1**,**u3X1**).

Merging sections

Now consider two contiguous mutually locally differentiable sections.

Each contiguous section shares two defining vectors, so that, for instance, if
D3[**x1**]∗(**r**(**x**,a)|a1;**dX1**) = **c1**∗**u1X1**+**c2**∗**u2X1**+**c3**∗**u3X1**
and
D3[**x1**]∗(**r**(**x**,a)|a1;**dX2**) = **c4**∗**u1X2**+**c5**∗**u2X2**+**c6**∗**u3X2**
are contiguous then two of the constituent vectors **uiXj** identical.

We now ask for a criterion for deciding when mutually locally differentiable contiguous sections may be merged.

Theoretical Physics: Derivatives

At observation *a1*, let

SECT(**x1**,**u1X1**,**u2X1**,**u3X1**)
with **r**(SECT(**x1**,**u1X1**,**u2X1**,**u3X1**))
= SECT(**r1**,**u1R1**,**u2R1**,**u3R1**)
with **D3[x1]**∗(**r**(**x**,a)|a1;**dX1**)
= **c1∗u1X1+c2∗u2X1+c3∗u3X1**
SECT(**x1**,**u1X1**,**u2X1**,**u4X1**)
with **r**(SECT(**x1**,**u1X1**,**u2X1**,**u4X1**))
= SECT(**r1**,**u1R1**,**u2R1**,**u4R1**)
with **D3[x1]**∗(**r**(**x**,a)|a1;**dX2**)
= **c1∗u1X1+c2∗u2X1+c4∗u4X1**
SECT(**x1**,**u1X1**,**u3X1**,**u4X1**)
with **r**(SECT(**x1**,**u1X1**,**u3X1**,**u4X1**))
= SECT(**r1**,**u1R1**,**u3R1**,**u4R1**)
with **D3[x1]**∗(**r**(**x**,a)|a1;**dX3**)
= **c1∗u1X1+c3∗u3X1+c4∗u4X1**
SECT(**x1**,**u2X1**,**u3X1**,**u4X1**)
with **r**(SECT(**x1**,**u2X1**,**u3X1**,**u4X1**))
= SECT(**r1**,**u2R1**,**u3R1**,**u4R1**)
with **D3[x1]**∗(**r**(**x**,a)|a1;**dX4**)
= **c2∗u2X1+c3∗u3X1+c4∗u4X1**

be four contiguous mutually locally differentiable sections where **u4X1** lies within SECT(**x1**,**u1X1**,**u2X1**,**u3X1**).

Then **c4** lies within SECT(**r1**,**u1R1**,**u2R1**,**u3R1**).

Clearly each sectional gradient differs from the others.

The possibility of merging
SECT(**x1**,**u1X1**,**u2X1**,**u4X1**),
SECT(**x1**,**u1X1**,**u3X1**,**u4X1**),
and
SECT(**x1**,**u2X1**,**u3X1**,**u4X1**)
into the section
SECT(**x1**,**u1X1**,**u2X1**,**u3X1**)
rests on the action of their gradients as transformations.

If
the transformation of any sectional vector in
SECT(**x1**,**u1X1**,**u2X1**,**u4X1**),
SECT(**x1**,**u1X1**,**u3X1**,**u4X1**),
or SECT(**x1**,**u2X1**,**u3X1**,**u4X1**)
into SECT(**r1**,**u1R1**,**u2R1**,**u4R1**),
SECT(**r1**,**u1R1**,**u3R1**,**u4R1**),
or SECT(**r1**,**u2R1**,**u3R1**,**u4R1**)
is identical to the same material vector represented in
SECT(**x1**,**u1X1**,**u2X1**,**u3X1**)
transformed into
SECT(**r1**,**u1R1**,**u2R1**,**u3R1**),
then
SECT(**x1**,**u1X1**,**u2X1**,**u4X1**),
SECT(**x1**,**u1X1**,**u3X1**,**u4X1**)
and SECT(**x1**,**u2X1**,**u3X1**,**u4X1**)
may be merged into
SECT(**x1**,**u1X1**,**u2X1**,**u3X1**).

This implies SECT(**r1**,**u1R1**,**u2R1**,**u4R1**), SECT(**r1**,**u1R1**,**u3R1**,**u4R1**) and SECT(**r1**,**u2R1**,**u3R1**,**u4R1**) may be merged into
SECT(**r1**,**u1R1**,**u2R1**,**u3R1**).

In such cases the more elaborate sectional notation merely offers a different, more complicated and unneeded representation of a simpler transformation.

Let
dX1=d11∗**u1X1**+d12∗**u2X1**+d13∗**u3X1**
be any vector in SECT(**0**,**u1X1**,**u2X1**,**u3X1**).

Then **dX1** also lies in either
SECT(**0**,**u1X1**,**u2X1**,**u4X1**),
SECT(**0**,**u1X1**,**u3X1**,**u4X1**)
or SECT(**0**,**u2X1**,**u3X1**,**u4X1**).

Say, for illustration, **dX1** lies in SECT(**0**,**u1X1**,**u2X1**,**u4X1**), so that **dX1** has a representation in SECT(**0**,**u1X1**,**u2X1**,**u4X1**) as
dX1 = d21∗**u1X1** + d22∗**u2X1** + d23∗**u4X1**
thus defining **UX2**.

Then,

$$\begin{aligned}
&\mathbf{dX1}\bullet[\mathbf{UX1}^{-1}]\bullet\mathbf{T}[\mathbf{D3}[\mathbf{x1}]*(\mathbf{r}(\mathbf{x},\mathrm{a})|\mathrm{a1};\mathbf{dX1})\bullet[\mathbf{UX1}^{-1}]]\\
&\qquad= \mathbf{dX1}\bullet[\mathbf{u1X1}*\mathbf{u1X1} + \mathbf{u2X1}*\mathbf{u2X1} + \mathbf{u3X1}*\mathbf{u3X1}]^{-1}\\
&\qquad\qquad\bullet(\mathbf{u1X1}*\mathbf{c1} + \mathbf{u2X1}*\mathbf{c2} + \mathbf{u3X1}*\mathbf{c3})\\
&\qquad= \mathrm{d11}*\mathbf{u1X1}+\mathrm{d12}*\mathbf{u2X1}+\mathrm{d13}*\mathbf{u3X1}\\
&\qquad\qquad\bullet((\mathbf{u2X1}\wedge\mathbf{u3X1})*(\mathbf{u2X1}\wedge\mathbf{u3X1})\\
&\qquad\qquad\qquad+ (\mathbf{u3X1}\wedge\mathbf{u1X1})*(\mathbf{u3X1}\wedge\mathbf{u1X1})\\
&\qquad\qquad\qquad+ (\mathbf{u1X1}\wedge\mathbf{u2X1})*(\mathbf{u1X1}\wedge\mathbf{u3X1}))\\
&\qquad\qquad\bullet(\mathbf{u1X1}*\mathbf{c1} + \mathbf{u2X1}*\mathbf{c2} + \mathbf{u3X1}*\mathbf{c3})\\
&\qquad\qquad\qquad\qquad/(\det[\mathbf{UX1}])^2\\
&\qquad= \mathrm{d11}*\mathbf{c1}+\mathrm{d12}*\mathbf{c2}+\mathrm{d13}*\mathbf{c3}
\end{aligned}$$

and

$$\begin{aligned}
&\mathbf{dX1}\bullet[\mathbf{UX2}^{-1}]\bullet\mathbf{T}[\mathbf{D3}[\mathbf{x1}]*(\mathbf{r}(\mathbf{x},\mathrm{a})|\mathrm{a1};\mathbf{dX2})\bullet[\mathbf{UX2}^{-1}]]\\
&\qquad= \mathrm{d21}*\mathbf{c1}+\mathrm{d22}*\mathbf{c2}+\mathrm{d23}*\mathbf{c4}.
\end{aligned}$$

When the image vectors coincide

$$\begin{aligned}
&\mathbf{dX1}\bullet[\mathbf{UX1}^{-1}]\bullet\mathbf{T}[\mathbf{D3}[\mathbf{x1}]*(\mathbf{r}(\mathbf{x},\mathrm{a})|\mathrm{a1};\mathbf{dX1})\bullet[\mathbf{UX1}^{-1}]]\\
&\qquad= \mathbf{dX1}\bullet[\mathbf{UX2}^{-1}]\bullet\mathbf{T}[\mathbf{D3}[\mathbf{x1}]*(\mathbf{r}(\mathbf{x},\mathrm{a})|\mathrm{a1};\mathbf{dX2})\bullet[\mathbf{UX2}^{-1}]]
\end{aligned}$$

so that

$$\begin{aligned}
\mathbf{dX1} &= \mathbf{dX1}\bullet[\mathbf{UX2}^{-1}]\bullet\mathbf{T}[\mathbf{D3}[\mathbf{x1}]*(\mathbf{r}(\mathbf{x},\mathrm{a})|\mathrm{a1};\mathbf{dX2})\bullet[\mathbf{UX2}^{-1}]]\\
&\quad\bullet[[\mathbf{UX1}^{-1}]\bullet\mathbf{T}[\mathbf{D3}[\mathbf{x1}]*(\mathbf{r}(\mathbf{x},\mathrm{a})|\mathrm{a1};\mathbf{dX1})\bullet[\mathbf{UX1}^{-1}]]]^{-1}
\end{aligned}$$

that is,

$$\begin{aligned}
&[\mathbf{UX1}^{-1}]\bullet\mathbf{T}[\mathbf{D3}[\mathbf{x1}]*(\mathbf{r}(\mathbf{x},\mathrm{a})|\mathrm{a1};\mathbf{dX1})\bullet[\mathbf{UX1}^{-1}]]\\
&\qquad\qquad= [\mathbf{UX2}^{-1}]\bullet\mathbf{T}[\mathbf{D3}[\mathbf{x1}]*(\mathbf{r}(\mathbf{x},\mathrm{a})|\mathrm{a1};\mathbf{dX2})\bullet[\mathbf{UX2}^{-1}]]. \qquad (1041)
\end{aligned}$$

For merging, this condition must hold for any vector **dX1** in
SECT(**x1**,**u1X1**,**u2X1**,**u3X1**)∩SECT(**x1**,**u1X1**,**u2X1**,**u4X1**).

The same argument holds for vectors shared by
SECT(**x1**,**u1X1**,**u2X1**,**u3X1**)
and
SECT(**x1**,**u1X1**,**u3X1**,**u4X1**)
or
SECT(**x1**,**u2X1**,**u3X1**,**u4X1**).

We have thus proven,

Theorem 12 (merging sections)
Given
SECT(**x1**,**u1X1**,**u2X1**,**u3X1**);
SECT(**x1**,**u1X1**,**u2X1**,**u4X1**),
a subset of SECT(**x1**,**u1X1**,**u2X1**,**u3X1**);
for
UX1 defined by SECT(**x1**,**u1X1**,**u2X1**,**u3X1**);
UX2 defined by SECT(**x1**,**u1X1**,**u2X1**,**u4X1**);
if

[**UX1**$^{-1}$]•**T**[**D3**[**x1**]∗(**r**(**x**,a)|a1;**dX1**)•[**UX1**$^{-1}$]]
= [**UX2**$^{-1}$]•**T**[**D3**[**x1**]∗(**r**(**x**,a)|a1;**dX2**)•[**UX2**$^{-1}$]] (1042)

then an increment expressed in and properly transformed from SECT(**x1**,**u1X1**,**u2X1**,**u4X1**) will have the same local image as the same increment expressed in and properly transformed from SECT(**x1**,**u1X1**,**u2X1**,**u3X1**).

Dually,
Given
SECT(**r1**,**u1R1**,**u2R1**,**u3R1**);
SECT(**r1**,**u1R1**,**u2R1**,**u4R1**); a subset of SECT(**r1**,**u1R1**,**u2R1**,**u3R1**);
for
UR1 defined by SECT(**r1**,**u1R1**,**u2R1**,**u3R1**);
UR2 defined by SECT(**r1**,**u1R1**,**u2R1**,**u4R1**);
if

[**UR1**$^{-1}$]•**T**[**D3**[**r1**]∗(**x**(**r**,a)|a1;**dR1**)•[**UR1**$^{-1}$]]
= [**UR2**$^{-1}$]•**T**[**D3**[**r1**]∗(**x**(**r**,a)|a1;**dR2**)•[**UR2**$^{-1}$]] (1043)

then an increment expressed in and properly transformed from SECT(**r1**,**u1R1**,**u2R1**,**u4R1**) will have the same material image as the same increment expressed in and properly transformed from SECT(**r1**,**u1R1**,**u2R1**,**u3R1**).

Theoretical Physics: Derivatives

Example: Let **u1X1**=**u1**, **u2X1**=**u2**, **u3X1**=**u3**, **u4X1**=(**u1**+**u2**+**u3**)/sqrt(3) and **r**(**x**)=**x** within a neighborhood of **r1**(**x1**,a1). Then **r**(SECT(**x1**,**u1**,**u2**,**u3**)) = SECT(**r1**,**u1**,**u2**,**u3**),

UX1 = **I** = **D3[x1]∗**(**r**(**x**,a)|a1;**dX1**)

the identity transformation so that

[**UX1**]$^{-1}$•**T**[**D3[x1]∗**(**r**(**x**,a)|a1;**dX1**)•[**UX1**]$^{-1}$] = **I**.

Consider SECT(**x1**,**u1**,**u2**,**u4X1**). Under **r**(**x**)=**x**, **u4R1**=(**u1**+**u2**+**u3**)/sqrt(3).

Then, **r**(SECT(**x1**,**u1**,**u2**,**u4X1**)) = SECT(**r1**,**u1**,**u2**,**u4R1**),

D3[x1]∗(**r**(**x**,a)|a1;**dX2**) = **u1∗u1**+**u2∗u2**+**u4R1∗u4X1**

UX2 = **u1∗u1**+**u2∗u2**+**u3∗u4X1**

and

[**UX2**$^{-1}$]•**T**[**D3[x1]∗**(**r**(**x**,a)|a1;**dX2**)•[**UX2**$^{-1}$]]

= ((**u2∧u4X1**)**∗**(**u2∧u4X1**)

+ (**u4X1∧u1**)**∗**(**u4X1∧u1**)

+(**u1∧u2**)**∗**(**u1∧u2**))

•(**u1∗u1**+**u2∗u2**+**u4X1∗u4R1**)

/(**u1**•(**u2∧u4X1**))^2.

Now, **u1**•(**u2∧u4X1**) = **u2**•(**u4X1∧u1**) = **u4X1**•(**u1∧u2**) = 1/sqrt(3).

Consequently,

[**UX2**$^{-1}$]•**T**[**D3[x1]∗**(**r**(**x**,a)|a1;**dX2**)•[**UX2**$^{-1}$]]

= ((**u2∧u4X1**)**∗u1** + (**u4X1∧u1**)**∗u2**

+(**u1∧u2**)**∗u4X1**)/(**u1**•(**u2∧u4X1**)).

Now,

$$(\mathbf{u2}\wedge\mathbf{u4X1})*\mathbf{u}1/\sqrt{3} = \begin{bmatrix} 1 & 0 & 0 \\ 0 & 0 & 0 \\ -1 & 0 & 0 \end{bmatrix}$$

$$(\mathbf{u4X1}\wedge\mathbf{u1})*\mathbf{u}2/\sqrt{3} = \begin{bmatrix} 0 & 0 & 0 \\ 0 & 1 & 0 \\ 0 & -1 & 0 \end{bmatrix}$$

$$(\mathbf{u1}\wedge\mathbf{u2})*\mathbf{u}4X1/\sqrt{3} = \begin{bmatrix} 0 & 0 & 0 \\ 0 & 0 & 0 \\ 1 & 1 & 1 \end{bmatrix}$$

Thus

[**UX2**$^{-1}$]•**T**[**D3[x1]∗**(**r**(**x**,a)|a1;**dX2**)•[**UX2**$^{-1}$]]

= **I**

= [**UX1**]$^{-1}$•**T**[**D3[x1]∗**(**r**(**x**,a)|a1;**dX1**)•[**UX1**]$^{-1}$]

even though **D3[x1]∗**(**r**(**x**,a)|a1;**dX2**) does not equal **D3[x1]∗**(**r**(**x**,a)|a1;**dX1**).

Under $[\mathbf{UX1}]^{-1}$•**T**[**D3**[**x1**]∗(**r**(**x**,a)|a1;**dX1**)•[**UX1**]$^{-1}$],
the vector **dX1** = **u1+u2+u3** is mapped into **dR1** = **u1+u2+u3**.
Under $[\mathbf{UX2}]^{-1}$•**T**[**D3**[**x1**]∗(**r**(**x**,a)|a1;**dX2**)•[**UX2**]$^{-1}$],
the same vector **dX1** = sqrt(3∗**u4X1** = **u1+u2+u3** is mapped into the same image vector

dR2 = sqrt(3)∗**u4R1** = **u1+u2+u3**.

Clearly under **r**(**x**) = **x** the creation of SECT(**x1**,**u1**,**u2**,**u4X1**) in SECT(**x1**,**u1**,**u2**,**u3**) merely introduces a different representation.

The Continuous Case of Gradients in Quadrant

What then is the case for a continuous gradient and further for the simply continuous gradient?

To see the continuous case, let SECT(**x1**,**±u1**,**±u2**,**±u3**) denote the eight material quadrants where the signs are taken appropriately for the selected quadrant. The material quadrants are associated similarly with eight local sections

SECT(**r1**,**±u1RXj**,**±u2RXj**,**±u3RXj**),

corresponding by sign with j = 1...8.

In the continuous case in each quadrant **r**(**x**,a1) is sectionally differentiable at **r1**(**x1**,a1).

The gradients for the first quadrant, for example, are

D3[**r1**]∗(**r**(**x**,a)|a1;**dX1**)
= b1∗**u1RX1**∗**u1**
+ b2∗**u2RX1**∗**u2**
+ b3∗**u3RX1**∗**u3** (1044)

where bi>0 and **uiRX1** are constant.

Each gradient may be more simply written as
D3[**r1**]∗(**r**(**x**,a)|a1;**dXj**)
= b1j∗**u1RXj**∗**u1**
+ b2j∗**u2RXj**∗**u2**
+ b3j∗**u3RXj**∗**u3**
≡ **b1j∗u1 + b2j∗u2 + b3j∗u3**. (1045)

Theoretical Physics: Derivatives

The vectors **bij** are derived from **r**(**x**,a1) as

$$(\mathbf{r}(\mathbf{x1}+di*\mathbf{ui}) - \mathbf{r}(\mathbf{x1}))/abs(di) \longrightarrow \mathbf{bij} = bij*\mathbf{uiRXj}. \quad (1046)$$

For the continuous case

$$\mathbf{UXj} = \mathbf{u1}*(\pm\mathbf{u1}) + \mathbf{u2}*(\pm\mathbf{u2}) + \mathbf{u3}*(\pm\mathbf{u3}), \quad (1047)$$

$$\mathbf{URXj} = \pm\mathbf{u1}*\mathbf{u1RXj} \pm \mathbf{u2}*\mathbf{u2RXj} \pm \mathbf{u3}*\mathbf{u3RXj} \quad (1048)$$

where again the signs are taken appropriately for the selected quadrant.

Note

$$\mathbf{UXj}\cdot\mathbf{T[UXj]} = \mathbf{I}, \qquad j = 1...8. \quad (1049)$$

$$[\mathbf{UXj}]^{-1} = \mathbf{T}[\mathbf{UXj}], \qquad j = 1...8 \quad (1050)$$

Consider now for illustration

$$\mathbf{dXj} = dx1*\mathbf{u1Xj} + dx2*\mathbf{u2Xj} + dx3*\mathbf{u3Xj}$$
$$= dx1*(-\mathbf{u1}) + dx2*\mathbf{u2} + dx3*\mathbf{u3},$$

a vector in SECT(−**u1**,**u2**,**u3**). Then

$$\mathbf{dXj}\cdot\mathbf{T}[\mathbf{D3}[\mathbf{r1}]*(\mathbf{r}(\mathbf{x},a)|a1;\mathbf{dXj})]$$
$$= \lim -dx1*\mathbf{u1}\cdot(-\mathbf{u1}*(\mathbf{r}(\mathbf{x1}-dx1*\mathbf{u1})-\mathbf{r1}))/dx1$$
$$+ dx2*\mathbf{u2}\cdot\mathbf{u2}*(\mathbf{r}(\mathbf{x1}+dx2*\mathbf{u2})-\mathbf{r1})/dx1$$
$$+ dx3*\mathbf{u3}\cdot\mathbf{u3}*(\mathbf{r}(\mathbf{x1}+dx3*\mathbf{u3})-\mathbf{r1})/dx1$$
$$= \mathbf{r}(\mathbf{x1}-dx1*\mathbf{u1})-\mathbf{r1}$$
$$+ \mathbf{r}(\mathbf{x1}+dx2*\mathbf{u2})-\mathbf{r1}$$
$$+ \mathbf{r}(\mathbf{x1}+dx3*\mathbf{u3})-\mathbf{r1}$$
$$\equiv \mathbf{dRXj}$$

in local section RXj.

Now consider

$$\mathbf{dXk} \equiv dx1*\mathbf{u1Xk} + dx2*\mathbf{u2Xk} + dx3*\mathbf{u3Xk}$$
$$\equiv dx1*\mathbf{u1} + dx2*(-\mathbf{u2}) + dx3*(-\mathbf{u3}),$$
$$= -\mathbf{dXj}.$$

Consequently

$$\mathbf{dXk}\cdot\mathbf{T}[\mathbf{D3}[\mathbf{r1}]*(\mathbf{r}(\mathbf{x},a)|a1;\mathbf{dXk})]$$
$$= \lim\ \mathbf{r}(\mathbf{x1}+dx1*\mathbf{u1})-\mathbf{r1}$$
$$+ \mathbf{r}(\mathbf{x1}-dx2*\mathbf{u2})-\mathbf{r1}$$
$$+ \mathbf{r}(\mathbf{x1}-dx3*\mathbf{u3})-\mathbf{r1}$$
$$\approx \mathbf{r}(\mathbf{x1} + \mathbf{dXk}) - \mathbf{r1}$$
$$\equiv \mathbf{dRXk}$$

in local section RXk.

Now in the *simply* continuous case

dRXj ≡ d1RXj∗**u1RXj** + d2RXj∗**u2RXj** + d3RXj∗**u3RXj**
and
dRXk ≡ d1RXk∗**u1RXk** + d2RXk∗**u2RXk** + d3RXk∗**u3RXk**

where

diRXj∗**uiRXj** = –diRXk∗**uiRXk**

and thus[72]

dRXj = –**dRXk**.

Accordingly, where **r**(**x**) is simply continuously locally differentiable, a straight material track through

x1 – **dXj**,
x1,

and

x1 + **dXj**

is transformed by the appropriate sectional gradients into a straight local trace through

r(**x1** – **dXj**) = **r1** – **dRXj**,
r1,

and

r(**x1** + **dXj**) = **r1** + **dRXj**.

In terms of an orthogonal description

D3[**r1**]∗(**r**(**x**,a)|a1;**dx**) ≡ **b1**∗**u1** + **b2**∗**u2** + **b3**∗**u3**.

Now for the above illustration in the positive definite convention,

dXj = (–dx1)∗**u1** + dx2∗**u2** + dx3∗**u3**,

and

dXk = dx1∗**u1** + (–dx2)∗**u2** + (–dx3)∗**u3**.

Thus

dXj•**T**[**D3**[**r1**]∗(**r**(**x**,a)|a1;**dx**)]
= –dx1∗**b1** + dx2∗**b2** + dx3∗**b2**
= **dRXj**

Theoretical Physics: Derivatives

while

dXk•T[D3[r1]∗(**r**(**x**,a)|a1;**dx**)]
= dx1∗**b1** – dx2∗**b2** – dx3∗**b2**
= **dRXk**

so that again

dRXj = –**dRXk**

under **D3[r1]**∗(**r**(**x**,a)|a1;**dx**).

An identical argument may be made for all the material quadrants. Thus in the simply continuous case, the eight material quadrants are associated under **D3[r1]**∗(**r**(**x**,a)|a1;**dx**) with eight local sections not necessarily quadrants. All straight material tracks through **x1** are carried into straight local traces through **r1**. However, the "density" of these material tracks may vary from local section to local section.

What about the dual?

The dual of the simply continuous case starts with the inverse relationship **x**(**r**) and eight local quadrants SECT(**r1**,**±u1**,**±u2**,**±u3**) where the signs are taken appropriately for the selected quadrant. These local quadrants are associated similarly with eight material sections
SECT(**x1**,**±u1XRj**,**±u2XRj**,**±u3XRj**),
corresponding by sign with j = 1...8.

Each gradient may be written as
D3[x1]∗(**x**(**r**,a)|a1;**dRj**)
= c1j∗**u1XRj**∗**u1**
+ c2j∗**u2XRj**∗**u2**
+ c3j∗**u3XRj**∗**u3**
≡ **c1j**∗**u1** + **c2j**∗**u2** + **c3j**∗**u3**. (1051)

The vectors **cij** are derived from **x**(**r**,a1) as
(**x**(**r1**+di∗**ui**) – **x**(**r1**))/abs(di) ⟶ **cij** = cij∗**uiXRj**. (1052)

Again dually for the continuous case
URj = **u1**∗(**±u1**) + **u2**∗(**±u2**) + **u3**∗(**±u3**), (1053)
UXRj = **±u1**∗**u1XRj** ± **u2**∗**u2XRj** ± **u3**∗**u3XRj** (1054)
where again the signs are taken appropriately for the selected quadrant.

URj•T[URj] = I,   j = 1...8.   (1055)
[**URj**]$^{-1}$ = **T**[**URj**],   j = 1...8   (1056)

Now in the *simply* continuous case

dXRj ≡ d1XRj∗**u1XRj** + d2XRj∗**u2XRj** + d3XRj∗**u3XRj**
and
dXRk ≡ d1XRk∗**u1XRk** + d2XRk∗**u2XRk** + d3XRk∗**u3XRk**

where
diXRj∗**uiXRj** = –diXRk∗**uiXRk**
and thus
dXRj = –**dXRk**
under **D3**[**x1**]∗(**x**(**r**,a)|a1;**dr**).

The same bijective transformation **r**(**x**) considered inversely as **x**(**r**) carries locally straight traces into straight material tracks at **x1**(**r1**) under
D3[**x1**]∗(**x**(**r**,a)|a1;**dr**).

Thus for the positive quadrant:

D3[**r1**]∗(**r**(**x**,a)|a1;**dx**) relates
material SECT(**x1**,**u1**,**u2**,**u3**)
to local SECT(**r1**,**u1RX1**,**u2RX1**,**u3RX1**)

D3[**x1**]∗(**x**(**r**,a)|a1;**dr**) relates
local SECT(**r1**,**u1**,**u2**,**u3**)
to material SECT(**x1**,**u1XR1**,**u2XR1**,**u3XR1**)

D3[**r1**]∗(**x**(**r**,a)|a1;**dRX1**) relates
local SECT(**r1**,**u1RX1**,**u2RX1**,**u3RX1**)
to material SECT(**x1**,**u1**,**u2**,**u3**)

D3[**x1**]∗(**r**(**x**,a)|a1;**dXR1**) relates
material SECT(**x1**,**u1XR1**,**u2XR1**,**u3XR1**)
to local SECT(**r1**,**u1**,**u2**,**u3**)
and similarly for the other quadrants.

Theoretical Physics: Derivatives

In the simply continuous case the gradients over the material quadrants are further related to the gradients over the local quadrants, a relationship we now investigate.

For

$$
\begin{aligned}
&\mathbf{D3}[\mathbf{r1}]*(\mathbf{r}(\mathbf{x},a)|a1;\mathbf{dx}) \\
&\quad = b1*\mathbf{u1RX1}*\mathbf{u1} \\
&\qquad + b2*\mathbf{u2RX1}*\mathbf{u2} \\
&\qquad + b3*\mathbf{u3RX1}*\mathbf{u3} \\
&\quad = \mathbf{u1RX1}*b1*\mathbf{u1} \\
&\qquad + \mathbf{u2RX1}*b2*\mathbf{u2} \\
&\qquad + \mathbf{u3RX1}*b3*\mathbf{u3} \\
&\quad = [\mathbf{u1RX1}*\mathbf{u1} + \mathbf{u2RX1}*\mathbf{u2} + \mathbf{u3RX1}*\mathbf{u3}]\bullet\mathbf{G}(\mathbf{b}) \\
&\quad = \mathbf{T}[\mathbf{URX1}]\bullet\mathbf{G}(\mathbf{b}) \\
&\quad = \mathbf{T}[\mathbf{G}(\mathbf{b})\bullet\mathbf{URX1}]
\end{aligned}
\qquad (1057)
$$

where **G** is the diagonal vector operator.

Moreover in the continuous case, for $bi>0$ any material vector

$$
\begin{aligned}
\mathbf{dx} &\equiv \mathbf{x}-\mathbf{x1} \\
&= dx1*\mathbf{u1} + dx2*\mathbf{u2} + dx3*\mathbf{u3}
\end{aligned}
$$

is transformed by $\mathbf{D3}[\mathbf{r1}]*(\mathbf{r}(\mathbf{x},a)|a1;\mathbf{dx})$ into its location according to

$$
\begin{aligned}
\mathbf{dx}\bullet\mathbf{T}[\mathbf{D3}[\mathbf{r1}]*(\mathbf{r}(\mathbf{x},a)|a1;\mathbf{dx})] &= \mathbf{dx}\bullet\mathbf{G}(\mathbf{b})\bullet\mathbf{URX1} \\
&= dx1*b1*\mathbf{u1RX1} \\
&\quad + dx2*b2*\mathbf{u2RX1} \\
&\quad + dx3*b3*\mathbf{u3RX1}.
\end{aligned}
\qquad (1058)
$$

In a neighborhood in which linearization holds,

$$
\begin{aligned}
&\mathbf{r}(\mathbf{x1}+\mathbf{dx}) - \mathbf{r}(\mathbf{x1}) \\
&\quad = dx1*b1*\mathbf{u1RX1} \\
&\qquad + dx2*b2*\mathbf{u2RX1} \\
&\qquad + dx3*b3*\mathbf{u3RX1} \\
&\quad \equiv dr1*\mathbf{u1} + dr2*\mathbf{u2} + dr3*\mathbf{u3} \\
&\quad \equiv \mathbf{dr}.
\end{aligned}
$$

Consequently the transpose of the continuous gradient of $\mathbf{r}(\mathbf{x})$, $\mathbf{T}[\mathbf{D3}[\mathbf{r1}]*(\mathbf{r}(\mathbf{x},a)|a1;\mathbf{dx})]$, describes the transformation of a material neighborhood into the local neighborhood it occupies.

Similarly

$$\mathbf{D3[x1]}*(\mathbf{x}(\mathbf{r},a)|a1;\mathbf{dr}) = \mathbf{T[UXR1]}\bullet\mathbf{G}(\mathbf{c}) = \mathbf{T}[\mathbf{G}(\mathbf{c})\bullet\mathbf{UXR1}] \qquad (1059)$$

describes the continuous transformation of a local neighborhood into the material neighborhood which occupies it under **x**(**r**) at **x1**(**r1**,a1).

For the simply continuous case the two transformations,
D3[**x1**]∗(**x**(**r**,a)|a1;**dr**) and **D3**[**r1**]∗(**r**(**x**,a)|a1;**dx**),
must be related through the relationship of **r**(**x**) to **x**(**r**) in a neighborhood of **r1**(**x1**).

Since in Theoretical Physics

$$\mathbf{r}(\mathbf{x}(\mathbf{r1})) = \mathbf{r1}$$
$$\mathbf{x}(\mathbf{r}(\mathbf{x1})) = \mathbf{x1}$$

and

$$\mathbf{x}(\mathbf{r}(\mathbf{x1}+\mathbf{dx})) - \mathbf{x}(\mathbf{r}(\mathbf{x1})) = \mathbf{dx}, \qquad (1060)$$

$$\begin{aligned}\mathbf{dx}\bullet\mathbf{T}[\mathbf{D3}[\mathbf{r1}]*(\mathbf{r}(\mathbf{x},a)|a1;\mathbf{dx})]\bullet\mathbf{T}[\mathbf{D3}[\mathbf{x1}]*(\mathbf{x}(\mathbf{r},a)|a1;\mathbf{dr})]\\ = \mathbf{dr}\bullet\mathbf{T}[\mathbf{D3}[\mathbf{x1}]*(\mathbf{x}(\mathbf{r},a)|a1;\mathbf{dr})]\\ = \mathbf{x}(\mathbf{r}(\mathbf{x1}+\mathbf{dx})) - \mathbf{x}(\mathbf{r}(\mathbf{x1}))\\ = \mathbf{dx}\end{aligned} \qquad (1061)$$

for all **dx**.

In the simply continuous case, then

$$\begin{aligned}\mathbf{T}[\mathbf{D3}[\mathbf{r1}]*(\mathbf{r}(\mathbf{x},a)|a1;\mathbf{dx})]\bullet\mathbf{T}[\mathbf{D3}[\mathbf{x1}]*(\mathbf{x}(\mathbf{r},a)|a1;\mathbf{dr})]\\ = \mathbf{I}\\ = [\mathbf{D3}[\mathbf{x1}]*(\mathbf{x}(\mathbf{r},a)|a1;\mathbf{dr})]\\ \bullet[\mathbf{D3}[\mathbf{r1}]*(\mathbf{r}(\mathbf{x},a)|a1;\mathbf{dx})]\end{aligned} \qquad (1062)$$

and similarly

$$\begin{aligned}\mathbf{T}[\mathbf{D3}[\mathbf{x1}]*(\mathbf{x}(\mathbf{r},a)|a1;\mathbf{dr})]\bullet\mathbf{T}[\mathbf{D3}[\mathbf{r1}]*(\mathbf{r}(\mathbf{x},a)|a1;\mathbf{dx})]\\ = \mathbf{I}\\ = [\mathbf{D3}[\mathbf{r1}]*(\mathbf{r}(\mathbf{x},a)|a1;\mathbf{dx})]\\ \bullet[\mathbf{D3}[\mathbf{x1}]*(\mathbf{x}(\mathbf{r},a)|a1;\mathbf{dr})].\end{aligned} \qquad (1063)$$

Theoretical Physics: Derivatives

Consequently for the simply continuous case in Theoretical Physics

D3[**x1**]∗(**x**(**r**,a)|a1;**dr**) and **D3**[**r1**]∗(**r**(**x**,a)|a1;**dx**)

are inverse transformations relating material and local neighborhoods.

In particular given

D3[**r1**]∗(**r**(**x**,a)|a1;**dx**) = **b1**∗**u1** + **b2**∗**u2** + **b3**∗**u3**,

D3[**x1**]∗(**x**(**r**,a)|a1;**dr**)
= [**D3**[**r1**]∗(**r**(**x**,a)|a1;**dx**)]$^{-1}$
= (**u1**∗(**b2**∧**b3**)
+ **u2**∗(**b3**∧**b1**)
+ **u3**∗(**b1**∧**b2**))/(**b1**•(**b2**∧**b3**))
= (**u1**∗**u2RX1**∧**u3RX1**/b1
+ **u2**∗(**u3RX1**∧**u1RX1**)/b2
+ **u3**∗(**u1RX1**∧**u2RX1**)/b3)
/(**u1RX1**•(**u2RX1**∧**u3RX1**)) (1064)
= [[**G**(**b**)]$^{-1}$]•**T**[**URX1**]$^{-1}$ (1065)

defines

D3[**x1**]∗(**x**(**r**,a)|a1;**dr**)
= **c1**∗**u1** + **c2**∗**u2** + **c3**∗**u3**
= **T**[**UXR1**]•**G**(**c**)
= **G**(**c**)•**UXR1**.

Consequently

G(**c**) = [**G**(**b**)]$^{-1}$ (1066)

UXR1 = **T**[**URX1**]$^{-1}$. (1067)

Putting

u2RX1∧**u3RX1** ≡ g11∗**u1** + g12∗**u2** + g13∗**u3**
u3RX1∧**u1RX1** ≡ g21∗**u1** + g22∗**u2** + g23∗**u3**
u1RX1∧**u2RX1** ≡ g31∗**u1** + g32∗**u2** + g33∗**u3** (1068)

c1 = (g11∗**u1**/b1
+ g21∗**u2**/b2
+ g31∗**u3**/b3)/det(**URX1**)
= c1∗**u1XR1**

c2 = (g12∗**u1**/b1
+ g22∗**u2**/b2
+ g32∗**u3**/b3)/det(**URX1**)
= c2∗**u2XR1**

c3 = (g13∗**u1**/b1
+ g23∗**u2**/b2
+ g33∗**u3**/b3)/det(**URX1**)
= c3∗**u3XR1**. (1069)

Dually, given **D3**[**x1**]∗(**x**(**r**,a)|a1;**dr**) = **c1∗u1 + c2∗u2 + c3∗u3**,

D3[**r1**]∗(**r**(**x**,a)|a1;**dx**)
= [**D3**[**x1**]∗(**x**(**r**,a)|a1;**dr**)]$^{-1}$
= (**u1∗(c2∧c3) + u2∗(c3∧c1) + u3∗(c1∧c2)**)
/(**c1**•(**c2∧c3**))
= (**u1∗u2XR1∧u3XR1**/c1
+ u2∗(**u3XR1∧u1XR1**)/c2
+ u3∗(**u1XR1∧u2XR1**)/c3)
/(**u1XR1**•(**u2XR1∧u3XR1**)) (1070)

defines

D3[**r1**]∗(**r**(**x**,a)|a1;**dx**)
≡ **G**(**b**)•**T**[**UXR1**]$^{-1}$
= **b1∗u1 + b2∗u2 + b3∗u3**.

Consequently

G(**b**) = [**G**(**c**)]$^{-1}$ (1071)
URX1 = **T**[**UXR1**]$^{-1}$. (1072)

Putting

u2XR1∧u3XR1 ≡ h11∗**u1** + h12∗**u2** + h13∗**u3**
u3XR1∧u1XR1 ≡ h21∗**u1** + h22∗**u2** + h23∗**u3**
u1XR1∧u2XR1 ≡ h31∗**u1** + h32∗**u2** + h33∗**u3** (1073)

b1 = (h11∗**u1**/c1 + h21∗**u2**/c2 + h31∗**u3**/c3)
/det[**UXR1**]
b2 = (h12∗**u1**/c1 + h22∗**u2**/c2 + h32∗**u3**/c3)
/det[**UXR1**]
b3 = (h13∗**u1**/c1 + h23∗**u2**/c2 + h33∗**u3**/c3)
/det[**UXR1**]. (1074)

In summary for the simply continuous case where **r**(**x**) and **x**(**r**) are locally differentiable at **r1**(**x1**,a1):

- *material and local quadrant gradients exist and may be succinctly described by **D3[r1]***(**r**(**x**,a)|a1;**dx**) or by **D3[x1]***(**x**(**r**,a)|a1;**dr**);*
- ***D3[r1]***(**r**(**x**,a)|a1;**dx**) and **D3[x1]***(**x**(**r**,a)|a1;**dr**) are inverses of each other;*
- *straight material tracks through **x1** are carried into straight local traces through **r1** and vice versa.*

Where these conditions hold, the gradients are **simply continuous** at **r1**(**x1**,a1).

The reader should note the opposite situation. Where **r**(**x**) is not simply continuously locally differentiable

- *material and local quadrant gradients may not be uniformly described by **D3[r1]***(**r**(**x**,a)|a1;**dx**) or by **D3[x1]***(**x**(**r**,a)|a1;**dr**);*
- *some straight material tracks are not be carried into straight local traces and vice versa.*

Example. Let SECT(**x1**,±**u1**,±**u2**,±**u3**) denote eight material quadrants (the signs are taken appropriately for the selected quadrant) with continuous sectional gradients. The eight sections can be denoted by six vectors. Under **r**(**x**,a1) let

u1 map into **u1RX1** = −**u1**
u2 map into **u2RX2** = −**u1** + **u2**
u3 map into **u3RX3** = **u3**
−**u1** map into −**u1RX4** = **u1**
−**u2** map into −**u2RX5** = **u1** − **u2**
−**u3** map into −**u3RX6** = −**u3**.

The local sections corresponding to the material sections are

1. SECT(**x1**,**+u1**,**+u2**,**+u3**) ⟶ SECT(r**1**,−**u1**,(**u2**−**u1**)/sqrt(2),**+u3**)
2. SECT(**x1**,−**u1**,**+u2**,**+u3**) ⟶ SECT(r**1**,**+u1**,(**u2**−**u1**)/sqrt(2),**+u3**)
3. SECT(**x1**,**+u1**,−**u2**,**+u3**) ⟶ SECT(r**1**,−**u1**,(**u1**−**u2**)/sqrt(2),**+u3**)
4. SECT(**x1**,−u1,−u2,+u3) ⟶ SECT(r**1**,+u1,(u1−u2)/sqrt(2),+u3)
5. SECT(**x1**,**+u1**,**+u2**,−**u3**) ⟶ SECT(r**1**,−**u1**,(**u2**−**u1**)/sqrt(2),−**u3**)
6. SECT(**x1**,−**u1**,**+u2**,−**u3**) ⟶ SECT(r**1**,**+u1**,(**u2**−**u1**)/sqrt(2),−**u3**)
7. SECT(**x1**,**+u1**,−**u2**,−**u3**) ⟶ SECT(r**1**,−**u1**,(**u1**−**u2**)/sqrt(2),−**u3**)
8. SECT(**x1**,−**u1**,−**u2**,−**u3**) ⟶ SECT(r**1**,**+u1**,(**u1**−**u2**)/sqrt(2),−**u3**).

Let **r**(**x**) be locally differentiable at **r1**(**x1**) with continuous sectional gradients,
D3[r1]*(**r**(**x**,a)|a1;**dX1**) = **−u1*u1 + (u2−u1)*u2 + u3*u3**
D3[r1]*(**r**(**x**,a)|a1;**dX2**) = **−u1*u1 + (u2−u1)*u2 + u3*u3**
D3[r1]*(**r**(**x**,a)|a1;**dX3**) = **−u1*u1 + (u2−u1)*u2 + u3*u3**
D3[r1]*(**r**(**x**,a)|a1;**dX4**) = **−u1*u1 + (u2−u1)*u2 + u3*u3**
D3[r1]*(**r**(**x**,a)|a1;**dX5**) = **−u1*u1 + (u2−u1)*u2 + u3*u3**
D3[r1]*(**r**(**x**,a)|a1;**dX6**) = **−u1*u1 + (u2−u1)*u2 + u3*u3**
D3[r1]*(**r**(**x**,a)|a1;**dX7**) = **−u1*u1 + (u2−u1)*u2 + u3*u3**
D3[r1]*(**r**(**x**,a)|a1;**dX8**) = **−u1*u1 + (u2−u1)*u2 + u3*u3**.

These eight gradients are adequately represented in the alternate convention by

D3[r1]*(**r**(**x**,a)|a1;**dx**) = **−u1*u1 + (u2−u1)*u2 + u3*u3**.

Thus,

b1 = **−u1** = +1*(**−u1**)
b2 = **u2** − **u1** = sqrt(2)***u**(**u2−u1**)
b3 = **u3** = +1*(**+u3**)

and

b = 1***u1** + sqrt(2)***u2** + 1***u3**.

Also

UXj = ±**u1*u1** ± **u2*u2** ± **u3*u3**, j= 1..8
[**UXj**]$^{-1}$ = ±**u1*u1** ± **u2*u2** ± **u3*u3**, j= 1..8.

Then,

URX1 = **−u1*u1** + **u2***(**u2−u1**)/sqrt(2) + **u3*u3**
URXj =±**u1*u1R1** ± **u2*u2R1** ± **u3*u3R1**, j= 1..8

so that

T[URX1]•G(**b**) = [**−u1*u1** + (**u2−u1**)***u2**/sqrt(2) + **u3*u3**]
• [1***u1*u1** + sqrt(2)***u2*u2** + 1***u3*u3**]
= **−u1*u1 + (u2−u1)*u2 + u3*u3**
= **D3[r1]***(**r**(**x**,a)|a1;**dx**).

The transformation **D3[r1]***(**r**(**x**,a)|a1;**dx**) has for inverse

[**D3[r1]***(**r**(**x**,a)|a1;**dx**)]$^{-1}$ = **u1*u1** +**u1*u2** − **u2*u2** − **u3*u3**.

The inverse sectional gradients are:
D3[x1]*(**x**(**r**,a)|a1;**dRX1**) = **−u1*(u1+u2) + u2*u2 + u3*u3**
D3[x1]*(**x**(**r**,a)|a1;**dRX2**) = **−u1*(u1+u2) + u2*u2 + u3*u3**
D3[x1]*(**x**(**r**,a)|a1;**dRX3**) = **−u1*(u1+u2) + u2*u2 + u3*u3**
D3[x1]*(**x**(**r**,a)|a1;**dRX4**) = **−u1*(u1+u2) + u2*u2 + u3*u3**
D3[x1]*(**x**(**r**,a)|a1;**dRX5**) = **−u1*(u1+u2) + u2*u2 + u3*u3**
D3[x1]*(**x**(**r**,a)|a1;**dRX6**) = **−u1*(u1+u2) + u2*u2 + u3*u3**
D3[x1]*(**x**(**r**,a)|a1;**dRX7**) = **−u1*(u1+u2) + u2*u2 + u3*u3**
D3[x1]*(**x**(**r**,a)|a1;**dRX8**) = **−u1*(u1+u2) + u2*u2 + u3*u3**.

Theoretical Physics: Derivatives

The inverse of **D3[r1]*(r(x,a)|a1;dx)** is

D3[x1]*(x(r,a)|a1;dr) = −u1*u1 + (u2−u1)*u2 + u3*u3.

Thus in this example **D3[r1]*(r(x,a)|a1;dx)** and **D3[x1]*(x(r,a)|a1;dr)** are identical matrices.

Look now at two vectors

dX1 = −**dX8** = dx11***u1** + dx12***u2** + dx13***u3**
= −(dx11*(−**u1**) + dx12*(−**u2**) + dx13*(−**u3**))
≡ **d**.

The vectors **x1**+**dX8**, **x1**, and **x1**+**dX1** determine a straight track of material. Then

dX1•T(D3X1*(r;x))= r(x1+dX1) − r1
= −dx11***u1** + dx12*(**u2**−**u1**) + dx13***u3**

and

dX8•T(D3X8*(r;x))= r(x1+dX8) − r1
= +dx11***u1** − dx12*(**u2**−**u1**) − dx13***u3**.

Thus **r(x1 + dX1) − r1 = r1 − r(x1 + dX8)**. It follows that **r(x1 + dX8)**, **r(x1)**, and **r(x1 + dX1)** are also collinear. Consequently we are dealing with the simply continuous case.

The local quadrants and corresponding material sections are:

1. SECT(r**1**,+**u1**,+**u2**,+**u3**) ⟶ SECT(**x1**,−**u1**,(**u2**−**u1**)/sqrt(2),+**u3**)
2. SECT(r**1**,−**u1**,+**u2**,+**u3**) ⟶ SECT(**x1**,+**u1**,(**u2**−**u1**)/sqrt(2),+**u3**)
3. SECT(r**1**,+**u1**,−**u2**,+**u3**) ⟶ SECT(**x1**,−**u1**,(**u1**−**u2**)/sqrt(2),+**u3**)
4. SECT(r**1**,−**u1**,−**u2**,+**u3**) ⟶ SECT(**x1**,+**u1**,(**u1**−**u2**)/sqrt(2),+**u3**)
5. SECT(r**1**,+**u1**,+**u2**,−**u3**) ⟶ SECT(**x1**,−**u1**,(**u2**−**u1**)/sqrt(2),−**u3**)
6. SECT(r**1**,−**u1**,+**u2**,−**u3**) ⟶ SECT(**x1**,+**u1**,(**u2**−**u1**)/sqrt(2),−**u3**)
7. SECT(r**1**,+**u1**,−**u2**,−**u3**) ⟶ SECT(**x1**,−**u1**,(**u1**−**u2**)/sqrt(2),−**u3**)
8. SECT(r**1**,−**u1**,−**u2**,−**u3**) ⟶ SECT(**x1**,+**u1**,(**u1**−**u2**)/sqrt(2),−**u3**)

with gradients

D3[x1]*(x(r,a)|a1;dR1) = −u1*u1 + (u2−u1)*u2 + u3*u3
D3[x1]*(x(r,a)|a1;dR2) = −u1*u1 + (u2−u1)*u2 + u3*u3
D3[x1]*(x(r,a)|a1;dR3) = −u1*u1 + (u2−u1)*u2 + u3*u3
D3[x1]*(x(r,a)|a1;dR4) = −u1*u1 + (u2−u1)*u2 + u3*u3
D3[x1]*(x(r,a)|a1;dR5) = −u1*u1 + (u2−u1)*u2 + u3*u3
D3[x1]*(x(r,a)|a1;dR6) = −u1*u1 + (u2−u1)*u2 + u3*u3
D3[x1]*(x(r,a)|a1;dR7) = −u1*u1 + (u2−u1)*u2 + u3*u3
D3[x1]*(x(r,a)|a1;dR8) = −u1*u1 + (u2−u1)*u2 + u3*u3.

This simply continuous case may thus be abbreviated as

D3[r1]*(r(x,a)|a1;dx) = −u1*u1 − u1*u2 + u2*u2 + u3*u3

and **D3[x1]*(x(r,a)|a1;dr) = −u1*u1 − u1*u2 + u2*u2 + u3*u3**

and thus

D3[r1]*(r(x,a)|a1;dx)•D3[x1]*(x(r,a)|a1;dr) = I.

Example. Let SECT(**x1**,±**u1**,±**u2**,±**u3**),denote eight material quadrants (the signs are taken appropriately for the selected quadrant) with continuous sectional gradients. The eight local sections into which the material quadrants are mapped under **r**(**x**,a1) can be denoted by six vectors. Let

u1X1 = **u1** map into −**u1** = **u1RX1**
u2X2 = **u2** map into −**u1** + **u2**= **u2RX2**
u3X3 = **u3** map into **u3** = **u3RX3**
u4X4 = −**u1** map into **u1** = **u4RX4**
u5X5 = −**u2** map into −**u2** = **u5RX5**
u6X6 = −**u3** map into −**u3** = **u6RX6**.

The local sections corresponding to material quadrants are

1. SECT(**x1**,+**u1**,+**u2**,+**u3**) ⟶ SECT(**r1**,−**u1**,(**u2**−**u1**)/sqrt(2),+**u3**)
2. SECT(**x1**,−**u1**,+**u2**,+**u3**) ⟶ SECT(**r1**,+**u1**,(**u2**−**u1**)/sqrt(2),+**u3**)
3. SECT(**x1**,+**u1**,−**u2**,+**u3**) ⟶ SECT(**r1**,−**u1**,−**u2**,+**u3**)
4. SECT(**x1**,−**u1**,−**u2**,+**u3**) ⟶ SECT(**r1**,+**u1**,−**u2**,+**u3**)
5. SECT(**x1**,+**u1**,+**u2**,−**u3**) ⟶ SECT(**r1**,−**u1**,(**u2**−**u1**)/sqrt(2),−**u3**)
6. SECT(**x1**,−**u1**,+**u2**,−**u3**) ⟶ SECT(**r1**,+**u1**,(**u2**−**u1**)/sqrt(2),−**u3**)
7. SECT(**x1**,+**u1**,−**u2**,−**u3**) ⟶ SECT(**r1**,−**u1**,−**u2**,−**u3**)
8. SECT(**x1**,−**u1**,−**u2**,−**u3**) ⟶ SECT(**r1**,+**u1**,−**u2**,−**u3**)

with continuous sectional gradients

D3[**r1**]∗(**r**(**x**,a)|a1;**dX1**) = −**u1**∗**u1** + (**u2**−**u1**)∗**u2** + **u3**∗**u3**
D3[**r1**]∗(**r**(**x**,a)|a1;**dX2**) = −**u1**∗**u1** + (**u2**−**u1**)∗**u2** + **u3**∗**u3**
D3[**r1**]∗(**r**(**x**,a)|a1;**dX3**) = −**u1**∗**u1** + **u2**∗**u2** + **u3**∗**u3**
D3[**r1**]∗(**r**(**x**,a)|a1;**dX4**) = −**u1**∗**u1** + **u2**∗**u2** + **u3**∗**u3**
D3[**r1**]∗(**r**(**x**,a)|a1;**dX5**) = −**u1**∗**u1** + (**u2**−**u1**)∗**u2** + **u3**∗**u3**
D3[**r1**]∗(**r**(**x**,a)|a1;**dX6**) = −**u1**∗**u1** + (**u2**−**u1**)∗**u2** + **u3**∗**u3**
D3[**r1**]∗(**r**(**x**,a)|a1;**dX7**) = −**u1**∗**u1** + **u2**∗**u2** + **u3**∗**u3**
D3[**r1**]∗(**r**(**x**,a)|a1;**dX8**) = −**u1**∗**u1** + **u2**∗**u2** + **u3**∗**u3**

whose inverses

[**D3**[**r1**]∗(**r**(**x**,a)|a1;**dX1**)]$^{-1}$ = −**u1**∗**u1** + (**u2**−**u1**)∗**u2** + **u3**∗**u3**
[**D3**[**r1**]∗(**r**(**x**,a)|a1;**dX2**)]$^{-1}$ = −**u1**∗**u1** + (**u2**−**u1**)∗**u2** + **u3**∗**u3**
[**D3**[**r1**]∗(**r**(**x**,a)|a1;**dX3**)]$^{-1}$ = −**u1**∗**u1** + **u2**∗**u2** + **u3**∗**u3**
[**D3**[**r1**]∗(**r**(**x**,a)|a1;**dX4**)]$^{-1}$ = −**u1**∗**u1** + **u2**∗**u2** + **u3**∗**u3**
[**D3**[**r1**]∗(**r**(**x**,a)|a1;**dX5**)]$^{-1}$ = −**u1**∗**u1** + (**u2**−**u1**)∗**u2** + **u3**∗**u3**
[**D3**[**r1**]∗(**r**(**x**,a)|a1;**dX6**)]$^{-1}$ = −**u1**∗**u1** + (**u2**−**u1**)∗**u2** + **u3**∗**u3**
[**D3**[**r1**]∗(**r**(**x**,a)|a1;**dX7**)]$^{-1}$ = −**u1**∗**u1** + **u2**∗**u2** + **u3**∗**u3**
[**D3**[**r1**]∗(**r**(**x**,a)|a1;**dX8**)]$^{-1}$ = −**u1**∗**u1** + **u2**∗**u2** + **u3**∗**u3**.

As in the previous example

D3[**r1**]∗(**r**(**x**,a)|a1;**dXj**) = **D3**[**x1**]∗(**x**(**r**,a)|a1;**dXj**)]$^{-1}$.

Such a mapping carries an entire material neighborhood into an entire local neighborhood at **r1**(**x1**,a1).

Theoretical Physics: Derivatives

Again,

UXj = ±u1*u1 ± u2*u2 ± u3*u3, j= 1..8
UXj^{-1} = UXj, j= 1..8,

but

URX1 = −u1*u1 + u2*(u2−u1)/sqrt(2) + u3*u3,
URX2 = +u1*u1 + u2*(u2−u1)/sqrt(2) + u3*u3,
URX3 = −u1*u1 − u2*u2 + u3*u3,
URX4 = +u1*u1 − u2*u2 + u3*u3,
URX5 = −u1*u1 + u2*(u2−u1)/sqrt(2) − u3*u3,
URX6 = +u1*u1 + u2*(u2−u1)/sqrt(2) − u3*u3,
URX7 = −u1*u1 − u2*u2 − u3*u3,
URX8 = +u1*u1 − u2*u2 − u3*u3.

In this case we cannot simply write D3[r1]*(r(x,a)|a1;dx) to designate the collection of D3[r1]*(r(x,a)|a1;dXj). Consequently we are dealing with gradients continuous in their respective sections, but not simply continuous.

Look now at the two vectors

dX1 = −dX8 = dx11*u1 + dx12*u2 + dx13*u3
= −(dx11*(−u1) + dx12*(−u2) + dx13*(−u3))
≡ d.

Consequently, the vectors x1+dX8, x1, and x1+dX1 determine a straight track of material.
However,

dX1•T[D3[r1]*(r(x,a)|a1;dX1)]
= r(x1 + dX1) − r1
= −dx11*u1 + dx12*(u2−u1) + dx13*u3
= −(dx11 + dx12)*u1 + dx12*u2 + dx13*u3

while

dX8•T[D3[r1]*(r(x,a)|a1;dX8)]
= r(x1 + dX8) − r1
= +dx11*u1 − dx12*u2 − dx13*u3.

Thus r(x1 + dX1)−r1 has a different direction than r1 − r(x1 + dX8). It follows that r(x1 + dX8), r(x1), and r(x1 + dX1) are not collinear.

For this non–simply continuous we revert to

D3[x1]*(x(r,a)|a1;dRXj) = [T[UXj]]•[UXj]
•$[D3[r1]*(r(x,a)|a1;dXj)]^{-1}$
•T[URXj]•[URXj]
= I•$[D3[r1]*(r(x,a)|a1;dXj)]^{-1}$
•[u1RXj*u1RXj
+ u2RXj*u2RXj
+ u3RXj*u3RXj].

Thus,

D3[x1]*(**x**(**r**,a)|a1;**dRX1**) = −**u1*****u1** − **u2*****u1**/2 + **u2*****u2**/2 + **u3*****u3**
D3[x1]*(**x**(**r**,a)|a1;**dRX2**) = −**u1*****u1** − **u2*****u1**/2 + **u2*****u2**/2 + **u3*****u3**
D3[x1]*(**x**(**r**,a)|a1;**dRX3**) = −**u1*****u1** − **u2*****u2** + **u3*****u3**
D3[x1]*(**x**(**r**,a)|a1;**dRX4**) = −**u1*****u1** − **u2*****u2** + **u3*****u3**
D3[x1]*(**x**(**r**,a)|a1;**dRX5**) = −**u1*****u1** − **u2*****u1**/2 + **u2*****u2**/2 + **u3*****u3**
D3[x1]*(**x**(**r**,a)|a1;**dRX6**) = −**u1*****u1** − **u2*****u1**/2 + **u2*****u2**/2 + **u3*****u3**
D3[x1]*(**x**(**r**,a)|a1;**dRX7**) = −**u1*****u1** − **u2*****u2** + **u3*****u3**
D3[x1]*(**x**(**r**,a)|a1;**dRX8**) = −**u1*****u1** − **u2*****u2** + **u3*****u3**.

For the transformations between the first material quadrant and its local image

D3[r1]*(**r**(**x**,a)|a1;**dX1**) = 1*(−**u1**)***u1**
+ sqrt(2)***u**(**u2**−**u1**)***u2**
+ 1***u3*****u3**

[**D3[r1]***(**r**(**x**,a)|a1;**dX1**)]$^{-1}$ = −**u1*****u1** + (**u2**−**u1**)***u2** + **u3*****u3**

D3[x1]*(**x**(**r**,a)|a1;**dRX1**) = **u1***(−**u1**)/1
+ **u2*****u**(**u2**−**u1**)/sqrt(2)
+ **u3*****u3**/1.

Thus, **D3[x1]***(**x**(**r**,a)|a1;**dRX1**) may be written from **D3[r1]***(**r**(**x**,a)|a1;**dX1**) directly.

Clearly the transformations **D3[x1]***(**x**(**r**,a)|a1;**dRXj**) may not be simply written **D3[x1]***(**x**(**r**,a)|a1;**dr**).

Let **dX1** = d1X1***u1** + d2X1***u2** + d3X1***u3**, diX1>0, be a material vector in SECT(**x1**,**u1**,**u2**,**u3**).

Then

dX1•[**UX1**$^{-1}$]•**T**[**UX1**$^{-1}$]•**T**[**D3[r1]***(**r**(**x**,a)|a1;**dX1**)]
= **dX1**•**I**•**I**•**T**[**D3[r1]***(**r**(**x**,a)|a1;**dX1**)]
= **dX1**•[−**u1*****u1** + **u2***(**u2**−**u1**) + **u3*****u3**]
= −(d1X1+d2X1)***u1** + d2X1***u2** + d3X1***u3**
= d1X1*(−**u1**)
+ (sqrt(2))*d2X1*(**u2**−**u1**)/sqrt(2)
+ d3X1***u3**
= **dRX1**,

but

dX1•[**UX1**$^{-1}$]•**T**[**UX1**$^{-1}$]•**D3[x1]***(**x**(**r**,a)|a1;**dRX1**)]
= **dX1**•[−**u1*****u1** − **u2*****u1**/2 + **u2*****u2**/2 + **u3*****u3**
= −(d1X1+d2X1/2)***u1** + d2X1***u2**/2 + d3X1***u3**
= (d1X1)*(−**u1**)
+ (d2X1/sqrt(2))*(**u2**−**u1**)/sqrt(2)
+ d3X1***u3**

a different vector in the same section.

Theoretical Physics: Derivatives

Suppose **dX1**, a vector in the first material quadrant is applied to a different sectional gradient, say

dX1•[**UX3**$^{-1}$]•T[**UX3**$^{-1}$]•T[**D3**[**r1**]∗(**r**(**x**,a)|a1;**dX3**)]
= **dX1**•[**u1**∗**u1**−**u2**∗**u2**+**u3**∗**u3**]
•[**u1**∗**u1**−**u2**∗**u2**+**u3**∗**u3**]
•[−**u1**∗**u1** + **u2**∗**u2** + **u3**∗**u3**]
= **dX1**•[−**u1**∗**u1** + **u2**∗**u2** + **u3**∗**u3**]
= −d1X1∗**u1** + d2X1∗**u2** + d3X1∗**u3**

a vector not in SECT(**r1**,−**u1**,−**u2**,+**u3**).
In contrast, for **dX3** = d1X1∗**u1** − d2X1∗**u2** + d3X1∗**u3**, a vector in **X3**,

dX3•[**UX3**$^{-1}$]•T[**UX3**$^{-1}$]•T[**D3**[**r1**]∗(**r**(**x**,a)|a1;**dX3**)]
= **dX3**•[−**u1**∗**u1** − **u2**∗**u2** + **u3**∗**u3**]
= −d1X1∗**u1** − d2X1∗**u2** + d3X1∗**u3**

a vector in SECT(**r1**,−**u1**,−**u2**,+**u3**).
In attempting to construct the dual, note first for the second local quadrant, SECT(**r1**,−**u1**,**u2**,**u3**)

dX2•[**UX2**$^{-1}$]•T[**UX2**$^{-1}$]•T[**D3**[**r1**]∗(**r**(**x**,a)|a1;**dX2**)
= **dX2**•T[**D3**[**r1**]∗(**r**(**x**,a)|a1;**dX2**)
= **u2**

defines the material vector in the second material section SECT(**x1**,−**u1**,+**u2**,+**u3**) which maps into the local vector **u2**. Solving then,

u2 = (−d1X2∗**u1** + d2X2∗**u2** + d3X2∗**u3**)
•[−**u1**∗**u1** + ∗**u2**∗(**u2**−**u1**) + **u3**∗**u3**]
= d1X2∗**u1** + d2X2∗(**u2**−**u1**)

gives

d1X2 = d2X2 = 1.

Consequently, for the inverse mapping

u1 in V3(**r**) maps into −**u1** in V3(**x**)
u2 in V3(**r**) maps into **u2**−**u1** in V3(**x**)
u3 in V3(**r**) maps into **u3** in V3(**x**)
−**u1** in V3(**r**) maps into −**u1** in V3(**x**)
−**u2** in V3(**r**) maps into −**u2** in V3(**x**)
−**u3** in V3(**r**) maps into −**u3** in V3(**x**).

Let SECT(**r1**,±**u1**,±**u2**,±**u3**), j= 1..8 be local quadrants. The following gradients may be defined for these quadrants:

D3[**r1**]∗(**x**(**r**,a)|a1;**dR1**) = −**u1**∗**u1** + (**u2**−**u1**)∗**u2** + **u3**∗**u3**
D3[**r1**]∗(**x**(**r**,a)|a1;**dR2**) = −**u1**∗**u1** + (**u2**−**u1**)∗**u2** + **u3**∗**u3**
D3[**r1**]∗(**x**(**r**,a)|a1;**dR3**) = −**u1**∗**u1** + **u2**∗**u2** + **u3**∗**u3**
D3[**r1**]∗(**x**(**r**,a)|a1;**dR4**) = −**u1**∗**u1** + **u2**∗**u2** + **u3**∗**u3**
D3[**r1**]∗(**x**(**r**,a)|a1;**dR5**) = −**u1**∗**u1** + (**u2**−**u1**)∗**u2** + **u3**∗**u3**
D3[**r1**]∗(**x**(**r**,a)|a1;**dR6**) = −**u1**∗**u1** + (**u2**−**u1**)∗**u2** + **u3**∗**u3**
D3[**r1**]∗(**x**(**r**,a)|a1;**dR7**) = −**u1**∗**u1** + **u2**∗**u2** + **u3**∗**u3**
D3[**r1**]∗(**x**(**r**,a)|a1;**dR8**) = −**u1**∗**u1** + **u2**∗**u2** + **u3**∗**u3**.

The corresponding material sections are
1. SECT(**r1**,**+u1**,**+u2**,**+u3**) ⟶ SECT(**x1**,**−u1**,(**u2**−**u1**)/sqrt(2),**+u3**) in V3(**x**)
2. SECT(**r1**,**−u1**,**+u2**,**+u3**) ⟶ SECT(**x1**,**+u1**,(**u2**−**u1**)/sqrt(2),**+u3**)
3. SECT(**r1**,**+u1**,**−u2**,**+u3**) ⟶ SECT(**x1**,**−u1**,**−u2**,**+u3**)
4. SECT(**r1**,**−u1**,**−u2**,**+u3**) ⟶ SECT(**x1**,**+u1**,**−u2**,**+u3**)
5. SECT(**r1**,**+u1**,**+u2**,**−u3**) ⟶ SECT(**x1**,**−u1**,(**u2**−**u1**)/sqrt(2),**−u3**)
6. SECT(**r1**,**−u1**,**+u2**,**−u3**) ⟶ SECT(**x1**,**+u1**,(**u2**−**u1**)/sqrt(2),**−u3**)
7. SECT(**r1**,**+u1**,**−u2**,**−u3**) ⟶ SECT(**x1**,**−u1**,**−u2**,**−u3**)
8. SECT(**r1**,**−u1**,**−u2**,**−u3**) ⟶ SECT(**x1**,**+u1**,**−u2**,**−u3**).

These gradients may not be locally differentiable. In this example, however, **T[D3[x1]**∗(**x**(**r**,a)|a1;**dRi**)]•**T[D3[r1]**∗(**r**(**x**,a)|a1;**dXj**)] = **I** for the following combinations:

i	j
1,2,5,6	1,2,5,6
3,4,7,8	3,4,7,8

Now

dRi•[**URi**$^{-1}$]•**T[URi**$^{-1}$]•**T[D3[x1]**∗(**x**(**r**,a)|a1;**dRi**)] = **dXj** for some j

and

dXj•[**UXj**$^{-1}$]•**T[UXj**$^{-1}$]•**T[D3[r1]**∗(**r**(**x**,a)|a1;**dXj**)] = **dRk** for some k.

If then, for every **i** there exists a proper **j** such that
dRi•**T[D3[x1]**∗(**x**(**r**,a)|a1;**dRi**)]•**T[D3[r1]**∗(**r**(**x**,a)|a1;**dXj**)] = **dRi**
the gradients **D3[x1]**∗(**x**(**r**,a)|a1;**dRi**) must be locally differentiable.

But the dual material sections are related to the original material sections as:
Dual *Original*
(1) SECT(−**u1**,(**u2**−**u1**)/sqrt(2),**+u3**)
is a subset of SECT(−**u1**,**+u2**,**+u3**) (2)
(2) SECT(**+u1**,(**u2**−**u1**)/sqrt(2),**+u3**)
is a subset of SECT(**+u1**,**+u2**,**+u3**)∪SECT(−**u1**,**+u2**,**+u3**) (1∪2)
(3) SECT(−**u1**,−**u2**,**+u3**)
is SECT(−**u1**,−**u2**,**+u3**) (4)
(4) SECT(**+u1**,−**u2**,**+u3**)
is SECT(**+u1**,−**u2**,**+u3**) (3)
(5) SECT(−**u1**,(**u2**−**u1**)/sqrt(2),−**u3**)
is a subset of SECT(−**u1**,**+u2**,−**u3**) (6)
(6) SECT(**+u1**,(**u2**−**u1**)/sqrt(2),−**u3**)
is a subset of SECT(**+u1**,**+u2**,−**u3**)∪SECT(−**u1**,**+u2**,−**u3**) (5∪6)

(7) SECT(−**u1**,−**u2**,−**u3**)
is SECT(−**u1**,−**u2**,−**u3**) (8)
(8) SECT(+**u1**,−**u2**,−**u3**)
is SECT(+**u1**,−**u2**,−**u3**) (7)

Consequently for every **i** there exists a proper **j** such that
T[D3[x1]∗(**x**(**r**,a)|a1;**dRi**)]•**T[D3[r1]**∗(**r**(**x**,a)|a1;**dXj**)] = **I**.

It follows **D3[r1]**∗(**x**;**dRi**), i=1..8 are locally differentiable.

In this example, then,

D3[x1]∗(**r**(**x**,a)|a1;**dXj**) exist as materially differentiable,
D3[r1]∗(**x**(**r**,a)|a1;**dRj**) exist as locally differentiable,
D3[x1]∗(**r**(**x**,a)|a1;**dx**) does not exist as a common form
for **D3[x1]**∗(**r**(**x**,a)|a1;**dXi**)
D3[r1]∗(**x**(**r**,a)|a1;**dr**) does not exist as a common form
for **D3[r1]**∗(**x**(**r**,a)|a1;**dRi**)

straight material tracks are not mapped into straight local traces at **r1**(**x1**,a1);
straight local traces are not mapped into straight material tracks at **x1**(**r1**,a1).

For any **r** and SECT(**r**,**u1Ri**,**u2Ri**,**u3Ri**)

D3[**r**]∗(**r**|a1;**dRi**)
= **u1Ri**∗**u1Ri** + **u2Ri**∗**u2Ri** + **u3Ri**∗**u3Ri**. (1075)

Then for any SECT(**r**,**u1Ri**,**u2Ri**,**u3Ri**)

[**URi**$^{-1}$]•**T**[**D3**[**r**]∗(**r**|a1;**dRi**)]•[**URi**$^{-1}$]] = **I**. (1076)

By theorem (12) any such section can be merged into a quadrant section, which is to say that **r** is continuous over **r**. Consequently

D3[**r1**]∗(**r**|a1;**dr**) = **I** (1077)

an otherwise obvious result.

Similarly, with the assumption of null matter,

D3[**x1**]∗(**x**|a1;**dx**) = **I** (1078)

exists as a simply continuous gradient.

Applying the definition of gradients in Theoretical Physics for a given SECT(**r1**,**u1Ri**,**u2Ri**,**u3Ri**) shows us

D3[**r1**]∗(**r**|a1;**dRi**)
= **D3**[**r1**]∗(**r**|a1+da;**dRi**)
= **D3**[**r2**]∗(**r**|a1;**dRi**)
= **D3**[**r2**]∗(**r**|a1+da;**dRi**)
= **D3**[**r0**]∗(**r**|a1;**dRi**)
= **D3**[**r0**]∗(**r**|a1−da;**dRi**). (1079)

Likewise[73],

D3[**x1**]∗(**x**|a1;**dXi**)
= **D3**[**x1**]∗(**x**|a1+da;**dXi**)
= **D3**[**x2**]∗(**x**|a1;**dXi**)
= **D3**[**x2**]∗(**x**|a1+da;**dXi**)
= **D3**[**x0**]∗(**x**|a1;**dXi**)
= **D3**[**x0**]∗(**x**|a1−da;**dXi**). (1080)

Different observations for related material/local sections may be acknowledged by using the convention for the indexed derivatives.

D3[**r1**]∗(**r**|a1;**dXi**)
= **T**[**URi**(**r1**,a1)]•[**URi**(**r1**,a1)]
•[**D3**[**x1**]∗(**x**|a1;**dRXi**)]$^{-1}$
•**T**[**UXi**(**x1**,a1)]•[**UXi**(**x1**,a1)]
D3[**r2**]∗(**r**|a1+da;**dXi**)
= **T**[**URi**(**r2**,a1+da)]•[**URi**(**r2**,a1+da)]
•[**D3**[**x1**]∗(**x**|a1+da;**dRXi**)]$^{-1}$
•**T**[**UXi**(**x1**,a1+da)]•[**UXi**(**x1**,a1+da)]
D3[**r2**]∗(**r**|a1;**dXi**)
= **T**[**URi**(**r2**,a1)]•[**URi**(**r2**,a1)]
•[**D3**[**y2**]∗(**x**|a1;**dRXi**)]$^{-1}$
•**T**[**UXi**(**y2**,a1)]•[**UXi**(**y2**,a1)]
D3[**r1**]∗(**r**|a1+da;**dXi**)
= **T**[**URi**(**r1**,a1+da)]•[**URi**(**r1**,a1+da)]
•[**D3**[**x2**]∗(**x**|a1;**dRXi**)]$^{-1}$
•**T**[**UXi**(**x2**,a1+da)]•[**UXi**(**x2**,a1+da)]
D3[**r0**]∗(**r**|a1−da;**dXi**)
= **T**[**URi**(**r0**,a1−da)]•[**URi**(**r0**,a1−da)]
•[**D3**[**x1**]∗(**x**|a1−da;**dRXi**)]$^{-1}$
•**T**[**UXi**(**x1**,a1−da)]•[**UXi**(**x1**,a1−da)]

Theoretical Physics: Derivatives

D3[**r0**]∗(**r**|a1;**dXi**)
= **T**[**URi**(**r0**,a1)]•[**URi**(**r0**,a1)]
•[**D3**[**y0**]∗(**x**|a1;**dRXi**)]$^{-1}$
•**T**[**UXi**(**y0**,a1)]•[**UXi**(**y0**,a1)]
D3[**r1**]∗(**r**|a1–da;**dXi**)
= **T**[**URi**(**r1**,a1–da)]•[**URi**(**r1**,a1–da)]
•[**D3**[**x0**]∗(**x**|a1–da;**dRXi**)]$^{-1}$
•**T**[**UXi**(**x0**,a1–da)]•[**UXi**(**x0**,a1–da)].

Derivatives of Generalized Functions

Derivatives of generalized functions in Theoretical Physics are somewhat more complicated than those of the primary reference functions, **r**(**x**,a) or **x**(**r**,a).

In Mathematics, {**r1** = **r2**} ⟶ {**f**(**r1**) = **f**(**r2**)}. Not so in Theoretical Physics, where **f**(**r1**(**x1**,a1)) does not necessarily equal **f**(**r1**(**x2**,a2)). For example, suppose **f** the density of a particle; then the density of particle **x1** at *a1* occupying location **r1** is not necessarily related to the density of particle **x2**, even if at observation *a2*, **x2** occupies location **r1**.

Rather, since **r1**(**x**,a1) = **r**(**x1**,a1) = **r1**(**x1**,a1),

f(**r1**,a1) = **f**(**x1**,a1) = **f**(**r1**(**x1**,a1)). (1081)

To discuss these functions expeditiously we will consider, without loss of generality, vector functions **f**(**r**(**x**,a)) = **f**(**x**(**r**,a)).

Change in the function **f**(**r**(**x**,a)) must reference one of three parameters. Each reference can be thought of as defining sub–functions of **f** since they reference specific subsets over which **f**(**r**(**x**,a)) may be defined.

Thus

f|**x1** ≡ **f**(**r**(**x1**,a)), **x1** fixed (1082)
f|**r1** ≡ **f**(**x**(**r1**,a)), **r1** fixed (1083)
f(**r**|a1) ≡ **f**(**r**(**x**,a1)), a1 fixed (1084)
f(**x**|a1) ≡ **f**(**x**(**r**,a1)), a1 fixed. (1085)

Clearly as sub–functions **f**|**x1** does not generally equal **f**|**r1**.

The assumption of null matter makes the continuity of the primary reference functions axiomatic (but not necessarily their derivatives.) The functions of Theoretical Physics, in contrast, are generalized functions with possible discontinuities.

Moreover, conditions in which two of the primary reference functions are stalled does not exclude change in the other primary function of Theoretical Physics. Under the stalled condition,

f(**r1**(**x1**,a))

becomes a function of the observational index.

Contrariwise, with reference to **r1**(**x1**,a1) the function

f(**r**(**x**,a1)) = **f**(**x**(**r**,a1))

may include possibly **branching** functional values.

Theoretical Physics does not exclude functions without derivatives or functions without a basic derivative. Nevertheless, when discussing derivatives of a function, it will usually be assumed to have a basic derivative.

In the sequel, the local/material reference at *a1* will be an open set, usually a neighborhood about **r1**(**x1**,a1) and the function under consideration will be assumed continuous throughout the open set except perhaps on a connected set of measure zero containing **r1**(**x1**,a1). In particular the function is assumed to have forward and backward *basic* derivatives which may not be equal at **r1**(**x1**,a1).

The general syntax used in Theoretical Physics for operators is:

Operator rank [**reference or interval**] **couple** (**f**|**condition**;**increment**)

(1086)

Theoretical Physics: Derivatives

Derivatives with respect to the Observational Index

Given the reference **r1**(**x1**,a1), the incremental function **diff** is defined in the directional case and in the positive–definite convention, consistent with the earlier convention, as

$$\mathbf{diff}_f(\mathbf{f}(\mathbf{r}|a1+da)) \equiv \mathbf{f}(\mathbf{r2}(\mathbf{x1},a1+da)) - \mathbf{f}(\mathbf{r1}(\mathbf{x2},a1+da)) \tag{1087}$$
$$\mathbf{diff}_f(\mathbf{f}(\mathbf{r}|a1)) \equiv \mathbf{f}(\mathbf{r2}(\mathbf{y2},a1)) - \mathbf{f}(\mathbf{r1}(\mathbf{x1},a1)) \tag{1088}$$
$$\mathbf{diff}_f(\mathbf{f}(\mathbf{r}|\mathbf{x1})) \equiv \mathbf{f}(\mathbf{r}(\mathbf{x1},a1 + da)) - \mathbf{f}(\mathbf{r}(\mathbf{x1},a1)) \tag{1089}$$
$$\mathbf{diff}_b(\mathbf{f}(\mathbf{r}|\mathbf{x1})) \equiv \mathbf{f}(\mathbf{r1}(\mathbf{x1},a1)) - \mathbf{f}(\mathbf{r0}(\mathbf{x1},a1 - da)) \tag{1090}$$
$$\mathbf{diff}_b(\mathbf{f}(\mathbf{r}|a1-da)) \equiv \mathbf{f}(\mathbf{r1}(\mathbf{x0},a1 - da)) - \mathbf{f}(\mathbf{r0}(\mathbf{x1},a1 - da)) \tag{1091}$$
$$\mathbf{diff}_b(\mathbf{f}(\mathbf{r}|a1)) \equiv \mathbf{f}(\mathbf{r1}(\mathbf{x1},a1)) - \mathbf{f}(\mathbf{r0}(\mathbf{y0},a1)). \tag{1092}$$

Dually:

$$\mathbf{diff}_f(\mathbf{f}(\mathbf{x}|\mathbf{r1})) \equiv \mathbf{f}(\mathbf{x2}(\mathbf{r1},a1+da)) - \mathbf{f}(\mathbf{x1}(\mathbf{r1},a1)) \tag{1093}$$
$$\mathbf{diff}_f(\mathbf{f}(\mathbf{x}|a1+da)) \equiv \mathbf{f}(\mathbf{x2}(\mathbf{r1},a1+da)) - \mathbf{f}(\mathbf{x1}(\mathbf{r2},a1+da)) \tag{1094}$$
$$\mathbf{diff}_f(\mathbf{f}(\mathbf{x}|a1)) \equiv \mathbf{f}(\mathbf{x2}(\mathbf{s2},a1)) - \mathbf{f}(\mathbf{x1}(\mathbf{r1},a1)) \tag{1095}$$
$$\mathbf{diff}_b(\mathbf{f}(\mathbf{x}|\mathbf{r1})) \equiv \mathbf{f}(\mathbf{x1}(\mathbf{r1},a1)) - \mathbf{f}(\mathbf{x0}(\mathbf{r1},a1 - da)) \tag{1096}$$
$$\mathbf{diff}_b(\mathbf{f}(\mathbf{x}|a1-da)) \equiv \mathbf{f}(\mathbf{x1}(\mathbf{r0},a1 - da)) - \mathbf{f}(\mathbf{x0}(\mathbf{r1},a1 - da)) \tag{1097}$$
$$\mathbf{diff}_b(\mathbf{f}(\mathbf{x}|a1)) \equiv \mathbf{f}(\mathbf{x1}(\mathbf{r1},a1)) - \mathbf{f}(\mathbf{x0}(\mathbf{s1},a1)). \tag{1098}$$

Note that while

$$\mathbf{diff}_f(\mathbf{r}|\mathbf{x1}) = \mathbf{diff}_f(\mathbf{r}|a1+da),$$

it does not follow that

$$\mathbf{diff}_f(\mathbf{f}(\mathbf{r}|\mathbf{x1})) \text{ equals } \mathbf{diff}_f(\mathbf{f}(\mathbf{r}|a1+da)).$$

The enumeration of functional increments can be simplified by noting that

$$\begin{aligned}\mathbf{f}(\mathbf{r}(\mathbf{x1},a1+da)) &= \mathbf{f}(\mathbf{x}(\mathbf{r2},a1+da))\\ &= \mathbf{f}(\mathbf{r2}(\mathbf{x1},a1+da))\\ &= \mathbf{f}(\mathbf{x1}(\mathbf{r2},a1+da))\end{aligned}$$

etc, so one may write

$$\mathbf{diff}_f(\mathbf{f}(\mathbf{r}|a1+da)) = -\mathbf{diff}_f(\mathbf{f}(\mathbf{x}|a1+da)) \tag{1099}$$
$$\mathbf{diff}_b(\mathbf{f}(\mathbf{r}|a1 - da)) = -\mathbf{diff}_b(\mathbf{f}(\mathbf{x}|a1 - da)). \tag{1100}$$

We will also abbreviate

$$\mathbf{diff}_f(\mathbf{f}(\mathbf{r}|\mathbf{x1})) \equiv \mathbf{diff}_f(\mathbf{f}|\mathbf{x1}) \tag{1101}$$
$$\mathbf{diff}_b(\mathbf{f}(\mathbf{r}|\mathbf{x1})) \equiv \mathbf{diff}_b(\mathbf{f}|\mathbf{x1}) \tag{1102}$$
$$\mathbf{diff}_f(\mathbf{f}(\mathbf{x}|\mathbf{r1})) \equiv \mathbf{diff}_f(\mathbf{f}|\mathbf{r1}) \tag{1103}$$
$$\mathbf{diff}_b(\mathbf{f}(\mathbf{x}|\mathbf{r1})) \equiv \mathbf{diff}_b(\mathbf{f}|\mathbf{r1}) \tag{1104}$$

keeping the full designation for the other differences.

For simple indexed derivatives the general syntax becomes

D[reference or interval](**f**|condition;increment).

A forward sense (a>a1) and backward sense (a<a1) is defined only with reference to the observational variable, for it alone is ordered.

Now consider: a forward change in the index can be related to any one of the *six* forward **diff** functions defined above. Similarly another *six* **diff** functions relate to a backward change in the index. Consequently there exists twelve possibilities for differentiation with respect to the observational index symbolized as:

D[**r1**,**r2**(**x1**)](**f**|**x1**;da)
≡ (**f**(**r2**(**x1**,a1+da)) – **f**(**r1**(**x1**,a1)))/da (1105)
D[**r1**,**r2**(**y2**)](**f**|a1;da)
≡ lim (**f**(**r2**(**y2**,a1)) – **f**(**r1**(**x1**,a1)))/da (1106)
D[**r1**(**x2**),**r2**](**f**|a1+da;da)
≡ lim (**f**(**r2**(**x1**,a1+da)) – **f**(**r1**(**x2**,a1+da)))/da (1107)
D[**r0**(**x1**),**r1**](**f**|**x1**;da)
≡ lim (**f**(**r1**(**x1**,a1)) – **f**(**r0**(**x1**,a1–da)))/da (1108)
D[**r0**(**y0**),**r1**](**f**|a1;da)
≡ lim (**f**(**r1**(**x1**,a1)) – **f**(**r0**(**y0**,a1)))/da (1109)
D[**r0**(**x1**),**r1**](**f**|a1–da;da)
≡ lim (**f**(**r1**(**x0**,a1–da)) – **f**(**r0**(**x1**,a1–da)))/da (1110)

D[**x1**,**x2**(**r1**)](**f**|**r1**;da)
≡ (**f**(**x2**(**r1**,a1+da)) – **f**(**x1**(**r1**,a1)))/da (1111)
D[**x1**,**x2**(**s2**)](**f**|a1;da)
≡ lim (**f**(**x2**(**s2**,a1)) – **f**(**x1**(**r1**,a1)))/da (1112)
D[**x1**(**r2**),**x2**](**f**|a1+da;da)
≡ lim (**f**(**x2**(**r1**,a1+da)) – **f**(**x1**(**r2**,a1+da)))/da (1113)
D[**x0**(**r1**),**r1**](**f**|**r1**;da)
≡ lim (**f**(**x1**(**r1**,a1)) – **f**(**x0**(**r1**,a1–da)))/da (1114)
D[**x0**(**s0**),**x1**](**f**|a1;da)
≡ lim (**f**(**x1**(**r1**,a1)) – **f**(**x0**(**s0**,a1)))/da (1115)
D[**x0**(**r1**),**x1**](**f**|a1–da;da)
≡ lim (**f**(**x1**(**r0**,a1–da)) – **f**(**x0**(**r1**,a1–da)))/da (1116)

as a1+da→a1 or a1–da→a1, which for the forward limits imply **r2→r1** and **x2→x1** and for the backward limits imply **r0→r1** and **x0 →x1**.

Theoretical Physics: Derivatives

For both forward and backward senses, da>0. Thus

$$D[\mathbf{r1},\mathbf{r2(y2)}](\mathbf{f}|a1;da) = -\,D[\mathbf{r2(y2)},\mathbf{r1}](\mathbf{f}|a1;da), \qquad (1117)$$

etc.

Somewhat ambiguously, the elaborated notation above is sometimes shortened with reference of **r1**(**x1**,a1) to:

$$D[\mathbf{r1},\mathbf{r2(x1)}](\mathbf{f}|\mathbf{x1};da) \equiv D[\mathbf{r1},\mathbf{r2}](\mathbf{f}|\mathbf{x1};da) \qquad (1118)$$

$$\equiv D[\mathbf{r1}](\mathbf{f}|\mathbf{x1};d_f a) \qquad (1119)$$

$$D[\mathbf{r1},\mathbf{r2(y2)}](\mathbf{f}|a1;da) \equiv D[\mathbf{r1},\mathbf{r2}](\mathbf{f}|a1;da) \qquad (1120)$$

$$D[\mathbf{r1(x2)},\mathbf{r2}](\mathbf{f}|a1+da;da) \equiv D[\mathbf{r1},\mathbf{r2}](\mathbf{f}|a1+da;da) \qquad (1121)$$

$$D[\mathbf{r0(x1)},\mathbf{r1}](\mathbf{f}|\mathbf{x1};da) \equiv D[\mathbf{r0},\mathbf{r1}](\mathbf{f}|\mathbf{x1};d_b a) \qquad (1122)$$

$$\equiv D[\mathbf{r1}](\mathbf{f}|\mathbf{x1};da) \qquad (1123)$$

$$D[\mathbf{r0(y0)},\mathbf{r1}](\mathbf{f}|a1;da) \equiv D[\mathbf{r0},\mathbf{r1}](\mathbf{f}|a1;da) \qquad (1124)$$

$$D[\mathbf{r0(x1)},\mathbf{r1}](\mathbf{f}|a1-da;da) \equiv D[\mathbf{r0},\mathbf{r1}](\mathbf{f}|a1-da;da) \qquad (1125)$$

$$D[\mathbf{x1},\mathbf{x2(r1)}](\mathbf{f}|\mathbf{r1};da) \equiv D[\mathbf{x1},\mathbf{x2}](\mathbf{f}|\mathbf{r1};da) \qquad (1126)$$

$$\equiv D[\mathbf{x1}](\mathbf{f}|\mathbf{r1};d_f a) \qquad (1127)$$

$$D[\mathbf{x1},\mathbf{x2(s2)}](\mathbf{f}|a1;da) \equiv D[\mathbf{x1},\mathbf{x2}](\mathbf{f}|a1;da) \qquad (1128)$$

$$D[\mathbf{x1(r2)},\mathbf{x2}](\mathbf{f}|a1+da;da) \equiv D[\mathbf{x1},\mathbf{x2}](\mathbf{f}|a1+da;da) \qquad (1129)$$

$$D[\mathbf{x0(xr)},\mathbf{r1}](\mathbf{f}|\mathbf{r1};da) \equiv D[\mathbf{x0},\mathbf{x1}](\mathbf{f}|\mathbf{r1};da) \qquad (1130)$$

$$\equiv D[\mathbf{x1}](\mathbf{f}|\mathbf{r1};d_b a) \qquad (1131)$$

$$D[\mathbf{x0(s0)},\mathbf{x1}](\mathbf{f}|a1;da) \equiv D[\mathbf{x0},\mathbf{x1}](\mathbf{f}|a1;da) \qquad (1132)$$

$$D[\mathbf{x0(r1)},\mathbf{x1}](\mathbf{f}|a1-da;da) \equiv D[\mathbf{x0},\mathbf{x1}](\mathbf{f}|a1-da;da). \qquad (1133)$$

It is immediately apparent

$$D[\mathbf{r1(x2)},\mathbf{r2}](\mathbf{f}|a1+da;da) = -D[\mathbf{x1(r2)},\mathbf{x2}](\mathbf{f}|a1+da;da) \qquad (1134)$$

and

$$D[\mathbf{r0(x1)},\mathbf{r1}](\mathbf{f}|a1-da;da) = -D[\mathbf{x0(r1)},\mathbf{x1}](\mathbf{f}|a1-da;da). \qquad (1135)$$

Directional Gradients

Now given suitable continuity, the *four* distinct, restricted increments of Theoretical Physics each approaches **0** as a1+da ⟶ a1 or a1−da ⟶ a1. Combined with the *twelve* index derivatives of Theoretical Physics, they can be used to define *forty–eight* distinct local/material directional gradients in V3 in the form of:

$$\mathbf{D1}[\mathbf{r1}(\mathbf{x1},a1),\mathbf{r2}(\mathbf{x1},a1+da)]*(\mathbf{f}|\mathbf{x1};\mathbf{d_b r}) \equiv \mathbf{D1}[\mathbf{r1},\mathbf{r2}]*(\mathbf{f}|\mathbf{x1};\mathbf{d_b r}) = \lim\,[\mathbf{diff_f}(\mathbf{f}|\mathbf{x1})*\mathbf{qd}(\mathbf{d_b}(\mathbf{r}|\mathbf{x1}))]. \quad (1136)$$

The collection will be symbolized as

$$\mathbf{D1}[\mathbf{p3},\mathbf{p4}]*(\mathbf{f}|\mathbf{p1};\mathbf{d_m p2})$$

for *m*, symbolizing either forward or backward sense; ***p1*** symbolizing ***x1***, ***r1***, *,a1*, *(a1+da)*, or *(a1−da)*; ***p2*** symbolizing ***r*** or ***x***, and [***p3***,***p4***] the appropriate material or local interval.

Where a mixture of senses is contemplated, it is assumed

$$(a1 + da) - a1 = a1 - (a1 - da). \quad (1137)$$

Since these gradients are outer products, their determinants equal zero.

Of the 48 directional gradients the following are most notable:

$$\mathbf{D1}[\mathbf{r1},\mathbf{r2}]*(\mathbf{f}|\mathbf{x1};\mathbf{d_f r})$$
$$\mathbf{D1}[\mathbf{x1},\mathbf{x2}]*(\mathbf{f}|\mathbf{r1};\mathbf{d_f x})$$
$$\mathbf{D1}[\mathbf{r1},\mathbf{r2}]*(\mathbf{f}|a1+da;\mathbf{d_f r})$$
$$\mathbf{D1}[\mathbf{x1},\mathbf{x2}]*(\mathbf{f}|a1+da\,;\mathbf{d_f x})$$
$$\mathbf{D1}[\mathbf{r0},\mathbf{r1}]*(\mathbf{f}|\mathbf{x1};\mathbf{d_b r})$$
$$\mathbf{D1}[\mathbf{x0},\mathbf{x1}]*(\mathbf{f}|\mathbf{r1};\mathbf{d_b x})$$
$$\mathbf{D1}[\mathbf{r0},\mathbf{r1}]*(\mathbf{f}|a1-da;\mathbf{d_f r})$$
$$\mathbf{D1}[\mathbf{x0},\mathbf{x2}]*(\mathbf{f}|a1-da;\mathbf{d_f r}).$$

Directional curls and divergences are similarly defined.

Sectional Gradients

For the sectional case, let **r1**(**x1**,a1) again be the reference.

Let **f**(**r**(**x**,a1)) be sectionally differentiable at observation *a1* in mutually related sections[74],

SECT(**r1**,**u1R1**,**u2R1**,**u3R1**)

and

SECT(**x1**,**u1X1**,**u2X1**,**u3X1**).

In Theoretical Physics, not only is the gradient

D3[**r1**]∗(**f**|a1;**dR1**)

defined over

SECT(**r1**,**u1R1**,**u2R1**,**u3R1**)

but also the gradient

D3[**x1**]∗(**f**|a1;**dX1**)

over

SECT(**x1**,**u1X1**,**u2X1**,**u3X1**).

The primary relationship **r**(**x**,a1) induces a relationship between

D3[**r1**]∗(**f**|a1;**dR1**) and **D3**[**x1**]∗(**f**|a1;**dX1**).

The function **f**(**r**(**x**,a1)) has corresponding values over the mutually related sections.

f(**r**(**x1**+**dX1**,a1)) = **f**(**r1**+**dR1**,a1)
= **f**(**r1**+**dR1**,**x1**+**dX1**,a1) (1138)

f(**r**(**x1**+diX1∗**uiX1**,a1))
= **f**(**r1**+diR1∗**uiR1**,**x1**+diX1∗**uiX1**,a1), i=1,2,3. (1139)

Many possibilities exist for the limits.

If

f(**r**(**x1**+diX1∗**uiX1**,a1)) as diX1⟶0

has no limit, then

f(**r1**+diR1∗**uiR1**,a1) as diR1⟶0

has no limit, either.

If

lim (**f**(**r**(**x1**+diX1∗**uiX1**,a1)) = **f0**

then

lim (**f**(**r1**+diR1∗**uiR1**,a1) = **f0**

also.

Thus the continuities and discontinuities of **f**(**r**(**x**,a1)) match those of **f**(**x**(**r**,a1)).

If

(**f**(**r**(**x1**+diX1∗**uiX1**,a1))–**f**(**r**(**x1**,a1)))/diX1

has no limit, then

(**f**(**r1**+diR1∗**uiR1**,a1)–**f**(**r1**))/diR1

has no limit, either.

The limits assume the diR1>0, while the **uiR1** may be any non–parallel directions.

If the limits exist then ***f*** has related sectional gradients in both related sections. These can be of any rank, but the rank of both related gradients will be the same. In this respect, the generalized functions of Theoretical Physics differ from those of the primary functions which, because of the principle of non–collocation, always have sectional gradients of rank 3.

Neither ***f***, even when endowed with sectional gradients, nor the primary functions need be sectionally continuous. However, if they are continuous in one section then they are continuous in the related section.

There are several distinct ways to define sectional gradients in Theoretical Physics. In addition to referencing the local/material variable, one can distinguish a sectional sense and the observational reference.

For example,

D3[**r1**]∗(**f**;**dR1**) = lim [(**f**(**x**(**r2**+d1R1∗**u1R1**,a1+da))
– **f**(**x1**(**r2**,a1+da)))∗**u1R1**
+ (**f**(**x**(**r2**+d2R1∗**u2R1**,a1+da))
– **f**(**x1**(**r2**,a1+da)))∗**u2R1**
+ (**f**(**x**(**r2**+d3R1∗**u3R1**,a1+da))
– **f**(**x1**(**r2**,a1+da)))∗**u3R1**]
•[**G**(**q**(**dR1**))]   as **dR1+⟶0**

may possibly differ from

D3[**r1**]∗(**f**;**dR1**) = lim[(**f**(**x**(**r1**+d1R1∗**u1R1**,a1))
– **f**(**x1**(**r1**,a1)))∗**u1R1**
+ (**f**(**x**(**r1**+d2R1∗**u2R1**,a1))
– **f**(**x1**(**r1**,a1)))∗**u2R1**
+ (**f**(**x**(**r1**+d3R1∗**u3R1**,a1))
– **f**(**x1**(**r1**,a1)))∗**u3R1**]
•[**G**(**q**(**dR1**))]   as **dR1+⟶0**.

Theoretical Physics: Derivatives

The indicated limiting processes may or may not exist, or both may exist but may not coincide. Both derivatives reference the particle **x1**--the first at **r2**(**x1**,a1+da), the second at **r1**(**x1**,a1).

To avoid ambiguity the first is symbolized as

D3[**r2**(**x1**)]∗(**f**|a1+da;**dR1**)

and the second as

D3[**r1**(**x1**)]∗(**f**|a1;**dR1**).

The first is often abbreviated to **D3**[**r2**]∗(**f**|a1+da;**dR1**); the second to **D3**[**r1**]∗(**f**|a1;**dR1**).

Of the many possibilities for Theoretical Physics, the following fourteen possibly distinct gradients are addressed here:

D3[**r1**]∗(**f**|a1;**dRi**)	**D3**[**r1**]∗(**f**|a1+da;**dRi**)	**D3**[**r1**]∗(**f**|a1−da;**dRi**)
D3[**r2**]∗(**f**|a1;**dRi**)	**D3**[**r2**]∗(**f**|a1+da;**dRi**)	
D3[**r0**]∗(**f**|a1;**dRi**)		**D3**[**r0**]∗(**f**|a1−da;**dRi**)
D3[**x1**]∗(**f**|a1;**dXi**)	**D3**[**x2**]∗(**f**|a1+da;**dXi**)	**D3**[**x0**]∗(**f**|a1−da;**dXi**)
D3[**x2**]∗(**f**|a1;**dXi**)	**D3**[**x1**]∗(**f**|a1+da;**dXi**)	
D3[**x0**]∗(**f**|a1;**dXi**)		**D3**[**x1**]∗(**f**|a1−da;**dXi**)

Sectional Gradients of Theoretical Physics

As a transformation, the sectional gradient of Theoretical Physics is a function V3⟶V3 where the units of the first set of vectors are those of **f**, say F, while the second set of vectors are those of F/L. Unlike the primary functions, both the domain and the image of ***f*** may possibly be only proper subsets of V3.

The definitions of Theoretical Physics allow the constitutive derivatives of the gradient to be impulsive or discontinuous. Where these various gradients coalesce into simply continuous gradients, the gradient is written simply as **D3**[**r1**]∗(**f**|a;**dr**) or **D3**[**x1**]∗(**f**|a;**dx**).

Derivatives: Mathematical versus Theoretical Physics

With the above enumeration the relationship between mathematical derivatives and those of Theoretical Physics can be seen more clearly. The following table illustrates some of the correspondences:

<table>
<tr><th>Mathematics</th><th>Theoretical Physics</th></tr>
<tr><td>D[r1]*(f;dr)</td><td>D3[r1](f|a1;dRi)</td></tr>
<tr><td></td><td>D3[r1]*(f|a1+da;dRi)</td></tr>
<tr><td></td><td>D3[r2]*(f|a1;dRi)</td></tr>
<tr><td></td><td>D3[r2]*(f|a1+da;dRi)</td></tr>
<tr><td></td><td>D3[r1]*(f|a1 – da;dRi)</td></tr>
<tr><td></td><td>D3[r0]*(f|a1;dRi)</td></tr>
<tr><td></td><td>D3[r0]*(f|a1 – da;dRi)</td></tr>
<tr><td></td><td></td></tr>
<tr><td>D[r1]*(f;dg(r))</td><td>D3[x1]*(f|a1;dXi)</td></tr>
<tr><td></td><td>D3[x1]*(f|a1+da;dXi)</td></tr>
<tr><td></td><td>D3[x2]*(f|a1;dXi)</td></tr>
<tr><td></td><td>D3[x2]*(f|a1+da;dXi)</td></tr>
<tr><td></td><td>D3[x1]*(f|a1 – da;dXi)</td></tr>
<tr><td></td><td>D3[x0]*(f|a1;dXi)</td></tr>
<tr><td></td><td>D3[x0]*(f|a1 – da;dXi)</td></tr>
<tr><td></td><td></td></tr>
<tr><td>D[r1]*(f;dr(x))</td><td>D1[r1(x1,a1)]*(f|x1;d_fr)</td></tr>
<tr><td></td><td>D1[r1(x1,a1)]*(f|r1;d_fr)</td></tr>
<tr><td></td><td>D1[r1(x),r(x1)]*(f|a1+da;d_fr</td></tr>
<tr><td></td><td>D1[x1(r),x(r1)]*(f|a1+da;d_fr)</td></tr>
<tr><td></td><td>D1[r1,r2]*(f|a1;d_fr)</td></tr>
<tr><td></td><td>D1[x1,x2]*(f|a1;d_fr)</td></tr>
<tr><td></td><td>D1[r1(x1,a1)]*(f|x1;d_br)</td></tr>
</table>

Theoretical Physics: Derivatives

Mathematics	Theoretical Physics
	D1[**r1**(**x1**,a1)]∗(**f**|**r1**;$\mathbf{d}_b\mathbf{r}$)
	D1[**r**(**x1**),**r1**(**x**)]∗(**f**|a1 – da;$\mathbf{d}_b\mathbf{r}$)
	D1[**x**(**r1**),**x1**(**r**)]∗(**f**|a1 – da;$\mathbf{d}_b\mathbf{r}$)
	D1(**r0**,**r1**)[**r1**(**x1**,a1)]∗(**f**|a1;$\mathbf{d}_b\mathbf{r}$)
	D1(**x0**,**x1**)[**r1**(**x1**,a1)]∗(**f**|a1;$\mathbf{d}_b\mathbf{r}$)
	D1[**x1**(**r1**,a1)]∗(**f**|**r1**;$\mathbf{d}_f\mathbf{x}$)
	D1[**x1**(**r1**,a1)]∗(**f**|**x1**;$\mathbf{d}_f\mathbf{x}$)
	D1[**x**(**r1**),**x1**(**r**)]∗(**f**|a1+da;$\mathbf{d}_f\mathbf{x}$)
	D1[**x1**(**r**),**x**(**r1**)]∗(**f**|a1+da;$\mathbf{d}_f\mathbf{x}$)
	D1[**r1**,**r2**]∗(**f**|a1;$\mathbf{d}_f\mathbf{x}$)
	D1[**x1**,**x2**)]∗(**f**|a1;$\mathbf{d}_f\mathbf{x}$)
	D1[**x1**(**r1**,a1)]∗(**f**|**x1**;$\mathbf{d}_b\mathbf{x}$)
	D1[**x1**(**r1**,a1)]∗(**f**|**r1**;$\mathbf{d}_b\mathbf{x}$)
	D1[**r**(**x1**),**r1**(**x**)]∗(**f**|a1 – da;$\mathbf{d}_b\mathbf{x}$)
	D1[**x**(**r1**),**x1**(**r**)]∗(**f**|a1 – da;$\mathbf{d}_b\mathbf{x}$)
	D1[**r0**,**r1**]∗(**f**|a1;$\mathbf{d}_b\mathbf{x}$)
	D1[**x0**,**x1**]∗(**f**|a1;$\mathbf{d}_b\mathbf{x}$)
D[a1](**f**(**r**(a);da))	D(**f**|**x1**;d_fa)
	D(**f**|**x1**;d_ba)
	D(**f**|**r1**;d_fa)
	D(**f**|**r1**;d_ba)
	D[**x1**,**x2**](**f**|a1;da)
	D[**x0**,**x1**](**f**|a1;da)
	D[**r1**,**r2**](**f**|a1;da)
	D[**r0**,**r1**](**f**|a1;da)

Mathematics	Theoretical Physics
	D[**x1**,**x2**](**f**|a1+da;da)
	D[**r1**,**r2**](**f**|a1+da;da)
	D[**x0**,**x1**](**f**|a1 – da;da)
	D[**r0**,**r1**](**f**|a1 – da;da)

Mathematical Derivatives vs Derivatives of Theoretical Physics

The entries in the table are made in terms of a vector function, **f**. Results for scalar functions, f, follow immediately. Curls and divergences are comprehended in **D3**[**r**]**∗**(**f**;**dr**)

The derivative operators of Theoretical Physics thus appear as restrictions followed by expansions of mathematical derivatives, analogous to the expansion and restriction of mathematical variables by units. They are distinctly new ideas, which nonetheless adhere to a logical development similar to mathematical ideas.

A numerical example: For **x**=z1∗**u1**+z2∗**u2**+z3∗**u3** let

$$\mathbf{r}(\mathbf{x},a) = (a/1_a)*((z1+2*z2)*\mathbf{u1}+(z2+3*z3)*\mathbf{u2}+(z1+z3)*\mathbf{u3})$$
$$\equiv p1*\mathbf{u1}+p2*\mathbf{u2}+p3*\mathbf{u3}, \text{ that is,}$$

$$\mathbf{r}(\mathbf{x},a) = (a/1_a)*(z1,z2,z3) \bullet \begin{bmatrix} 1 & 0 & 1 \\ 2 & 1 & 0 \\ 0 & 3 & 1 \end{bmatrix}$$

so that

$$\mathbf{x}(\mathbf{r},a) = (1_a/(7*a))*(p1,p2,p3) \bullet \begin{bmatrix} 1 & 3 & -1 \\ -2 & 1 & 2 \\ 6 & -3 & 1 \end{bmatrix}$$

where 1_a is a unit constant with dimension of a.

Theoretical Physics: Derivatives

Then for the particle $\mathbf{x1}=\mathbf{x}(\mathbf{r1},a1) = (\mathbf{u1}+\mathbf{u2}+\mathbf{u3})$,
$\mathbf{r}(\mathbf{x1},a)=(a/1_a)*(3*\mathbf{u1}+4*\mathbf{u2}+2*\mathbf{u3})$,
from which

$\mathbf{r0} = ((a1 - da)/1_a)*(3*\mathbf{u1}+4*\mathbf{u2}+2*\mathbf{u3})$,
$\mathbf{r1} = (a1/1_a)*(3*\mathbf{u1}+4*\mathbf{u2}+2*\mathbf{u3})$,
$\mathbf{r2} = ((a1+da)/1_a)*(3*\mathbf{u1}+4*\mathbf{u2}+2*\mathbf{u3})$,
$\mathbf{x0} = (a1/(a1 - da))*\mathbf{x1}$,
$\mathbf{x2} = (a1/(a1+da))*\mathbf{x1}$,
$\mathbf{s0} = ((a1 - da)/1_a)*(3*\mathbf{u1}+4*\mathbf{u2}+2*\mathbf{u3})$,
$\mathbf{s2} = ((a1+da)/1_a)*(3*\mathbf{u1}+4*\mathbf{u2}+2*\mathbf{u3})$,
$\mathbf{y0} = ((a1 - da)/a1)*\mathbf{x1}$
$\mathbf{y2} = ((a1+da)/a1)*\mathbf{x1}$.

Thus

$D[a1](\mathbf{r}|\mathbf{x1};da) = (3*\mathbf{u1}+4*\mathbf{u2}+2*\mathbf{u3})/1_a$.

$$\mathbf{D3}[\mathbf{x1}]*(\mathbf{r}|a1;\mathbf{dx}) = (a1/1_a)*\begin{bmatrix}1 & 2 & 0\\ 0 & 1 & 3\\ 1 & 0 & 1\end{bmatrix}$$

$\mathbf{D3}[\mathbf{x1}]*(\mathbf{r}|a1+da;\mathbf{dx}) = (a1+da/a1)*\mathbf{D3}[\mathbf{x1}]*(\mathbf{r}|a1;\mathbf{dx})$
$\mathbf{D3}[\mathbf{y2}]*(\mathbf{r}|a1;\mathbf{dx}) = \mathbf{D3}[\mathbf{x1}]*(\mathbf{r}|a1;\mathbf{dx})$
$\mathbf{D3}[\mathbf{x2}]*(\mathbf{r}|a1+da;\mathbf{dx}) = ((a1+da)/a1)*\ \mathbf{D3}[\mathbf{x1}]*(\mathbf{r}|a1;\mathbf{dx})$
$D[a1](\mathbf{x}|\mathbf{r1};da) = a1*(\mathbf{u1}+\mathbf{u2}+\mathbf{u3})/a^2$

$$\mathbf{D3}[\mathbf{r1}]*(\mathbf{x}|a1;\mathbf{dr}) = (1_a/(7*a1))*\begin{bmatrix}1 & -2 & 6\\ 3 & 1 & -3\\ -1 & 2 & 1\end{bmatrix}$$

$\mathbf{D3}[\mathbf{r1}]*(\mathbf{x}|a1+da;\mathbf{dr}) = (a1/(a1+da))*\mathbf{D3}[\mathbf{r1}]*(\mathbf{x}|a1;\mathbf{dr})$
$\mathbf{D3}[\mathbf{r2}]*(\mathbf{x}|a1;\mathbf{dr}) = \mathbf{D3}[\mathbf{r1}]*(\mathbf{x}|a1;\mathbf{dr})$
$\mathbf{D3}[\mathbf{r2}]*(\mathbf{x}|a1+da;\mathbf{dr}) = (a1/(a1+da))*\mathbf{D3}[\mathbf{r1}]*(\mathbf{x}|a1;\mathbf{dr})$.

In this case, not only does $\mathbf{r}(\mathbf{x},a)$ have continuous derivatives at $\mathbf{r1}(\mathbf{x1},a1)$, but also $\mathbf{r0}=\mathbf{s0}$ and $\mathbf{r2}=\mathbf{s2}$.

Let $\mathbf{f}(\mathbf{r}(\mathbf{x},a))=\mathbf{r}(\mathbf{x},a)\wedge\mathbf{x}(\mathbf{r},a)$. The derivatives of this function are also continuous at $\mathbf{r1}(\mathbf{x1},a1)$, so that forward and backward senses agree.

Then

$\mathbf{f}(\mathbf{r1}(\mathbf{x1},a1)) = (a1/1_a)*(2*\mathbf{u1} - \mathbf{u2} - \mathbf{u3})$
$D[a1](\mathbf{f}|\mathbf{x1};da) = 2*\mathbf{u1} - \mathbf{u2} - \mathbf{u3}/1_a$
$D[a1](\mathbf{f}|\mathbf{r1};da) = -\ (a1/a)*(2*\mathbf{u1} - \mathbf{u2} - \mathbf{u3})/1_a$
$\longrightarrow -\ D[a1](\mathbf{f}|\mathbf{x1};da)$
$D[a1](\mathbf{f}(\mathbf{r})|a1;da) = ((a1+da+a1)/a1)*(2*\mathbf{u1} - \mathbf{u2} - \mathbf{u3})/1_a$
$\longrightarrow 2*D[a1](\mathbf{f}|\mathbf{x1};da)$
$D[a1](\mathbf{f}(\mathbf{x}|a1);da) = 0$
$D[a1](\mathbf{f}(\mathbf{r}|a1+da);da) = ((a1+da+a1)/(a1+da))*(2*\mathbf{u1} - \mathbf{u2} - \mathbf{u3})/1_a$
$\longrightarrow 2*D[a1](\mathbf{f}|\mathbf{x1};da)$

D[a1](**f**(**x**|a1+da);da) = − ((a1+da+a1)/a1+da)∗(2∗**u1** − **u2** − **u3**)/1_a

$$\mathbf{D3[r1]}(\mathbf{f}(\mathbf{r}|a1);\mathbf{dr}) = 1/7* \begin{bmatrix} -10 & 13 & 3 \\ -2 & -10 & 16 \\ 12 & 4 & -33 \end{bmatrix}$$

$$\mathbf{D3[r1]}(\mathbf{f}(\mathbf{r}|a1+da);\mathbf{dr}) = (a1/(7*a1+da))* \begin{bmatrix} -10 & 13 & 3 \\ -2 & -10 & 16 \\ 12 & 4 & -33 \end{bmatrix}$$

= (a1/(a1+da))∗**D3[r1]**(**f**|a1;**dr**)

D3[r2](**f**(**r**|a1);**dr**) = ((a1+da)/a1)∗**D3[r1]**(**f**|a1;**dr**)

D3[r2](**f**(**r**|a1+da);**dr**) = (1_a/a1)∗**D3[r1]**(**f**|a1;**dr**)

$$\mathbf{D3[x1]}(\mathbf{f}(\mathbf{x}|a1);\mathbf{dx}) = (a1/1_a)* \begin{bmatrix} -1 & -1 & 6 \\ 2 & -2 & -2 \\ -3 & 4 & -3 \end{bmatrix}$$

D3[x1](**f**(**x**|a1+da);**dx**) = ((a1+da)/1_a)∗**D3[x1]**(**f**|a1;**dx**)

D3[y2](**f**(**x**|a1);**dx**) = (a1+da/1_a)∗(a1∗1_a)∗**D3[x1]**(**f**|a1;**dx**)

D3[x2](**f**(**x**|a1+da);**dx**) = (a1/1_a)∗**D3[x1]**(**f**|a1;**dx**).

Relationships between the Derivatives of Theoretical Physics

The derivatives of Theoretical Physics arise from functions of the one given set of observations, **r**(**x**,a), alternately stated as **x**(**r**,a). Consequently one might expect related observations to generate relationships among derivatives.

Let **diff**(**f**|q) be the change in ***f*** with *q* symbolizing ***r1, x1*** or *a* held fixed. Then beginning with D[a1](**f**|**x1**;da),

D[a1](**f**|**x1**;d_fa)
= lim (**f**(**r**(**x1**,a1+da)) − **f**(**r**(**x1**,a1)))/da
= lim **diff**$_f$(**f**|**x1**)/da
= lim (**d**$_f$(**r**|**x1**)•**qd**(**d**$_f$(**r**|**x1**)))∗**diff**$_f$(**f**|**x1**)/da
= lim (**d**$_f$(**r**|**x1**)/da)
•**qd**(**d**$_f$(**r**|**x1**))∗**diff**(**f**|**x1**)
= D[a1](**r**|**x1**;d_fa)
•**T**[**D1**[**r1**(**x1**,a1),**r2**(**x1**,a2)]∗(**f**|**x1**;**dr**)].

(1140)

Theoretical Physics: Derivatives

Likewise, by the same argument

$$\begin{aligned}&D[a1](\mathbf{f}|\mathbf{r1};d_f a)\\ &\qquad = D[a1](\mathbf{x}|\mathbf{r1};d_f a)\bullet\mathbf{T}[\mathbf{D1}[\mathbf{x1},\mathbf{x2}]*(\mathbf{f}|\mathbf{r1};\mathbf{dx})]\end{aligned} \tag{1141}$$

$$\begin{aligned}&D[\mathbf{r1}(\mathbf{x1},a1),\mathbf{r2}(\mathbf{x},a1)](\mathbf{f}|a1;da)\\ &\qquad = D[a1](\mathbf{r}|\mathbf{x1};d_f a)\bullet\mathbf{T}[\mathbf{D1}[\mathbf{r1}(\mathbf{x},a1),\mathbf{r2}(\mathbf{x},a1)]*(\mathbf{f}|a1;\mathbf{dr})]\end{aligned} \tag{1142}$$

$$\begin{aligned}&D[\mathbf{r1}(\mathbf{x},a1+da),\mathbf{r2}(\mathbf{x1},a1+da)](\mathbf{f}|a1+da;da)\\ &\qquad = D[a1](\mathbf{r}|\mathbf{x1};d_f a)\\ &\qquad\qquad \bullet\mathbf{T}[\mathbf{D1}[\mathbf{r1}(\mathbf{x},a1+da),\mathbf{r}(\mathbf{x1},a1+da)]*(\mathbf{f}|a1+da;\mathbf{dr})]\end{aligned} \tag{1143}$$

$$\begin{aligned}&D[\mathbf{x1}(\mathbf{r1},a1),\mathbf{x2}(\mathbf{r},a1)](\mathbf{f}|a1;da)\\ &\qquad = D[a1](\mathbf{x}|\mathbf{r1};d_f a)\bullet\mathbf{T}[\mathbf{D1}[\mathbf{x1}(\mathbf{r1},a1),\mathbf{x2}(\mathbf{s2},a1)]*(\mathbf{f}|a1;\mathbf{dx})]\end{aligned} \tag{1144}$$

$$\begin{aligned}&D[\mathbf{x1}(\mathbf{r},a1+da),\mathbf{x2}(\mathbf{r1},a1+da)](\mathbf{f}|a1+da;da)\\ &\qquad = D[a1](\mathbf{x}|\mathbf{r1};d_f a)\\ &\qquad\qquad \bullet\mathbf{T}[\mathbf{D1}[\mathbf{x1}(\mathbf{r},a1+da),\mathbf{x}(\mathbf{r1},a1+da)]*(\mathbf{f}|a1+da;\mathbf{dx})].\end{aligned} \tag{1145}$$

Note mixed combinations may be well–defined, for example,

$$D[a1](\mathbf{f}|\mathbf{x1};d_f a) = D[a1](\mathbf{r}|\mathbf{x1};d_b a)\bullet\mathbf{T}[\mathbf{D1}[\mathbf{r1},\mathbf{r2}]*(\mathbf{f}|\mathbf{x1};\mathbf{d_b r})] \tag{1146}$$

$$D[a1](\mathbf{f}|\mathbf{x1};d_b a) = D[a1](\mathbf{r}|\mathbf{x1};d_f a)\bullet\mathbf{T}[\mathbf{D1}[\mathbf{r0},\mathbf{r1}]*(\mathbf{f}|\mathbf{x1};\mathbf{d_f r})] \tag{1147}$$

$$D[a1](\mathbf{f}|\mathbf{x1};d_f a) = D[a1](\mathbf{r}|\mathbf{x1};d_b a)\bullet\mathbf{T}[\mathbf{D1}[\mathbf{r1},\mathbf{r2}]*(\mathbf{f}|\mathbf{x1};\mathbf{d_b r})] \tag{1148}$$

$$D[a1](\mathbf{f}|\mathbf{x1};d_b a) = D[a1](\mathbf{r}|\mathbf{x1};d_b a)\bullet\mathbf{T}[\mathbf{D1}[\mathbf{r0},\mathbf{r1}]*(\mathbf{f}|\mathbf{x1};\mathbf{d_b r})] \tag{1149}$$

and related variations.

By combining equations, other equations may be written of the form:

$$\begin{aligned}&D[a1](\mathbf{r}|\mathbf{x1};d_n a)\bullet\mathbf{T}[\mathbf{D1}[\mathbf{r1},\mathbf{r2}]*(\mathbf{f}|\mathbf{x1};\mathbf{d_n r})]\\ &\qquad\qquad = D[a1](\mathbf{x}|\mathbf{r1};d_m a)\bullet\mathbf{T}[\mathbf{D1}[\mathbf{r1},\mathbf{r2}]*(\mathbf{f}|\mathbf{x1};\mathbf{d_m x})]\end{aligned} \tag{1150}$$

for *m* and *n* symbolizing either forward or backward senses.

The results for indexed derivatives have their counterparts in directional gradients. For example,

$$\begin{aligned}&\mathbf{D1}[\mathbf{r1},\mathbf{r2}]*(\mathbf{f}|\mathbf{x1};\mathbf{d_f r})\\ &\qquad = \lim\ \mathbf{diff_f}(\mathbf{f}|\mathbf{x1})*\mathbf{qd}(\mathbf{d_f}(\mathbf{r}|\mathbf{x1}))\\ &\qquad = \lim\ (\mathbf{diff_f}(\mathbf{f}|\mathbf{x1})*\mathbf{qd}(\mathbf{d_f}(\mathbf{x}|\mathbf{r1})))\bullet(\mathbf{d_f}(\mathbf{x}|\mathbf{r1})*\mathbf{qd}(\mathbf{d_f}(\mathbf{r}|\mathbf{x1})))\\ &\qquad = \mathbf{D1}[\mathbf{r1},\mathbf{r2}]*(\mathbf{f}|\mathbf{x1};\mathbf{d_f x})\bullet\mathbf{D1}[\mathbf{x1},\mathbf{x2}]*(\mathbf{x}|\mathbf{r1};\mathbf{d_f r}).\end{aligned} \tag{1151}$$

Thus again compactly,

$$\mathbf{D1}[\mathbf{r1},\mathbf{r2}]*(\mathbf{f}|\mathbf{x1};\mathbf{d}_n p1) = \mathbf{D1}[\mathbf{r1},\mathbf{r2}]*(\mathbf{f}|\mathbf{x1};\mathbf{d}_m p2)\bullet\mathbf{D1}[p3,p4]*(p2;\mathbf{d}_n p1) \quad (1152)$$

$$\mathbf{D1}[\mathbf{x1},\mathbf{x2}]*(\mathbf{f}|\mathbf{r1};\mathbf{d}_n p1) = \mathbf{D1}[\mathbf{x1},\mathbf{x2}]*(\mathbf{f}|\mathbf{r1};\mathbf{d}_m p2)\bullet\mathbf{D1}[p3,p4]*(p2;\mathbf{d}_n p1) \quad (1153)$$

and similar variations for *m*, *n* symbolizing either forward or backward senses or intervals; *p1* symbolizing ***x1***, ***r1*** *,a1*, *(a1+da)*, or *(a1−da)*; *p2* symbolizing ***r*/*x1*** and ***x*/*r1*** in either a forward or backward sense, and [*p3,p4*] the appropriate material or local interval.

Furthermore from their definition, directional gradients are related to indexed derivatives. With reference to **r1**(**x1**,a1)

$$\begin{aligned}\mathbf{D1}[\mathbf{r1},\mathbf{r2}]*(\mathbf{f}|\mathbf{x1};\mathbf{d}_f\mathbf{r}) &= \lim \mathbf{diff}_f(\mathbf{f}|\mathbf{x1})*\mathbf{qd}(\mathbf{d}_f(\mathbf{r}|\mathbf{x1}))\\ &= \lim \mathbf{diff}_f(\mathbf{f}|\mathbf{x1})*\mathbf{qd}(\mathbf{d}_f(\mathbf{r}|\mathbf{x1}))/(da/da)\\ &= D[a1](\mathbf{f}|\mathbf{x1};d_f a)*\mathbf{qd}(D[a1](\mathbf{r}|\mathbf{x1},d_f a)). \quad (1154)\end{aligned}$$

By a similar argument again in compact notation

$$\mathbf{D1}[\mathbf{r1},\mathbf{r2}]*(\mathbf{f}|p1;dnp2) = D[a1](\mathbf{f}|p1;d_f a)*\mathbf{qd}(D[a1](p2;d_n a)) \quad (1155)$$

for *n* symbolizing either forward or backward sense; *p1* symbolizing ***x1***, ***r1***, *a1*, *(a1+da)*, *(a1 − da)*; *p2* symbolizing an increment in ***r*/*x1*** or ***x*/*r1*** in either a forward or backward sense.

In view of this latter decomposition, directional gradients are often written somewhat ambiguously but more simply as

$$\mathbf{D1}[\mathbf{r1},\mathbf{r2}]*(\mathbf{f}|\mathbf{x1};\mathbf{d}_f\mathbf{r}) \equiv D[a1](\mathbf{f}|\mathbf{x1};d_f a)*\mathbf{qd}(D[a1](\mathbf{r}|\mathbf{x1},d_f a)) \quad (1156)$$

or even

$$\mathbf{D1}[\mathbf{r1}]*(\mathbf{f}|\mathbf{x1};\mathbf{d}_f\mathbf{r}) \equiv D[a1](\mathbf{f}|\mathbf{x1};d_f a)*\mathbf{qd}(D[a1](\mathbf{r}|\mathbf{x1},d_f a)) \quad (1157)$$

and similar variations. Whenever this abbreviated symbolism is used, both the functional and incremental differences are assumed taken in the same sense.

Theoretical Physics: Derivatives

Let t(a) be a strictly monotonic function (without a derivative equaling 0).

Then

$$\begin{aligned}
&\mathbf{D1}[\mathbf{r1}(\mathbf{x1},a1),\mathbf{r2}(\mathbf{x1},a2)]*(\mathbf{f}|\mathbf{x1};\mathbf{d_f r})\\
&\qquad = D[a1](\mathbf{f}|\mathbf{x1};d_f a)*\mathbf{qd}(D[a1](\mathbf{r}|\mathbf{x1};d_f a))\\
&\qquad = D[t(a1)](\mathbf{f}|\mathbf{x1};d_f t)*D[a1](t(a);d_f a)\\
&\qquad\qquad *\mathbf{qd}(D[t(a1)](\mathbf{r}|\mathbf{x1};d_f t)*D[a1](t(a);d_f a))\\
&\qquad = D[t(a1)](\mathbf{f}|\mathbf{x1};d_f t)*D[a1](t(a);d_f a)\\
&\qquad\qquad *\mathbf{qd}(D[t(a1)](\mathbf{r}|\mathbf{x1};d_f t))/D[a](t(a);d_f a)\\
&\qquad = D[t(a1)](\mathbf{f}|\mathbf{x1};d_f t)*\mathbf{qd}(D[t(a1)](\mathbf{r}|\mathbf{x1};d_f t))\\
&\qquad = \mathbf{D1}[\mathbf{r1}(\mathbf{x1},t(a1)),\mathbf{r2}(\mathbf{x1},t(a2))]*(\mathbf{f}|\mathbf{x1};\mathbf{d_f r} && (1158)
\end{aligned}$$

even for nonlinear movement.

Similar results hold for the backward sense and for duals.

Directional gradients are thus *intrinsic* to their traces or tracks.

Results may also be written in terms of an intrinsic variable,

$$\begin{aligned}
&\mathbf{D1}[\mathbf{r1},\mathbf{r2}]*(\mathbf{f}|\mathbf{x1};\mathbf{d_f r})\\
&\qquad\qquad = D[\mathbf{r1}](\mathbf{f}|\mathbf{x1};srf(\mathbf{x1}))*\mathbf{qd}(\mathbf{utrf}(\mathbf{x1})) && (1159)
\end{aligned}$$

where again similar results hold for the backward sense and for duals.

How are these results for directional gradients related to directional divergences and curls?

$$\begin{aligned}
&D1[\mathbf{r1},\mathbf{r2}]\bullet(\mathbf{f}|\mathbf{x1};\mathbf{d_f r})\\
&\qquad = \lim \mathbf{diff_f}(\mathbf{f}|\mathbf{x1})\bullet\mathbf{qd}(\mathbf{d_f}(\mathbf{r}|\mathbf{x1}))\\
&\qquad = D[a1](\mathbf{f}|\mathbf{x1};d_f a)\bullet\mathbf{qd}(D[a1](\mathbf{r}|\mathbf{x1},d_f a)) && (1160)\\
&\mathbf{D1}[\mathbf{r1},\mathbf{r2}]\wedge(\mathbf{f}|\mathbf{x1};\mathbf{d_f r})\\
&\qquad = \lim \mathbf{diff_f}(\mathbf{f}|\mathbf{x1})\wedge\mathbf{qd}(\mathbf{d_f}(\mathbf{r}|\mathbf{x1}))\\
&\qquad = D[a1](\mathbf{f}|\mathbf{x1};d_f a)\wedge\mathbf{qd}(D[a1](\mathbf{r}|\mathbf{x1},d_f a)). && (1161)
\end{aligned}$$

Similarly,

$$\begin{aligned}
&D1[\mathbf{r1},\mathbf{r2}]\bullet(\mathbf{f}|\mathbf{x1};\mathbf{d_f r})\\
&\qquad = D[a1](\mathbf{x}|\mathbf{r1};d_f a)\bullet\mathbf{T}[\mathbf{D1_f}[\mathbf{r1},\mathbf{r2}]*(\mathbf{f}|\mathbf{x1};\mathbf{d_f x})]\\
&\qquad\qquad\qquad \bullet\mathbf{qd}(D[a1](\mathbf{r}|\mathbf{x1};d_f a)) && (1162)
\end{aligned}$$

D1[**r1**,**r2**]∧(**f**|**x1**;**d**$_f$**r**)
= (D[a1](**x**|**r1**;d$_f$a)•**T**[**D1**$_f$[**r1**,**r2**]∗(**f**|**x1**;**d**$_f$**x**)])
∧**qd**(D[a1](**r**|**x1**;d$_f$a)). (1163)

As the illustration shows, the many directional material/local gradients are related to indexed derivatives through D[a1](**r**|**x1**;da) and dually through D[a1](**x**|**r1**;da).

Turning now to relationships with sectional gradients of generalized functions, consider **f**(**r**(**x**,a1)) at **r1**(**x1**,a1).

The relationship between mutually related gradients is given in the following theorem.

Theorem 13 (relationship between mutually related gradients)
Given
r(**x**,a1) and **x**(**r**,a1), an observation of particles
and their location at observation a1;
mutually related sections
SECT(**r1**,**u1R1**,**u2R1**,**u3R1**) and
SECT(**x1**,**u1X1**,**u2X1**,**u3X1**);
dX1 = d1X1∗**u1X1**+d2X1∗**u2X1**+d3X1∗**u3X1**;
f(**r**(**x**,a1), a function
with mutually related gradients in the sections;
for
UR1 = **u1**∗**u1R1** + **u2**∗**u2R1** +**u3**∗**u3R1;**
UX1 = **u1**∗**u1X1** + **u2**∗**u2X1** +**u3**∗**u3X1;**
r(**x1** + d1X1∗**u1X1,**a1)–**r1**= d1R1∗**u1R1;**
r(**x1** + d2X1∗**u2X1,**a1)–**r1**= d2R1∗**u2R1;**
r(**x1** + d3X1∗**u3X1,**a1)–**r1**= d3R1∗**u3R1;**
dR1 = d1R1∗**u1R1**+d2R1∗**u2R1**+d3R1∗**u3R1**;
p = lim (d1X1/d1R1)∗**u1**
+ (d2X1/d2R1)∗**u2**
+ (d3X1/d3R1)∗**u3**;
then

D3[**r1**]∗(**f**(**r**(**x**,a1))|a1;**dR1**)
= **D3**[**x1**]∗(**f**(**x**(**r**,a1))|a1;**dX1**)•[**UX1**$^{-1}$]•[**G**$^{-1}$(**p**)]•**UR1** (1164)
= **D3**[**x1**]∗(**f**(**x**(**r**,a1))|a1;**dX1**)
•[**UX1**$^{-1}$]•**T**[**UX1**$^{-1}$]•**D3**[**r1**]∗(**x**(**r**,a1)|a1;**dR1**). (1165)

Theoretical Physics: Derivatives

$$\begin{aligned}
&\mathbf{D3[x1]}\ast(\mathbf{f}(\mathbf{x}(\mathbf{r},\mathrm{a1}))|\mathrm{a1};\mathbf{dX1})\\
&\quad= \mathbf{D3[r1]}\ast(\mathbf{f}(\mathbf{r}(\mathbf{x},\mathrm{a1}))|\mathrm{a1};\mathbf{dR1})\bullet[\mathbf{UR1}^{-1}]\bullet[\mathbf{G}(\mathbf{p})]\bullet\mathbf{UX1} && (1166)\\
&\quad= \mathbf{D3[r1]}\ast(\mathbf{f}(\mathbf{r}(\mathbf{x},\mathrm{a1}))|\mathrm{a1};\mathbf{dR1})\\
&\qquad\qquad\bullet[\mathbf{UR1}^{-1}]\bullet\mathbf{T}[\mathbf{UR1}^{-1}]\bullet\mathbf{D3[x1]}\ast(\mathbf{r}(\mathbf{x},\mathrm{a1})|\mathrm{a1};\mathbf{dX1}). && (1167)
\end{aligned}$$

Proof:

$$\begin{aligned}
&\mathbf{D3[x1]}\ast(\mathbf{f}|\mathrm{a1};\mathbf{dX1})\bullet[\mathbf{UX1}^{-1}]\bullet[\mathbf{G}^{-1}(\mathbf{p})]\bullet\mathbf{UR1}\\
&\quad= \lim\ (\mathbf{f}(\mathbf{x1}+\mathrm{d1X1}\ast\mathbf{u1X1},\mathrm{a1}) - \mathbf{f}(\mathbf{x1},\mathrm{a1}))\ast\mathbf{u1X1}/\mathrm{d1X1}\\
&\qquad\quad + (\mathbf{f}(\mathbf{x1}+\mathrm{d2X1}\ast\mathbf{u2X1},\mathrm{a1}) - \mathbf{f}(\mathbf{x1},\mathrm{a1}))\ast\mathbf{u2X1}/\mathrm{d2X1}\\
&\qquad\quad + (\mathbf{f}(\mathbf{x1}+\mathrm{d3X1}\ast\mathbf{u3X1},\mathrm{a1}) - \mathbf{f}(\mathbf{x1},\mathrm{a1}))\ast\mathbf{u3X1}/\mathrm{d3X1}\\
&\qquad\qquad\bullet[\mathbf{UX1}^{-1}]\bullet[\mathbf{G}^{-1}(\mathbf{p})]\bullet\mathbf{UR1}\\
&\quad= \lim\ (\mathbf{f}(\mathbf{x1}+\mathrm{d1X1}\ast\mathbf{u1X1},\mathrm{a1}) - \mathbf{f}(\mathbf{x1},\mathrm{a1}))\ast\mathbf{u1}/\mathrm{d1X1}\\
&\qquad\quad + (\mathbf{f}(\mathbf{x1}+\mathrm{d2X1}\ast\mathbf{u2X1},\mathrm{a1}) - \mathbf{f}(\mathbf{x1},\mathrm{a1}))\ast\mathbf{u2}/\mathrm{d2X1}\\
&\qquad\quad + (\mathbf{f}(\mathbf{x1}+\mathrm{d3X1}\ast\mathbf{u3X1},\mathrm{a1}) - \mathbf{f}(\mathbf{x1},\mathrm{a1}))\ast\mathbf{u3}/\mathrm{d3X1}\\
&\qquad\qquad\bullet[\mathbf{G}(\mathbf{p})]\bullet\mathbf{UR1}\\
&\quad= \lim\ (\mathbf{f}(\mathbf{x1}+\mathrm{d1X1}\ast\mathbf{u1X1},\mathrm{a1}) - \mathbf{f}(\mathbf{x1},\mathrm{a1}))\ast\mathbf{u1}/\mathrm{d1R1}\\
&\qquad\quad + (\mathbf{f}(\mathbf{x1}+\mathrm{d2X1}\ast\mathbf{u2X1},\mathrm{a1}) - \mathbf{f}(\mathbf{x1},\mathrm{a1}))\ast\mathbf{u2}/\mathrm{d2R1}\\
&\qquad\quad + (\mathbf{f}(\mathbf{x1}+\mathrm{d3X1}\ast\mathbf{u3X1},\mathrm{a1}) - \mathbf{f}(\mathbf{x1},\mathrm{a1}))\ast\mathbf{u3}/\mathrm{d3R1}\\
&\qquad\qquad\bullet\mathbf{UR1}\\
&\quad= \lim\ (\mathbf{f}(\mathbf{x1}+\mathrm{d1X1}\ast\mathbf{u1X1},\mathrm{a1}) - \mathbf{f}(\mathbf{x1},\mathrm{a1}))\ast\mathbf{u1R1}/\mathrm{d1R1}\\
&\qquad\quad + (\mathbf{f}(\mathbf{x1}+\mathrm{d2X1}\ast\mathbf{u2X1},\mathrm{a1}) - \mathbf{f}(\mathbf{x1},\mathrm{a1}))\ast\mathbf{u2R1}/\mathrm{d2R1}\\
&\qquad\quad + (\mathbf{f}(\mathbf{x1}+\mathrm{d3X1}\ast\mathbf{u3X1},\mathrm{a1}) - \mathbf{f}(\mathbf{x1},\mathrm{a1}))\ast\mathbf{u3R1}/\mathrm{d3R1}\\
&\quad= \mathbf{D3[r1]}\ast(\mathbf{f}|\mathrm{a1};\mathbf{dR1}).
\end{aligned}$$

This proves the first proposition.
Now from the theorem on the gradients of primary functions

$$\mathbf{D3[r1]}\ast(\mathbf{x}(\mathbf{r},\mathrm{a1});\mathbf{dR1}) = \mathbf{T}[\mathbf{UX1}]\bullet[\mathbf{G}(\mathbf{p})]^{-1}\bullet\mathbf{UR1}$$

that is,

$$\mathbf{G}^{-1}(\mathbf{p})\bullet\mathbf{UR1} = \mathbf{T}[\mathbf{UX1}]^{-1}\bullet[\mathbf{D3[r1]}\ast(\mathbf{x}(\mathbf{r},\mathrm{a1});\mathbf{dR1}).$$

Thus

$$\begin{aligned}
&\mathbf{D3[x1]}\ast(\mathbf{f}|\mathrm{a1};\mathbf{dX1})\bullet[\mathbf{UX1}^{-1}]\bullet[\mathbf{G}^{-1}(\mathbf{p})]\bullet\mathbf{UR1}\\
&\quad= \mathbf{D3[x1]}\ast(\mathbf{f}|\mathbf{a1};\mathbf{dX1})\\
&\qquad\qquad\bullet[\mathbf{UX1}^{-1}]\bullet\mathbf{T}[\mathbf{UX1}]^{-1}\bullet\mathbf{D3[r1]}\ast(\mathbf{x}(\mathbf{r},\mathrm{a1});\mathbf{dR1})].
\end{aligned}$$

This proves the second proposition.
The dual is proven similarly. qed

Discontinuities of **f** along **r**(**x1**,a) may bear no relationship to possible discontinuities of **f** along **x**(**r1**+d1 ∗ **u1**,a+da) so that in general a relationship between the sectional gradients of ***f*** and the positive quadrant gradient of ***f*** may not exist. Indeed, the trace of **r**(**x1**,a) may not even fall in the positive quadrant where **D3[r1]**∗(**f**;**dr**) is defined.

The conditions under which a sectional gradient is related to a positive quadrant gradient is given in theorem 2 restated here in dual form.

For a function with a *simply continuous* gradient at **r1**(**x1**,a1)

D3[**r1**]∗(**f**|a1;**dR1**)
 = **D3**[**r1**]∗(**f**|a1;**dr**)
 •[**u1R1**∗**u1R1** + **u2R1**∗**u2R1** + **u3R1**∗**u3R1**] (1168)
 = [**D**[**r1**]∗(**f**|a1;**dr**)]•**T**[**UR1**]•**UR1**. (1169)

Dually,

D3[**x1**]∗(**f**|a1;**dX1**)
 = **D3**[**x1**]∗(**f**|a1;**dx**)
 •[**u1X1**∗**u1X1** + **u2X1**∗**u2X1** + **u3X1**∗**u3X1**] (1170)
 = [**D**[**x1**]∗(**f**|a1;**dx**)]•**T**[**UX1**]•**UX1**. (1171)

Then in combination with the theorem 13

D3[**r1**]∗(**f**(**r**(**x**,a1))|a1;**dR1**)
 = **D3**[**x1**]∗(**f**(**x**(**r**,a1))|a1;**dx**)•**D3**[**r1**]∗(**x**(**r**,a1)|a1;**dR1**) (1172)

and dually

D3[**x1**]∗(**f**(**x**(**r**,a1))|a1;**dX1**)
 = **D3**[**r1**]∗(**f**(**r**(**x**,a1))|a1;**dr**)•**D3**[**x1**]∗(**r**(**x**,a1)|a1;**dX1**). (1173)

When SECT(**r1**,**u1R1**,**u2R1**,**u3R1**) and SECT(**x1**,**u1X1**,**u2X1**,**u3X1**) are positive quadrants

D3[**r1**]∗(**f**(**r**(**x**,a1))|a1;**dr**)
 = **D3**[**x1**]∗(**f**(**x**(**r**,a1))|a1;**dx**)•**D3**[**r1**]∗(**x**(**r**,a1)|a1;**dr**) (1174)

and

D3[**x1**]∗(**f**(**x**(**r**,a1))|a1;**dx**)
 = **D3**[**r1**]∗(**f**(**r**(**x**,a1))|a1;**dr**)•**D3**[**x1**]∗(**r**(**x**,a1)|a1;**dx**). (1175)

By setting **f**(**x**(**r**,a1)) = **x**(**r**,a1) or **f**(**r**(**x**,a1)) = **r**(**x**,a1), these results confirm equation (1062).

Theoretical Physics: Derivatives

For other closely related observations

D3[**r1**]∗(**f**|a1;**dRi**)
= **D3**[**x1**]∗(**f**|a1;**dXi**)
•[**UXi**(**x1**|a1)$^{-1}$]•**T**[**UXi**(**x1**|a1)$^{-1}$]•**D3**[**r1**]∗(**x**|a1;**dRi**) (1176)

D3[**r1**]∗(**f**|a1+da;**dRi**)
= **D3**[**x2**]∗(**f**|a1+da;**dXi**)
•[**UXi**(**x2**|a1+da)$^{-1}$]•**T**[**UXi**(**x2**|a1+da)$^{-1}$]
•**D3**[**r1**]∗(**x**|a1+da;**dRi**) (1177)

D3[**r1**]∗(**f**|a1–da;**dRi**)
= **D3**[**x0**]∗(**f**|a1–da;**dXi**)
•[**UXi**(**x0**|a1–da)$^{-1}$[•**T**[**UXi**(**x0**|a1–da)$^{-1}$]
•**D3**[**r1**]∗(**x**|a1–da;**dRi**) (1178)

D3[**r2**]∗(**f**|a1;**dRi**)
= **D3**[**y2**]∗(**f**|a1;**dXi**)
•[**UXi**(**y2**|a1)$^{-1}$[•**T**[**UXi**(**y2**|a1)$^{-1}$]
•**D3**[**r2**]∗(**x**|a1;**dRi**) (1179)

D3[**r2**]∗(**f**|a1+da;**dRi**)
= **D3**[**x1**]∗(**f**|a1+da;**dXi**)
•[**UXi**(**x1**|a1+da)$^{-1}$]•**T**[**UXi**(**x1**|a1+da)$^{-1}$]
•**D3**[**r2**]∗(**x**|a1+da;**dRi**) (1180)

D3[**r0**]∗(**f**|a1;**dRi**)
= **D3**[**y0**]∗(**f**|a1;**dXi**)
•[**UXi**(**y0**|**a1**)$^{-1}$[•**T**[**UXi**(**y0**|**a1**)$^{-1}$]
•**D3**[**r0**]∗(**x**|a1;**dRi**) (1181)

D3[**r0**]∗(**f**|a1–da;**dRi**)
= **D3**[**x1**]∗(**f**|a1–da;**dXi**)
•[**UXi**(**x1**|a1–da)$^{-1}$[•**T**[**UXi**(**x1**|a1–da)$^{-1}$]
•**D3**[**r0**]∗(**x**|a1–da;**dRi**) (**1182**)

D3[**x1**]∗(**f**|a1;**dXi**)
= **D3**[**r1**]∗(**f**|a1;**dRi**)
•[**URi**(**r1**|a1)$^{-1}$]•T[**URi**(**r1**|a1)$^{-1}$]
•**D3**[**x1**]∗(**r**|a1;**dXi**) (1183)

D3[**x2**]∗(**f**|a1+da;**dXi**)
= **D3**[**r1**]∗(**f**|a1+da;**dRi**)
•[**URi**(**r1**|a1+da)$^{-1}$[•**T**[**URi**(**r1**|a1+da)$^{-1}$]
•**D3**[**x2**]∗(**r**|a1+da;**dXi**) (1184)

D3[**x0**]∗(**f**|a1−da;**dXi**)
 = **D3**[**r1**]∗(**f**|a1−da;**dRi**)
 •[**URi**(**r1**|a1−da)$^{-1}$[•**T**[**URi**(**r1**|a1−da)$^{-1}$]
 •**D3**[**x0**]∗(**r**|a1−da;**dXi**) (1185)

D3[**y2**]∗(**f**|a1;**dXi**)
 = **D3**[**r2**]∗(**f**|a1;**dRi**)
 •[**URi**(**r2**|a1)$^{-1}$[•**T**[**URi**(**r2**|a1)$^{-1}$]
 •**D3**[**y2**]∗(**r**|a1;**dXi**) (1186)

D3[**x1**]∗(**f**|a1+da;**dXi**)
 = **D3**[**r2**]∗(**f**|a1+da;**dRi**)
 •[**URi**(**r2**|a1+da)$^{-1}$[•**T**[**URi**(**r2**|a1+da)$^{-1}$]
 •**D3**[**x1**]∗(**r**|a1+da;**dXi**) (1187)

D3[**y0**]∗(**f**|a1;**dXi**)
 = **D3**[**r0**]∗(**f**|a1;**dRi**)
 •[**URi**(**r0**|a1)$^{-1}$[•**T**[**URi**(**r0**|a1)$^{-1}$]
 •**D3**[**y0**]∗(**r**|a1;**dXi**) (1188)

D3[**x1**]∗(**f**|a1−da;**dXi**)
 = **D3**[**r0**]∗(**f**|a1−da;**dRi**)
 •[**URi**(**r0**|a1−da)$^{-1}$[•**T**[**URi**(**r0**|a1−da)$^{-1}$]
 •**D3**[**x1**]∗(**r**|a1−da;**dXi**). (1189)

Now let a trace **r**(**x1**,a) be given such that, as above,

r1 = **r**(**x1**,a1) = **r**(**x0**,a1 – da) = **r**(**x2**,a1+da);
r2 = **r**(**x1**,a1+da) = **r**(**y2**,a1);
r0 = **r**(**x1**,a1 – da) = **r**(**y0**,a1);

and **f**(**r**(**x**,a)) continuously differentiable with respect to each of the three primary variables, except perhaps at **r1**(**x1**,a1) itself, within a section of the open neighborhood of **r1**(**x1**,a1) containing the local curve.

Theorem 14 (the relationship between local derivatives and gradients)
Given

r1(**x1**,a1), the reference location, particle and observation;
SECT(**r1**,**u1Rf**,**u2Rf**,**u3Rf**) a local section containing **r**(**x1**,a1+da);
SECT(**x1**,**u1Xf**,**u2Xf**,**u3Xf**) a material section containing **x**(**r1**,a1+da);
f(**r**(**x**,a)), a function mutually locally differentiable in SECT(**r1**,u1Rf,u2Rf,u3Rf) and SECT(**x1**,u1Xf,u2Xf,u3Xf);

then

D[**r1**(**x2**),**r2**](**f**|a1+da;da)
= D[a1](**r**|**x1**;d$_f$a)
•**T**[**URf**(a1+da)$^{-1}$]]•[**URf**(a1+da)$^{-1}$]
•**T**[**D3**[**r1**]∗(**f**|a1+da;**dRf**)] (1190)

and

D[**x1**(**r2**),**x2**](**f**|a1+da;da)
= D[a1](**x**|**r1**;d$_f$a)
•**T**[**UXf**(a1+da)$^{-1}$]]•[**UXf**(a1+da)$^{-1}$]
•**T**[**D3**[**x1**]∗(**f**|a1+da;**dXf**)]. (1191)

Proof:

Recall **r2** = **r**(**x1**,a1+da) and **x2** = **x**(**r1**,a1+da).
Then call
r2 − **r1** ≡ **dRf**
= dr1∗**u1Rf**+dr2∗**u2Rf**+dr3∗**u3Rf**
= (dr1∗**u1**+dr2∗**u2**+dr3∗**u3**)
•(**u1**∗**u1Rf**+**u2**∗**u2Rf**+**u3**∗**u3Rf**)
≡ **dr**•**URf**.

Now since **f** is differentiable in Rf,
f(**r2**(**x1**,a1+da)) − **f**(**r1**(**x2**,a1+da))
= **f**(**r1**+**dRf**,a1+da) − **f**(**r1**,a1+da)
≈ **f**(**r1**+dr1∗**u1Rf**,a1+da) − **f**(**r1**,a1+da)
+ **f**(**r1**+dr2∗**u2Rf**,a1+da) − **f**(**r1**,a1+da)
+ **f**(**r1**+dr3∗**u3Rf**,a1+da) − **f**(**r1**,a1+da)
= (dr1∗**u1Rf**+dr2∗**u2Rf**+dr3∗**u3Rf**)
•**T**[**URf**(a1+da)$^{-1}$]]•[**URf**(a1+da)$^{-1}$]
•[**u1Rf** ∗(**f**(**r1**+dr1∗**u1Rf**,a1+da) − **f**(**r1**,a1+da))
/dr1
+ **u2Rf** ∗(**f**(**r1**+dr2∗**u2Rf**,a1+da) − **f**(**r1**,a1+da))
/dr2
+ **u3Rf** ∗(**f**(**r1**+dr3∗**u3Rf**,a1+da) − **f**(**r1**,a1+da))
/dr3]
= (**r2**−**r1**)•**T**[**URf**(a1+da)$^{-1}$]]•[**URf**(a1+da)$^{-1}$]
•**T**[**D3**[**r1**]∗(**f**|a1+da;**dRf**)]

that is,

D[**r1**(**x2**),**r2**](**f**|a1+da;da)
= D[**r1**(**x2**),**r2**](**r**|a1+da;da)
•**T**[**URf**(a1+da)$^{-1}$]]•[**URf**(a1+da)$^{-1}$]
•**T**[**D3**[**r1**]∗(**f**|a1+da;**dRf**)]
= D[a1](**r**|**x1**;d$_f$a)
•**T**[**URf**(a1+da)$^{-1}$]]
•[**URf**(a1+da)$^{-1}$]•**T**[**D3**[**r1**]∗(**f**|a1+da;**dRf**)].

Similarly, since **f** is also differentiable in Xf, call

x2–**x1** ≡ **dXf**
= dx1 ∗ **u1Xf**+dx2 ∗ **u2Xf**+dx3 ∗ **u3Xf**
= (dx1 ∗ **u1**+dx2 ∗ **u2**+dx3 ∗ **u3**)
•(**u1** ∗ **u1Xf**+**u2** ∗ **u2Xf**+**u3** ∗ **u3Xf**)
≡ **dx**•**UXf**.

Then

f(**x2**(**r1**,a1+da))–**f**(**x1**(**r2**,a1+da))
= **f**(**x1**+**dXf**,a1+da) – **f**(**x1**,a1+da)
≈ **f**(**x1**+dx1 ∗ **u1Xf**,a1+da) – **f**(**x1**,a1+da)
+ **f**(**x1**+dx2 ∗ **u2Xf**,a1+da) – **f**(**x1**,a1+da)
+ **f**(**x1**+dx3 ∗ **u3Xf**,a1+da) – **f**(**x1**,a1+da)
= (dx1 ∗ **u1Xf**+dx2 ∗ **u2Xf**+dx3 ∗ **u3Xf**)
•**T**[**UXf**(a1+da)$^{-1}$]]•[**UXf**(a1+da)$^{-1}$]
•[**u1Xf** ∗ (**f**(**r1**+dx1 ∗ **u1Xf**,a1+da) – **f**(**x1**,a1+da))
/dx1
+ **u2Xf** ∗ (**f**(**r1**+dx2 ∗ **u2Xf**,a1+da)– **f**(**x1**,a1+da))
/dx2
+ **u3Xf** ∗ (**f**(**r1**+dx3 ∗ **u3Xf**,a1+da) – **f**(**x1**,a1+da))
/dx3]
= (**x2**–**x1**)•**T**[**UXf**(a1+da)$^{-1}$]]•[**UXf**(a1+da)$^{-1}$]
•**T**[**D3**[**x1**] ∗ (**f**|a1+da;**dXf**)]

that is,

D[**x1**(**r2**),**x2**](**f**|a1+da;da)
= D[**x1**(**r2**),**x2**](**x**|a1+da;da)
•**T**[**UXf**(a1+da)$^{-1}$]]•[**UXf**(a1+da)$^{-1}$]
•**T**[**D3**[**x1**] ∗ (**f**|a1+da;**dXf**)]
= D[a1](**x**|**r1**;d_fa)
•**T**[**UXf**(a1+da)$^{-1}$]]•[**UXf**(a1+da)$^{-1}$]
•**T**[**D3**[**x1**] ∗ (**f**|a1+da;**dXf**)].

qed

Corollary

D[a1](**r**|**x1**;d_fa)•**T**[**URf**(a1+da)$^{-1}$]]•[**URf**(a1+da)$^{-1}$]
•**T**[**D3**[**r1**] ∗ (**f**|a1+da;**dRf**)]
= – D[a1](**x**|**r1**;d_fa)
•**T**[**UXf**(a1+da)$^{-1}$]]•[**UXf**(a1+da)$^{-1}$]
•**T**[**D3**[**x1**] ∗ (**f**|a1+da;**dXf**)]. (1192)

Proof:

Since **f** is mutually differentiable, the implication follows from equation (1134).

Theoretical Physics: Derivatives

Corollary

Given

r1(**x1**,a1), the reference location, particle and observation;
SECT(**r1**,**u1Rb**,**u2Rb**,**u3Rb**) a local section containing
r(**x1**,a1−da);
SECT(**x1**,**u1Xb**,**u2Xb**,**u3Xb**) a material section containing
x(**r1**,a1−da);
f(**r**(**x**,a)), a function mutually locally differentiable in
SECT(**r1**,**u1Rb**,**u2Rb**,**u3Rb**)
and SECT(**x1**,**u1Xb**,**u2Xb**,**u3Xb**);

then

D[**r0**(**x1**),**r1**](**f**|a1−da;da)
= D[a1](**r**|**x1**;d_ba)
•**T**[**URb**(a1−da)$^{-1}$]]•[**URb**(a1−da)$^{-1}$]
•**T**[**D3**[**r1**]∗(**f**|a1−da;**dRb**)] (1193)

and

D[**x0**(**r1**),**x1**](**f**|a1−da;da)
= D[a1](**x**|**r1**;d_ba)
•**T**[**UXb**(a1−da)$^{-1}$]]•[**UXb**(a1−da)$^{-1}$]
•**T**[**D3**[**x1**]∗(**f**|a1−da;**dXbf**)]. (1194)

Corollary

D[a1](**r**|**x1**;d_ba)•**T**[**URb**(a1−da)$^{-1}$]]•[**URb**(a1−da)$^{-1}$]
•**T**[**D3**[**r1**]∗(**f**|a1−da;**dRb**)]
= −D[a1](**x**|**r1**;d_ba)
•**T**[**UXb**(a1−da)$^{-1}$]]•[**UXb**(a1−da)$^{-1}$]
•**T**[**D3**[**x1**]∗(**f**|a1−da;**dXb**)]. (1195)

These results cannot be appropriated to the *a1* observation completely.

For the *a1* observation

D[**r1**,**r2**(**y2**)](**f**|a1;da) = lim (**f**(**r2**(**y2**,a1)) – **f**(**r1**(**x1**,a1)))/da. (1196)

The vector **r2**–**r1** lies in SECT(**u1Rf**(a1),**u2Rf**(a1),**u3Rf**(a1)) defined by the sectional continuity of ***f***. It follows then

f(**r2**(**y2**,a1)) – **f**(**r1**(**x1**,a1))
= **f**(**r1** + **dR1**,a1) – **f**(**r1**,a1)
≈ (**r2**–**r1**)•**T**[**URf**(a1)$^{-1}$]]•[**URf**(a1)$^{-1}$]•**T**[**D3**[**r1**]∗(**f**|a1;**dRf**)]

that is,

$$
\begin{aligned}
&D[\mathbf{r1},\mathbf{r2}(\mathbf{y2})](\mathbf{f}|a1;da)\\
&\quad = D[\mathbf{r1},\mathbf{r2}](\mathbf{r}|a1;da)\cdot\mathbf{T}[\mathbf{URf}(a1)^{-1}]]\cdot[\mathbf{URf}(a1)^{-1}]\\
&\qquad\qquad \cdot\mathbf{T}[\mathbf{D3}[\mathbf{r1}]*(\mathbf{f}|a1;\mathbf{dRf})]
\end{aligned}
\tag{1197}
$$

$$
\begin{aligned}
&\quad = D(\mathbf{r}|\mathbf{x1};d_f a)\cdot\mathbf{T}[\mathbf{URf}(a1)^{-1}]]\cdot[\mathbf{URf}(a1)^{-1}]\\
&\qquad\qquad \cdot\mathbf{T}[\mathbf{D3}[\mathbf{r1}]*(\mathbf{f}|a1;\mathbf{dRf})].
\end{aligned}
\tag{1198}
$$

However

$$D[\mathbf{r1},\mathbf{r2}(\mathbf{y2})](\mathbf{f}|a1;da)$$

does not necessarily equal

$$-D[\mathbf{x1},\mathbf{x}(\mathbf{s2})](\mathbf{f}|a1;da).$$

Likewise,

$$
\begin{aligned}
&D[\mathbf{x1},\mathbf{x}(\mathbf{s2})](\mathbf{f}|a1;da)\\
&\quad = D(\mathbf{x}|\mathbf{r1};d_f a)\cdot\mathbf{T}[\mathbf{UXf}(a1)^{-1}]]\cdot[\mathbf{UXf}(a1)^{-1}]\\
&\qquad\qquad \cdot\mathbf{T}[\mathbf{D3}[\mathbf{x1}]*(\mathbf{f}|a1;\mathbf{dXf})]
\end{aligned}
\tag{1199}
$$

$$
\begin{aligned}
&D[\mathbf{r}(\mathbf{y0}),\mathbf{r1}](\mathbf{f}|a1;da)\\
&\quad = D(\mathbf{r}|\mathbf{x1};d_b a)\cdot\mathbf{T}[\mathbf{URb}(a1)^{-1}]]\cdot[\mathbf{URb}(a1)^{-1}]\\
&\qquad\qquad \cdot\mathbf{T}[\mathbf{D3}[\mathbf{r1}]*(\mathbf{f}|a1;\mathbf{dRf})]
\end{aligned}
\tag{1200}
$$

$$
\begin{aligned}
&D[\mathbf{x}(\mathbf{s0}),\mathbf{x1}](\mathbf{f}|a1;da)\\
&\quad = D(\mathbf{x}|\mathbf{r1};d_b a)\cdot\mathbf{T}[\mathbf{UXb}(a1)^{-1}]]\cdot[\mathbf{UXb}(a1)^{-1}]\\
&\qquad\qquad \cdot\mathbf{T}[\mathbf{D3}[\mathbf{x1}]*(\mathbf{f}|a1;\mathbf{dXb})].
\end{aligned}
\tag{1201}
$$

Consequently,

$$
\begin{aligned}
&D[\mathbf{r}(\mathbf{x2}),\mathbf{r2}](\mathbf{f}|a1+da;da) - D[\mathbf{r1},\mathbf{r}(\mathbf{y2})](\mathbf{f}|a1;da)\\
&\quad = D(\mathbf{r}|\mathbf{x1};d_f a)\cdot[\mathbf{T}[\mathbf{URf}(a1+da)^{-1}]\cdot[\mathbf{URf}(a1+da)^{-1}]\\
&\qquad\qquad \cdot\mathbf{T}[\mathbf{D3}[\mathbf{r1}]*(\mathbf{f}|a1+da;\mathbf{dRf})]\\
&\qquad\qquad\quad -\mathbf{T}[\mathbf{URf}(a1)^{-1}]\cdot[\mathbf{URf}(a1)^{-1}]\\
&\qquad\qquad\qquad \cdot\mathbf{T}[\mathbf{D3}[\mathbf{r1}]*(\mathbf{f}|a1;\mathbf{dRf})]]
\end{aligned}
\tag{1202}
$$

$$
\begin{aligned}
&D[\mathbf{x}(\mathbf{r2}),\mathbf{x2}](\mathbf{f}|a1+da;da) - D[\mathbf{x1},\mathbf{x}(\mathbf{s2})](\mathbf{f}|a1;da)\\
&\quad = D(\mathbf{x}|\mathbf{r1};d_f a)\cdot[\mathbf{T}[\mathbf{UXf}(a1+da)^{-1}]]\cdot[\mathbf{UXf}(a1+da)^{-1}]\\
&\qquad\qquad \cdot\mathbf{T}[\mathbf{D3}[\mathbf{x1}]*(\mathbf{f}|a1+da;\mathbf{dXf})]\\
&\qquad\qquad\quad -\mathbf{T}[\mathbf{UXf}(a1)^{-1}]]\cdot[\mathbf{UXf}(a1)^{-1}]\\
&\qquad\qquad\qquad \cdot\mathbf{T}[\mathbf{D3}[\mathbf{x1}]*(\mathbf{f}|a1;\mathbf{dXf})]
\end{aligned}
\tag{1203}
$$

$$
\begin{aligned}
&D[\mathbf{r}(\mathbf{y0}),\mathbf{r1}](\mathbf{f}|a1;da)- D[\mathbf{r}(\mathbf{x1}),\mathbf{r1}](\mathbf{f}|a1-da;da)\\
&\quad = D(\mathbf{r}|\mathbf{x1};d_b a)\cdot[\mathbf{T}[\mathbf{URb}(a1)^{-1}]]\cdot[\mathbf{URb}(a1)^{-1}]\\
&\qquad\qquad \cdot\mathbf{T}[\mathbf{D3}[\mathbf{r1}]*(\mathbf{f}|a1;\mathbf{dRb})]\\
&\qquad\qquad\quad -\mathbf{T}[\mathbf{URb}(a1-da)^{-1}]]\cdot[\mathbf{URb}(a1-da)^{-1}]\\
&\qquad\qquad\qquad \cdot\mathbf{T}[\mathbf{D3}[\mathbf{r1}]*(\mathbf{f}|a1-da;\mathbf{dRb})]
\end{aligned}
\tag{1204}
$$

Theoretical Physics: Derivatives

$$\begin{aligned}&D[\mathbf{x}(\mathbf{s0}),\mathbf{r1}](\mathbf{f}|a1;da) - D[\mathbf{x}(\mathbf{r1}),\mathbf{x1}](\mathbf{f}|a1-da;da)\\ &\quad = D(\mathbf{x}|\mathbf{r1};d_b a)\bullet[\mathbf{T}[\mathbf{UXb}(a1)^{-1}]]\bullet[\mathbf{UXb}(a1)^{-1}]\\ &\qquad\bullet\mathbf{T}[\mathbf{D3}[\mathbf{x1}]*(\mathbf{f}|a1;\mathbf{dXb})]\\ &\qquad -\mathbf{T}[\mathbf{UXb}(a1-da)^{-1}]]\bullet[\mathbf{UXb}(a1-da)^{-1}]\\ &\qquad\quad\bullet\mathbf{T}[\mathbf{D3}[\mathbf{x1}]*(\mathbf{f}|a1-da;\mathbf{dXb})].\end{aligned} \tag{1205}$$

The above results with implied reference of **r1**(**x1**,a1) are summarized in the following tables.

Derivatives	Rank 1 Decomposition
D[**r0**,**r1**](**f**|a1;da)	D(**r**|**x1**;d_ba)•**T**[**D1**[**r0**(**x**,a1),**r1**(**x1**,a1)]*(**f**|a1;$\mathbf{d_b}$**r**)]
D[**r0**,**r1**] (**f**|a1−da;da)	D(**r**|**x1**;d_ba)•**T**[**D1**[**r0**(**x1**),**r1**(**x**,a1−da)]*(**f**;$\mathbf{d_b}$**r**)]
D[**r1**,**r2**](**f**|a1+da;da)	D(**r**|**x1**;d_fa)•**T**[**D1**[**r1**(**x**,a1+da),**r2**(**x1**)]*(**f**;$\mathbf{d_f}$**r**)]
D[(**r1**,**r2**](**f**|a1;da)	D(**r**|**x1**;d_fa)•**T**[**D1**[**r1**(**x1**),**r2**(**x**,a1)]*(**f**|a1;$\mathbf{d_f}$**r**)]
D[**x0**,**x1**](**f**|a1;da)	D(**x**|**r1**;d_ba)•**T**[**D1**[**x0**(**r**,a1),**x1**(**r1**)]*(**f**;$\mathbf{d_b}$**x**)]
D[**x0**,**x1**](**f**|a1−da;da)	D(**x**|**r1**;d_ba)•**T**[**D1**(**x0**(**r1**),**x1**(**r1**,a1−da)]*(**f**;$\mathbf{d_b}$**x**)]
D[**x1**,**x2**](**f**|a1+da;da)	D(**x**|**r1**;d_fa)•**T**[**D1**(**x1**(**r**,a1+da),**x2**(**r1**)]*(**f**|a1+da;$\mathbf{d_f}$**x**)]
D[**x1**,**x2**](**f**|a1;da)	D(**x**|**r1**;d_fa)•**T**[**D1**[**x1**(**r1**),**x2**(**r**,a1)]*(**f**|a1;$\mathbf{d_f}$**x**)]

Decomposition of Indexed Derivatives

<table>
<tr><th>Derivatives</th><th>Rank 3 Decomposition</th></tr>
<tr><td>D[r0,r1](f|a1;da)</td><td>D(r|x1;d_ba)•T[URb(a1)$^{-1}$]•[URb(a1)^{1}]
•T[D3[r1]*(f|a1;dRb)]</td></tr>
<tr><td>D[r0,r1](f|a1−da;da)</td><td>D(r|x1;d_ba)•T[URb(a1−da)$^{-1}$]•[URb(a1−da)$^{-1}$]
•T[D3[r1]*(f|a1−da;dRb)]</td></tr>
<tr><td>D[r1,r2](f|a1+da;da)</td><td>D(r|x1;d_fa)•T[URf(a1+da)$^{-1}$]•[URf(a1+da)$^{-1}$]
•T[D3[r1]*(f|a1+da;dRf)]</td></tr>
<tr><td>D[r1,r2](f|a1;da)</td><td>D(r|x1;d_fa)•T[URf(a1)$^{-1}$]•[URf(a1)$^{-1}$]
•T[D3[r1]*(f|a1;dRf)]</td></tr>
<tr><td>D[x0,x1](f|a1;da)</td><td>D(x|r1;d_ba)•T[UXb(a1)$^{-1}$]•[UXb(a1)$^{-1}$]
•T[D3[x1]*(f|a1;dXb)</td></tr>
<tr><td>D[x0,x1](f|a1−da;da)</td><td>D(x|r1;d_ba)•T[UXb(a1−da)$^{-1}$]•[UXb(a1−da)$^{-1}$]
•T[D3[x1]*(f|a1−da;dXb)]</td></tr>
<tr><td>D[x1,x2](f|a1+da;da)</td><td>D(x|r1;d_fa)•T[UXf(a1+da)$^{-1}$]•[UXf(a1+da)$^{-1}$]
•T[D3[x1]*(f|a1+da;dXf)]</td></tr>
<tr><td>D[x1,x2](f|a1;da)</td><td>D(x|r1;d_fa)•T[UXf(a1)$^{-1}$]•[UXf(a1)$^{-1}$]
•T[D3[x1]*(f|a1;dXf)]</td></tr>
</table>

Decomposition of Indexed Derivatives

Where ***f*** has a simply continuous gradient at **r1**(**x1**,a1), **uiRf** may be chosen as one of the three arbitrary orthogonal directions. Then,

UXf^{-1}]•[**UXf**$^{-1}$] = **I**
D3[**r1**]*(**f**|a1+da;**dRf**)] = **D3**[**r1**]*(**f**|a1−da;**dRb**)
= **D3**[**r1**]*(**f**|a1;**dr**). (1206)

Results for the simply continuous case are

D3[**r1**]*(**f**|a1;**dr**)
= **D3**[**x1**]*(**f**|a1;**dx**)•**D3**[**r1**]*(**x**|a1;**dr**) (1207)
D3[**r1**]*(**f**|a1+da;**dr**)
= **D3**[**x2**]*(**f**|a1+da;**dx**)•**D3**[**r1**]*(**x**|a1+da;**dr**) (1208)
D3[**r1**]*(**f**|a1−da;**dr**)
= **D3**[**x0**]*(**f**|a1−da;**dx**)•**D3**[**r1**]*(**x**|a1−da;**dr**) (1209)
D3[**r2**]*(**f**|a1;**dr**)
= **D3**[**y2**]*(**f**|a1;**dx**)•**D3**[**r2**]*(**x**|a1;**dr**) (1210)

Theoretical Physics: Derivatives

D3[**r2**]∗(**f**|a1+da;**dr**)
  = **D3**[**x1**]∗(**f**|a1+da;**dx**)•**D3**[**r2**]∗(**x**|a1+da;**dr**) (1211)
D3[**r0**]∗(**f**|a1;**dr**)
  = **D3**[**y0**]∗(**f**|a1;**dx**)•**D3**[**r0**]∗(**x**|a1;**dr**) (1212)
D3[**r0**]∗(**f**|a1−da;**dr**)
  = **D3**[**x1**]∗(**f**|a1−da;**dx**)•**D3**[**r0**]∗(**x**|a1−da;**dr**) (1213)

D3[**x1**]∗(**f**|a1;**dx**)
  = **D3**[**r1**]∗(**f**|a1;**dr**)•**D3**[**x1**]∗(**r**|a1;**dx**) (1214)
D3[**x2**]∗(**f**|a1+da;**dx**)
  = **D3**[**r1**]∗(**f**|a1+da;**dr**)•**D3**[**x2**]∗(**r**|a1+da;**dx**) (1215)
D3[**x0**]∗(**f**|a1−da;**dx**)
  = **D3**[**r1**]∗(**f**|a1−da;**dr**)•**D3**[**x0**]∗(**r**|a1−da;**dx**) (1216)
D3[**y2**]∗(**f**|a1;**dx**)
  = **D3**[**r2**]∗(**f**|a1;**dr**)•**D3**[**y2**]∗(**r**|a1;**dx**) (1217)
D3[**x1**]∗(**f**|a1+da;**dx**)
  = **D3**[**r2**]∗(**f**|a1+da;**dr**)•**D3**[**x1**]∗(**r**|a1+da;**dx**) (1218)
D3[**y0**]∗(**f**|a1;**dx**)
  = **D3**[**r0**]∗(**f**|a1;**dr**)•**D3**[**y0**]∗(**r**|a1;**dx**) (1219)
D3[**x1**]∗(**f**|a1−da;**dx**)
  = **D3**[**r0**]∗(**f**|a1−da;**dr**)•**D3**[**x1**]∗(**r**|a1−da;**dx**) (1220)

D[**r1**(**x2**,a1+da),**r2**(**x1,**a1+da)](**f**|a1+da;da)
  = D(**r**|**x1**;d_fa)•**T**[**D3**[**r1**]∗(**f**|a1+da;**dr**)] (1221)
  = −D[**x1**(**r**,a1+da),**x2**(**r1**)](**f**|a1+da;da) (1222)
D[**r0**(**x1**),**r1**(**x**,a1−da)](**f**|a1−da;da)
  = D(**r**|**x1**;d_ba)•**T**[**D3**[**r1**]∗(**f**|a1−da;**dr**)] (1223)
  = −D[**x0**(**r1**),**x1**(**r0**,a1−da)](**f**|a1−da;da)
D[**x1**(**r2**),**x2**(**r1**,a1+da)](**f**|a1+da;da)
  = D(**x**|**r1**;d_fa)•**T**[**D3**[**x1**]∗(**f**|a1+da;**dx**)] (1224)
  = −D[**r1**(**x2**,a1+da),**r2**(**x1**)](**f**|a1+da;da) (1225)
D[**x**(**r1**),**x1**](**f**|a1−da;da)
  = D(**x**|**r1**;d_ba)•**T**[**D3**[**x1**]∗(**f**|a1−da;**dx**)] (1226)
  = −D[**r**(**x1**),**r1**](**f**|a1−da;da) (1227)
D[**r1**,**r**(**y2**)](**f**|a1;da)
  = D(**r**|**x1**;d_fa)•**T**[**D3**[**r1**]∗(**f**|a1;**dr**)] (1228)
D[**x1**,**x**(**s2**)](**f**|a1;da)
  = D(**x**|**r1**;d_fa)•**T**[**D3**[**x1**]∗(**f**|a1;**dx**)] (1229)
D[**r**(**y0**),**r1**](**f**|a1;da)
  = D(**r**|**x1**;d_ba)•**T**[**D3**[**r1**]∗(**f**|a1;**dr**)] (1230)

$$D[\mathbf{x}(\mathbf{s0}),\mathbf{x1}](\mathbf{f}|a1;da) = D(\mathbf{x}|\mathbf{r1};d_b a)\cdot\mathbf{T}[\mathbf{D3}[\mathbf{x1}]*(\mathbf{f}|a1;\mathbf{dx})] \quad (1231)$$

$$D(\mathbf{r}|\mathbf{x1};d_f a)\cdot\mathbf{T}[\mathbf{D3}[\mathbf{r1}]*(\mathbf{f}|a1+da;\mathbf{dr})] = -D(\mathbf{x}|\mathbf{r1};d_f a)\cdot\mathbf{T}[\mathbf{D3}[\mathbf{x1}]*(\mathbf{f}|a1+da;\mathbf{dx})] \quad (1232)$$

$$D(\mathbf{r}|\mathbf{x1};d_b a)\cdot\mathbf{T}[\mathbf{D3}[\mathbf{r1}]*(\mathbf{f}|a1-da;\mathbf{dr})] = -D(\mathbf{x}|\mathbf{r1};d_b a)\cdot\mathbf{T}[\mathbf{D3}[\mathbf{x1}]*(\mathbf{f}|a1-da;\mathbf{dx})]. \quad (1233)$$

On the other hand, when **uiRf⟶ur**, i=1,2,3

$$\begin{aligned}\mathbf{URf} &= \mathbf{u1}*\mathbf{u1Rf}+\mathbf{u2}*\mathbf{u2Rf}+\mathbf{u3}*\mathbf{u3Rf}\\ &= \mathbf{u1}*\mathbf{ur}+\mathbf{u2}*\mathbf{ur}+\mathbf{u3}*\mathbf{ur}\\ &= (\mathbf{u1}+\mathbf{u2}+\mathbf{u3})*\mathbf{ur}\end{aligned}$$

is no longer a rank 3 transformation. Then for **URf** defined by

$$D(\mathbf{r}|\mathbf{x1};da) = d*\mathbf{ur} = d1*\mathbf{u1}+d2*\mathbf{u2}+d3*\mathbf{u3}$$

$$\begin{aligned}&(\mathbf{f}(\mathbf{r1}+d*\mathbf{ur})-\mathbf{f}(\mathbf{r1}))*\mathbf{u1}/d1\\ &\quad+(\mathbf{f}(\mathbf{r1}+d*\mathbf{ur})-\mathbf{f}(\mathbf{r1}))*\mathbf{u2}/d2\\ &\quad+(\mathbf{f}(\mathbf{r1}+d*\mathbf{ur})-\mathbf{f}(\mathbf{r1}))*\mathbf{u3}/d3\\ &\qquad= (\mathbf{f}(\mathbf{r1}+d*\mathbf{ur})-\mathbf{f}(\mathbf{r1}))*\mathbf{qd}(d*\mathbf{ur})\\ &\qquad\longrightarrow \mathbf{D1}[\mathbf{r1}(\mathbf{x1},a1)]*(\mathbf{f}|\mathbf{x1};\mathbf{dr}).\end{aligned}$$

Accordingly as the section narrows, the sectional gradient collapses into a directional gradient.

Sectional gradients in Theoretical Physics, as in Mathematics, occupy an intermediate position between the continuous gradients and directional gradients.

Numerical example (continued). In the previous numerical example on page 295,

$$\begin{aligned}D[\mathbf{r1},\mathbf{r2}](\mathbf{f}|a1+da;da) &= ((a1+da+a1)/a1+da)*(2*\mathbf{u1}-\mathbf{u2}-\mathbf{u3})/1_a\\ &= -D[\mathbf{x1},\mathbf{x2}](\mathbf{f}|a1+da;da);\\ D[a1](\mathbf{r}|\mathbf{x1};da) &= (3*\mathbf{u1}+4*\mathbf{u2}+2*\mathbf{u3})/1_a;\\ D[a1](\mathbf{x}|\mathbf{r1};da) &= -(\mathbf{u1}+\mathbf{u2}+\mathbf{u3})/a\end{aligned}$$

$$\mathbf{T}[\mathbf{D3}[\mathbf{r1}]*(\mathbf{f}|a1+da;\mathbf{dr})] = (a1/(7*a1+da)) * \begin{bmatrix}-10 & -2 & 12\\ 13 & -10 & 4\\ 3 & 16 & -33\end{bmatrix}$$

$$\mathbf{T}[\mathbf{D3}[\mathbf{x1}]*(\mathbf{f}|a1+da;\mathbf{dx})] = (a1+da/1_a) * \begin{bmatrix} -1 & 2 & -3 \\ -1 & -2 & 4 \\ 6 & -2 & -3 \end{bmatrix}$$

from which may be verified
D[a1](**r**|**x1**;da)•**T**(**D3**[**r1**]*(**f**|a1+da;**dr**)) = (4***u1**−2***u2**−2***u2**)/1_a
and
D[a1](x|**r1**;da)•**T**(**D3**[**x1**](**f**|a1+da;**dx**)) = −(4***u1**−2***u2**−2***u2**)/1_a
as a1+da⟶a1.

Continuing now with directional local derivatives and assuming sectional continuity in SECT(**r1**, **u1Rf**,**u2Rf**,**u3Rf**) which contains the directional increment **r**(**x1**,a1+da)−**r**(**x1**,a1), consider

D3[**r1**]*(**f**|a1+da;**dRf**)•**T**[**URf**(a1+da)]$^{-1}$•[**URf**(a1+da)$^{-1}$]
•[D[a1](**r**|**x1**;d_fa)***qd**(D[a1](**r**|**x1**;d_fa))]
= (D[a1](**r**|**x1**;d_fa)
•**T**[**D3**[**r1**]*(**f**|a1+da;**dRf**)
•**T**[**URf**(a1+da)]$^{-1}$•[**URf**(a1+da)$^{-1}$]])
***qd**(D[a1](**r**|**x1**;d_fa))]
= D[a1](**r**|**x1**;d_fa)•[**T**[**URf**(a1+da)$^{-1}$]]•[**URf**(a1+da)$^{-1}$]
•**T**[**D3**[**r1**]*(**f**|a1+da;**dRf**)]***qd**(D[a1](**r**|**x1**;d_fa))]
= D[**r1**,**r**(**x1**)](**f**|a1+da;da)***qd**(D[a1](**r**|**x1**;d_fa))
= **D1**[**r1**,**r2**]*(**f**|a1+da;$\mathbf{d_f r}$)
by Theorem 14.

Thus,

D1[**r1**,**r2**]*(**f**|a1+da;$\mathbf{d_f r}$)
= **D3**[**r1**]*(**f**|a1+da;**dRf**)
•**T**[**URf**(a1+da)]$^{-1}$•[**URf**(a1+da)$^{-1}$]
•**D1**[**r1**,**r2**]*(**r**|**x1**;$\mathbf{d_f r}$) (1234)
= −**D1**[**x1**(**r2**,a1+da),**x**(**r1**,a1+da)]*(**f**|a1+da;$\mathbf{d_f r}$). (1235)

Dually, for the related section SECT(**x1**,**u1Xf**,**u2Xf**,**u3Xf**)

D1[**x1**(**r**,a1+da),**x2**(**r1**,a1+da)]*(**f**|a1+da;$\mathbf{d_n q}$)
= **D3**[**x1**]*(**f**|a1+da;**dXf**)
•**T**[**UXf**(a1+da)]$^{-1}$•[**UXf**(a1+da)$^{-1}$]
•**D1**[**x1**,**x2**]*(**x**|**r1**;$\mathbf{d_n q}$) (1236)
= −**D1**[**r1**(**x**,a1+da),**r**(**x1**,a1+da)]*(**f**|a1+da;$\mathbf{d_n q}$). (1237)

Similarly for SECT(**r1**,**u1Rb**,**u2Rb**,**u3Rb**) spanning the backward increment

D1[**r**(**x1**,a1–da),**r1**(**x**,a1+da)]∗(**f**|a1–da;$\mathbf{d_n q}$)
= **D3**[**r1**]∗(**f**|a1–da;**dRb**)
•**T**[**URb**(a1–da)]$^{-1}$•[**URb**(a1–da)$^{-1}$]
•**D1**[**r0**,**r1**]∗(**r**|**x1**;$\mathbf{d_n q}$) (1238)
= –**D1**[**x**(**r1**,a1–da),**r1**(**x**,a1–da)]∗(**f**|a1–da;$\mathbf{d_n q}$) (1239)

D1[**x**(**r1**,a1–da),**x1**(**r**,a1–da)]∗(**f**|a1–da;$\mathbf{d_n q}$)
= **D3**[**x1**]∗(**f**|a1–da;**dXb**)
•**T**[**UXb**(a1–da)]$^{-1}$•[**URb**(a1–da)$^{-1}$]
•**D1**[**x0**,**x1**]∗(**x**|**r1**;$\mathbf{d_n q}$) (1240)
= –**D1**[**r**(**x1**,a1–da),**r1**(**x**,a1–da)]∗(**f**|a1–da;$\mathbf{d_n q}$) (1241)

where SECT(**x1**,**u1Xb**,**u2Xb**,**u3Xb**) is the material section corresponding to SECT(**r1**,**u1Rb**,**u2Rb**,**u3Rb**) and $\mathbf{d_b q}$ symbolizes either $\mathbf{d_f r}$ or $\mathbf{d_b r}$ or $\mathbf{d_f x}$ or $\mathbf{d_b x}$.

Also

D1[**r1**(**x1**,a1),**r2**(**y2**,a1)]∗(**f**|a1;$\mathbf{d_n q}$)
= **D3**[**r1**]∗(**f**|a1;**dRf**)
•**T**[**URf**(a1)]$^{-1}$•[**URf**(a1)$^{-1}$]
•**D1**[**r1**,**r2**]∗(**r**|**x1**;$\mathbf{d_n q}$) (1242)

D1[**x1**(**r1**,a1),**x2**(**s2**,a1)]∗(**f**|a1;$\mathbf{d_n q}$)
= **D3**[**x1**]∗(**f**|a1;**dXf**)
•**T**[**UXf**(a1)]$^{-1}$•[**UXf**(a1)$^{-1}$]
•**D1**[**x1**,**x2**]∗(**x**|**r1**;$\mathbf{d_n q}$) (1243)

D1[**r0**(**y0**,a1),**r1**(**x1**,a1)]∗(**f**|a1;$\mathbf{d_n q}$)
= **D3**[**r1**]∗(**f**|a1;**dRb**)
•**T**[**URb**(a1)]$^{-1}$•[**URb**(a1)$^{-1}$]
•**D1**[**r0**,**r1**]∗(**r**|**x1**;$\mathbf{d_n q}$) (1244)

D1[**x0**(**s0**,a1),**x1**(**r1**,a1)]∗(**f**|a1;$\mathbf{d_n q}$)
= **D3**[**x1**]∗(**f**|a1;**dXb**)
•**T**[**UXb**(a1)]$^{-1}$•[**UXb**(a1)$^{-1}$]
•**D1**[**x0**,**x1**]∗(**x**|**r1**;$\mathbf{d_n q}$) (1245)

where
SECT(**r1**,**u1Rf**(a1),**u2Rf**(a1),**u3Rf**(a1))
may not equal
SECT(**r1**,**u1Rf**(a1+da),**u2Rf**(a1+da),**u3Rf**(a1+da))

nor

SECT(**r1**,**u1Rb**(a1),**u2Rb**(a1),**u3Rb**(a1))

equal

SECT(**r1**,**u1Rf**(a1–da),**u2Rf**(a1–da),**u3Rf**(a1–da)).

Results for Primary Functions

Applied to the primary functions **r** and **x**, the general results provide a convenient summary which includes previous results.

D[**r1**](**r**|**x1**;d$_f$a)

= D[**r1**,**r2**](**r**|a1+da;da)

= D[**r1**,**r2**](**r**|a1;da)

= –D[**x1**,**x2**](**r**|a1+da;da)

= D[**x1**](**x**|**r1**;d$_f$a)

•**T**[**D1**[**r1**,**r2**]∗(**r**|**x1**;**d**$_f$**x**)]

= D[**x1**](**x**|**r1**;d$_b$a)

•**T**[**D1**[**r1**,**r2**]∗(**r**|**x1**;**d**$_b$**x**)]

= D[**r1**](**r**|**x1**;d$_b$a)

•**T**[**D1**[**r1**,**r2**]∗(**r**|**x1**;**d**$_b$**r**)]

= D[a1](**x**|**r1**;d$_f$a)

•**T**[**UXf**(a1+da)$^{-1}$]]•[**UXf**(a1+da)$^{-1}$]

•**T**[**D3**[**x1**]∗(**f**|a1+da;**dXf**)] (1246)

D[**r1**](**r**|**x1**;d$_b$a)

= D[**r0**,**r1**](**r**|a1–da;da)

= D[**r0**,**r1**](**r**|a1;da)

= –D[**x0**,**x1**](**r**|a1–da;da)

= D[**x1**](**x**|**r1**;d$_f$a)

•**T**[**D1**[**r0**,**r1**]∗(**r**|**x1**;**d**$_f$**x**)]

= D[**x1**](**x**|**r1**;d$_b$a)

•**T**[**D1**[**r0**,**r1**]∗(**r**|**x1**;**d**$_b$**x**)]

= D[**r1**](**r**|**x1**;d$_f$a)

•**T**[**D1**[**r0**,**r1**]∗(**r**|**x1**;**d**$_f$**r**)]

= D[a1](**x**|**r1**;d$_b$a)

•**T**[**UXb**(a1–da)$^{-1}$]]•[**UXb**(a1–da)$^{-1}$]

•**T**[**D3**[**x1**]∗(**f**|a1–da;**dXf**)] (1247)

$$
\begin{aligned}
&D[\mathbf{x1}](\mathbf{x}|\mathbf{r1};d_f a)\\
&\quad = D[\mathbf{x1},\mathbf{x2}](\mathbf{x}|a1+da;da)\\
&\quad = D[\mathbf{x1},\mathbf{x2}](\mathbf{x}|a1;da)\\
&\quad = -D[\mathbf{r1},\mathbf{r2}](\mathbf{x}|a1+da;da)\\
&\quad = D[\mathbf{r1}](\mathbf{r}|\mathbf{x1};d_f a)\\
&\qquad\quad \bullet\mathbf{T}[\mathbf{D1}[\mathbf{x1},\mathbf{x2}]*(\mathbf{x}|\mathbf{r1};\mathbf{d_f r})]\\
&\quad = D[[\mathbf{r1}](\mathbf{r}|\mathbf{x1};d_b a)\\
&\qquad\quad \bullet\mathbf{T}[\mathbf{D1}[\mathbf{x1},\mathbf{x2}]*(\mathbf{x}|\mathbf{r1};\mathbf{d_b r})]\\
&\quad = D[\mathbf{x1}](\mathbf{x}|\mathbf{r1};d_b a)\\
&\qquad\quad \bullet\mathbf{T}[\mathbf{D1_f}[\mathbf{x1},\mathbf{x2}]*(\mathbf{x}|\mathbf{r1};\mathbf{d_b x})]\\
&\quad = D[a1](\mathbf{r}|\mathbf{x1};d_f a)\\
&\qquad\quad \bullet\mathbf{T}[\mathbf{URf}(a1+da)^{-1}]]\bullet[\mathbf{URf}(a1+da)^{-1}]\\
&\qquad\qquad\qquad \bullet\mathbf{T}[\mathbf{D3}[\mathbf{r1}]*(\mathbf{f}|a1+da;\mathbf{dRf})]
\end{aligned}
\tag{1248}
$$

$$
\begin{aligned}
&D[\mathbf{x1}](\mathbf{x}|\mathbf{r1};d_b a)\\
&\quad = D[\mathbf{x0},\mathbf{x1}](\mathbf{x}|a1-da;da)\\
&\quad = D[\mathbf{x0},\mathbf{x1}](\mathbf{x}|a1;da)\\
&\quad = -D[\mathbf{r0},\mathbf{r1}](\mathbf{x}|a1-da;da)\\
&\quad = D[\mathbf{r1}](\mathbf{r}|\mathbf{x1};d_f a)\\
&\qquad\quad \bullet\mathbf{T}[\mathbf{D1}[\mathbf{x0},\mathbf{x1}]*(\mathbf{x}|\mathbf{r1};\mathbf{d_f r})]\\
&\quad = D[\mathbf{r1}](\mathbf{r}|\mathbf{x1};d_b a)\\
&\qquad\quad \bullet\mathbf{T}[\mathbf{D1}[\mathbf{x0},\mathbf{x1}]*(\mathbf{x}|\mathbf{r1};\mathbf{d_b r})]\\
&\quad = D[\mathbf{x1}](\mathbf{x}|\mathbf{r1};d_f a)\\
&\qquad\quad \bullet\mathbf{T}[\mathbf{D1}[\mathbf{x0},\mathbf{x1}]*(\mathbf{x}|\mathbf{r1};\mathbf{d_f x})]\\
&\quad = D[a1](\mathbf{r}|\mathbf{x1};d_b a)\\
&\qquad\quad \bullet\mathbf{T}[\mathbf{URb}(a1-da)^{-1}]]\bullet[\mathbf{URb}(a1-da)^{-1}]\\
&\qquad\qquad\qquad \bullet\mathbf{T}[\mathbf{D3}[\mathbf{r1}]*(\mathbf{f}|a1-da;\mathbf{dRb})]
\end{aligned}
\tag{1249}
$$

$$
\begin{aligned}
&\mathbf{D1}[\mathbf{r1},\mathbf{r2}]*(\mathbf{r}|\mathbf{x1};\mathbf{d_n q})\\
&\quad = \mathbf{D1}[\mathbf{r1},\mathbf{r2}]*(\mathbf{r}|a1+da;\mathbf{d_n q})\\
&\quad = \mathbf{D1}[\mathbf{r1},\mathbf{r2}]*(\mathbf{r}|a1;\mathbf{d_n q})\\
&\quad = -\mathbf{D1}[\mathbf{x1},\mathbf{x2}]*(\mathbf{r}|a1+da;\mathbf{d_n q})\\
&\quad = D[a1](\mathbf{r}|\mathbf{x1};\mathbf{d_f a})*\mathbf{qd}(D[a1](\mathbf{q};d_n a))\\
&\quad = \mathbf{D1}[\mathbf{r1},\mathbf{r2}]*(\mathbf{r}|\mathbf{x1};\mathbf{d_m q1})\\
&\qquad\qquad\qquad \bullet[D[a1](\mathbf{q1};d_m a)*\mathbf{qd}(D(\mathbf{q};d_n a))]
\end{aligned}
\tag{1250}
$$

$$
\begin{aligned}
&\mathbf{D1}[\mathbf{r0},\mathbf{r1}]*(\mathbf{r}|\mathbf{x1};\mathbf{d_n q})\\
&\quad = \mathbf{D1}[\mathbf{r0},\mathbf{r1}]*(\mathbf{r}|a1-da;\mathbf{d_n q})\\
&\quad = \mathbf{D1}[\mathbf{r0},\mathbf{r1}]*(\mathbf{r}|a1;\mathbf{d_n q})\\
&\quad = -\mathbf{D1}[\mathbf{x0},\mathbf{x1}]*(\mathbf{r}|a1-da;\mathbf{d_n q})\\
&\quad = D[a1](\mathbf{r}|\mathbf{x1};d_b a)*\mathbf{qd}(D[a1](\mathbf{q};d_n a))\\
&\quad = \mathbf{D1}[\mathbf{r0},\mathbf{r1}]*(\mathbf{r}|\mathbf{x1};\mathbf{d_m q1})\\
&\qquad\qquad\qquad \bullet[D[a1](\mathbf{q1};d_m a)*\mathbf{qd}(D(\mathbf{q};d_n a))]
\end{aligned}
\tag{1251}
$$

$$\begin{aligned}
&\mathbf{D1}[\mathbf{x1},\mathbf{x2}]*(\mathbf{x}|\mathbf{r1};\mathbf{d}_n\mathbf{q})\\
&\quad = \mathbf{D1}[\mathbf{x1},\mathbf{x2}]*(\mathbf{x}|a1+da;\mathbf{d}_n\mathbf{q})\\
&\quad = \mathbf{D1}[\mathbf{x1},\mathbf{x2}]*(\mathbf{x}|a1;\mathbf{d}_n\mathbf{q})\\
&\quad = -\mathbf{D1}[\mathbf{r1},\mathbf{r2}]*(\mathbf{x}|a1+da;\mathbf{d}_n\mathbf{q})\\
&\quad = \mathbf{D1}[\mathbf{x1},\mathbf{x2}]*(\mathbf{x}|\mathbf{r1};\mathbf{d}_m\mathbf{q1})\\
&\qquad\qquad \bullet[D[a1](\mathbf{q1};d_m a)*\mathbf{qd}(D(\mathbf{q};d_n a))]
\end{aligned} \tag{1252}$$

$$\begin{aligned}
&\mathbf{D1}[\mathbf{x0},\mathbf{x1}]*(\mathbf{x}|\mathbf{r1};\mathbf{d}_n\mathbf{q})\\
&\quad = \mathbf{D1}[\mathbf{x0},\mathbf{x1}]*(\mathbf{x}|a1-da;\mathbf{d}_n\mathbf{q})\\
&\quad = \mathbf{D1}[\mathbf{x0},\mathbf{x1}]*(\mathbf{x}|a1;\mathbf{d}_n\mathbf{q})\\
&\quad = -\mathbf{D1}[\mathbf{r0},\mathbf{r1}]*(\mathbf{x}|a1-da;\mathbf{d}_n\mathbf{q})\\
&\quad = D[a1](\mathbf{x}|\mathbf{r1};d_b a)*\mathbf{qd}(D(\mathbf{q};d_n a))\\
&\quad = \mathbf{D1}[\mathbf{x0},\mathbf{x1}]*(\mathbf{x}|\mathbf{r1};\mathbf{d}_m\mathbf{q1})\\
&\qquad\qquad \bullet[D[a1](\mathbf{q1};d_m a)*\mathbf{qd}(D(\mathbf{q};d_n a))]
\end{aligned} \tag{1253}$$

$$\begin{aligned}
&\mathbf{D3}[\mathbf{r1}]*(\mathbf{r}|a1;\mathbf{dR1})\\
&\quad = \mathbf{D3}[\mathbf{x1}]*(\mathbf{r}|a1;\mathbf{dX1})\\
&\qquad \bullet[\mathbf{UX1}^{-1}]\bullet[\mathbf{T}[\mathbf{UX1}^{-1}]]\\
&\qquad\qquad \bullet\mathbf{D3}[\mathbf{r1}]*(\mathbf{x}|a1;\mathbf{dR1})
\end{aligned} \tag{1254}$$

$$\begin{aligned}
&\mathbf{D3}[\mathbf{x1}]*(\mathbf{x}|a1;\mathbf{dX1})\\
&\quad = \mathbf{D3}[\mathbf{r1}]*(\mathbf{x}|a1;\mathbf{dR1})\\
&\qquad \bullet[\mathbf{UR1}^{-1}]\bullet[\mathbf{T}[\mathbf{UR1}^{-1}]]\\
&\qquad\qquad \bullet\mathbf{D3}[\mathbf{x1}]*(\mathbf{r}|a1;\mathbf{dX1})
\end{aligned} \tag{1255}$$

for m and n symbolizing either forward or backward senses; **q** symbolizing **dr** or **dx** in either a forward or backward sense; **q1** symbolizing **r**(**x1**), **x**(**r1**), **x**|a1+da, **r**|a1+da, **x**|a1 or **r**|a1; where SECT(**x1**,**u1X1**,**u2X1**,**u3X1**) and SECT(**r1**,**u1R1**,**u2R1**,**u3R1**), SECT(**x1**,**u1Xf**,**u2Xf**,**u3Xf**) and SECT(**r1**,**u1Rf**,**u2Rf**,**u3Rf**), and SECT(**x1**,**u1Xb**,**u2Xb**,**u3Xb**) and SECT(**r1**,**u1Rb**,**u2Rb**,**u3Rb**) are mutually related sections.

Note

$$\mathbf{D3}[\mathbf{r1}]*(\mathbf{r}|a;\mathbf{dR1}) = \mathbf{T}[\mathbf{UR1}(a)]\bullet\mathbf{UR1}(a) \tag{1256}$$

for any section SECT(**r1**,**u1R1**,**u2R1**,**u3R1**). Consequently since,

$$\mathbf{T}[\mathbf{UR1}^{-1}]\bullet\mathbf{T}[\mathbf{D3}[\mathbf{r1}]*(\mathbf{r};\mathbf{dR1})]\bullet[\mathbf{UR1}^{-1}] = \mathbf{I} \tag{1257}$$

any chosen SECT(**r1**,**u1R1**,**u2R1**,**u3R1**) may be reassembled and merged into quadrants, as one should expect since **r** in V3(**r**) is continuous. Similarly for **x** in V3(**x**). This case may be written simply[75] as

$$\mathbf{D3}[\mathbf{r1}]*(\mathbf{r};\mathbf{dr}) = \mathbf{D3}[\mathbf{x1}]*(\mathbf{x};\mathbf{dx}) = \mathbf{I}. \tag{1258}$$

Some Fallacies.

While the above equations may appear exhaustive, some similarly appearing statements are fallacies to be avoided.

Please note carefully

D[a1](**f**|**x1**;d_fa)
= D[a1](**r**|**x1**;d_fa)•**T**(**D3**[**r1**]∗**f**|a1;**dRf**)
= −D[a1](**x**|**r1**;d_fa)•**T**(**D3**[**x1**]∗**f**|a1;**dXf**)

D[a1](**f**|**x1**;d_ba)
= D[a1](**r**|**x1**;d_ba)•**T**(**D3**[**r1**]∗**f**|a1;**dRb**)
= −D[a1](**x**|**r1**;d_ba)•**T**(**D3**[**x1**]∗**f**|a1;**dXb**)

D[a1](**f**|**r1**;d_fa)
= −D[a1](**x**|**r1**;d_fa)•**T**(**D3**[**x1**]∗**f**|a1;**dXf**)
= D[a1](**r**|**x1**;d_fa)•**T**(**D3**[**r1**]∗**f**|a1;**dRf**)

D[a1](**f**|**r1**;d_ba)
= D[a1](**r**|**x1**;d_ba)•**T**(**D3**[**r1**]∗**f**|a1;**dRb**)
= −D[a1](**x**|**r1**;d_ba)•**T**(**D3**[**x1**]∗**f**|a1;**dXb**)

D1[**r1**(**x1**,a1),**r2**(**x1**,a1+da)]∗(**f**|**x1**;**dr**)
= **D3**[**r1**]∗(**f**|a1+da;**dRf**)
•**D1**[**r1**(**x1**,a1),**r2**(**x1**,a1+da),]∗(**r**|**x1**;**dr**)

D1[**x1**(**r1**,a1),**x2**(**r1**,a1+da)]∗(**f**|**r1**;**dx**)
= **D3**[**x1**]∗(**f**|a1+da;**dXf**)
•**D1**[**x1**(**r1**,a1),**x2**(**r1**,a1+da)]∗(**x**|**r1**;**dx**)

and similar statements are *generally false* even when **f** is locally differentiable.

Further, the following:

D3[**r1**]∗(D[a1](**f**|**x1**;d_na)|a1;**dRf**)
= D[a1](**D3**[**r1**]∗(**f**|a1;**dRf**)|**x1**;d_na)

D3[**x1**]∗(D[a1](**f**|**r1**;d_na)|a1;**dXf**)
= D[a1](**D3**[**x1**]∗(**f**|a1;**dXf**)|**r1**;d_na)

in either forward or backward sense are also *generally false*.

In addition, such statements as

D3[**x1**]∗(D[a1](**f**|**x1**;d_fa)|a1;**dXf**)
= D[a1](**D1**[**r1**,**r2**]∗(**f**|**x1**;$\mathbf{d_f x}$)|**x1**;d_fa)

and

D[a1](**D3**[**x1**]∗(**f**|a1;**dXf**)|**x1**;d_fa)
= **D1**[**r1**,**r2**]∗(D[a1](**f**|**x1**;d_fa)|**x1**;$\mathbf{d_f x}$).

are also *generally false*.

Theoretical Physics: Step Functions

Step functions of Theoretical Physics

The derivatives of step functions give rise to impulse functions. Theoretical Physics supplies a variety of such functions.

Definition 53 (observational step functions)

Given

$\{\mathbf{r}(\mathbf{x},a)\}$, the set of ordered observations;

for

a1, a reference observation

then

$$\begin{aligned} ua(\mathbf{r}(\mathbf{x},a)-\mathbf{r}(\mathbf{x},a1)) &\equiv 1,\ a>a1 \\ &\equiv 0\ \ a\le a1 \end{aligned} \tag{1259}$$

and

$$\begin{aligned} va(\mathbf{r}(\mathbf{x},a)-\mathbf{r}(\mathbf{x},a1)) &\equiv 1,\ a\ge a1 \\ &\equiv 0\ \ a<a1 \end{aligned} \tag{1260}$$

are the **unit observational step functions**.

end of definition

The unit observational step functions, may be used to designate the start of an interval of observations. The functions 1–ua or 1–va may be used to designate the end of the interval.

The unit observational step functions are generic step functions which may be specified either to a particle or to a location as

$$\begin{aligned} uax(\mathbf{r}(\mathbf{x1},a)-\mathbf{r}(\mathbf{x1},a1)) &\equiv 1,\ a>a1 \\ &\equiv 0\ \ a\le a1 \end{aligned} \tag{1261}$$

and

$$\begin{aligned} vax(\mathbf{r}(\mathbf{x1},a)-\mathbf{r}(\mathbf{x1},a1)) &\equiv 1,\ a\ge a1 \\ &\equiv 0\ \ a<a1 \end{aligned} \tag{1262}$$

called the **local unit observational step functions with reference to the particle x1** or

$$\begin{aligned}\mathrm{uar}(\mathbf{x}(\mathbf{r1},a)-\mathbf{x}(\mathbf{r1},a1)) &\equiv 1,\ a>a1\\ &\equiv 0\ \ a\le a1 \qquad (1263)\end{aligned}$$

and

$$\begin{aligned}\mathrm{var}(\mathbf{x}(\mathbf{r1},a)-\mathbf{x}(\mathbf{r1},a1)) &\equiv 1,\ a\ge a1\\ &\equiv 0\ \ a<a1 \qquad (1264)\end{aligned}$$

called the **material unit observational step functions with reference to location r1**.

As functions of observations, the function of *specific* unit observational step functions, vax(**r**(**x1**,a)–**r**(**x1**,a1)), singles out that portion of the local trace **r**(**x1**,a) starting at **r1**(**x1**,a1), while var(**x**(**r1**,a)–**x**(**r1**,a1)) singles out that portion of the material track **x**(**r1**,a) starting at **x1**(**r1**,a1).

Indexed derivatives of these step functions are impulsive.

Generically,

$$\begin{aligned}&D[a](\mathrm{ua}(\mathbf{r}(\mathbf{x},a)-\mathbf{r}(\mathbf{x},a1));d_f a)\\ &\quad= \lim\ (\mathrm{ua}(\mathbf{r}(\mathbf{x},a+da)-\mathbf{r}(\mathbf{x},a1)) - \mathrm{ua}(\mathbf{r}(\mathbf{x},a)-\mathbf{r}(\mathbf{x},a1)))/da\\ &\quad= \lim\ 1/da, \qquad a=a1\\ &\quad= 0 \qquad\qquad \text{otherwise} \qquad (1265)\end{aligned}$$

$$\begin{aligned}&D[a](\mathrm{va}(\mathbf{r}(\mathbf{x},a)-\mathbf{r}(\mathbf{x},a1));d_f a)\\ &\quad= \lim\ (\mathrm{va}(\mathbf{r}(\mathbf{x},a+da)-\mathbf{r}(\mathbf{x},a1)) - \mathrm{va}(\mathbf{r}(\mathbf{x},a)-\mathbf{r}(\mathbf{x},a1)))/da\\ &\quad= 0, \qquad \text{for all } a. \qquad (1266)\end{aligned}$$

In the backward sense

$$\begin{aligned}&D[a](\mathrm{ua}(\mathbf{r}(\mathbf{x},a)-\mathbf{r}(\mathbf{x},a1));d_b a)\\ &\quad= \lim\ (\mathrm{ua}(\mathbf{r}(\mathbf{x},a)-\mathbf{r}(\mathbf{x},a1)) - \mathrm{ua}(\mathbf{r}(\mathbf{x},a-da)-\mathbf{r}(\mathbf{x},a1)))/da\\ &\quad= 0, \qquad \text{for all } a. \qquad (1267)\end{aligned}$$

$$\begin{aligned}&D[a](\mathrm{va}(\mathbf{r}(\mathbf{x},a)-\mathbf{r}(\mathbf{x},a1));d_b a)\\ &\quad= \lim\ (\mathrm{va}(\mathbf{r}(\mathbf{x},a)-\mathbf{r}(\mathbf{x},a1)) - \mathrm{va}(\mathbf{r}(\mathbf{x},a-da)-\mathbf{r}(\mathbf{x},a1)))/da\\ &\quad= \lim\ 1/da, \qquad a=a1\\ &\quad= 0 \qquad\qquad \text{otherwise.} \qquad (1268)\end{aligned}$$

Clearly,

$$D[a](\mathrm{ua}(\mathbf{r}(\mathbf{x},a)-\mathbf{r}(\mathbf{x},a1));d_f a) = D[a](\mathrm{va}(\mathbf{r}(\mathbf{x},a)-\mathbf{r}(\mathbf{x},a1));d_b a) \qquad (1269)$$

$$D[a](\mathrm{ua}(\mathbf{r}(\mathbf{x},a)-\mathbf{r}(\mathbf{x},a1));d_b a) = D[a](\mathrm{va}(\mathbf{r}(\mathbf{x},a)-\mathbf{r}(\mathbf{x},a1));d_f a). \qquad (1270)$$

Theoretical Physics: Step Functions

Dually,

$$\begin{aligned}
&D[a](ua(\mathbf{x}(\mathbf{r},a)-\mathbf{x}(\mathbf{r},a1));d_f a)\\
&\qquad = \lim\ (ua(\mathbf{x}(\mathbf{r},a+da)-\mathbf{x}(\mathbf{r},a1)) - ua(\mathbf{x}(\mathbf{r},a)-\mathbf{x}(\mathbf{x},a1)))/da\\
&\qquad = \lim 1/da, \quad a=a1\\
&\qquad = 0, \quad \text{otherwise}
\end{aligned} \tag{1271}$$

$$\begin{aligned}
&D[a](va(\mathbf{x}(\mathbf{r},a)-\mathbf{x}(\mathbf{r},a1));d_f a)\\
&\qquad = \lim\ (va(\mathbf{x}(\mathbf{r},a+da)-\mathbf{x}(\mathbf{r},a1)) - ua(\mathbf{x}(\mathbf{r},a)-\mathbf{x}(\mathbf{x},a1)))/da\\
&\qquad = 0, \quad \text{for all } a.
\end{aligned} \tag{1272}$$

In the backward sense

$$\begin{aligned}
&D[a](ua(\mathbf{x}(\mathbf{r},a)-\mathbf{x}(\mathbf{r},a1));d_b a)\\
&\qquad = \lim\ (ua(\mathbf{x}(\mathbf{r},a)-\mathbf{x}(\mathbf{r},a1)) - ua(\mathbf{x}(\mathbf{r},a-da)-\mathbf{x}(\mathbf{r},a1)))/da\\
&\qquad = 0, \quad \text{for all } a
\end{aligned} \tag{1273}$$

$$\begin{aligned}
&D[a](va(\mathbf{x}(\mathbf{r},a)-\mathbf{x}(\mathbf{r},a1));d_b a)\\
&\qquad = \lim\ (va(\mathbf{x}(\mathbf{r},a)-\mathbf{x}(\mathbf{r},a1)) - va(\mathbf{x}(\mathbf{r},a-da)-\mathbf{x}(\mathbf{r},a1)))/da\\
&\qquad = \lim 1/da, \quad a=a1\\
&\qquad = 0, \quad \text{otherwise.}
\end{aligned} \tag{1274}$$

Clearly,

$$D[a](ua(\mathbf{x}(\mathbf{r},a)-\mathbf{x}(\mathbf{r},a1));d_f a) = D[a](va(\mathbf{x}(\mathbf{r},a)-\mathbf{x}(\mathbf{r},a1));d_b a) \tag{1275}$$

$$D[a](ua(\mathbf{x}(\mathbf{r},a)-\mathbf{x}(\mathbf{r},a1));d_b a) = D[a](va(\mathbf{x}(\mathbf{r},a)-\mathbf{x}(\mathbf{r},a1));d_f a) \tag{1276}$$

and

$$\begin{aligned}
&D[a](ua(\mathbf{r}(\mathbf{x},a) - \mathbf{r}(\mathbf{x},a1));d_f a)\\
&\qquad = D[a](ua(\mathbf{x}(\mathbf{r},a) - \mathbf{x}(\mathbf{r},a1));d_f a)\\
&\qquad = D[a](va(\mathbf{r}(\mathbf{x},a) - \mathbf{r}(\mathbf{x},a1));d_b a)\\
&\qquad = D[a](va(\mathbf{x}(\mathbf{r},a) - \mathbf{x}(\mathbf{r},a1));d_b a)
\end{aligned} \tag{1277}$$

$$\begin{aligned}
&D[a](va(\mathbf{r}(\mathbf{x},a) - \mathbf{r}(\mathbf{x},a1));d_f a)\\
&\qquad = D[a](va(\mathbf{x}(\mathbf{r},a) - \mathbf{x}(\mathbf{r},a1));d_f a)\\
&\qquad = D[a](ua(\mathbf{r}(\mathbf{x},a) - \mathbf{r}(\mathbf{x},a1));d_b a)\\
&\qquad = D[a](ua(\mathbf{x}(\mathbf{r},a) - \mathbf{x}(\mathbf{r},a1));d_b a).
\end{aligned} \tag{1278}$$

The first group of these derivatives selects the *a1* observation.

Similarly, in a specific sense,

D[a](uax(**r**(**x1**,a) – **r**(**x1**,a1))|**x1**;d_fa)
= D[a](uar(**x**(**r1**,a) – **x**(**r1**,a1))|**r1**;d_fa)
= D[a](vax(**r**(**x1**,a) – **r**(**x1**,a1))|**x1**;d_ba)
= D[a](var(**x**(**r1**,a) – **x**(**r1**,a1))|**r1**;d_ba) (1279)

D[a](vax(**r**(**x1**,a) – **r**(**x1**,a1))|**x1**;d_fa)
= D[a](var(**x**(**r1**,a) – **x**(**r1**,a1))|**r1**;d_fa)
= D[a](uax(**r**(**x1**,a) – **r**(**x1**,a1))|**x1**;d_ba)
= D[a](uar(**x**(**r1**,a) – **x**(**r1**,a1))|**r1**;d_ba). (1280)

These derivatives all select **r1**(**x1**,a1) on the local trace of ***x1*** or **x1**(**r1**,a1) on the material track passing through **r1**.

Other indexed step functions may be defined at **r1**(**x1**,a1) involving both track and trace there,

ua(**r**(**x1**,a)–**r1**(**x**,a)) ≡ 1, a>a1
≡ 0 a≤a1 (1281)
va(**r**(**x1**,a)–**r1**(**x**,a)) ≡ 1, a≥a1
≡ 0, a<a1 (1282)

and

ua(**x**(**r1**,a)–**x1**(**r**,a)) ≡ 1, a>a1
≡ 0 a≤a1 (1283)
va(**x**(**r1**,a)–**x1**(**r**,a)) ≡ 1, a≥a1
≡ 0, a<a1 (1284)

called the **specific unit observational step functions with reference to observation** a1.

From their definitions,

ua(**r**(**x1**,a)–**r1**(**x**,a)) = ua(**x**(**r1**,a)–**x1**(**r**,a)) (1285)

and

va(**r**(**x1**,a)–**r1**(**x**,a)) = va(**x**(**r1**,a)–**x1**(**r**,a)). (1286)

Their derivatives are also impulsive.

D[a1](ua(**r**(**x1**,a)–**r1**(**x**,a)))a1+da;da)
= lim (ua(**r**(**x1**,a1+da)–**r1**(**x**,a1+da))
– ua(**r**(**x1**,a1)–**r1**(**x**,a1)))/da
= lim 1/da, da>0
= 0, otherwise (1287)

Theoretical Physics: Step Functions

D[a1](va(**r**(**x1**,a)–**r1**(**x**,a))}a1+da;da)
= lim (va(**r**(**x1**,a1+da)–**r1**(**x**,a1+da))
– va(**r**(**x1**,a1)–**r1**(**x**,a1)))/da
= lim 1/da, da≥0
= 0, otherwise (1288)

D[a1](ua(**x**(**r1**,a)–**x1**(**r**,a))}a1+da;da)
= lim (ua(**x**(**r1**,a1+da)–**x1**(**r**,a1+da))
– ua(**x**(**r1**,a1)–**x1**(**r**,a1)))/da
= lim 1/da, da>0
= 0, otherwise (1289)
D[a1](va(**x**(**r1**,a)–**x1**(**x**,a))}a1+da;da)
= lim (va(**x**(**r1**,a1+da)–**x1**(**r**,a1+da))
– va(**x**(**r1**,a1)–**x1**(**r**,a1)))/da
= lim 1/da, da≥0
= 0, otherwise. (1290)

Where the trace or track is not stalled, the indexed derivatives are the basis for directional gradients.

D1[r1,**r2]**∗(uax(**r**(**x**,a)–**r**(**x**,a1))|**x1**;**d**$_f$**r**)
= lim **qd**(**d**$_f$(**r**|**x1**)) at **r**(**x1**,a1)
= **0**, otherwise (1291)
D1[r1,**r2]**∗(vax(**r**(**x**,a)–**r**(**x**,a1))|**x1**;**d**$_f$**r**)
= **0** (1292)
D1[x1,**x2]**∗(uar(**x**(**r**,a)–**x**(**r**,a1))|**r1**;**d**$_f$**x**)
= lim **qd**(**d**$_f$(**x**|**r1**)) at **x**(**r1**,a1)
= **0**, otherwise. (1293)
D1[x1,**x2]**∗(var(**x**(**r**,a)–**x**(**r**,a1))|**r1**;**d**$_f$**x**)
= **0** (1294)
D1[r1,**r2]**∗(ua(**r**(**x1**,a)–**r1**(**x**,a))|a1+da;**d**$_f$**r**)
= lim **qd**(**d**$_f$(**r**|**x1**)) at **r**(**x1**,a1)
= **0**, otherwise. (1295)
D1[r1,**r2]**∗(va(**r**(**x1**,a)–**r1**(**x**,a))|a1+da;**d**$_f$**r**)
= **0** (1296)
D1[x1,**x2]**∗(uar(**x**(r**1**,a)–**x1**(**r**,a))|a1+da;**d**$_f$**x**)
= lim **qd**(**d**$_f$(**x**|**r1**)) at **r**(**x1**,a1)
= **0**, otherwise. (1297)
D1[x1,**x2]**∗(va(**x**(**r1**,a)–**x1**(**r**,a))|a1+da;**d**$_f$**x**)
= **0** (1298)

D1[r0,**r1]**∗(uax(**r**(**x**,a)–**r**(**x**,a1))|**x1**;$\mathbf{d_b r}$)
    = **0** (1299)
D1[r0,**r1]**∗(vax(**r**(**x**,a)–**r**(**x**,a1))|**x1**;$\mathbf{d_b r}$)
    = **0** . (1300)
D1[x0,**x1]**∗(uar(**x**(**r**,a)–**x**(**r**,a1))|**r1**;$\mathbf{d_b x}$)
    = lim **qd**($\mathbf{d_b}$(**x**|**r1**)) at **x**(**r1**,a1)
    = **0**, otherwise (1301)
D1[x0,**x1]**∗(uar(**x**(**r**,a)–**x**(**r**,a1))|**r1**;$\mathbf{d_b x}$)
    = **0** (1302)
D1[r0,**r1]**∗(uax(**r**(**x1**,a)–**r1**(**x**,a))|a1–da;$\mathbf{d_b r}$)
    = **0** (1303)
D1[r0,**r1]**∗(vax(**r**(**x1**,a)–**r1**(**x**,a))|a1–da;$\mathbf{d_b r}$)
    = lim **qd**($\mathbf{d_b}$(**r**|**x1**)) at **r**(**x1**,a1)
    = **0**, otherwise (1304)
D1[x0,**x1]**∗(uax(**x**(r**1**,a)–**x1**(**r**,a))|a1–da;$\mathbf{d_b x}$)
    = **0** (1305)
D1[x0,**x1]**∗(vax(**x**(**r1**,a)–**x1**(**r**,a))|a1–da;$\mathbf{d_b x}$)
    = lim **qd**($\mathbf{d_b}$(**x**|**r1**)) at **r**(**x1**,a1)
    = **0**, otherwise. (1306)

Clearly,

D1[r1,**r2]**∗(vax(**r**(**x**,a)–**r**(**x**,a1))|**x1**;$\mathbf{d_f r}$)
    = **D1[x1**,**x2]**∗(uar(**x**(**r**,a)–**x**(**r**,a1))|**r1**;$\mathbf{d_f x}$)
    = **D1[r0**,**r1]**∗(uax(**r**(**x**,a)–**r**(**x,**a1))|**x1**;$\mathbf{d_b r}$)
    = **D1[x0**,**x1]**∗(uar(**x**(**r**,**a**)–**x**(**r**,**a1**))|**r1**;$\mathbf{d_b x}$)
    = **D1[r1**,**r2]**∗(va(**r**(**x1**,a)–**r1**(**x**,a))|a1+da;$\mathbf{d_f r}$)
    = **D1[r0**,**r1]**∗(ua(**r**(**x1**,a)–**r1**(**x**,a))|a1–da;$\mathbf{d_b r}$)
    = **D1[x0**,**x1]**∗(ua(**x**(r**1**,a)–**x1**(**r**,a))|a1–da;$\mathbf{d_b x}$)
    = **0** (1307)

but the non–zero directional gradients may differ.

Theoretical Physics: Step Functions

At observation *a1*, let MX be a measurable set of particles in V3(**x**) and SMX its surface; let MR be the set of locations of particles in MX and SMR the set of locations of particles in SMX[76].

Definition 54 (local step functions in V3)

Given

r(**x**,a1), a single observation;
MX, a measurable set of particles in V3(**x**);
SMX, the surface of MX;

then

$$\begin{aligned} ux(\mathbf{x},a1)|MX &\equiv 0, \quad \mathbf{x} \text{ in } MX\mathbf{u}SMX \\ &\equiv 1, \quad \text{otherwise} \end{aligned} \qquad (1308)$$

$$\begin{aligned} vx(\mathbf{x},a1)|MX &\equiv 0, \quad \mathbf{x} \text{ in } MX\text{–}SMX \\ &\equiv 1, \quad \text{otherwise.} \end{aligned} \qquad (1309)$$

Dually,

Given

MR, a measurable set of locations in V3(**r**);
SMR, the surface of MR;

then

$$\begin{aligned} ur(\mathbf{r},a1)|MR &\equiv 0, \quad \mathbf{r} \text{ in } MR\mathbf{u}SMR \\ &\equiv 1, \quad \text{otherwise} \end{aligned} \qquad (1310)$$

$$\begin{aligned} vr(\mathbf{r},a1)|MR &\equiv 0, \quad \mathbf{r} \text{ in } MR\text{–}SMR \\ &\equiv 1, \quad \text{otherwise.} \end{aligned} \qquad (1311)$$

end of definition

Corollary

Given MR and MX as mutually corresponding sets under **r**(**x**,a1), that is,

MX = **x**(MR,a1), the set of particles at locations in MR;
SMX = **x**(SMR,a1), the set of particles at locations in SMX;
MR = **r**(MX,a1), the set of locations of particles in MX;
SMR = **r**(SMX,a1), the set of locations of particles in SMX;

then

$$ur(\mathbf{r},a1)|MR = ux(\mathbf{x},a1)|MX \qquad (1312)$$

$$vr(\mathbf{r},a1)|MR = vr(\mathbf{r},a1)|MX. \qquad (1313)$$

Proof:

Let **x** in V3(**x**) be in MX**u**SMX; then **r**(**x**,a1) is in MR**u**SMR.
Let **x** in V3(**x**) be in V3(**x**)–(MX**u**SMX) then **r**(**x**,a1) is in V3(**r**)–(MR**u**SMR).
Thus

$$ur(\mathbf{r},a1)|MR = ux(\mathbf{x},a1)|MX.$$

Similarly,
let **x** in V3(x) be in MX−SMX; then **r**(**x**,a1) is in MR−SMR.
Let **x** in V3(**x**) be in V3(**x**)−(MX−SMX) then **r**(**x**,a1) is in V3(**r**)−(MR−SMR).
Thus

vr(**r**,a1)|MR = vx(**x**,a1)|MX.

qed

Now[77] let

dX1 ≡ d1X1∗**u1X1** + d2X1∗**u2X1** + d3X1∗**u3X1** (1314)

and

dR1 = **r**(**dX1**)
≡ **dX1•[UX1**$^{-1}$**]•T[UX1**$^{-1}$**]•T[D3[r1]∗**(**r**(**x**,a)|a1;**dX1**)] (1315)

imply two mutually differentiable sections.

Then

D3[**x**]∗(ux(**x**,a1)|MX|a1;**dX1**) = **0**, **x** in MX−SMX (1316)
D3[**x**]∗(ux(**x**,a1)|MX|a1;**dX1**) = **0**, **x** in V3−(MX**u**SMX) (1317)
D3[**x**]∗(ux(**x**,a1)|MX|a1;**dX1**)
= lim (ux(**x**+d1X1∗**u1X1**,a1)|MX)∗**u1X1**/d1X1
+ (ux(**x**+d2X1∗**u2X1**,a1)|MX)∗**u2X1**/d2X1
+ (ux(**x**+d3X1∗**u3X1**,a1)|MX)∗**u3X1**/d3X1,
x in SMX. (1318)

For ***x*** in SMX then, **D3**[**x**]∗(ux(**x**,a1)|MX;**dX1**) may equal **0**, or may be impulsive in one, two, or all three sectional directions.

D3[**x**]∗(vx(**x**,a1)|MX|a1;**dX1**) = **0**, **x** in MX−SMX (1319)
D3[**x**]∗(vx(**x**,a1)|MX|a1;**dX1**) = **0**, **x** in V3−(MX**u**SMX) (1320)
D3[**x**]∗(vx(**x**,a1)|MX|a1;**dX1**)
= lim ((vx(**x**+d1X1∗**u1X1**,a1)|MX)−1)∗**u1X1**/d1X1
+ ((vx(**x**+d2X1∗**u2X1**,a1)|MX)−1)∗**u2X1**/d2X1
+ ((vx(**x**+d3X1∗**u3X1**,a1)|MX)−1)∗**u3X1**/d3X1,
x in SMX. (1321)

For ***x*** in SMX then, **D3**[**x**]∗(vx(**x**,a1)|MX;**dX1**)r may equal **0**, or may be negatively impulsive in one, two, or all three sectional directions.

Theoretical Physics: Step Functions

Dually,

D3[**r**]∗(ur(**r**,a1)|MR|a1;**dR1**) = **0**, **r** in MR–SMR (1322)
D3[**r**]∗(ur(**r**,a1)|MR|a1;**dR1**) = **0**, **r** in V3–(MRuSMR) (1323)
D3[**r**]∗(ur(**r**,a1)|MR|a1;**dR1**)
= lim (ur(**r**+d1R1∗**u1R1**,a1)|MR)∗**u1R1**/d1R1
+ (ur(**r**+d2R1∗**u2R1**,a1)|MR)∗**u2R1**/d2R1
+ (ur(**r**+d3R1∗**u3R1**,a1)|MR)∗**u3R1**/d3R1,
r in SMR (1324)

D3[**r**]∗(vr(**r**,a1)|MR|a1;**dR1**) = **0**, **r** in MR–SMR (1325)
D3[**r**]∗(vr(**r**,a1)|MR|a1;**dR1**) = **0**, **r** in V3–(MRuSMR) (1326)
D3[**r**]∗(vr(**r**,a1)|MR|a1;**dR1**)
= lim ((vr(**r**+d1R1∗**u1R1**,a1)|MR)–1)∗**u1R1**/d1R1
+ ((vr(**r**+d2R1∗**u2R1**,a1)|MR)–1)∗**u2R1**/d2R1
+ ((vr(**r**+d3R1∗**u3R1**,a1)|MR)–1)∗**u3R1**/d3R1,
r in SMR. (1327)

Now let **f**(**r**(**x**,a1)) be a generalized function over V3(**x**) and the sections restricted by the differentiability of ***f***. Then

D3[**x**]∗(**f**(**x**,a1)∗ur(**r**(**x**,a1))|MX|a1;**dX1**)
= **D3**[**x**]∗(**f**(**x**,a1)|a1;**dX1**), **x** in V3(**x**)–(MXuSMX)
= lim **f**(**x**+d1X1∗**u1X1**)∗ur(**r**+d1X1∗**u1X1**)∗**u1X1**/d1X1
+ f(**x**+d2X1∗**u2X1**)∗ur(**x**+d2X1∗**u2X1**)∗**u2X1**/d2X1
+ f(**x**+d3X1∗**u3X1**)∗ur(**x**+d3X1∗**u3X1**)∗**u3X1**/d3X1,
x in SMX
= [**0**], **x** in MX–SMX (1328)
f(**x**,a1)∗**D3**[**r**]∗(ur(**r**(**x**,a1))|MX|a1;**dX1**)
= [**0**], **x** not in SM
= **f**(**r**)∗ lim (ur(**x**+d1X1∗**u1X1**)∗**u1X1**/d1X1
+ ur(**x**+d2X1∗**u2X1**)∗**u2X1**/d2X1
+ ur(**x**+d3X1∗**u3X1**)∗**u3X1**/d3X1),
x in SMX (1329)
(ur(**r**(**x**,a1))|MX)∗**D3**[**x**]∗(**f**|a1;**dX1**)
= **D3**[**x**]∗(**f**|a1;**dX1**), **x** in V3–(MXuSMX)
= [**0**], **x** in SMX
= [**0**], **x** in MX–SMX (1330)

D3[**r**]∗(**f**(**r**)∗v(**r**)|M|a1;**dR1**
    = **D3**[**r**]∗(**f**|a1;**dR1**),   **r** in V3–(M∪SM)
    = lim (**f**(**r**+d1R1∗**u1R1**)∗v(**r**+d1R1∗**u1R1**)∗**u1R1**–**f**(**r**))
        /d1R1
      +(**f**(**r**+d2R1∗**u2R1**)∗v(**r**+d2R1∗**u2R1**)∗**u2R1**–**f**(**r**))
        /d2R1
      +(**f**(**r**+d3R1∗**u3R1**)∗v(**r**+d3R1∗**u3R1**)∗**u3R1**–**f**(**r**))
        /d3R1,
    r in SM
    = [**0**],   **r** in M–SM   (1331)

f(**r**)∗**D3**[**r**]∗(v(**r**)|M|a1;**dR1**)
    = [**0**],   **r** not in SM
    = **f**(**r**)∗ lim (v(**r**+d1R1∗**u1R1**)–1)∗**u1R1**/d1R1
        + (v(**r**+d2R1∗**u2R1**)–1)∗**u2R1**/d2R1
        + (v(**r**+d3R1∗**u3R1**)–1)∗**u3R1**/d3R1),
    r in SM   (1332)

(v(**r**)|M)∗**D3**[**r**]∗(**f**|a1;**dR1**)
    = **D3**[**r**]∗(**f**|a1;**dR1**),   **r** in V3–(M∪SM)
    = **D3**[**r**]∗(**f**|a1;**dR1**),   **r** in SM
    = [**0**],   **r** in M–SM.   (1333)

Similar formulations may be extended for other observational references.

Example: Let

f(**r**(**x**,a)) = c,   **x** = **x1**
    = 0   otherwise.

for non–zero c. Thus f is a choice function which selects the particle **x1** in its trace, **r**(**x1**,a).

Then at the reference **r1**(**x1**,a1),

D[a1](f|**x1**;d_fa)= 0
D[a1](f|**x1**;d_ba) = 0
D[a1](f|**r1**;d_fa) = –c∗D[a](uar(**x**(**r1**,a) – **x**(**r1**,a1))|**r1**;d_fa)
D[a1](f|**r1**;d_ba) = c∗D[a](var(**x**(**r1**,a) – **x**(**r1**,a1))|**r1**;d_ba)
D[**r1**,**r2**](f|a1;da) = –c∗D[a1](ua(**r**(**x1**,a) – **r1**(**x**,a))|a1;d_fa)
D[**x1**,**x2**](f|a1;da) = –c∗D[a1](ua(**x**(**r1**,a) – **x1**(**r**,a))|a1;d_fa)
D[**r0**,**r1**](f|a1;da) = c∗D[a1](va(**r**(**x1**,a) – **r1**(**x**,a))|a1;d_ba)
D[**x0**,**x1**](f|a1;da) = c∗D[a1](va(**x**(**r1**,a) – **x1**(**x**,a))|a1;d_ba)
D[**r1**,**r2**](f|a1+da;da) = c∗D[a1](ua(**r**(**x1**,a) – **r1**(**x**,a))|a1+da;d_fa)
D[**x1**,**x2**](f|a1+da;da) = –c∗D[a1](ua(**x**(**r1**,a) – **x1**(**r**,a))|a1+da;d_fa)
D[**r0**,**r1**](f|a1–da;da) = –c∗D[a1](va(**r**(**x1**,a) – **r1**(**x**,a))|a1–da ;d_ba)
D[**x0**,**x1**](f|a1–da;da) = c∗D[a1](va(**x**(**r1**,a) – **x1**(**x**,a))|a1–da;d_ba).

Theoretical Physics: Step Functions

Each of these derivatives has a related gradient too.

At **r1**(**x1**,a1) for mutually related sections SECT(**r1**,**u1R1**,**u2R1**,**u3R1**) and SECT(**x1**,**u1X1**,**u2X1**,**u3X1**)

D3[r1]∗(f|a1;**dR1**) = – lim c∗(**u1R1**/d1R1+**u2R1**/d2R1+**u3R1**/d3R1)
D3[x1]∗(f|a1;**dX1**) = – lim c∗(**u1X1**/d1X1+**u2X1**/d2X1+**u3X1**/d3X1)
D3[r1]∗(f|a1+da;**dR1**)= **0**
= **D3[r1]**∗(f|a1–da;**dR1**)
= **D3[r2]**∗(f|a1;**dR1**)
= **D3[r0]**∗(f|a1;**dR1**)
= **D3[y2]**∗(f|a1;**dX1**)
= **D3[y0]**∗(f|a1;**dX1**)
= **D3[x2]**∗(f|a1+da;**dX1**)
= **D3[x0]**∗(f|a1–da;**dX1**).

At **r2**(**x1**,a1+da) for mutually related sections SECT(**r2**,**u1R1**,**u2R1**,**u3R1**) and SECT(**x1**,**u1X1**,**u2X1**,**u3X1**)

D3[r2]∗(f|a1+da;**dR1**) = – lim c∗(**u1R1**/d1R1+**u2R1**/d2R1+**u3R1**/d3R1)
D3[x1]∗(f|a1+da;**dX1**) = – lim c∗(**u1X1**/d1X1+**u2X1**/d2X1+**u3X1**/d3X1).

At **r0**(**x1**,a1–da) for mutually related sections SECT(**r0**,**u1R1**,**u2R1**,**u3R1**) and SECT(**x1**,**u1X1**,**u2X1**,**u3X1**)

D3[r0]∗(f|a1–da;**dR1**) = – lim c∗(**u1R1**/d1R1+**u2R1**/d2R1+**u3R1**/d3R1)
D3[x1]∗(f|a1–da;**dX1**) = – lim c∗(**u1X1**/d1X1+**u2X1**/d2X1+**u3X1**/d3X1).

The isolated particle of classical mechanics is thus placed in the richer context of Theoretical Physics.

INTEGRALS OF THEORETICAL PHYSICS

Theoretical Physics conceives an integral to match each of its derivatives. For each it defines a measure, the conditions for interior cancellation, the integration of impulse functions. From one viewpoint, these new integrals are extensions of the fundamental theorem of integral calculus; in Theoretical Physics they also are seen as defining an extended class of functions suitable for describing, analyzing, and explaining physical reality.

Observational Integration

Let pt>0 be a real number called the **partition scalar**. For integers n>0, let Pn be the collection of the following sets of observations *a*,

$$Pni \equiv \{a|(i-1)*pt/n \leq a < i*pt/n\} \quad \text{for } i=-(n^2-1),\ldots,-1,0,1,2,3,\ldots n^2\}$$
$$\text{and } Pn\infty \text{ otherwise.} \qquad (1334)$$

Any observation may be placed into one and only one of the subsets of Pn. Pn is thus a partition of observations into $2*n^2+1$ disjoint subsets Pni, called **observational cells**.

For each cell define a central observation

$$a(Pni) \equiv (i-1/2)*pt/n \qquad (1335)$$
$$a(Pn\infty) = 0$$

For a measurable set M of observations in a given partition Pn, let *bn(M∩Pn)* be the number of cells for which M∩Pni is non–empty.

For each Pni of Pn and j=i+1 let

$$mn(Pni) \equiv pt*(a(Pnj)-a(Pni))$$
$$= pt/n \qquad (1336)$$
$$\text{with } mn(Pn\infty) \equiv 0. \qquad (1337)$$

Then define the **measure of M with respect to the partition Pn** as the following sum

$$mn(M\cap Pn]) \equiv S[bn(M\cap Pn)](mn(Pni))$$
$$= (pt/n)*S[bn(M\cap Pn)]. \qquad (1338)$$

and the **measure of M** as

$$m(M) \equiv \lim(mn(M \cap Pn)) \text{ as } n \rightarrow \infty \quad (1339)$$
$$= pt * \lim(S[M \cap Pn]/n). \quad (1340)$$

For M1 and M2 measurable and disjoint,

$$m(M1 \cup M2) = m(M1) + m(M2). \quad (1341)$$

Moreover, for M = Pni

$$m(Pni) = pt/n \quad (1342)$$

since for

$$mn(Pni) = pt/n$$
$$m2n(Pni) = (pt/(2*n))*2$$
$$= mn(Pni)$$

Given the partition scalar, then, the measure of each subset of any given partition is known immediately.

Definition 55 (observational integrals)

Given

{**r**(**x**,a)} the set of ordered observations;

f(**r**(**x**,a)), a generalized function of the observations;

Pn a set of partitions of the observations based on the partition scalar pt;

m(Pni), the measure of a cell in the Pn partition;

M a measurable set of the observations;

for

bn(M∩Pn) the number of cells of Pn intersecting M;

S[bn(M∩Pn)], a summation over cells indicated by bn;

then

$$\mathrm{I}[M](\mathbf{f}(\mathbf{r}(\mathbf{x},a));da)$$
$$\equiv \lim S[bn(M \cap Pn)](\mathbf{f}(a(Pni)) * m(Pni))$$
$$= pt * \lim S[bn(M \cap Pn)](\mathbf{f}(a(Pni))/n) \quad \text{as } n \rightarrow \infty. \quad (1343)$$

end of definition

When M is an interval [a0,an], the integral is written as

I[a0,an](**f**(**r**(**x**,a));da)
≡ lim S[bn(Pnn[a0,an])](**f**(**r**(**x**,a0+i∗da))∗da) (1344)

where

a(Pni) = a0 + i∗pt/n (1345)
da ≡ pt/n (1346)
= a(Pn(i+1))–a(Pni). (1347)

Just as the indexed derivative D[a1](**f**;da) may be considered a value (for fixed a1) or a function of a1 (for variable a1), so the indexed integral I[a0,a2](**f**;da) may be considered a value (for fixed a2) or a function of a2 (for variable a2). To emphasize the indexed integral as an indexed function, it is sometimes written as

g(a) = **I**[a0,a](**f**;dt) (1348)

where *t* is a dummy variable of *a*.

Because the addends in the summation commute it is unnecessary to define a forward or backward sense for the integral.

In Theoretical Physics integration with respect to the observational index may be applied either generically or specifically[78] as for example,

Specifically: **I**[](**f**(**r**(**x1**,a))|**x1**;da)
Generically: **I**[](**f**(**r**(**x**,a))|**x**;da).

In the generic case, **r**(**x**,a), represents continuous observations of locations of material particles throughout the whole universe, or more modestly, over a measurable region of physical locations. Observational integration yields specific results in the first case or results for the region in the second.

A local trace, **r**(**x1**,a) denotes the location of a material particle ***x1*** at observation *a*. Specific observational integration yields a result for particle ***x1***. Such integrals are symbolized

I[](**f**|**x1**;da) ≡ **I**[](**f**(**r**(**x1**,a));da). (1349)

Theoretical Physics: Integration

Likewise material tracks, **x**(**r1**,a) represent observations of the particles occupying location ***r1***. Specific observational integration yields a result for location ***r1***. Such integrals may be symbolized

$$\mathbf{I}[](\mathbf{f}|\mathbf{r1};da) \equiv \mathbf{I}[](\mathbf{f}(\mathbf{x}(\mathbf{r1},a));da). \tag{1350}$$

Functions over given traces or tracks may be continuously differentiable over all observations without being so everywhere unrestrictedly.

Theorem 15 (integration[79] over the observational index)
Generically
Given
f(**r**(**x**,a)) a generalized function of the observational index;
pt any partition scalar;
then

$$\mathbf{I}[](\mathbf{f}(\mathbf{r}(\mathbf{x},a)) \ast D[a](ua(\mathbf{r}(\mathbf{x},a)-\mathbf{r}(\mathbf{x},a1));d_f a);da) = \mathbf{f}(\mathbf{r}(\mathbf{x},a1)) \tag{1351}$$

$$\mathbf{I}[](\mathbf{f}(\mathbf{r}(\mathbf{x},a)) \ast D[a](ua(\mathbf{r}(\mathbf{x},a)-\mathbf{r}(\mathbf{x},a1));d_b a);da) = \mathbf{0} \tag{1352}$$

$$\mathbf{I}[](\mathbf{f}(\mathbf{r}(\mathbf{x},a)) \ast D[a](va(\mathbf{r}(\mathbf{x},a)-\mathbf{r}(\mathbf{x},a1));d_f a);da) = \mathbf{0} \tag{1353}$$

$$\mathbf{I}[](\mathbf{f}(\mathbf{r}(\mathbf{x},a)) \ast D[a](va(\mathbf{r}(\mathbf{x},a)-\mathbf{r}(\mathbf{x},a1));d_b a);da) = \mathbf{f}(\mathbf{r}(\mathbf{x},a1)). \tag{1354}$$

Proof:

$$\begin{aligned}
&\mathbf{I}[](\mathbf{f}(\mathbf{r}(\mathbf{x},a)) \ast D[a](ua(\mathbf{r}(\mathbf{x},a)-\mathbf{r}(\mathbf{x},a1));d_f a);da)\\
&\quad = pt \ast \lim S[bn(R \cap Pn)](\mathbf{f}(\mathbf{r}(\mathbf{x},a(Pni)))\\
&\qquad\qquad \ast D[a(Pni)](ua(\mathbf{r}(\mathbf{x},a)-\mathbf{r}(\mathbf{x},a1));d_f a))/n\\
&\quad = pt \ast \lim S[bn(R \cap Pn)](\mathbf{f}(\mathbf{r}(\mathbf{x},a(Pni)))\\
&\qquad \ast\ ua(\mathbf{r}(\mathbf{x},a(Pni)+da)-\mathbf{r}(\mathbf{x},a1)) - ua(\mathbf{r}(\mathbf{x},a(Pni))-\mathbf{r}(\mathbf{x},a1)))/(da \ast n)\\
&\quad = pt \ast \lim\ (\mathbf{f}(a1) \ast ua(\mathbf{r}(\mathbf{x},a1+da)-\mathbf{r}(\mathbf{x},a1)))/(da \ast n).
\end{aligned}$$

For da and n such that $da \ast n < pt$

$$= \mathbf{f}(\mathbf{r}(\mathbf{x},a1)).$$

This proves the first proposition.

$$\begin{aligned}
&\mathbf{I}[](\mathbf{f}(\mathbf{r}(\mathbf{x},a)) \ast D[a](ua(\mathbf{r}(\mathbf{x},a)-\mathbf{r}(\mathbf{x},a1));d_b a);da)\\
&\quad = pt \ast \lim S[bn(R \cap Pn)](\mathbf{f}(\mathbf{r}(\mathbf{x},a(Pni)))\\
&\qquad \ast\ ua(\mathbf{r}(\mathbf{x},a(Pni))-\mathbf{r}(\mathbf{x},a1)) - ua(\mathbf{r}(\mathbf{x},a(Pni)-da)-\mathbf{r}(\mathbf{x},a1)))\\
&\qquad\qquad /(da \ast n)\\
&\quad = \mathbf{0}.
\end{aligned}$$

This proves the second proposition.

$$\begin{aligned}
&\mathbf{I}[](\mathbf{f}(\mathbf{r}(\mathbf{x},a)) \ast D[a](va(\mathbf{r}(\mathbf{x},a)-\mathbf{r}(\mathbf{x},a1));d_f a);da)\\
&\quad = pt \ast \lim S[bn(R \cap Pn)](\mathbf{f}(\mathbf{r}(\mathbf{x},a(Pni)))\\
&\qquad\qquad \ast D[a(Pni)](va(\mathbf{r}(\mathbf{x},a)-\mathbf{r}(\mathbf{x},a1));d_f a))/n\\
&\quad = pt \ast \lim S[bn(R \cap Pn)](\mathbf{f}(\mathbf{r}(\mathbf{x},a(Pni)))\\
&\qquad \ast\ va(\mathbf{r}(\mathbf{x},a(Pni)+da)-\mathbf{r}(\mathbf{x},a1)) - va(\mathbf{r}(\mathbf{x},a(Pni))-\mathbf{r}(\mathbf{x},a1)))\\
&\qquad\qquad /(da \ast n)\\
&\quad = \mathbf{0}.
\end{aligned}$$

So is proved the third proposition.
I[](**f**(**r**(**x**,a))∗D[a](va(**r**(**x**,a)−**r**(**x**,a1));d_ba);da)
= pt∗lim S[bn(R∩Pn)](**f**(**r**(**x**,a(Pni)))
∗ va(**r**(**x**,a(Pni))−**r**(**x**,a1)) − va(**r**(**x**,a(Pni)−da)−**r**(**x**,a1)))
/(da∗n)
= pt∗lim (**f**(a1)∗va(**r**(**x**,a1)−**r**(**x**,a1)))/(da∗n)
For da and n such that da∗n < pt
= **f**(**r**(**x**,a1)).
This proves the fourth proposition.

qed

Corollary
Specifically
for the trace **r**(**x1**,a)

I[](**f**(**r**(**x1**,a))∗D[a](uax(**r**(**x1**,a)−**r**(**x1**,a1));d_fa);da)
= **f**(**r**(**x1**,a1)) (1355)
I[](**f**(**r**(**x1**,a))∗D[a](uax(**r**(**x1**,a)−**r**(**x1**,a1));d_ba);da)
= **0** (1356)
I[](**f**(**r**(**x1**,a))∗D[a](vax(**r**(**x1**,a)−**r**(**x1**,a1));d_fa);da)
= **0** (1357)
I[](**f**(**r**(**x1**,a))∗D[a](vax(**r**(**x1**,a)−**r**(**x1**,a1));d_ba);da)
= **f**(**r**(**x1**,a1)) (1358)

for the track **x**(**r1**,a)

I[](**f**(**x**(**r1**,a))∗D[a](uar(**x**(**r1**,a)−**x**(**r1**,a1));d_fa);da)
= **f**(**x**(**r1**,a1)) (1359)
I[](**f**(**x**(**r1**,a))∗D[a](uar(**x**(**r1**,a)−**x**(**r1**,a1));d_ba);da)
= **0** (1360)
I[](**f**(**x**(**r1**,a))∗D[a](var(**x**(**r1**,a)−**x**(**r1**,a1));d_fa);da)
= **0** (1361)
I[](**f**(**x**(**r1**,a))∗D[a](var(**x**(**r1**,a)−**x**(**r1**,a1));d_ba);da)
= **f**(**x**(**r1**,a1)). (1362)

Theoretical Physics: Integration

Theorem 16 (integration over the observational index)
Generically
Given

f(**r**(**x**,a)) a continuous function of the observational index;
D[a](**f**(**r**(**x**,t));d_ft) a basic observational derivative;
pt any partition scalar;

for

I[](**f**(**r**(**x**,a));da) bounded;

then

I[](ua(**r**(**x**,a)–**r**(**x**,a1))*D[a](**f**(**r**(**x**,t));d_ft);da) = –**f**(**r**(**x**,a1)) (1363)
I[](ua(**r**(**x**,a)–**r**(**x**,a1))*D[a](**f**(**r**(**x**,t));d_bt);da) = –**f**(**r**(**x**,a1)) (1364)
I[](va(**r**(**x**,a)–**r**(**x**,a1))*D[a](**f**(**r**(**x**,t));d_ft);da) = –**f**(**r**(**x**,a1)) (1365)
I[](va(**r**(**x**,a)–**r**(**x**,a1))*D[a](**f**(**r**(**x**,t));d_ft);da) = –**f**(**r**(**x**,a1)). (1366)

Proof:
I[](ua(**r**(**x**,a)–**r**(**x**,a1))*D[a](**f**(**r**(**x**,t));d_ft);da)
= pt*lim S[bn(R∩Pn)](ua(**r**(**x**,a(Pni))–**r**(**x**,a1))
*D[a(Pni)](**f**;d_ft))/n
= pt*lim S[bn((a1,)∩Pn)](**f**(a(Pni)+dt)–**f**(a(Pni)))
/(dt*n)
= lim (... + **f**(a1+2*da+dt) – **f**(a1+2*da)
+ **f**(a1+da+dt) – **f**(a1+da))
= lim – **f**(a1+da)
= –**f**(**r**(**x**,a1)), since f is continuous.
This proves the first proposition.
I[](ua(**r**(**x**,a)–**r**(**x**,a1))*D[a](**f**(**r**(**x**,t));d_ba);da)
= pt*lim S[bn(R∩Pn)](ua(**r**(**x**,a(Pni))–**r**(**x**,a1))
*D[a(Pni)](**f**;d_ba))/n
= pt*lim S[bn((a1,)∩Pn)](**f**(a(Pni))–**f**(a(Pni)–dt))
/(dt*n)
= lim (... + **f**(a1+2*da) – **f**(a1+2*da–dt)
+ **f**(a1+da) – **f**(a1+da–dt))
= lim – **f**(a1+da–dt).
= –**f**(**r**(**x**,a1)).
This proves the second proposition.
I[](va(**r**(**x**,a)–**r**(**x**,a1))*D[a](**f**(**r**(**x**,t));d_fa);da)
= pt*lim S[bn(R∩Pn)](va(**r**(**x**,a(Pni))–**r**(**x**,a1))
*D[a(Pni)](**f**;d_fa))/n
= pt*lim S[bn([a1,]∩Pn)](**f**(a(Pni)+da)–**f**(a(Pni)))
/(da*n)
= lim (... + **f**(a1+da+dt) – **f**(a1+da)
+ **f**(a1+dt) – **f**(a1))
= –**f**(**r**(**x**,a1)).

So is proved the third proposition.

I[](va(**r**(**x**,a)−**r**(**x**,a1))∗D[a](**f**(**r**(**x**,t));d_ba);da)
= pt∗lim S[bn(R∩Pn)](va(**r**(**x**,a(Pni))−**r**(**x**,a1))
∗D[a(Pni)](**f**;d_ba))/n
= pt∗lim S[bn([a1,]∩Pn)](**f**(a(Pni))−**f**(a(Pni)−da))
/(da∗n)
= lim (... + **f**(a1+da) − **f**(a1+da−dt)
+ **f**(a1) − **f**(a1−dt))
= − **f**(a1−dt).
= −**f**(**r**(**x**,a1)).

This proves the fourth proposition. qed

Corollary

Specifically

for the trace **r**(**x1**,a)

I[](uax(**r**(**x1**,a)−**r**(**x1**,a1))∗D[a](**f**(**r**(**x1**,t));d_fa);da)
= −**f**(**r**(**x1**,a1)) (1367)

I[](uax(**r**(**x1**,a)−**r**(**x1**,a1))∗D[a](**f**(**r**(**x1**,t));d_ba);da)
= −**f**(**r**(**x1**,a1)) (1368)

I[](vax(**r**(**x1**,a)−**r**(**x1**,a1))∗D[a](**f**(**r**(**x1**,t));d_fa);da)
= − **f**(**r**(**x1**,a1)) (1369)

I[](vax(**r**(**x1**,a)−**r**(**x1**,a1))∗D[a](**f**(**r**(**x1**,t));d_fa);da)
= −**f**(**r**(**x1**,a1)) (1370)

for the track **x**(**r1**,a)

I[](uar(**x**(**r1**,a)−**x**(**r1**,a1))∗D[a](**f**(**x**(**r1**,t));d_fa);da)
= −**f**(**x**(**r1**,a1)) (1371)

I[](uar(**x**(**r1**,a)−**x**(**r1**,a1))∗D[a](**f**(**x**(**r1**,t));d_ba);da)
= −**f**(**x**(**r1**,a1)) (1372)

I[](var(**x**(**r1**,a)−**x**(**r1**,a1))∗D[a](**f**(**x**(**r1**,t));d_fa);da)
= −**f**(**x**(**r1**,a1)) (1373)

I[](var(**x**(**r1**,a)−**x**(**r1**,a1))∗D[a](**f**(**x**(**r1**,t));d_fa);da)
=− **f**(**x**(**r1**,a1)). (1374)

Theoretical Physics: Integration

Theorem 17 (integration over the observational index)
Generically
Given
$\mathbf{f}(\mathbf{r}(\mathbf{x},a))$ a continuous function of the observational index;
pt any partition scalar;
for
$\mathrm{I}[](\mathbf{f}(\mathbf{r}(\mathbf{x},a));da)$ bounded;
then

$$\mathrm{I}[](D[a](\mathbf{f}(\mathbf{r}(\mathbf{x},a))\ast ua(\mathbf{r}(\mathbf{x},a)-\mathbf{r}(\mathbf{x},a1));d_f a);da) = \mathbf{0} \tag{1375}$$

$$\mathrm{I}[](D[a](\mathbf{f}(\mathbf{r}(\mathbf{x},a))\ast ua(\mathbf{r}(\mathbf{x},a)-\mathbf{r}(\mathbf{x},a1));d_b a);da) = \mathbf{0} \tag{1376}$$

$$\mathrm{I}[](D[a](\mathbf{f}(\mathbf{r}(\mathbf{x},a))\ast va(\mathbf{r}(\mathbf{x},a)-\mathbf{r}(\mathbf{x},a1));d_f a);da) = \mathbf{0} \tag{1377}$$

$$\mathrm{I}[](D[a](\mathbf{f}(\mathbf{r}(\mathbf{x},a))\ast va(\mathbf{r}(\mathbf{x},a)-\mathbf{r}(\mathbf{x},a1));d_b a);da) = \mathbf{0}. \tag{1378}$$

Proof:

$$\begin{aligned}
&\mathrm{I}[](D[a](\mathbf{f}(\mathbf{r}(\mathbf{x},a))\ast ua(\mathbf{r}(\mathbf{x},a)-\mathbf{r}(\mathbf{x},a1));d_f a);da)\\
&\quad = pt\ast \lim S[bn(R\cap Pn)]D[a(Pni)](\mathbf{f}(\mathbf{r}(\mathbf{x},a(Pni)))\\
&\qquad\qquad \ast ua(\mathbf{r}(\mathbf{x},a)-\mathbf{r}(\mathbf{x},a1));d_f a)/n\\
&\quad = pt\ast \lim S[bn(R\cap Pn)](\mathbf{f}(\mathbf{r}(\mathbf{x},a(Pni)+da))\\
&\qquad\qquad \ast\ ua(\mathbf{r}(\mathbf{x},a(Pni)+da)-\mathbf{r}(\mathbf{x},a1))\\
&\qquad\qquad - \mathbf{f}(\mathbf{r}(\mathbf{x},a(Pni)))\ast ua(\mathbf{r}(\mathbf{x},a(Pni))-\mathbf{r}(\mathbf{x},a1)))\\
&\qquad\qquad /(da\ast n)\\
&\quad = \lim\ (\mathbf{f}(\mathbf{r}(\mathbf{x},a(Pnj)+da))\\
&\qquad\qquad + \mathbf{f}(\mathbf{r}(\mathbf{x},a(Pn(j+1))+da))-\mathbf{f}\ (\mathbf{r}(\mathbf{x},a(Pn(j+1))))\\
&\qquad\qquad + \ldots)
\end{aligned}$$

where a1 falls in cell Pnj.
Now

$$a(Pn(j+1)) = a(Pnj) + pt/n \longrightarrow a(Pnj) + da$$

and since f is continuous,

$$\mathbf{f}(\mathbf{r}(\mathbf{x},a(Pnj)+da)) \longrightarrow \mathbf{f}(\mathbf{r}(\mathbf{x},a(Pn(j+1)))).$$

Further since **f** must be asymptotic because its integral is bounded,

$$\begin{aligned}
&\mathrm{I}[](D[a](\mathbf{f}(\mathbf{r}(\mathbf{x},a))\ast ua(\mathbf{r}(\mathbf{x},a)-\mathbf{r}(\mathbf{x},a1));d_f a);da)\\
&\quad = \lim\ \mathbf{f}(\mathbf{r}(\mathbf{x},a(Pn(j+m))+da)) \text{ as } m \longrightarrow \infty\\
&\quad = \mathbf{0}.
\end{aligned}$$

This proves the first proposition.

$$\begin{aligned}
&\mathrm{I}[](D[a](\mathbf{f}(\mathbf{r}(\mathbf{x},a))\ast ua(\mathbf{r}(\mathbf{x},a)-\mathbf{r}(\mathbf{x},a1));d_b a);da)\\
&\quad = pt\ast \lim S[bn(R\cap Pn)](\mathbf{f}(\mathbf{r}(\mathbf{x},a(Pni)))\ast\ ua(\mathbf{r}(\mathbf{x},a(Pni))-\mathbf{r}(\mathbf{x},a1))\\
&\qquad\qquad -\mathbf{f}(\mathbf{r}(\mathbf{x},a(Pni)-da))\ast ua(\mathbf{r}(\mathbf{x},a(Pni)-da)-\mathbf{r}(\mathbf{x},a1)))\\
&\qquad\qquad /(da\ast n)\\
&\quad = \lim\ (\mathbf{f}(\mathbf{r}(\mathbf{x},a(Pn(j+1))))-\mathbf{f}(\mathbf{r}(\mathbf{x},a(Pn(j+1))-da))\\
&\qquad\qquad + \mathbf{f}(\mathbf{r}(\mathbf{x},a(Pn(j+2))+da))-\mathbf{f}\ (\mathbf{r}(\mathbf{x},a(Pn(j+2))-da))\\
&\qquad\qquad + \ldots)\\
&\quad = \mathbf{0} \text{ since the function is continuous.}
\end{aligned}$$

This proves the second proposition.

I[](D[a](**f**(**r**(**x**,a))∗va(**r**(**x**,a)−**r**(**x**,a1));d_fa);da)
 = pt∗lim S[bn(R∩Pn)](**f**(**r**(**x**,a(Pni)+da))
 ∗va(**r**(**x**,a(Pni)+da)−**r**(**x**,a1))
 − **f**(**r**(**x**,a(Pni)))∗va(**r**(**x**,a(Pni))−**r**(**x**,a1)))
 /(da∗n)
 = lim (**f**(**r**(**x**,a(Pnj+da))) − **f**(**r**(**x**,a(Pnj)))
 + ...)
 = **0** since the function is continuous.

This proves the third proposition.

I[](D[a](**f**(**r**(**x**,a))∗va(**r**(**x**,a)−**r**(**x**,a1));d_ba);da)
 =lim **f**(**r**(**x**,a(Pn(j+m)))) as m ⟶ ∞
 = **0**.

This proves the fourth proposition.

qed

Corollary

Specifically

for the trace **r**(**x1**,a)

I[](D[a](**f**(**r**(**x1**,a))∗uax(**r**(**x1**,a)−**r**(**x1**,a1));d_fa);da) = **0** (1379)

I[](D[a](**f**(**r**(**x1**,a))∗uax(**r**(**x1**,a)−**r**(**x1**,a1));d_ba);da) = **0** (1380)

I[](D[a](**f**(**r**(**x1**,a))∗vax(**r**(**x1**,a)−**r**(**x1**,a1));d_fa);da) = **0** (1381)

I[](D[a](**f**(**r**(**x1**,a))∗vax(**r**(**x1**,a)−**r**(**x1**,a1));d_ba);da) = **0** (1382)

for the track **x**(**r1**,a)

I[](D[a](**f**(**x**(**r1**,a))∗uar(**x**(**r1**,a)−**x**(**r1**,a1));d_fa);da) = **0** (1383)

I[](D[a](**f**(**x**(**r1**,a))∗uar(**x**(**r1**,a)−**x**(**r1**,a1));d_ba);da) = **0** (1384)

I[](D[a](**f**(**x**(**r1**,a))∗var(**x**(**r1**,a)−**x**(**r1**,a1));d_fa);da) = **0** (1385)

I[](D[a](**f**(**x**(**r1**,a))∗var(**x**(**r1**,a)−**x**(**r1**,a1));d_ba);da) = **0**. (1386)

Theorem 18 (integration over the observational index)

Generically

given

f(**r**(**x**,a)) a continuous function of the observational index;

D[a](**f**(**r**(**x**,t));dt) a basic observational derivative;

pt any partition scalar;

for

I[](**f**(**r**(**x**,a));da) bounded

then

I[](D[a](**f**(**r**(**x**,t));da);da) = **0** (1387)

for only

I[a0,](**f**(**r**(**x**,t));da) bounded.

then

I[](va(**r**(**x**,a)−**r**(**x**,a0))∗D[a](**f**(**r**(**x**,t));da);da)
= −**f**(**r**(**x**,a0)) (1388)
= I[a0,](D[a](**f**(**r**(**x**,t));da);da)

for only

I[,a2](**f**(**r**(**x**,t));da) bounded

then

I[](1−ua(**r**(**x**,a)−**r**(**x**,a2))∗D[a](**f**(**r**(**x**,t));da);da)
= **f**(**r**(**x**,a2)) (1389)
= I[,a2](D[a](**f**(**r**(**x**,t));da);da)

for only

f(**r**(**x**,t)) bounded[80] only in the interval [a0,a2]

then

I[]((va(**r**(**x**,a)−**r**(**x**,a0))−ua(**r**(**x**,a)−**r**(**x**,a2)))
∗D[a](**f**(**r**(**x**,t));da);da)
= **f**(**r**(**x**,a2))−**f**(**r**(**x**,a0)) (1390)
= I[a0,a2](D[a](**f**(**r**(**x**,t));da);da).

Proof:

I[](D[a](**f**;dt);da)
= pt∗lim S[bn(R∩Pn)](D[a](**f**;da)/n)
= pt∗lim S[bn(R∩Pn)]((**f**(a+pt/n)−**f**(a))/(pt/n))/n
= lim S[bn(R∩Pn)](**f**(a+pt/n)−**f**(a))
= (... + **f**(a+2∗pt/n) − **f**(a+pt/n)
+ **f**(a+pt/n) − **f**(a)
+ **f**(a) − **f**(a−pt/n) + ...)
= **0**.

This proves the first proposition.

I[](va(**r**(**x**,a)−**r**(**x**,a0))∗D[a](**f**(**r**(**x**,t));da);da)
= va(**r**(**x**,a)−**r**(**x**,a0))∗pt∗lim S[bn(R∩Pn)](D[a](**f**;da)/n)
= lim S[bn((a0,)∩Pn)](**f**(a+pt/n)−**f**(a))
= (... + **f**(a0+2∗pt/n) − **f**(a0+pt/n)
+ **f**(a0+pt/n)
+ **f**(a0+pt/n) − **f**(a0))
= −**f**(a0).

This proves the second proposition.

The third and fourth propositions are proven similarly.

qed

Corollary

Specifically

for the trace **r**(**x1**,a)

I[](D[a](**f**(**r**(**x1**,t))|**x1**;da)|**x1**;da)
= **0** (1391)

I[](vax(**r**(**x1**,a)–**r**(**x1**,a0))∗D[a](**f**(**r**(**x1**,t));da);da)
= –**f**(**r**(**x1**,a0)) (1392)
= I[a0,](D[a](**f**(**r**(**x1**,t));da);da)

I[](1–uax(**r**(**x1**,a)–**r**(**x1**,a2))∗D[a](**f**(**r**(**x1**,t));da);da)
= **f**(**r**(**x1**,a2)) (1393)
= I[,a2](D[a](**f**(**r**(**x1**,t));da);da)

I[](vax(**r**(**x1**,a)–**r**(**x1**,a0))–uax(**r**(**x1**,a)–**r**(**x1**,a2))
∗D[a](**f**(**r**(**x1**,t));da);da)
= **f**(**r**(**x1**,a2))–**f**(**r**(**x1**,a0)) (1394)

for the track **x**(**r1**,a)

I[](D[a](**f**(**x**(**r1**,a));da);da)
= **0** (1395)

I[](var(**x**(**r1**,a)–**x**(**r1**,a0))∗D[a](**f**(**x**(**r1**,t));da);da)
= –**f**(**x**(**r1**,a0)) (1396)
= I[a0,](D[a](**f**(**x**(**r1**,a));da);da)

I[](1–uar(**x**(**r1**,a)–**x**(**r1**,a2))∗D[a](**f**(**x**(**r1**,a));da);da)
= **f**(**x**(**r1**,a2)) (1397)
= I[,a2](D[a](**f**(**x**(**r1**,a));da);da)

I[]((var(**x**(**r1**,a)–**x**(**r1**,a0))–uar(**x**(**r1**,a)–**x**(**r1**,a2)))
∗D[a](**f**(**x**(**r1**,t));da);da)
= **f**(**x**(**r1**,a2))–**f**(**x**(**r1**,a0)) (1398)
= I[a0,a2](D[a](**f**(**x**(**r1**,a));da);da).

Equation 1398 is the **fundamental theorem of integral calculus in Theoretical Physics extended to vectorial functions over observations**.

Theoretical Physics: Integration

Now for, $a0<a1<a2$, let

$$\mathbf{f}(\mathbf{r}(\mathbf{x},a)) \equiv \mathbf{f1}(\mathbf{r}(\mathbf{x},a)) + \mathbf{f0} * ua(\mathbf{r}(\mathbf{x},a)-\mathbf{r}(\mathbf{x},a1)) \qquad (1399)$$

where ***f1*** enjoys interior cancellation and **f0** is constant. Such a function has a vectorial step discontinuity, **f0**, at observation a1.

Then generically

$$D[a](\mathbf{f};d_f a) = D[a](\mathbf{f1};da) + \mathbf{f0} * D[a](ua(\mathbf{r}(\mathbf{x},a)-\mathbf{r}(\mathbf{x},a1));d_f a) \qquad (1400)$$

and

$$I[a0,a2](D[a](\mathbf{f};d_f a);da) = \mathbf{f1}(a2) - \mathbf{f1}(a0) + \mathbf{f0}. \qquad (1401)$$

Similarly, for

$$\mathbf{f}(\mathbf{r}(\mathbf{x},a)) \equiv \mathbf{f1}(\mathbf{r}(\mathbf{x},a)) + \mathbf{f0} * va(\mathbf{r}(\mathbf{x},a)-\mathbf{r}(\mathbf{x},a1)) \qquad (1402)$$

$$D[a](\mathbf{f};d_b a) = D[a](\mathbf{f1};da) + \mathbf{f0} * D[a](va(\mathbf{r}(\mathbf{x},a)-\mathbf{r}(\mathbf{x},a1));d_b a) \qquad (1403)$$

and

$$I[a0,a2](D[a](\mathbf{f};d_b a);da) = \mathbf{f1}(a2) - \mathbf{f1}(a0) + \mathbf{f0}. \qquad (1404)$$

Specifically for the track $\mathbf{x}(\mathbf{r1},a)$

$$\begin{aligned} D[a](\mathbf{f}(\mathbf{x}(\mathbf{r1},a))|\mathbf{r1};d_f a) &= D[a](\mathbf{f1}(\mathbf{x}(\mathbf{r1},a))|\mathbf{r1};da) \\ &\quad + \mathbf{f0} * D[a](uar(\mathbf{x}(\mathbf{r1},a)-\mathbf{x}(\mathbf{r1},a1));d_f a) \end{aligned} \qquad (1405)$$

and

$$\begin{aligned} I[a0,a2](D[a](\mathbf{f}(\mathbf{x}(\mathbf{r1},a))|\mathbf{r1};d_f a);da) \\ = \mathbf{f1}(\mathbf{x}(\mathbf{r1},a2)) - \mathbf{f1}(\mathbf{x}(\mathbf{r1},a0)) + \mathbf{f0}. \end{aligned} \qquad (1406)$$

These equations are extensions in Theoretical Physics to the **fundamental theorem of integral calculus to vectorial functions with a vectorial step discontinuity over observations**.

Let us symbolize

$$D^{(-1)}(ua(a-ai);da) \equiv I[](ua(\mathbf{r}(\mathbf{x},a)-\mathbf{r}(\mathbf{x},a1));da) \quad (1407)$$
$$D^{(0)}(ua(a-ai);da) \equiv ua(\mathbf{r}(\mathbf{x},a)-\mathbf{r}(\mathbf{x},a1)) \quad (1408)$$
$$D^{(1)}(ua(a-ai);da) \equiv D[a](ua(\mathbf{r}(\mathbf{x},a)-\mathbf{r}(\mathbf{x},a1));d_f a) \quad (1409)$$
$$D^{(2)}(ua(a-ai);da) \equiv D[a](D[a](ua(\mathbf{r}(\mathbf{x},a)-\mathbf{r}(\mathbf{x},a1));d_f a);d_f a) \quad (1410)$$

...

and

$$D^{(-1)}(va(a-ai);da) \equiv I[](va(\mathbf{r}(\mathbf{x},a)-\mathbf{r}(\mathbf{x},a1));da) \quad (1411)$$
$$D^{(0)}(va(a-ai);da) \equiv va(\mathbf{r}(\mathbf{x},a)-\mathbf{r}(\mathbf{x},a1)) \quad (1412)$$
$$D^{(1)}(va(a-ai);da) \equiv D[a](va(\mathbf{r}(\mathbf{x},a)-\mathbf{r}(\mathbf{x},a1));d_b a) \quad (1413)$$
$$D^{(2)}(va(a-ai);da) \equiv D[a](D[a](va(\mathbf{r}(\mathbf{x},a)-\mathbf{r}(\mathbf{x},a1));d_b a);d_b a) \quad (1414)$$

...

An observational function, discontinuous at *a1* but otherwise continuous, may have any or no value at the discontinuity. Regardless of the value of the function there, it will also have a limit from above and from below.

Consequently, it is often appropriate to deal with the discontinuity in terms of these limits rather than with the functional value there.

Functions of **r**(**x**,a) with a finite number of observational discontinuities may always be written in terms of other functions which are identical almost everywhere with the original function.

For instance where the functional values at the discontinuities are its limits from below:

$$\mathbf{f}(\mathbf{r}(\mathbf{x},a)) = \mathbf{f1}(\mathbf{r}(\mathbf{x},a))$$
$$+ S[1,n]\ \mathbf{ci} * I[](ua(\mathbf{r}(\mathbf{x},a)-\mathbf{r}(\mathbf{x},ai));da)$$
$$+ S[1,n0]\ \mathbf{c0i} * (ua(\mathbf{r}(\mathbf{x},a)-\mathbf{r}(\mathbf{x},ai)))$$
$$+ S[1,n1]\ \mathbf{c1i} * D[a](ua(\mathbf{r}(\mathbf{x},a)-\mathbf{r}(\mathbf{x},ai));d_f a)$$
$$+ S[1,n2]\ \mathbf{c2i} * D[a](D[a](ua(\mathbf{r}(\mathbf{x},a)-\mathbf{r}(\mathbf{x},a1));d_f a);d_f a)$$
$$+ \ldots \quad (1415)$$
$$= \mathbf{f1}(\mathbf{r}(\mathbf{x},a)) + S[-1,](S[1,nj](\mathbf{cji} * D^{(j)}(ua(a-ai);d_f a))) \quad (1416)$$

where **f1** enjoys interior cancellation, and

$$\mathbf{ci} \equiv D[ai]((\mathbf{f}-\mathbf{f1});da) \quad (1417)$$
$$\mathbf{c0i} \equiv (\mathbf{f}-\mathbf{f1}) \text{ at ai} \quad (1418)$$
c1i, the impulse constant. (1419)

...

The physical units of **cji**$*D^{(j)}$(ua(a–ai);da) are always the physical units of **f** and the **cji** are all constant.

Theoretical Physics: Integration

The representation of **f**(**r**(**x**,a)), in terms of a similar function almost everywhere identical, may likewise be made in terms of the step function va(**r**(**x**,a)–**r**(**x**,ai)) and specifically for

uax(**r**(**x1**,a)–**r**(**x1**,ai)),
vax(**r**(**x1**,a)–**r**(**x1**,ai)),
uar(**x**(**r1**,a)–**x**(**r1**,ai)),
var(**x**(**r1**,a)–**x**(**r1**,ai)).

Directional Integration of Directional Gradients

Here one is given a local trace **r**(**x1**,a) or a material track **x**(**r1**,a). Because of the principle of non–collocation, traces or tracks are continuous curved lines with possible discontinuities in D[a](**r**|**x1**;da) or D[a](**x**|**r1**;da). Consideration is restricted in the following to curves with no more than a finite number of such discontinuities.

Traces or tracks may overlap, that is for a1<a2

r(**x1**,a1) = **r**(**x1**,a2)
and **x**(**r1**,a1) = **x**(**r1**,a2)

are possible observations.

Theoretical Physics considers 48 gradients of the form

$$\mathbf{D1}[\mathbf{r1},\mathbf{r2}]*(\mathbf{f}|p1;\mathbf{d}_n p2) = D[a1](\mathbf{f}|p1;d_m a)*\mathbf{qd}(D[a1](p2,d_n a)) \qquad (1420)$$

for m and n symbolizing either forward or backward senses; p1 symbolizing **x1**, **r1**, a1, (a1+da), (a1–da); p2 symbolizing an increment in **r**|**x1** or **x**|**r1** in either a forward or backward sense.

Associated with these gradients are curls and divergences. For these many ideas, Theoretical Physics creates corresponding invergences, incurls and ingradients.

Consider first **D1**[**r**(**x1**,a),**r**(**x1**,a+da)]∗(**f**|**x1**;$\mathbf{d_f r}$) and the trace **r**(**x1**,a) under materially unstalled conditions.

Let pt>0 be a chosen real number called the partition scalar. Let Prn(**x1**) be the collection of the following sets of observations *a*,

$$\begin{aligned} &\text{Prni}(\mathbf{x1}) = \{\mathbf{r}(\mathbf{x1},a)|(i-1)*pt/n \le a < i*pt/n\} \\ &\qquad\qquad \text{for } i = -(n^2-1),\ldots,-1,0,1,2,3,\ldots n^2\} \\ \text{and} \quad &\text{Prn}\infty(\mathbf{x1}) \text{ otherwise.} \end{aligned} \qquad (1421)$$

Any observation of the trace **r**(**x1**,a) may be placed into one and only one of the subsets of Prn(**x1**). Prn(**x1**) is thus a partition of the trace **r**(**x1**,a) into $2*n^2+1$ disjoint cells Prni(**x1**).

For each cell define a reference location

$$\begin{aligned} &\mathbf{r}(\text{Prni}(\mathbf{x1})) \equiv \mathbf{r}(\mathbf{x1},i*pt/n) \qquad (1422) \\ &\mathbf{r}(\text{Prn}\infty(\mathbf{x1})) = \mathbf{0}. \end{aligned}$$

For each **r**(Prni(**x1**)) define

$$\begin{aligned} &\mathbf{m1_f n}(\text{Prni}(\mathbf{x1})) \equiv \mathbf{r}(\mathbf{x1},(i+1)*pt/n) - \mathbf{r}(\mathbf{x1},i*pt/n) \qquad (1423) \\ &\mathbf{m1_b n}(\text{Prni}(\mathbf{x1})) \equiv \mathbf{r}(\mathbf{x1},i*pt/n) - \mathbf{r}(\mathbf{x1},(i-1)*pt/n) \qquad (1424) \\ \text{with } &\mathbf{m1n}(\text{Prn}\infty(\mathbf{x1})) \equiv \mathbf{0}. \qquad (1425) \end{aligned}$$

For a measurable set M of observations of the trace in a given partition Prn(**x1**) let bn(M∩Prn(**x1**)) be the number of cells for which {M∩Prn(**x1**)} is non–empty.

Then define the **measure of** M **with respect to the partition Px1n as**

$$\begin{aligned} &\mathbf{m1_f n}(M\cap\text{Prn}(\mathbf{x1})) \equiv \text{bn}(M\cap\text{Prn}(\mathbf{x1}))*(\mathbf{m1_f n}(\text{Prni}(\mathbf{x1}))) \qquad (1426) \\ &\mathbf{m1_b n}(M\cap\text{Prn}(\mathbf{x1})) \equiv \text{bn}(M\cap\text{Prn}(\mathbf{x1}))*(\mathbf{m1_b n}(\text{Prni}(\mathbf{x1}))) \qquad (1427) \end{aligned}$$

and the **measure of** M as

$$\begin{aligned} \mathbf{m1}(M) &\equiv \lim(\mathbf{m1_f n}(M\cap\text{Prn}(\mathbf{x1}))) \text{ as } n\rightarrow\infty \qquad (1428) \\ &\equiv \lim(\mathbf{m1_b n}(M\cap\text{Prn}(\mathbf{x1}))) \text{ as } n\rightarrow\infty \qquad (1429) \end{aligned}$$

noting

$$\mathbf{m1_f n}(M\cap\text{Prn}(\mathbf{x1})) - \mathbf{m1_b n}(M\cap\text{Prn}(\mathbf{x1})) \rightarrow \mathbf{0} \text{ as } n\rightarrow\infty.$$

For M1 and M2 measurable and disjoint,

$$\mathbf{m1}(\mathrm{M1}\cup\mathrm{M2}) = \mathbf{m1}(\mathrm{M1}) + \mathbf{m1}(\mathrm{M2}). \tag{1430}$$

Moreover, for M = Prni(**x1**)

$$\mathbf{m1}(\mathrm{Prni}(\mathbf{x1})) = \mathbf{m2n}(\mathrm{Prni}(\mathbf{x1})) \tag{1431}$$

since for

$$\begin{aligned}
\mathbf{m1n}(\mathrm{Prni}(\mathbf{x1})) &= \mathbf{r}(\mathbf{x1},i*pt/n) - \mathbf{r}(\mathbf{x1},(i-1)*pt/n)\\
\mathbf{m2n}(\mathrm{Prni}(\mathbf{x1})) &= \mathbf{r}(\mathbf{x1},2*i*pt/(2*n))\\
&\quad - \mathbf{r}(\mathbf{x1},(2*i-1)*pt/(2*n))\\
&\quad + \mathbf{r}(\mathbf{x1},(2*i-1)*pt/(2*n))\\
&\quad - \mathbf{r}(\mathbf{x1},(2*i-2)*pt/(2*n))\\
&= \mathbf{r}(\mathbf{x1},i*pt/n) - \mathbf{r}(\mathbf{x1},(i-1)*pt/n)\\
&= \mathbf{m1n}(\mathrm{Prni}(x1)).
\end{aligned}$$

Dually, for partition

$$\begin{aligned}
&\mathrm{Pxni}(\mathbf{r1}) = \{\mathbf{x}(\mathbf{r1},a) \mid (i-1)*pt/n \le a < i*pt/n\}\\
&\qquad \text{for } i=-(n^2-1),\ldots,-1,0,1,2,3,\ldots n^2\}\\
\text{and}\quad &\mathrm{Pxn}\infty(\mathbf{r1}) = 0. \quad \text{otherwise.}
\end{aligned} \tag{1432}$$

Any observation of the track **x**(**r1**,a) may be placed into one and only one of the subsets of Prn(**r1**). Prn(**r1**) is thus a partition of the track **x**(**r1**,a) into $2*n^2+1$ disjoint cells Prni(**r1**).

For each cell define a reference location

$$\mathbf{x}(\mathrm{Pxni}(\mathbf{r1})) \equiv \mathbf{x}(\mathbf{r1},(i)*pt/n) \tag{1433}$$

$$\mathbf{x}(\mathrm{Pxn}\infty(\mathbf{x1})) = \mathbf{0}$$

and a measure of each cell

$$\mathbf{m1_fn}(\mathbf{x}(\mathrm{Pxni}(\mathbf{r1}))) \equiv \mathbf{x}(\mathbf{r1},(i+1)*pt/n) - \mathbf{x}(\mathbf{r1},i*pt/n) \tag{1434}$$

$$\mathbf{m1_bn}(\mathbf{x}(\mathrm{Pxni}(\mathbf{r1}))) \equiv \mathbf{x}(\mathbf{r1},i*pt/n) - \mathbf{x}(\mathbf{r1},(i-1)*pt/n) \tag{1435}$$

$$\text{with}\quad \mathbf{m1n}(\mathrm{Pxn}\infty(\mathbf{r1})) \equiv \mathbf{0}. \tag{1436}$$

For a measurable set M of observations of the track in a given partition Pxn(**r1**) with bn(M∩Pxn(**r1**)) the number of cells for which {M∩Pxn(**r1**)} is non–empty the **measure of** M **with respect to the partition Pxn**(**r1**) is

$$\mathbf{m1_fn}(\mathrm{M}\cap\mathrm{Pxn}(\mathbf{r1})) \equiv \mathrm{bn}(\mathrm{M}\cap\mathrm{Pxn}(\mathbf{r1}))*(\mathbf{m1_fn}(\mathrm{Pxni}(\mathbf{r1}))) \tag{1437}$$

$$\mathbf{m1_bn}(\mathrm{M}\cap\mathrm{Pxn}(\mathbf{r1})) \equiv \mathrm{bn}(\mathrm{M}\cap\mathrm{Pxn}(\mathbf{r1}))*(\mathbf{m1_bn}(\mathrm{Pxni}(\mathbf{r1}))). \tag{1438}$$

Definition 56 (integrals of a measurable set in a trace or in a track)

Given

a non–stalled trace $\mathbf{r}(\mathbf{x1},\mathrm{a})$;
$\mathbf{f}(\mathbf{r}(\mathbf{x1},\mathrm{a}))$ a measurable function;
M measurable set in the trace;
pt a partition scalar;
$\mathrm{Prn}(\mathbf{x1})$ a set of partitions defined from pt;
$\mathbf{mn_f1}$ the forward measure of partition cells $\mathrm{Prni}(\mathbf{x1})$;
$\mathbf{mn_b1}$ the backward measure of partition cells $\mathrm{Prni}(\mathbf{x1})$;

then

the invergence, incurl and ingradient of $\mathbf{f}$ over $\mathbf{r}(\mathbf{x1},\mathrm{a})$ with respect to $\mathbf{m1}$ are defined as

$$\mathrm{I1}[\mathrm{M}]\cdot(\mathbf{f}|\mathbf{x1};\mathbf{d_f r}) \equiv \lim(\mathrm{S}[\mathrm{bn}(\mathrm{M}\cap\mathrm{Prn}(\mathbf{x1}))](\mathbf{f}(\mathbf{r}(\mathrm{Prni}(\mathbf{x1})))\cdot\mathbf{mn_f1}(\mathrm{Prni}(\mathbf{x1})))) \quad (1439)$$

$$\mathbf{I1}[\mathrm{M}]\wedge(\mathbf{f}|\mathbf{x1};\mathbf{d_f r}) \equiv \lim(\mathrm{S}[\mathrm{bn}(\mathrm{M}\cap\mathrm{Prn}(\mathbf{x1}))](\mathbf{f}(\mathbf{r}(\mathrm{Prni}(\mathbf{x1})))\wedge\mathbf{mn_f1}(\mathrm{Prni}(\mathbf{x1})))) \quad (1440)$$

$$\mathbf{I1}[\mathrm{M}]*(\mathbf{f}|\mathbf{x1};\mathbf{d_f r}) \equiv \lim(\mathrm{S}[\mathrm{bn}(\mathrm{M}\cap\mathrm{Prn}(\mathbf{x1}))](\mathbf{f}(\mathbf{r}(\mathrm{Prni}(\mathbf{x1})))*\mathbf{mn_f1}(\mathrm{Prni}(\mathbf{x1})))) \quad (1441)$$

as $\mathrm{n}\longrightarrow\infty$.

$$\mathrm{I1}[\mathrm{M}]\cdot(\mathbf{f}|\mathbf{x1};\mathbf{d_b r}) \equiv \lim(\mathrm{S}[\mathrm{bn}(\mathrm{M}\cap\mathrm{Prn}(\mathbf{x1}))](\mathbf{f}(\mathbf{r}(\mathrm{Prni}(\mathbf{x1})))\cdot\mathbf{mn_b1}(\mathrm{Prni}(\mathbf{x1})))) \quad (1442)$$

$$\mathbf{I1}[\mathrm{M}]\wedge(\mathbf{f}|\mathbf{x1};\mathbf{d_b r}) \equiv \lim(\mathrm{S}[\mathrm{bn}(\mathrm{M}\cap\mathrm{Prn}(\mathbf{x1}))](\mathbf{f}(\mathbf{r}(\mathrm{Prni}(\mathbf{x1})))\wedge\mathbf{mn_b1}(\mathrm{Prni}(\mathbf{x1})))) \quad (1443)$$

$$\mathbf{I1}[\mathrm{M}]*(\mathbf{f}|\mathbf{x1};\mathbf{d_b r}) \equiv \lim(\mathrm{S}[\mathrm{bn}(\mathrm{M}\cap\mathrm{Prn}(\mathbf{x1}))](\mathbf{f}(\mathbf{r}(\mathrm{Prni}(\mathbf{x1})))*\mathbf{mn_b1}(\mathrm{Prni}(\mathbf{x1})))) \quad (1444)$$

as $\mathrm{n}\longrightarrow\infty$.

Given

a non–stalled track $\mathbf{x}(\mathbf{r1},\mathrm{a})$;
$\mathbf{f}(\mathbf{x}(\mathbf{r1},\mathrm{a}))$ a measurable function;
M measurable set in the trace;
pt a partition scalar;
$\mathrm{Prn}(\mathbf{r1})$ a set of partitions defined from pt;
$\mathbf{mn_f1}$ the forward measure of partition cells $\mathrm{Prni}(\mathbf{r1})$;
$\mathbf{mn_b1}$ the backward measure of partition cells $\mathrm{Prni}(\mathbf{r1})$;

then

the invergence, incurl and ingradient of **f** over **x**(**r1**,a) with respect to **m1** are defined as

I1[M]•(**f**|**r1**;**d**$_f$**x**)
≡ lim(S[bn(M∩Pxn(**r1**))](**f**(**x**(Pxni(**r1**)))•**mn**$_f$**1**(Pxni(**r1**)))) (1445)

I1[M]∧(**f**|**r1**;**d**$_f$**x**)
≡ lim(S[bn(M∩Pxn(**r1**))](**f**(**x**(Pxni(**r1**)))∧**mn**$_f$**1**(Pxni(**r1**)))) (1446)

I1[M]∗(**f**|**r1**;**d**$_f$**x**)
≡ lim(S[bn(M∩Pxn(**r1**))](**f**(**x**(Pxni(**r1**)))∗**mn**$_f$**1**(Pxni(**r1**)))) (1447)

as n⟶∞.

I1[M]•(**f**|**r1**;**d**$_b$**x**)
≡ lim(S[bn(M∩Pxn(**r1**))](**f**(**x**(Pxni(**r1**)))•**mn**$_b$**1**(Pxni(**r1**)))) (1448)

I1[M]∧(**f**|**r1**;**d**$_b$**x**)
≡ lim(S[bn(M∩Pxn(**r1**))](**f**(**x**(Pxni(**r1**)))∧**mn**$_b$**1**(Pxni(**r1**)))) (1449)

I1[M]∗(**f**|**r1**;**d**$_b$**x**)
≡ lim(S[bn(M∩Pxn(**r1**))](**f**(**x**(Pxni(**r1**)))∗**mn**$_b$**1**(Pxni(**r1**)))) (1450)

as n⟶∞.

end of definition

Since **mn1**(Prni(**x1**)) is a vector, abs(**mn1**(Prni(**x1**))):V3⟶R enables the following definitions:

I1[M](**f**|**x1**;d$_f$r)
≡ lim(S[bn(M∩Prn(**x1**))](**f**(**r**(Prni(**x1**)))
∗abs(**mn**$_f$**1**(Prni(**x1**))))) (1451)

I1[M](**f**|**x1**;d$_b$r)
≡ lim(S[bn(M∩Prn(**x1**))](**f**(**r**(Prni(**x1**)))
∗abs(**mn**$_b$**1**(Prni(**x1**))))) (1452)

I1[M](**f**|**r1**;d$_f$x)
≡ lim(S[bn(M∩Pxn(**r1**))](**f**(**x**(Pxni(**r1**)))
∗abs(**mn**$_f$**1**(Pxni(**r1**))))) (1453)

I1[M](**f**|**r1**;d$_b$x)
≡ lim(S[bn(M∩Pxn(**r1**))](**f**(**x**(Pxni(**r1**)))
∗abs(**mn**$_b$**1**(Pxni(**r1**))))). (1454)

The directional integration of a transformation, **A**(**r**(**x1**,a)), is defined as

I1[M]•(**A**|**x1**;$\mathbf{d_f r}$)
 ≡ lim (S[bn(M∩Prn(**x1**))](**mn**$_\mathbf{f}$**1**(Prni(**x1**))•**T**[**A**(**r**(Prni(**x1**)))])) 1455)

I1[M]•(**A**|**x1**;$\mathbf{d_b r}$)
 ≡ lim (S[bn(M∩Prn(**x1**))](**mn**$_\mathbf{b}$**1**(Prni(**x1**))•**T**[**A**(**r**(Prni(**x1**)))])) (1456)

I1[M]•(**A**|**r1**;$\mathbf{d_f x}$)
 ≡ lim (S[bn(M∩Pxn(**r1**))](**mn**$_\mathbf{f}$**1**(Pxni(**r1**))•**T**[**A**(**r**(Pxni(**r1**)))])) (1457)

I1[M]•(**A**|**r1**;$\mathbf{d_b x}$)
 ≡ lim (S[bn(M∩Pxn(**r1**))](**mn**$_\mathbf{b}$**1**(Pxni(**r1**))•**T**[**A**(**r**(Pxni(**r1**)))])).(1458)

Theorem 19 (integration of directional gradients over traces and tracks)

Given

- **r**(**x1**,a), a non–stalled trace of the particle **x1**;
- **f**(**r**(**x1**,a)) a generalized function over the trace;
- pt any partition scalar;
- Prn(**x1**), a set of partitions of **r**(**x1**,a) based on pt;

then

I1[]•(**f**(**r**(**x1**,a))∗**D1**[**r**(**x1**,a),**r**(**x1**,a+da)]
 ∗(uax(**r**(**x**,a)−**r**(**x**,a1))|**x1**;$\mathbf{d_f r}$)|**x1**;$\mathbf{d_f r}$)
 = **f**(**r**(**x1**,a1)) (1459)

I1[]•(**f**(**r**(**x1**,a))∗**D1**[**r**(**x1**,a−da),**r**(**x1**,a)]
 ∗(uax(**r**(**x**,a)−**r**(**x**,a1))|**x1**;$\mathbf{d_b r}$)|**x1**;$\mathbf{d_b r}$)
 = **0** (1460)

I1[]•(**f**(**r**(**x1**,a))∗**D1**[**r**(**x1**,a),**r**(**x1**,a+da)]
 ∗(vax(**r**(**x**,a)−**r**(**x**,a1))|**x1**;$\mathbf{d_f r}$)|**x1**;$\mathbf{d_f r}$)
 = **0** (1461)

I1[]•(**f**(**r**(**x1**,a))∗**D1**[**r**(**x1**,a−da),**r**(**x1**,a)]
 ∗(vax(**r**(**x**,a)−**r**(**x**,a1))|**x1**;$\mathbf{d_b r}$)|**x1**;$\mathbf{d_b r}$)
 = **f**(**r**(**x1**,a1)). (1462)

Dually given

- **x**(**r1**,a), a non–stalled trace of the particle **r1**;
- **f**(**x**(**r1**,a)) a generalized function over the trace;
- pt any partition scalar;
- Pxn(**r1**), a set of partitions of **x**(**r1**,a) based on pt;

Theoretical Physics: Integration

then

I1[]•(f(x(r1,a))∗D1[x(r1,a),x(r1,a+da)]
∗(uar(x(r,a)−x(r,a1))|r1;d_fx)|r1;d_fx)
= f(x(r1,a1)) (1463)

I1[]•(f(x(r1,a))∗D1[x(r1,a−da),x(r1,a)]
∗(uar(x(r,a)−x(r,a1))|r1;d_bx)|r1;d_bx)
= 0 (1464)

I1[]•(f(x(r1,a))∗D1[x(r1,a),x(r1,a+da)]
∗(var(x(r,a)−x(r,a1))|r1;d_fx)|r1;d_fx)
= 0 (1465)

I1[]•(f(x(r1,a))∗D1[x(r1,a−da),x(r1,a)]
∗(var(x(r,a)−x(r,a1))|r1;d_bx)|r1;d_bx)
= f(x(r1,a1)). (1466)

Proof:
I1[]•(f(r(x1,a))∗D1[r(x1,a),r(x1,a+da)]∗(uax(r(x,a)−r(x,a1))
|x1;d_fr)|x1;d_fr)
= lim S[bn(Prn(x1))](f(r(Prni(x1)))
∗uax(r(Prni(x1)+da)−r(x,a1))−uax(r(Prni(x1))−r(x,a1))
∗qd(r(Prni(x1)+da)−r(Prni(x1)))
•(r(x1,(i+1)∗pt/n) − r(x1,(i∗pt/n)))).

Now da ≤ pt/n and r(x1,i∗pt/n) = r(Prni(x1)).
Thus,
qd(r(Prni(x1)+da)−r(Prni(x1)))•(r(x1,(i+1)∗pt/n) − r(x1,i∗pt/n))
=qd(r(x1,(i+1)∗pt/n) − r(x1,i∗pt/n))•
(r(x1,(i+1)∗pt/n) − r(x1,i∗pt/n))
= 1.

Consequently,
I1[]•(f(r(x1,a))∗D1[r(x1,a),r(x1,a+da)]∗(uax(r(x,a)−r(x,a1))
|x1;d_fr)|x1;d_fr)
= lim f(r(Porn(x1)))∗uax(r(Prn(j+1)(x1))−r(x,a1))
where Prnj(x1) contains r(x,a1)
= f(r(x1,a1)).

This proves the first proposition.
I1[]•(f(r(x1,a))∗D1[r(x1,a−da),r(x1,a)]∗(uax(r(x,a)−r(x,a1))
|x1;d_br)|x1;d_br)
= 0
I1]•(f(r(x1,a))∗D1[r(x1,a),r(x1,a+da)]∗(vax(r(x,a)−r(x,a1))|x1;d_fr)
|x1;d_fr)
= 0
since the gradients equal 0 everywhere along the trace.
This proves the second and third propositions.

I1]•(**f**(**r**(**x1**,a))∗**D1**[**r**(**x1**,a−da),**r**(**x1**,a)]∗(vax(**r**(**x**,a)−**r**(**x**,a1))
|**x1**;$\mathbf{d}_b\mathbf{r}$)|**x1**;$\mathbf{d}_b\mathbf{r}$)
= lim S[bn(Prn(**x1**))](**f**(**r**(Prni(**x1**)))
∗(vax(**r**(Prni(**x1**))−**r**(**x**,a1))−vax(**r**(Prn(i−1)(**x1**))−**r**(**x**,a1)))
∗**qd**(**r**(Prni(**x1**))−**r**(Prni(**x1**)−da))
•(**r**(**x1**,i∗pt/n) − **r**(**x1**,i−1)))
= **f**(**r**(**x1**,a1)).

This proves the fourth proposition.
The duals are proven similarly. qed

Theorem 20 (integration of directional gradients over traces and tracks)

Given
- **r**(**x1**,a), a non−stalled trace of the particle **x1**;
- **f**(**r**(**x1**,a)) a generalized function over the trace;
- pt any partition scalar;
- Prn(**x1**), a set of partitions of **r**(**x1**,a) based on pt;

for
- **f** continuous at **r**(**x1**,a1);
- I[](**f**(**r**(**x1**,a));da) bounded;

then

I1[]•(uax(**r**(**x1**,a)−**r**(**x1**,a1))
∗**D1**[**r**(**x1**,a),**r**(**x1**,a+da)]∗(**f**(**r**(**x**,a))|**x1**;$\mathbf{d}_f\mathbf{r}$)|**x1**;$\mathbf{d}_f\mathbf{r}$)
= −**f**(**r**(**x1**,a1)) (1467)

I1[]•(uax(**r**(**x1**,a)−**r**(**x1**,a1))
∗**D1**[**r**(**x1**,a),**r**(**x1**,a+da)]∗(**f**(**r**(**x**,a))|**x1**;$\mathbf{d}_b\mathbf{r}$)|**x1**;$\mathbf{d}_b\mathbf{r}$)
= −**f**(**r**(**x1**,a1)) (1468)

I1[]•(vax(**r**(**x1**,a)−**r**(**x1**,a1))
∗**D1**[**r**(**x1**,a),**r**(**x1**,a+da)]∗(**f**(**r**(**x**,a))|**x1**;$\mathbf{d}_f\mathbf{r}$)|**x1**;$\mathbf{d}_f\mathbf{r}$)
= −**f**(**r**(**x1**,a1)) (1469)

I1[]•(vax(**r**(**x1**,a)−**r**(**x1**,a1))
∗**D1**[**r**(**x1**,a),**r**(**x1**,a+da)]∗(**f**(**r**(**x**,a))|**x1**;$\mathbf{d}_b\mathbf{r}$)|**x1**;$\mathbf{d}_b\mathbf{r}$)
= −**f**(**r**(**x1**,a1)). (1470)

Dually given
- **x**(**r1**,a), a non−stalled track of particles passing **r1**;
- **f**(**x**(**r1**,a)) a generalized function over the track;
- pt any partition scalar;
- Pxn(**r1**), a set of partitions of **x**(**r1**,a) based on pt;

for

f continuous at **x**(**r1**,a1);

I[](**f**(**x**(**r1**,a));da) bounded;

then

I1[]•(uar(**x**(**r1**,a)−**x**(**r1**,a1))
∗**D1**[**x**(**r1**,a),**x**(**r1**,a+da)]∗(**f**(**x**(**x**,a))|**r1**;**d**$_f$**x**)|**r1**;**d**$_f$**x**)
= −**f**(**x**(**r1**,a1)) (1471)

I1[]•(uar(**x**(**r1**,a)−**x**(**r1**,a1))
∗**D1**[**x**(**r1**,a),**x**(**r1**,a+da)]∗(**f**(**x**(**x**,a))|**r1**;**d**$_b$**x**)|**r1**;**d**$_b$**x**)
= −**f**(**x**(**r1**,a1)) (1472)

I1[]•(var(**x**(**r1**,a)−**x**(**r1**,a1))
∗**D1**[**x**(**r1**,a),**x**(**r1**,a+da)]∗(**f**(**x**(**x**,a))|**r1**;**d**$_f$**x**)|**r1**;**d**$_f$**x**)
= −**f**(**x**(**r1**,a1)) (1473)

I1[]•(var(**x**(**r1**,a)−**x**(**r1**,a1))
∗**D1**[**x**(**r1**,a),**x**(**r1**,a+da)]∗(**f**(**x**(**x**,a))|**r1**;**d**$_b$**x**)|**r1**;**d**$_b$**x**)
= −**f**(**x**(**r1**,a1)). (1474)

Proof:

I1[]•(uax(**r**(**x1**,a)−**r**(**x1**,a1))
∗**D1**[**r**(**x1**,a),**r**(**x1**,a+da)]∗(**f**(**r**(**x**,a))|**x1**;**d**$_f$**r**)|**x1**;**d**$_f$**r**)
=lim S[bn({**r**(**x1**,a)}∩Prn)](uax(**r**(**x1**,a(Prni))−**r**(**x1**,a1))
∗(**r**(**x1**,(i+1)∗pt/n) − **r**(**x1**,(i∗pt/n)))
•**T**[**D1**[**r**(**x1**,a(Prni)),**r**(**x1**,a(Prni)+da)]∗(**f**(**r**(**x**,a))|**x1**;**d**$_f$**r**)])
= lim S[bn(R+∩Prn)](**r**(**x1**,(i+1)∗pt/n) − **r**(**x1**,(i∗pt/n)))
•**qd**(**r**(**x1**,a(Prni)+da)−**r**(**x1**,a(Prni)))
∗(**f**(**r**(**x1**,a(Prni)+da))−**f**(**r**(**x1**,a(Prni))))

where R+ = [**r**(**x1**,a)|a>a1}
= lim **f**(**r**(**x1**,a(Prn(j+2)))−**f**(**r**(**x1**,a(Prn(j+1)))
+ **f**(**r**(**x1**,a(Prn(j+3))))−**f**(**r**(**x1**,a(Prn(j+2))))
+ ...

where Prnj contains **r**(**x1**,a1)
= lim −**f**(**r**(**x1**,a(Prn(j+1))))

since **f**(**r**(**x1**,a))⟶ **0** as a ⟶∞.
= −**f**(**r**(**x1**,a1))

since a(Prn(j+1)) ⟶ a1+da
and **f**(**r**(**x1**,a)) is continuous at **r**(**x1**,a1).
This proves the first proposition.

I1[]•(uax(**r**(**x1**,a)−**r**(**x1**,a1))
∗**D1**[**r**(**x1**,a),**r**(**x1**,a+da)]∗(**f**(**r**(**x**,a))|**x1**;**d**$_b$**r**)|**x1**;**d**$_b$**r**)
=lim S[bn({**r**(**x1**,a)}∩Prn)](uax(**r**(**x1**,a(Prni))−**r**(**x1**,a1))
∗(**r**(**x1**,(i+1)∗pt/n) − **r**(**x1**,(i∗pt/n)))
•**T**[**D1**[**r**(**x1**,a(Prni)),**r**(**x1**,a(Prni)+da)]∗(**f**(**r**(**x**,a))|**x1**;**d**$_b$**r**)])

= lim S[bn(R+∩Prn)](**r**(**x1**,i*pt/n) − **r**(**x1**,(i−1)*pt/n))
•**qd**(**r**(**x1**,a(Prni))−**r**(**x1**,a(Prni)−da))
*(**f**(**r**(**x1**,a(Prni)))−**f**(**r**(**x1**,a(Prni)−da))))
= lim **f**(**r**(**x1**,a(Prn(j+1))))−**f**(**r**(**x1**,a(Prnj)))
+ **f**(**r**(**x1**,a(Prn(j+2))))−**f**(**r**(**x1**,a(Prn(j+1))))
+ ...
where Prnj contains **r**(**x1**,a1)
= −**f**(**r**(**x1**,a1)).
This proves the second proposition.
I1[]•(vax(**r**(**x1**,a)−**r**(**x1**,a1))
***D1**[**r**(**x1**,a),**r**(**x1**,a+da)]*(**f**(**r**(**x**,a))|**x1**;$\mathbf{d_f r}$)|**x1**;$\mathbf{d_f r}$)
= lim S[bn(R+∩Prn)](**r**(**x1**,(i+1)*pt/n) − **r**(**x1**,(i*pt/n)))
•**qd**(**r**(**x1**,a(Prni)+da)−**r**(**x1**,a(Prni)))
*(**f**(**r**(**x1**,a(Prni)+da)))−**f**(**r**(**x1**,a(Prni)))
= lim **f**(**r**(**x1**,a(Prn(j+1))))−**f**(**r**(**x1**,a(Prnj)))
+ **f**(**r**(**x1**,a(Prn(j+21))))−**f**(**r**(**x1**,a(Prn(j+1))))
+ ...
= lim −**f**(**r**(**x1**,a(Prnj)))
= −**f**(**r**(**x1**,a1)).
This proves the third proposition.
I1[]•(vax(**r**(**x1**,a)−**r**(**x1**,a1))
***D1**[**r**(**x1**,a),**r**(**x1**,a+da)]*(**f**(**r**(**x**,a))|**x1**;$\mathbf{d_b r}$)|**x1**;$\mathbf{d_b r}$)
= lim S[bn(R+∩Prn)](**r**(**x1**,i*pt/n) − **r**(**x1**,(i−1)*pt/n)))
•**qd**(**r**(**x1**,a(Prni))−**r**(**x1**,a(Prni)−da))
*(**f**(**r**(**x1**,a(Prni)))−**f**(**r**(**x1**,a(Prni)−da)))
= lim **f**(**r**(**x1**,a(Prnj)))−**f**(**r**(**x1**,a(Prn(j−1))))
+ **f**(**r**(**x1**,a(Prn(j+1))))−**f**(**r**(**x1**,a(Prnj)))
+ ...
= lim −**f**(**r**(**x1**,a(Prn(j−1))))
= −**f**(**r**(**x1**,a1)).
This proves the fourth proposition.
The duals are proven similarly.

qed

Theoretical Physics: Integration

Theorem 21 (integration of directional gradients over traces and tracks)

Given

r(**x1**,a), a non–stalled trace of the particle **x1**;
f(**r**(**x1**,a)) a generalized function over the trace;
pt any partition scalar;
Prn(**x1**), a set of partitions of **r**(**x1**,a) based on pt;

then

$$\mathbf{I1}[]\bullet\mathbf{D1}[\mathbf{r}(\mathbf{x1},a),\mathbf{r}(\mathbf{x1},a+da)]*(\mathbf{f}(\mathbf{r}(\mathbf{x1},a))*(\mathrm{uax}(\mathbf{r}(\mathbf{x},a)-\mathbf{r}(\mathbf{x},a1))|\mathbf{x1};\mathbf{d_f r})|\mathbf{x1};\mathbf{d_f r}) = \mathbf{0} \tag{1475}$$

$$\mathbf{I1}[]\bullet\mathbf{D1}[\mathbf{r}(\mathbf{x1},a-da),\mathbf{r}(\mathbf{x1},a)]*(\mathbf{f}(\mathbf{r}(\mathbf{x1},a))*(\mathrm{uax}(\mathbf{r}(\mathbf{x},a)-\mathbf{r}(\mathbf{x},a1))|\mathbf{x1};\mathbf{d_b r})|\mathbf{x1};\mathbf{d_b r}) = \mathbf{0} \tag{1476}$$

$$\mathbf{I1}[]\bullet\mathbf{D1}[\mathbf{r}(\mathbf{x1},a),\mathbf{r}(\mathbf{x1},a+da)]*(\mathbf{f}(\mathbf{r}(\mathbf{x1},a))*(\mathrm{vax}(\mathbf{r}(\mathbf{x},a)-\mathbf{r}(\mathbf{x},a1))|\mathbf{x1};\mathbf{d_f r})|\mathbf{x1};\mathbf{d_f r}) = \mathbf{0} \tag{1477}$$

$$\mathbf{I1}[]\bullet\mathbf{D1}[\mathbf{r}(\mathbf{x1},a-da),\mathbf{r}(\mathbf{x1},a)]*(\mathbf{f}(\mathbf{r}(\mathbf{x1},a))*(\mathrm{vax}(\mathbf{r}(\mathbf{x},a)-\mathbf{r}(\mathbf{x},a1))|\mathbf{x1};\mathbf{d_b r})|\mathbf{x1};\mathbf{d_b r}) = \mathbf{0}. \tag{1478}$$

Dually

Given

x(**r1**,a), a non–stalled trace of particles passing **r1**;
f(**x**(**r1**,a)) a generalized function over the track;
pt any partition scalar
Pxn(**r1**), a set of partitions of **x**(**r1**,a) based on pt

then

$$\mathbf{I1}[]\bullet\mathbf{D1}[\mathbf{x}(\mathbf{r1},a),\mathbf{x}(\mathbf{r1},a+da)](\mathbf{f}(\mathbf{x}(\mathbf{r1},a))*(\mathrm{uar}(\mathbf{x}(\mathbf{r},a)-\mathbf{x}(\mathbf{r},a1))|\mathbf{r1};\mathbf{d_f x})|\mathbf{r1};\mathbf{d_f x}) = \mathbf{0} \tag{1479}$$

$$\mathbf{I1}[]\bullet\mathbf{D1}[\mathbf{x}(\mathbf{r1},a-da),\mathbf{x}(\mathbf{r1},a)](\mathbf{f}(\mathbf{x}(\mathbf{r1},a))*(\mathrm{uar}(\mathbf{x}(\mathbf{r},a)-\mathbf{x}(\mathbf{r},a1))|\mathbf{r1};\mathbf{d_b x})|\mathbf{r1};\mathbf{d_b x}) = \mathbf{0} \tag{1480}$$

$$\mathbf{I1}[]\bullet\mathbf{D1}[\mathbf{x}(\mathbf{r1},a),\mathbf{x}(\mathbf{r1},a+da)](\mathbf{f}(\mathbf{x}(\mathbf{r1},a))*(\mathrm{var}(\mathbf{x}(\mathbf{r},a)-\mathbf{x}(\mathbf{r},a1))|\mathbf{r1};\mathbf{d_f x})|\mathbf{r1};\mathbf{d_f x}) = \mathbf{0} \tag{1481}$$

$$\mathbf{I1}[]\bullet\mathbf{D1}[\mathbf{x}(\mathbf{r1},a-da),\mathbf{x}(\mathbf{r1},a)](\mathbf{f}(\mathbf{x}(\mathbf{r1},a))*(\mathrm{var}(\mathbf{x}(\mathbf{r},a)-\mathbf{x}(\mathbf{r},a1))|\mathbf{r1};\mathbf{d_b x})|\mathbf{r1};\mathbf{d_b x}) = \mathbf{0}. \tag{1482}$$

Proof:

I1[]•(D1[r(x1,a),r(x1,a+da)]*(f(r(x1,a))*uax(r(x,a)−r(x,a1))
|x1;d_fr)|x1;d_fr)
= lim S[bn(Prn(x1))]
((f(r(Prni(x1)+da))*uax(r(Prni(x1)+da)−r(x,a1))
−f(r(Prni(x1))*uax(r(Prni(x1)))−r(x,a1)))
*qd(r(Prni(x1)+da)−r(Prni(x1)))
•(r(x1,(i+1)*pt/n) − r(x1,(i*pt/n))))
= lim f(r(x1,a(Prn(j+1))))
+ f(r(x1,a(Prn(j+2))))−f(r(x1,a(Prn(j+1))))
+ ...
= 0.

This proves the first proposition.

I1[]•(D1[r(x1,a−da),r(x1,a)]*(f(r(x1,a))*uax(r(x,a)−r(x,a1))
|x1;d_br)|x1;d_br)
= lim S[bn(Prn(x1))]
((f(r(Prni(x1)))*uax(r(Prni(x1))−r(x,a1))
−f(r(Prni(x1)−da)*uax(r(Prni(x1)−da))−r(x,a1)))
*qd(r(Prni(x1))−r(Prni(x1)−da))
•(r(x1,i*pt/n) − r(x1,(i−1)*pt/n))))
= lim f(r(x1,a(Prn(j+1))))
+ f(r(x1,a(Prn(j+2))))−f(r(x1,a(Prn(j+1))))
+ ...
= 0.

This proves the second proposition.

I1[]•(D1[r(x1,a),r(x1,a+da)]*(f(r(x1,a))*vax(r(x,a)−r(x,a1))
|x1;d_fr)|x1;d_fr)
= lim S[bn(Prn(x1))]
((f(r(Prn(i+1)(x1)))*vax(r(Prni(x1)+da)−r(x,a1))
−f(r(Prni(x1))*vax(r(Prni(x1)))−r(x,a1)))
*qd(r(Prni(x1)+da)−r(Prni(x1)))
•(r(x1,(i+1)*pt/n) − r(x1,(i*pt/n))))
= lim f(r(x1,a(Prnj)))
+ f(r(x1,a(Prn(j+1))))−f(r(x1,a(Prnj)))
+ ...
= 0.

This proves the third proposition.

I1[]•(D1[r(x1,a−da),r(x1,a)]*(f(r(x1,a))*vax(r(x,a)−r(x,a1))
|x1;d_br)|x1;d_br)
= lim S[bn(Prn(x1))]
((f(r(Prni(x1)))*vax(r(Prni(x1))−r(x,a1))
−f(r(Prn(i−1)(x1))*vax(r(Prn(i−1)(x1)))−r(x,a1)))
*qd(r(Prni(x1))−r(Prn(i−1)(x1)))
•(r(x1,i*pt/n) − r(x1,(i−1)*pt/n))))

= lim **f**(**r**(**x1**,a(Prnj)))
+ **f**(**r**(**x1**,a(Prn(j+1))))−**f**(**r**(**x1**,a(Prnj)))
+ ...
= **0**.

This proves the fourth proposition.
The duals are proven similarly.

qed

Theorem 22 (integration of directional gradients over traces and tracks)

Given
r(**x1**,a), a non–stalled trace of the particle **x1**;
f(**r**(**x1**,a)) a continuous function over the trace;
D1[**r**(**x1**,a),**r**(**x1**,a+da)]∗(**f**(**r**(**x1**,a))|**x1**;**d**f**r**)
a basic directional gradient;
pt any partition scalar;
Prn(**x1**), a set of partitions of **r**(**x1**,a) based on pt;

for
I[](**f**(**r**(**x1**,a));da) bounded

then

I1[]•(**D1**[**r**(**x1**,a),**r**(**x1**,a+da)]∗(**f**(**r**(**x1**,a))|**x1**;**d**f**r**)|**x1**;**d**f**r**)
= **0** (1483)

for only
I[a0,](**f**(**r**(**x1**,t));da) bounded.

then

I1[]•(vax(**r**(**x1**,a)−**r**(**x1**,a0))
∗**D1**[**r**(**x1**,a),**r**(**x1**,a+da)]∗(**f**(**r**(**x1**,a))|**x1**;**d**f**r**)|**x1**;**d**f**r**)
= −**f**(**r**(**x1**,a0)) 1484)
= **I1**[**r**(**x1**,a0),]•(**D1**[**r**(**x1**,a),**r**(**x1**,a+da)]∗(**f**(**r**(**x1**,a))
|**x1**;**d**f**r**)|**x1**;**d**f**r**)

for only
I[,a2](**f**(**r**(**x1**,t));da) bounded

then

I1[]•(1−uax(**r**(**x1**,a)−**r**(**x1**,a2))
∗**D1**[**r**(**x1**,a),**r**(**x1**,a+da)]∗(**f**(**r**(**x1**,a))|**x1**;**d**f**r**)|**x1**;**d**f**r**)
= **f**(**r**(**x1**,a2)) (1485)
= **I1**[,**r**(**x1**,a2)]•(**D1**[**r**(**x1**,a),**r**(**x1**,a+da)]∗(**f**(**r**(**x**,a))
|**x1**;**d**f**r**)|**x1**;**d**f**r**)

for only

r(x1,t)) bounded only in the interval [r(x1,a0),r(x1,a2)]

then

I1[]•((vax(r(x1,a)–r(x1,a0))–uax(r(x1,a)–r(x1,a2)))
D1[r(x1,a),r(x1,a+da)](f(r(x,a))|x1;d$_f$r)|x1;d$_f$r)
= f(r(x1,a2))–f(r(x1,a0)) (1486)
= I1[r(x1,a0),r(x1,a2)]•(D1[r(x1,a),r(x1,a+da)]*(f(r(x1,a))
|x1;d$_f$r)|x1;d$_f$r).

Dually given

x(r1,a), a non–stalled track of particles at location r1;
f(x(r1,a)) a continuous function over the track;
D1[x(r1,a),x(r1,a+da)]*(f(x(r1,a))|r1;d$_f$x)
a basic directional gradient;
pt any partition scalar;
Pxn(r1), a set of partitions of x(r1,a) based on pt;

for

I[](f(x(r1,a));da) bounded

then

I1[]•(D1[x(r1,a),x(r1,a+da)]*(f(x(r1,a))|r1;d$_f$x)|r1;d$_f$x)
= 0 (1487)

for only

I[a0,](f(x(r1,t));da) bounded.

then

I1[]•((vax(x(r1,a)–x(r1,a0))
D1[x(r1,a),x(r1,a+da)](f(x(r1,a))|r1;d$_f$x)|r1;d$_f$x)
= –f(x(r1,a0)) (1488)
= I1[x(r1,a0),]•(D1[x(r1,a),x(r1,a+da)]*(f(x(r1,a))
|r1;d$_f$x)|r1;d$_f$x)

for only

I[,a2](f(x(r1,t));da) bounded

then

I1[]•(1–uax(x(r1,a)–x(r1,a2))
D1[x(r1,a),x(r1,a+da)](f(x(r1,a))|r1;d$_f$x)|r1;d$_f$x)
= f(x(r1,a2)) (1489)
= I1[,x(r1,a2)]•(D1[x(r1,a),x(r1,a+da)]*(f(x(x,a))
|r1;d$_f$x)|r1;d$_f$x) (1490)

Theoretical Physics: Integration

for only
  f(**x**(**r1**,t)) bounded only in the interval [**x**(**r1**,a0),**x**(**r1**,a2)]
then

I1[]•((vax(**x**(**r1**,a)–**x**(**r1**,a0))–uax(**x**(**r1**,a)–**x**(**r1**,a2)))
  ∗**D1**[**x**(**r1**,a),**x**(**r1**,a+da)]∗(**f**(**x**(**x**,a))|**r1**;**d**$_f$**x**)|**r1**;**d**$_f$**x**)
 = **f**(**x**(**r1**,a2))–**f**(**x**(**r1**,a0)) (1491)
 = **I1**[**x**(**r1**,a0),**x**(**r1**,a2)]•(**D1**[**x**(**r1**,a),**x**(**r1**,a+da)]∗(**f**(**x**(**r1**,a))
  |**r1**;**d**$_f$**x**)|**r1**;**d**$_f$**x**). (1492)

Proof:
  I1[]•(**D1**[**r**(**x1**,a),**r**(**x1**,a+da)]∗(**f**(**r**(**x1**,a))|**x1**;**d**$_f$**r**)|**x1**;**d**$_f$**r**)
   = lim S[bn(Prn(**x1**))](**f**(**r**(Prni(**x1**),a+da))–**f**(**r**(Prni(**x1**),a)))
    ∗**qd**(**r**(Prni(x1),a+da)–**r**(Prni(**x1**),a))
     •(**r**(**x1**,(i+1)∗pt/n) – **r**(**x1**,(i∗pt/n)))
   = lim ... +
    + **f**(**r**(Prni(**x1**),a))–**f**(**r**(Prni(**x1**),a–da))
    + **f**(**r**(Prni(**x1**),a+da))–**f**(**r**(Prni(**x1**),a))
    + ...
   = **0**.
  This proves the first proposition.
  I1[]•(vax(**r**(**x1**,a)–**r**(**x1**,a0))
    ∗**D1**[**r**(**x1**,a),**r**(**x1**,a+da)]∗(**f**(**r**(**x1**,a))|**x1**;**d**$_f$**r**)|**x1**;**d**$_f$**r**)
   = –**f**(**r**(**x1**,a0))
  by the fourth proposition of theorem 20
   = **I1**[**r**(**x1**,a0),]•(**D1**[**r**(**x1**,a),**r**(**x1**,a+da)]∗(**f**(**r**(**x1**,a))
    |**x1**;**d**$_f$**r**)|**x1**;**d**$_f$**r**).
  This proves the second proposition.
  I1[]•(1–uax(**r**(**x1**,a)–**r**(**x1**,a2))
    ∗**D1**[**r**(**x1**,a),**r**(**x1**,a+da)]∗(**f**(**r**(**x1**,a))|**x1**;**d**$_f$**r**);**d**$_f$**r**)
   = lim S[bn(Prn(**x1**))]
    (**f**(**r**(Prni(**x1**),+da))–**f**(**r**(Prni(**x1**),a)))
     ∗(1–uax(**r**(Prni(**x1**),a)–**r**(**x**,a2)))
    ∗**qd**(**r**(Prni(**x1**),a+da)–**r**(Prni(**x1**),a))
     •(**r**(**x1**,(i+1)∗pt/n) – **r**(**x1**,i∗pt/n))
   = lim ... +
    + **f**(**r**(Prni(**x1**),a))–**f**(**r**(Prni(**x1**),a–da))
    + **f**(**r**(Prni(**x1**),a+da))–**f**(**r**(Prni(**x1**),a))
    + ...
    + **f**(**r**(Prni(**x1**),a2))–**f**(**r**(Prni(**x1**),a2–da))
   = **f**(**r**(**x1**,a2))
   = **I1**[,**r**(**x1**,a2)]•(**D1**[**r**(**x1**,a),**r**(**x1**,a+da)]∗(**f**(**r**(**x**,a))
    |**x1**;**d**$_f$**r**)|**x1**;**d**$_f$**r**).
  This proves the third proposition.

I1[]•((vax(**r**(**x1**,a)−**r**(**x1**,a0))−uax(**r**(**x1**,a)−**r**(**x1**,a2)))
∗**D1**[**r**(**x1**,a),**r**(**x1**,a+da)]∗(**f**(**r**(**x**,a))|**x1**;**d**$_f$**r**)|**x1**;**d**$_f$**r**)
= lim S[bn(Prn(**x1**))]
(**f**(**r**(Prni(**x1**),a+da))−**f**(**r**(Prni(**x1**),a)))
∗(vax(**r**(Prni(**x1**),a)−**r**(**x**,a0))−uax(**r**(Prni(**x1**),a)−**r**(**x**,a2)))
∗**qd**(**r**(Prni(**x1**),a+da)−**r**(Prni(**x1**),a))
•(**r**(**x1**,(i+1)∗pt/n) − **r**(**x1**,i∗pt/n))
= lim **f**(**r**(Prni(**x1**),a0+da)−**f**(**r**(Prni(**x1**),a0)
+ **f**(**r**(Prni(**x1**),a0+2∗da)−**f**(**r**(Prni(**x1**),a+da)
+ ...
+ **f**(**r**(Prni(**x1**),a2−da)−**f**(**r**(Prni(**x1**),a2−2∗da)
+ **f**(**r**(Prni(**x1**),a2)−**f**(**r**(Prni(**x1**),a2−da)
= **f**(**r**(**x1**,a2))−**f**(**r**(**x1**,a0))
= **I1**[**r**(**x1**,a0),**r**(**x1**,a2)]
•(**D1**[**r**(**x1**,a),**r**(**x1**,a+da)]∗(**f**(**r**(**x1**,a))|**x1**;**d**$_f$**r**)|**x1**;**d**$_f$**r**).

This proves the fourth proposition.
The duals are proven similarly.

qed

Analogous results may be made for the invergences of directional gradients in the backward sense.

Results for directional invergences of directional gradients extend to other combinations.

Corollary

Given the conditions of the theorems,

I1[]•(**D1**[**r**(**x1**,a),**r**(**x1**,a+da)]∧(**f**|**x1**;**d**$_f$**r**)|**x1**;**d**$_f$**r**)
= 0 (1493)

I**1**[]•(**f**(**r**(**x1**,a))∧**D1**[**r**(**x1**,a),**r**(**x1**,a+da)]
∗(uax(**r**(**x**,a)−**r**(**x**,a1))|**x1**;**d**$_f$**r**)|**x1**;**d**$_f$**r**)
= 0 (1494)

I**1**[]•(**f**(**r**(**x1**,a))∧**D1**[**r**(**x1**,a−da),**r**(**x1**,a)]
∗(vax(**r**(**x**,a)−**r**(**x**,a1))|**x1**;**d**$_b$**r**)|**x1**;**d**$_b$**r**)
= 0 (1495)

I**1**[]•(uax(**r**(**x1**,a)−**r**(**x1**,a1))
∗**D1**[**r**(**x1**,a),**r**(**x1**,a+da)]∧(**f**(**r**(**x**,a))|**x1**;**d**$_f$**r**)|**x1**;**d**$_f$**r**)
= 0 (1496)

Theoretical Physics: Integration

$$\mathbf{I1}[]\cdot(\mathrm{vax}(\mathbf{r}(\mathbf{x1},a)-\mathbf{r}(\mathbf{x1},a1)) * \mathbf{D1}[\mathbf{r}(\mathbf{x1},a),\mathbf{r}(\mathbf{x1},a+da)]\wedge(\mathbf{f}(\mathbf{r}(\mathbf{x},a))|\mathbf{x1};\mathbf{d_b r})|\mathbf{x1};\mathbf{d_b r}) = 0 \tag{1497}$$

$$\mathbf{I1}[]\cdot\mathbf{D1}[\mathbf{r}(\mathbf{x1},a),\mathbf{r}(\mathbf{x1},a+da)]\wedge(\mathbf{f}(\mathbf{r}(\mathbf{x1},a)) * (\mathrm{uax}(\mathbf{r}(\mathbf{x},a)-\mathbf{r}(\mathbf{x},a1)) |\mathbf{x1};\mathbf{d_f r})|\mathbf{x1};\mathbf{d_f r}) = 0 \tag{1498}$$

$$\mathbf{I1}[]\cdot\mathbf{D1}[\mathbf{r}(\mathbf{x1},a-da),\mathbf{r}(\mathbf{x1},a)]\wedge(\mathbf{f}(\mathbf{r}(\mathbf{x1},a)) * (\mathrm{vax}(\mathbf{r}(\mathbf{x},a) * \mathbf{r}(\mathbf{x},a1)) |\mathbf{x1};\mathbf{d_b r})|\mathbf{x1};\mathbf{d_b r}) = 0 \tag{1499}$$

$$\mathbf{I1}[]\wedge(\mathbf{D1}[\mathbf{r}(\mathbf{x1},a),\mathbf{r}(\mathbf{x1},a+da)]\wedge(\mathbf{f}|\mathbf{x1};\mathbf{d_f r})|\mathbf{x1};\mathbf{d_f r}) = \lim \mathrm{S}[\mathrm{bn}(\mathrm{Prn}(\mathbf{x1}))] (\mathbf{f}(\mathbf{r}(\mathrm{Prni}(\mathbf{x1}),a+da))-\mathbf{f}(\mathbf{r}(\mathrm{Prni}(\mathbf{x1}),a)))) \cdot[\mathbf{utr}(\mathbf{x1},a) * \mathbf{utr}(\mathbf{x1},a)-\mathbf{I}] \tag{1500}$$

$$\mathbf{I1}[]\wedge(\mathbf{f}(\mathbf{r}(\mathbf{x1},a))\wedge\mathbf{D1}[\mathbf{r}(\mathbf{x1},a),\mathbf{r}(\mathbf{x1},a+da)] * (\mathrm{uax}(\mathbf{r}(\mathbf{x},a)-\mathbf{r}(\mathbf{x},a1)) |\mathbf{x1};\mathbf{d_f r})|\mathbf{x1};\mathbf{d_f r}) = \mathbf{f}(\mathbf{r}(\mathbf{x1},a1)\cdot[\mathbf{utr}(\mathbf{x1},a1) * \mathbf{utr}(\mathbf{x1},a1)-\mathbf{I}]) \tag{1501}$$

$$\mathbf{I1}[]\wedge(\mathbf{f}(\mathbf{r}(\mathbf{x1},a))\wedge\mathbf{D1}[\mathbf{r}(\mathbf{x1},a),\mathbf{r}(\mathbf{x1},a+da)] * (\mathrm{vax}(\mathbf{r}(\mathbf{x},a)-\mathbf{r}(\mathbf{x},a1)) |\mathbf{x1};\mathbf{d_b r})|\mathbf{x1};\mathbf{d_b r}) = \mathbf{f}(\mathbf{r}(\mathbf{x1},a1)\cdot[\mathbf{utr}(\mathbf{x1},a1) * \mathbf{utr}(\mathbf{x1},a1)-\mathbf{I}]) \tag{1502}$$

$$\mathbf{I1}[]\wedge(\mathrm{uax}(\mathbf{r}(\mathbf{x1},a)-\mathbf{r}(\mathbf{x1},a1)) * \mathbf{D1}[\mathbf{r}(\mathbf{x1},a),\mathbf{r}(\mathbf{x1},a+da)]\wedge(\mathbf{f}(\mathbf{r}(\mathbf{x},a)) |\mathbf{x1};\mathbf{d_f r})|\mathbf{x1};\mathbf{d_f r}) = \lim (\mathbf{f}(\mathbf{r}(\mathrm{Prni}(\mathbf{x1}),a1+2*da))-\mathbf{f}(\mathbf{r}(\mathrm{Prni}(\mathbf{x1}),a1+da))) \cdot[\mathbf{utr}(\mathbf{x1},a1+da) * \mathbf{utr}(\mathbf{x1},a1+da)-\mathbf{I}] + \ldots \tag{1503}$$

$$\mathbf{I1}[]\wedge(\mathrm{vax}(\mathbf{r}(\mathbf{x1},a)-\mathbf{r}(\mathbf{x1},a1)) * \mathbf{D1}[\mathbf{r}(\mathbf{x1},a),\mathbf{r}(\mathbf{x1},a+da)]\wedge(\mathbf{f}(\mathbf{r}(\mathbf{x},a))|\mathbf{x1};\mathbf{d_b r})|\mathbf{x1};\mathbf{d_b r}) = \lim (\mathbf{f}(\mathbf{r}(\mathrm{Prni}(\mathbf{x1}),a1))-\mathbf{f}(\mathbf{r}(\mathrm{Prni}(\mathbf{x1}),a1-da)))) \cdot[\mathbf{utr}(\mathbf{x1},a1) * \mathbf{utr}(\mathbf{x1},a1)-\mathbf{I}] + \ldots \tag{1504}$$

$$\mathbf{I1}[]\wedge\mathbf{D1}[\mathbf{r}(\mathbf{x1},a),\mathbf{r}(\mathbf{x1},a+da)]\wedge(\mathbf{f}(\mathbf{r}(\mathbf{x1},a)) * (\mathrm{uax}(\mathbf{r}(\mathbf{x},a)-\mathbf{r}(\mathbf{x},a1)) |\mathbf{x1};\mathbf{d_f r})|\mathbf{x1};\mathbf{d_f r}) = \lim \mathbf{f}(\mathbf{r}(\mathbf{x1},a1+da))\cdot[\mathbf{utr}(\mathbf{x1},a1) * \mathbf{utr}(\mathbf{x1},a1)-\mathbf{I}] + (\mathbf{f}(\mathbf{r}(\mathrm{Prni}(\mathbf{x1}),a1+2*da))-\mathbf{f}(\mathbf{r}(\mathrm{Prni}(\mathbf{x1}),a1+da))) \cdot[\mathbf{utr}(\mathbf{x1},a1+da) * \mathbf{utr}(\mathbf{x1},a1+da)-\mathbf{I}] + \ldots \tag{1505}$$

I1[]∧D1[r(x1,a−da),r(x1,a)]∧(f(r(x1,a))∗(vax(r(x,a)−r(x,a1))

|x1;d_b r)|x1;d_b r)

= lim f(r(x1,a1)•[utr(x1,a1)∗utr(x1,a1)−I])

+ f(r(Prni(x1),a1+da))−f(r(Prni(x1),a1))

•[utr(x1,a1+da)∗utr(x1,a1+da)−I]

+ ... (1506)

I1[]∗(D1[r(x1,a),r(x1,a+da)]∧(f|x1;d_f r)|x1;d_f r)

= lim S[bn(Prn(x1))]C(f(r(Prni(x1),a+da))−f(r(Prni(x1),a)))

•[utr(x1,a)∗utr(x1,a)] (1507)

I1[]∗(f(r(x1,a))∧D1[r(x1,a),r(x1,a+da)]∗(uax(r(x,a)−r(x,a1))

|x1;d_f r)|x1;d_f r)

= C(f(r(x1,a1)))•[utr(x1,a1)∗utr(x1,a1)] (1508)

I1[]∗(f(r(x1,a))∧D1[r(x1,a),r(x1,a+da)]∗(vax(r(x,a)−r(x,a1))

|x1;d_b r)|x1;d_b r)

= C(f(r(x1,a1)))•[utr(x1,a1)∗utr(x1,a1)] (1509)

I1[]∗(uax(r(x1,a)−r(x1,a1)∧D1[r(x1,a),r(x1,a+da)]∧(f(r(x,a))

|x1;d_f r)|x1;d_f r)

= lim C(f(r(Prni(x1),a1+2∗da))−f(r(Prni(x1),a1+da)))

•[utr(x1,a1+da)∗utr(x1,a1+da)]

+ ... (1510)

I1[]∗(vax(r(x1,a)−r(x1,a1))

∧D1[r(x1,a),r(x1,a+da)]∧(f(r(x,a))|x1;d_b r)|x1;d_b r)

= lim C(f(r(Prni(x1),a1))−f(r(Prni(x1),a1−da)))

•[utr(x1,a1)∗utr(x1,a1)]

+ ... (1511)

I1[]∗D1[r(x1,a),r(x1,a+da)]∧(f(r(x1,a))∗(uax(r(x,a)−r(x,a1))

|x1;d_f r)|x1;d_f r)

= lim C(f(r(x1,a1+da)))•[utr(x1,a1)∗utr(x1,a1)]

+ C(f(r(Prni(x1),a1+2∗da))−f(r(Prni(x1),a1+da)))

•[utr(x1,a1+da)∗utr(x1,a1+da)]

+ ... (1512)

I1[]∗D1[r(x1,a−da),r(x1,a)]∧(f(r(x1,a))∗(vax(r(x,a)−r(x,a1))

|x1;d_b r)|x1;d_b r)

= lim f(r(x1,a1))•[utr(x1,a1)∗utr(x1,a1)−I]

+ C(f(r(Prni(x1),a1+da))−f(r(Prni(x1),a1)))

•[utr(x1,a1+da)∗utr(x1,a1+da)]

+ ... (1513)

Theoretical Physics: Integration

I1[]*(D1[r(x1,a),r(x1,a+da)]•(f|x1;d_fr)|x1;d_fr)
= lim S[bn(Prn(x1))]
(f(r(Prni(x1),a+da))–f(r(Prni(x1),a)))
•[utr(x1,a)*utr(x1,a)] (1514)

I1[]*(f(r(x1,a))•D1[r(x1,a),r(x1,a+da)]*(uax(r(x,a)–r(x,a1))
|x1;d_fr)|x1;d_fr)
= f(r(x1,a1))•[utr(x1,a1)*utr(x1,a1)] (1515)

I1[]*(f(r(x1,a))•D1[r(x1,a),r(x1,a+da)]*(vax(r(x,a)–r(x,a1))
|x1;d_br)|x1;d_br)
= f(r(x1,a1))•[utr(x1,a1)*utr(x1,a1)] (1516)

I1[]*(uax(r(x1,a)–r(x1,a1))•D1[r(x1,a),r(x1,a+da)]∧(f(r(x,a))
|x1;d_fr)|x1;d_fr)
= lim (f(r(Prni(x1),a1+2*da))–f(r(Prni(x1),a1+da)))
•[utr(x1,a1+da)*utr(x1,a1+da)]
+ ... (1517)

I1[]*(vax(r(x1,a)–r(x1,a1)).•D1[r(x1,a),r(x1,a+da)]∧(f(r(x,a))
|x1;d_br)|x1;d_br)
= lim (f(r(Prni(x1),a1))–f(r(Prni(x1),a1–da)))
•[utr(x1,a1)*utr(x1,a1)])
+ ... (1518)

I1[]*D1[r(x1,a),r(x1,a+da)]•(f(r(x1,a))*(uax(r(x,a)–r(x,a1))
|x1;d_fr)|x1;d_fr)
= lim f(r(x1,a1+da))•[utr(x1,a1)*utr(x1,a1)])
+ (f(r(Prni(x1),a1+2*da))–f(r(Prni(x1),a1+da)))
•[utr(x1,a1+da)*utr(x1,a1+da)]
+ ... (1519)

I1[]*D1[r(x1,a–da),r(x1,a)]•(f(r(x1,a))*(vax(r(x,a)–r(,a1))
|x1;d_br)|x1;d_br)
= lim f(r(x1,a1)•[utr(x1,a1)*utr(x1,a1)])
+ (f(r(Prni(x1),a1+da))–f(r(Prni(x1),a1)))
•[utr(x1,a1+da)*utr(x1,a1+da)]
+ ... (1520)

For scalar functions

$$\mathbf{I1}[]\cdot(\mathbf{D1}[\mathbf{r}(\mathbf{x1},a),\mathbf{r}(\mathbf{x1},a+da)]*(f|\mathbf{x1};\mathbf{d}_f\mathbf{r})|\mathbf{x1};\mathbf{d}_f\mathbf{r}) = 0 \tag{1521}$$

$$\mathbf{I1}[]\cdot(f(\mathbf{r}(\mathbf{x1},a))*\mathbf{D1}[\mathbf{r}(\mathbf{x1},a),\mathbf{r}(\mathbf{x1},a+da)]*(uax(\mathbf{r}(\mathbf{x},a)-\mathbf{r}(\mathbf{x},a1))|\mathbf{x1};\mathbf{d}_f\mathbf{r})|\mathbf{x1};\mathbf{d}_f\mathbf{r}) = f(\mathbf{r}(\mathbf{x1},a1)) \tag{1522}$$

$$\mathbf{I1}[]\cdot(f(\mathbf{r}(\mathbf{x1},a))*\mathbf{D1}[\mathbf{r}(\mathbf{x1},a),\mathbf{r}(\mathbf{x1},a+da)]*(vax(\mathbf{r}(\mathbf{x},a)-\mathbf{r}(\mathbf{x},a1))|\mathbf{x1};\mathbf{d}_b\mathbf{r})|\mathbf{x1};\mathbf{d}_b\mathbf{r}) = f(\mathbf{r}(\mathbf{x1},a1)) \tag{1523}$$

$$\mathbf{I1}[]\cdot(uax(\mathbf{r}(\mathbf{x1},a)-\mathbf{r}(\mathbf{x1},a1))*\mathbf{D1}[\mathbf{r}(\mathbf{x1},a),\mathbf{r}(\mathbf{x1},a+da)]*(f(\mathbf{r}(\mathbf{x},a))|\mathbf{x1};\mathbf{d}_f\mathbf{r})|\mathbf{x1};\mathbf{d}_f\mathbf{r}) = -f(\mathbf{r}(\mathbf{x1},a1)) \tag{1524}$$

$$\mathbf{I1}[]\cdot(vax(\mathbf{r}(\mathbf{x1},a)-\mathbf{r}(\mathbf{x1},a1))*\mathbf{D1}[\mathbf{r}(\mathbf{x1},a),\mathbf{r}(\mathbf{x1},a+da)]*(f(\mathbf{r}(\mathbf{x},a))|\mathbf{x1};\mathbf{d}_b\mathbf{r})|\mathbf{x1};\mathbf{d}_b\mathbf{r}) = -f(\mathbf{r}(\mathbf{x1},a1)) \tag{1525}$$

$$\mathbf{I1}[]\cdot\mathbf{D1}[\mathbf{r}(\mathbf{x1},a),\mathbf{r}(\mathbf{x1},a+da)]*(f(\mathbf{r}(\mathbf{x1},a))*(uax(\mathbf{r}(\mathbf{x},a)-\mathbf{r}(\mathbf{x},a1))|\mathbf{x1};\mathbf{d}_f\mathbf{r})|\mathbf{x1};\mathbf{d}_f\mathbf{r}) = 0 \tag{1526}$$

$$\mathbf{I1}[]\cdot\mathbf{D1}[\mathbf{r}(\mathbf{x1},a-da),\mathbf{r}(\mathbf{x1},a)]*(f(\mathbf{r}(\mathbf{x1},a))*(vax(\mathbf{r}(\mathbf{x},a)-\mathbf{r}(\mathbf{x},a1))|\mathbf{x1};\mathbf{d}_b\mathbf{r})|\mathbf{x1};\mathbf{d}_b\mathbf{r}) = 0 \tag{1527}$$

$$\mathbf{I1}[]\wedge(\mathbf{D1}[\mathbf{r}(\mathbf{x1},a),\mathbf{r}(\mathbf{x1},a+da)]*(f|\mathbf{x1};\mathbf{d}_f\mathbf{r})|\mathbf{x1};\mathbf{d}_f\mathbf{r}) = 0 \tag{1528}$$

$$\mathbf{I1}[]\wedge(f(\mathbf{r}(\mathbf{x1},a))*\mathbf{D1}[\mathbf{r}(\mathbf{x1},a),\mathbf{r}(\mathbf{x1},a+da)]*(uax(\mathbf{r}(\mathbf{x},a)-\mathbf{r}(\mathbf{x},a1))|\mathbf{x1};\mathbf{d}_f\mathbf{r})|\mathbf{x1};\mathbf{d}_f\mathbf{r}) = 0 \tag{1529}$$

$$\mathbf{I1}[]\wedge(f(\mathbf{r}(\mathbf{x1},a))*\mathbf{D1}[\mathbf{r}(\mathbf{x1},a),\mathbf{r}(\mathbf{x1},a+da)]*(vax(\mathbf{r}(\mathbf{x},a)-\mathbf{r}(\mathbf{x},a1))|\mathbf{x1};\mathbf{d}_b\mathbf{r})|\mathbf{x1};\mathbf{d}_b\mathbf{r}) = 0 \tag{1530}$$

$$\mathbf{I1}[]\wedge(uax(\mathbf{r}(\mathbf{x1},a)-\mathbf{r}(\mathbf{x1},a1))*\mathbf{D1}[\mathbf{r}(\mathbf{x1},a),\mathbf{r}(\mathbf{x1},a+da)]\wedge(f(\mathbf{r}(\mathbf{x},a))|\mathbf{x1};\mathbf{d}_f\mathbf{r})|\mathbf{x1};\mathbf{d}_f\mathbf{r}) = 0. \tag{1531}$$

$$\mathbf{I1}[]\wedge(vax(\mathbf{r}(\mathbf{x1},a)-\mathbf{r}(\mathbf{x1},a1))*\mathbf{D1}[\mathbf{r}(\mathbf{x1},a),\mathbf{r}(\mathbf{x1},a+da)]\wedge(f(\mathbf{r}(\mathbf{x},a))|\mathbf{x1};\mathbf{d}_b\mathbf{r})|\mathbf{x1};\mathbf{d}_b\mathbf{r}) = 0 \tag{1532}$$

$$\begin{aligned}&\mathbf{I1}[\,]\wedge\mathbf{D1}[\mathbf{r}(\mathbf{x1},a),\mathbf{r}(\mathbf{x1},a+da)]\ast(f(\mathbf{r}(\mathbf{x1},a))\ast(\mathrm{uax}(\mathbf{r}(\mathbf{x},a)-\mathbf{r}(\mathbf{x},a1))\\&\qquad\qquad|\mathbf{x1};\mathbf{d_f r})|\mathbf{x1};\mathbf{d_f r})\\&\quad=0\end{aligned}\tag{1533}$$

$$\begin{aligned}&\mathbf{I1}[\,]\wedge\mathbf{D1}[\mathbf{r}(\mathbf{x1},a-da),\mathbf{r}(\mathbf{x1},a)]\ast(f(\mathbf{r}(\mathbf{x1},a))\ast(\mathrm{vax}(\mathbf{r}(\mathbf{x},a)-\mathbf{r}(\mathbf{x},a1))\\&\qquad\qquad|\mathbf{x1};\mathbf{d_b r})|\mathbf{x1};\mathbf{d_b r})\\&\quad=0\end{aligned}\tag{1534}$$

$$\begin{aligned}&\mathbf{I1}[\,]\ast(\mathbf{D1}[\mathbf{r}(\mathbf{x1},a),\mathbf{r}(\mathbf{x1},a+da)]\ast(f|\mathbf{x1};\mathbf{d_f r})|\mathbf{x1};\mathbf{d_f r})\\&\quad=\lim S[\mathrm{bn}(\mathrm{Prn}(\mathbf{x1}))]\\&\qquad\qquad(f(\mathbf{r}(\mathrm{Prni}(\mathbf{x1}),a+da))-f(\mathbf{r}(\mathrm{Prni}(\mathbf{x1}),a)))\\&\qquad\qquad\quad\ast[\mathbf{utr}(\mathbf{x1},a)\ast\mathbf{utr}(\mathbf{x1},a)]\\&\qquad+\ldots\end{aligned}\tag{1535}$$

$$\begin{aligned}&\mathbf{I1}[\,]\ast(f(\mathbf{r}(\mathbf{x1},a))\ast\mathbf{D1}[\mathbf{r}(\mathbf{x1},a),\mathbf{r}(\mathbf{x1},a+da)]\ast(\mathrm{uax}(\mathbf{r}(\mathbf{x},a)-\mathbf{r}(\mathbf{x},a1))\\&\qquad\qquad|\mathbf{x1};\mathbf{d_f r})|\mathbf{x1};\mathbf{d_f r})\\&\quad=f(\mathbf{r}(\mathbf{x1},a1))\ast[\mathbf{utr}(\mathbf{x1},a1)\ast\mathbf{utr}(\mathbf{x1},a1)]\end{aligned}\tag{1536}$$

$$\begin{aligned}&\mathbf{I1}[\,]\ast(f(\mathbf{r}(\mathbf{x1},a))\ast\mathbf{D1}[\mathbf{r}(\mathbf{x1},a),\mathbf{r}(\mathbf{x1},a+da)]\ast(\mathrm{vax}(\mathbf{r}(\mathbf{x},a)-\mathbf{r}(\mathbf{x},a1))\\&\qquad\qquad|\mathbf{x1};\mathbf{d_b r})|\mathbf{x1};\mathbf{d_b r})\\&\quad=f(\mathbf{r}(\mathbf{x1},a1))\ast[\mathbf{utr}(\mathbf{x1},a1)\ast\mathbf{utr}(\mathbf{x1},a1)]\end{aligned}\tag{1537}$$

$$\begin{aligned}&\mathbf{I1}[\,]\ast(\mathrm{uax}(\mathbf{r}(\mathbf{x1},a)-\mathbf{r}(\mathbf{x1},a1))\ast\mathbf{D1}[\mathbf{r}(\mathbf{x1},a),\mathbf{r}(\mathbf{x1},a+da)]\wedge(f(\mathbf{r}(\mathbf{x},a))\\&\qquad\qquad|\mathbf{x1};\mathbf{d_f r})|\mathbf{x1};\mathbf{d_f r})\\&\quad=\lim\,(f(\mathbf{r}(\mathrm{Prni}(\mathbf{x1}),a1+2\ast da))-f(\mathbf{r}(\mathrm{Prni}(\mathbf{x1}),a1+da)))\\&\qquad\qquad\ast[\mathbf{utr}(\mathbf{x1},a1+da)\ast\mathbf{utr}(\mathbf{x1},a1+da)]\\&\qquad+\ldots\end{aligned}\tag{1538}$$

$$\begin{aligned}&\mathbf{I1}[\,]\ast(\mathrm{vax}(\mathbf{r}(\mathbf{x1},a)-\mathbf{r}(\mathbf{x1},a1))\ast\mathbf{D1}[\mathbf{r}(\mathbf{x1},a),\mathbf{r}(\mathbf{x1},a+da)]\wedge(f(\mathbf{r}(\mathbf{x},a))\\&\qquad\qquad|\mathbf{x1};\mathbf{d_b r})|\mathbf{x1};\mathbf{d_b r})\\&\quad=\lim\,(f(\mathbf{r}(\mathrm{Prni}(\mathbf{x1}),a1))-f(\mathbf{r}(\mathrm{Prni}(\mathbf{x1}),a1-da)))\\&\qquad\qquad\ast[\mathbf{utr}(\mathbf{x1},a1)\ast\mathbf{utr}(\mathbf{x1},a1)]\\&\qquad+\ldots\end{aligned}\tag{1539}$$

$$\begin{aligned}&\mathbf{I1}[\,]\ast\mathbf{D1}[\mathbf{r}(\mathbf{x1},a),\mathbf{r}(\mathbf{x1},a+da)]\ast(f(\mathbf{r}(\mathbf{x1},a))\ast(\mathrm{uax}(\mathbf{r}(\mathbf{x},a)-\mathbf{r}(\mathbf{x},a1))\\&\qquad\qquad|\mathbf{x1};\mathbf{d_f r})|\mathbf{x1};\mathbf{d_f r})\\&\quad=\lim f(\mathbf{r}(\mathbf{x1},a1+da))\\&\qquad+(f(\mathbf{r}(\mathrm{Prni}(\mathbf{x1}),a1+2\ast da))-f(\mathbf{r}(\mathrm{Prni}(\mathbf{x1}),a1+da)))\\&\qquad\qquad\ast[\mathbf{utr}(\mathbf{x1},a+da)\ast\mathbf{utr}(\mathbf{x1},a+da)]\\&\qquad+\ldots\end{aligned}\tag{1540}$$

$$\begin{aligned}&\mathbf{I1}[\,]\bullet\mathbf{D1}[\mathbf{r}(\mathbf{x1},a-da),\mathbf{r}(\mathbf{x1},a)]\ast(f(\mathbf{r}(\mathbf{x1},a))\ast(\mathrm{vax}(\mathbf{r}(\mathbf{x},a)-\mathbf{r}(\mathbf{x},a1))\\&\qquad\qquad|\mathbf{x1};\mathbf{d_b r})|\mathbf{x1};\mathbf{d_b r})\\&\quad=\lim f(\mathbf{r}(\mathbf{x1},a1))\\&\qquad+(f(\mathbf{r}(\mathrm{Prni}(\mathbf{x1}),a1+da))-f(\mathbf{r}(\mathrm{Prni}(\mathbf{x1}),a1)))\\&\qquad\quad\ast[\mathbf{utr}(\mathbf{x1},a1)\ast\mathbf{utr}(\mathbf{x1},a1)]\\&\qquad+\ldots\end{aligned}\tag{1541}$$

These results also have dual expressions.

Corollary

Given the conditions of the theorems,

$$\mathrm{I1}[]\bullet(\mathbf{D1}[\mathbf{x}(\mathbf{r1},a),\mathbf{x}(\mathbf{r1},a+da)]\wedge(\mathbf{f}|\mathbf{r1};\mathbf{d_f x})|\mathbf{r1};\mathbf{d_f x}) = 0 \qquad (1542)$$

$$\mathbf{I1}[]\bullet(\mathbf{f}(\mathbf{x}(\mathbf{r1},a))\wedge\mathbf{D1}[\mathbf{x}(\mathbf{r1},a),\mathbf{x}(\mathbf{r1},a+da)]*(\mathrm{uar}(\mathbf{x}(\mathbf{r},a)-\mathbf{x}(\mathbf{r},a1))|\mathbf{r1};\mathbf{d_f x})|\mathbf{r1};\mathbf{d_f x}) = 0 \qquad (1543)$$

$$\mathbf{I1}[]\bullet(\mathbf{f}(\mathbf{x}(\mathbf{r1},a))\wedge\mathbf{D1}[\mathbf{x}(\mathbf{r1},a-da),\mathbf{x}(\mathbf{r1},a)]*(\mathrm{var}(\mathbf{x}(\mathbf{r},a)-\mathbf{x}(\mathbf{r},a1))|\mathbf{r1};\mathbf{d_b x})|\mathbf{r1};\mathbf{d_b x}) = 0 \qquad (1544)$$

$$\mathbf{I1}[]\bullet(\mathrm{uar}(\mathbf{x}(\mathbf{r1},a)-\mathbf{x}(\mathbf{r1},a1))*\mathbf{D1}[\mathbf{x}(\mathbf{r1},a),\mathbf{x}(\mathbf{r1},a+da)]\wedge(\mathbf{f}(\mathbf{x}(\mathbf{r},a))|\mathbf{r1};\mathbf{d_f x})|\mathbf{r1};\mathbf{d_f x}) = 0 \qquad (1545)$$

$$\mathbf{I1}[]\bullet(\mathrm{var}(\mathbf{x}(\mathbf{r1},a)-\mathbf{x}(\mathbf{r1},a1))*\mathbf{D1}[\mathbf{x}(\mathbf{r1},a),\mathbf{x}(\mathbf{r1},a+da)]\wedge(\mathbf{f}(\mathbf{x}(\mathbf{r},a))|\mathbf{r1};\mathbf{d_b x})|\mathbf{r1};\mathbf{d_b x}) = 0 \qquad (1546)$$

$$\mathbf{I1}[]\bullet\mathbf{D1}[\mathbf{x}(\mathbf{r1},a),\mathbf{x}(\mathbf{r1},a+da)]\wedge(\mathbf{f}(\mathbf{x}(\mathbf{r1},a))*(\mathrm{uar}(\mathbf{x}(\mathbf{r},a)-\mathbf{x}(\mathbf{r},a1))|\mathbf{r1};\mathbf{d_f x})|\mathbf{r1};\mathbf{d_f x}) = 0 \qquad (1547)$$

$$\mathbf{I1}[]\bullet\mathbf{D1}[\mathbf{x}(\mathbf{r1},a-da),\mathbf{x}(\mathbf{r1},a)]\wedge(\mathbf{f}(\mathbf{x}(\mathbf{r1},a))*(\mathrm{var}(\mathbf{x}(\mathbf{x},a)*\mathbf{x}(\mathbf{x},a1))|\mathbf{r1};\mathbf{d_b x})|\mathbf{r1};\mathbf{d_b x}) = 0 \qquad (1548)$$

$$\mathbf{I1}[]\wedge(\mathbf{D1}[\mathbf{x}(\mathbf{r1},a),\mathbf{x}(\mathbf{r1},a+da)]\wedge(\mathbf{f}|\mathbf{r1};\mathbf{d_f x})|\mathbf{r1};\mathbf{d_f x}) = \lim S[\mathrm{bn}(\mathrm{Pxn}(\mathbf{r1}))](\mathbf{f}(\mathbf{x}(\mathrm{Pxni}(\mathbf{r1}),a+da))-\mathbf{f}(\mathbf{x}(\mathrm{Pxni}(\mathbf{r1}),a)))\bullet[\mathbf{utx}(\mathbf{r1},a)*\mathbf{utx}(\mathbf{r1},a)-\mathbf{I}] \qquad (1549)$$

$$\mathbf{I1}[]\wedge(\mathbf{f}(\mathbf{x}(\mathbf{r1},a))\wedge\mathbf{D1}[\mathbf{x}(\mathbf{r1},a),\mathbf{x}(\mathbf{r1},a+da)]*(\mathrm{uar}(\mathbf{x}(\mathbf{r},a)-\mathbf{x}(\mathbf{r},a1))|\mathbf{r1};\mathbf{d_f x})|\mathbf{r1};\mathbf{d_f x}) = \mathbf{f}(\mathbf{x}(\mathbf{r1},a1))\bullet[\mathbf{utx}(\mathbf{r1},a1)*\mathbf{utx}(\mathbf{r1},a1)-\mathbf{I}] \qquad (1550)$$

$$\mathbf{I1}[]\wedge(\mathbf{f}(\mathbf{x}(\mathbf{r1},a))\wedge\mathbf{D1}[\mathbf{x}(\mathbf{r1},a),\mathbf{x}(\mathbf{r1},a+da)]*(\mathrm{var}(\mathbf{x}(\mathbf{r},a)-\mathbf{x}(\mathbf{r},a1))|\mathbf{r1};\mathbf{d_b x})|\mathbf{r1};\mathbf{d_b x}) = \mathbf{f}(\mathbf{x}(\mathbf{r1},a1))\bullet[\mathbf{utx}(\mathbf{r1},a1)*\mathbf{utx}(\mathbf{r1},a1)-\mathbf{I}] \qquad (1551)$$

$$\mathbf{I1}[]\wedge(\mathrm{uar}(\mathbf{x}(\mathbf{r1},a)-\mathbf{x}(\mathbf{r1},a1))*\mathbf{D1}[\mathbf{x}(\mathbf{r1},a),\mathbf{x}(\mathbf{r1},a+da)]\wedge(\mathbf{f}(\mathbf{x}(\mathbf{r},a))|\mathbf{r1};\mathbf{d_f x})|\mathbf{r1};\mathbf{d_f x}) = \lim (\mathbf{f}(\mathbf{x}(\mathrm{Pxni}(\mathbf{r1}),a1+2*da))-\mathbf{f}(\mathbf{x}(\mathrm{Pxni}(\mathbf{r1}),a1+da)))\bullet[\mathbf{utx}(\mathbf{r1},a1+da)*\mathbf{utx}(\mathbf{r1},a1+da)-\mathbf{I}] + \ldots \qquad (1552)$$

Theoretical Physics: Integration

I1[]∧(var(x(r1,a)−x(r1,a1))∗D1[x(r1,a),x(r1,a+da)]∧(f(x(r,a))
|r1;d_bx)|r1;d_bx)
= lim (f(x(Pxni(r1),a1))−f(x(Pxni(r1),a1−da)))
•[utx(r1,a1)∗utx(r1,a1)−I]
+ ... (1553)

I1[]∧D1[x(r1,a),x(r1,a+da)]∧(f(x(r1,a))∗(uar(x(r,a)−x(r,a1))
|r1;d_fx)|r1;d_fx)
= lim f(x(r1,a1+da))•[utx(r1,a1)∗utx(r1,a1)−I]
+ f(x(Pxni(r1),a1+2∗da))−f(x(Pxni(r1),a1+da))
•[utx(r1,a1+da)∗utx(r1,a1+da)−I]
+ ... (1554)

I1[]∧D1[x(r1,a−da),x(r1,a)]∧(f(x(r1,a))∗(var(x(r,a)−x(r,a1))
|r1;d_bx)|r1;d_bx)
= lim f(x(r1,a1))•[utx(r1,a1)∗utx(r1,a1)−I]
+ f(x(Pxni(r1),a1+da))−f(x(Pxni(r1),a1))
•[utx(r1,a1+da)∗utx(r1,a1+da)−I]
+ ... (1555)

I1[]∗(D1[x(r1,a),x(r1,a+da)]∧(f|r1;d_fx)|r1;d_fx)
= lim S[bn(Pxn(r1))]C(f(x(Pxni(r1),a+da))−f(x(Pxni(r1),a)))
•[utx(r1,a)∗utx(r1,a)] (1556)

I1[]∗(f(x(r1,a))∧D1[x(r1,a),x(r1,a+da)]
∗(uar(x(r,a)−x(r,a1))|r1;d_fx)|r1;d_fx)
= C(f(x(r1,a1)))•[utx(r1,a1)∗utx(r1,a1)] (1557)

I1[]∗(f(x(r1,a))∧D1[x(r1,a),x(r1,a+da)]
∗(var(x(r,a)−x(r,a1))|r1;d_bx)|r1;d_bx)
= C(f(x(r1,a1)))•[utx(r1,a1)∗utx(r1,a1)] (1558)

I1[]∗(uar(x(r1,a)−x(r1,a1))∧D1[x(r1,a),x(r1,a+da)]∧(f(x(r,a))
|r1;d_fx)|r1;d_fx)
= lim C(f(x(Pxni(r1),a1+2∗da))−f(x(Pxni(r1),a1+da))
•[utx(r1,a1+da)∗utx(r1,a1+da)])
+ ... (1559)

I1[]∗(var(x(r1,a)−x(r1,a1))∧D1[x(r1,a),x(r1,a+da)]∧(f(x(r,a))
|r1;d_bx)|r1;d_bx)
= lim C(f(x(Pxni(r1),a1))−f(x(Pxni(r1),a1−da)))
•[utx(r1,a1)∗utx(r1,a1)]
+ ... (1560)

I1[]∗D1[x(r1,a),x(r1,a+da)]∧(f(x(r1,a))∗(uar(x(r,a)–x(r,a1))
|r1;d_fx)|r1;d_fx)
= lim C(f(x(r1,a1+da)))•[utx(r1,a1)∗utx(r1,a1)]
+ C(f(x(Pxni(r1),a1+2∗da))–f(x(Pxni(r1),a1+da)))
•[utx(r1,a1+da)∗utx(r1,a1+da)]
+ ... (1561)

I1[]∗D1[x(r1,a–da),x(r1,a)]∧(f(x(r1,a))∗(var(x(r,a)–x(r,a1))
|r1;d_bx)|r1;d_bx)
= lim f(x(r1,a1))•[utx(r1,a1)∗utx(r1,a1)–I]
+ C(f(x(Pxni(r1),a1+da))–f(x(Pxni(r1),a1)))
•[utx(r1,a1+da)∗utx(r1,a1+da)]
+ ... (1562)

I1[]∗(D1[x(r1,a),x(r1,a+da)]•(f|r1;d_fx)|r1;d_fx)
= lim S[bn(Pxn(r1))](f(x(Pxni(r1),a+da))–f(x(Pxni(r1),a)))
•[utx(r1,a)∗utx(r1,a)] (1563)

I1[]∗(f(x(r1,a))•D1[x(r1,a),x(r1,a+da)]∗(uar(x(r,a)–x(r,a1))
|r1;d_fx)|r1;d_fx)
= f(x(r1,a1)))•[utx(r1,a1)∗utx(r1,a1)] (1564)

I1[]∗(f(x(r1,a))•D1[x(r1,a),x(r1,a+da)]∗(var(x(r,a)–x(r,a1))
|r1;d_bx)|r1;d_bx)
= f(x(r1,a1))•[utx(r1,a1)∗utx(r1,a1)] (1565)

I1[]∗(uar(x(r1,a)–x(r1,a1))•D1[x(r1,a),x(r1,a+da)]∧(f(x(r,a))
|r1;d_fx)|r1;d_fx)
= lim (f(x(Pxni(r1),a1+2∗da))–f(x(Pxni(r1),a1+da)))
•[utx(r1,a1+da)∗utx(r1,a1+da)]
+ ... (1566)

I1[]∗(var(x(r1,a)–x(r1,a1))•D1[x(r1,a),x(r1,a+da)]∧(f(x(r,a))
|r1;d_bx)|r1;d_bx)
= lim (f(x(Pxni(r1),a1))–f(x(Pxni(r1),a1–da)))
•[utx(r1,a1)∗utx(r1,a1)]
+ ... (1567)

I1[]∗D1[x(r1,a),x(r1,a+da)]•(f(x(r1,a))∗(uar(x(r,a)–x(r,a1))
|r1;d_fx)|r1;d_fx)
= lim f(x(r1,a1+da))•[utx(r1,a1)∗utx(r1,a1)]
+ (f(x(Pxni(r1),a1+2∗da))–f(x(Pxni(r1),a1+da)))
•[utx(r1,a1+da)∗utx(r1,a1+da)]
+ ... (1568)

Theoretical Physics: Integration

I1[]∗**D1**[**x**(**r1**,a–da),**x**(**r1**,a)]•(**f**(**x**(**r1**,a))∗(var(**x**(**r**,a)–**x**(**r**,a1))
|**r1**;$\mathbf{d_b}$**x**)|**r1**;$\mathbf{d_b}$**x**)
= lim **f**(**x**(**r1**,a1))•[**utx**(**r1**,a1)∗**utx**(**r1**,a1)]
+ (**f**(**x**(Pxni(**r1**),a1+da))–**f**(**x**(Pxni(**r1**),a1)))
•[**utx**(**r1**,a1+da)∗**utx**(**r1**,a1+da)]
+ ... (1569)

For scalar functions

I1[]•(**D1**[**x**(**r1**,a),**x**(**r1**,a+da)]∗(f|**r1**;$\mathbf{d_f}$**x**)|**r1**;$\mathbf{d_f}$**x**)
= 0 (1570)

I1[]•(f(**x**(**r1**,a))∗**D1**[**x**(**r1**,a),**x**(**r1**,a+da)]∗(uar(**x**(**r**,a)–**x**(**r**,a1))
|**r1**;$\mathbf{d_f}$**x**)|**r1**;$\mathbf{d_f}$**x**)
= f(**x**(**r1**,a1)) (1571)

I1[]•(f(**x**(**r1**,a))∗**D1**[**x**(**r1**,a),**x**(**r1**,a+da)]∗(var(**x**(**r**,a)–**x**(**r**,a1))
|**r1**;$\mathbf{d_b}$**x**)|**r1**;$\mathbf{d_b}$**x**)
= f(**x**(**r1**,a1)) (1572)

I1[]•(uar(**x**(**r1**,a)–**x**(**r1**,a1))∗**D1**[**x**(**r1**,a),**x**(**r1**,a+da)]∗(f(**x**(**r**,a))
|**r1**;$\mathbf{d_f}$**x**)|**r1**;$\mathbf{d_f}$**x**)
= –f(**x**(**r1**,a1)) (1573)

I1[]•(var(**x**(**r1**,a)–**x**(**r1**,a1))∗**D1**[**x**(**r1**,a),**x**(**r1**,a+da)]∗(f(**x**(**r**,a))
|**r1**;$\mathbf{d_b}$**x**)|**r1**;$\mathbf{d_b}$**x**)
= –f(**x**(**r1**,a1)) (1574)

I1[]•**D1**[**x**(**r1**,a),**x**(**r1**,a+da)]∗(f(**x**(**r1**,a))∗(uar(**x**(**r**,a)–**x**(**r**,a1))
|**r1**;$\mathbf{d_f}$**x**)|**r1**;$\mathbf{d_f}$**x**)
= 0 (1575)

I1[]•**D1**[**x**(**r1**,a–da),**x**(**r1**,a)]∗(f(**x**(**r1**,a))∗(var(**x**(**r**,a)–**x**(**r**,a1))
|**r1**;$\mathbf{d_b}$**x**)|**r1**;$\mathbf{d_b}$**x**)
= 0 (1576)

I1[]∧(**D1**[**x**(**r1**,a),**x**(**r1**,a+da)]∗(f|**r1**;$\mathbf{d_f}$**x**)|**r1**;$\mathbf{d_f}$**x**)
= 0 (1577)

I1[]∧(f(**x**(**r1**,a))∗**D1**[**x**(**r1**,a),**x**(**r1**,a+da)]∗(uar(**x**(**r**,a)–**x**(**r**,a1))
|**r1**;$\mathbf{d_f}$**x**)|**r1**;$\mathbf{d_f}$**x**)
= 0 (1578)

I1[]∧(f(**x**(**r1**,a))∗**D1**[**x**(**r1**,a),**x**(**r1**,a+da)]∗(var(**x**(**r**,a)–**x**(**r**,a1))
|**r1**;$\mathbf{d_b}$**x**)|**r1**;$\mathbf{d_b}$**x**)
= 0 (1579)

I1[]∧(uar(x(r1,a)−x(r1,a1))∗D1[x(r1,a),x(r1,a+da)]∧(f(x(r,a))
|r1;d_fx)|r1;d_fx)
= 0 (1580)

I1[]∧(var(x(r1,a)−x(r1,a1))∗D1[x(r1,a),x(r1,a+da)]∧(f(x(r,a)
|r1;d_bx)|r1;d_bx)
= 0 (1581)

I1[]∧D1[x(r1,a),x(r1,a+da)]∗(f(x(r1,a))∗(uar(x(r,a)−x(r,a1))
|r1;d_fx)|r1;d_fx)
= 0 (1582)

I1[]∧D1[x(r1,a−da),x(r1,a)]∗(f(x(r1,a))∗(var(x(r,a)−x(r,a1))
|r1;d_bx)|r1;d_bx)
= 0 (1583)

I1[]∗(D1[x(r1,a),x(r1,a+da)]∗(f|r1;d_fx)|r1;d_fx)
= lim S[bn(Pxn(r1))](f(x(Pxni(r1),a+da))−f(x(Pxni(r1),a)))
∗[utx(r1,a)∗utx(r1,a)] (1584)

I1[]∗(f(x(r1,a))∗D1[x(r1,a),x(r1,a+da)]∗(uar(x(r,a)−x(r,a1))
|r1;d_fx)|r1;d_fx)
= f(x(r1,a1))∗[utx(r1,a1)∗utx(r1,a1)] (1585)

I1[]∗(f(x(r1,a))∗D1[x(r1,a),x(r1,a+da)]∗(var(x(r,a)−x(r,a1))
|r1;d_bx)|r1;d_bx)
= f(x(r1,a1))∗[utx(r1,a1)∗utx(r1,a1)] (1586)

I1[]∗(uar(x(r1,a)−x(r1,a1))∗D1[x(r1,a),x(r1,a+da)]∧(f(x(r,a))
|r1;d_fx)|r1;d_fx)
= lim (f(x(Pxni(r1),a1+2∗da))−f(x(Pxni(r1),a1+da)))
∗[utx(r1,a1+da)∗utx(r1,a1+da)]
+ ... (1587)

I1[]∗(var(x(r1,a)−x(r1,a1))∗D1[x(r1,a),x(r1,a+da)]∧(f(x(r,a))
|r1;d_bx|r1;d_bx)
= lim (f(x(Pxni(r1),a1))−f(x(Pxni(r1),a1−da)))
∗[utx(r1,a1)∗utx(r1,a1)]
+ ... (1588)

I1[]∗D1[x(r1,a),x(r1,a+da)]∗(f(x(r1,a))∗(uar(x(r,a)−x(r,a1))
|r1;d_fx)|r1;d_fx)
= lim f(x(r1,a1+da))
+ (f(x(Pxni(r1),a1+2∗da))−f(x(Pxni(r1),a1+da)))
∗[utx(r1,a+da)∗utx(r1,a+da)]
+ ... (1589)

I1[]•**D1**[**x**(**r1**,a−da),**x**(**r1**,a)]∗(f(**x**(**r1**,a))∗(var(**x**(**r**,a)−**x**(**r**,a1))
|**r1**;$\mathbf{d_b x}$)|**r1**;$\mathbf{d_b x}$)
= lim f(**x**(**r1**,a1))
+ (f(**x**(Pxni(**r1**),a1+da))−f(**x**(Pxni(**r1**),a1)))
∗[**utx**(**r1**,a1)∗**utx**(**r1**,a1)]
+ ... (1590)

Now let CR ≡ {**r**(**x1**,a):a0≤a<a2}be the section of the curve **r**(**x1**,a) designated by the interval

vax(**r**(**x1**,a)–**r**(**x1**,a0)) – uax(**r**(**x1**,a)–**r**(**x1**,a2)).

Theorem 23 (extended fundamental theorem)

Given

- **r**(**x1**,a), the trace of the particle **x1**;
- CX ≡ {**r**(**x1**,a):a0≤a≤a2};
- **f**(**r**(**x1**,a)) a continuous function over the trace;
- **D1**[**r**(**x1**,a),**r**(**x1**,a+da)]∗(**f**|**x1**;**dr**) a basic directional gradient;
- pt any partition scalar;

then

I1[]•((vax(**r**(**x1**,a)–**r**(**x1**,a0))–uax(**r**(**x1**,a)–**r**(**x1**,a2)))
∗**D1**[**r**(**x1**,a),**r**(**x1**,a+da)]∗(**f**|**x1**;$\mathbf{d}_f\mathbf{r}$)|**x1**;$\mathbf{d}_f\mathbf{r}$)
= **f**(**r**(**x1**,a2))–**f**(**r**(**x1**,a0)) (1591)
= **I1**[CX]•(**D1**[**r**(**x1**,a),**r**(**x1**,a+da)]∗(**f**|**x1**;$\mathbf{d}_f\mathbf{r}$)]|**x1**;$\mathbf{d}_f\mathbf{r}$).

Given

- **x**(**r1**,a), the track of particles at location **r1**;
- CR ≡ {**x**(**r1**,a):a0≤a≤a2};
- **f**(**x**(**r1**,a)) a continuous function over the track;
- **D1**[**x**(**r1**,a),**x**(**r1**,a+da)]∗(**f**|**r1**;**dx**) a basic directional gradient;
- pt any partition scalar;

then

I1[]•((var(**x**(**r1**,a)–**x**(**r1**,a0))–uar(**x**(**r1**,a)–**x**(**r1**,a2)))
∗**D1**[**x**(**r1**,a),**x**(**r1**,a+da)]∗(**f**|**r1**;$\mathbf{d}_f\mathbf{x}$)|**r1**;$\mathbf{d}_f\mathbf{x}$)
= **f**(**x**(**r1**,a2))–**f**(**x**(**r1**,a0)) (1592)
= **I1**[CR]•(**D1**[**x**(**r1**,a),**x**(**r1**,a+da)]∗(**f**|**r1**;$\mathbf{d}_f\mathbf{x}$)]|**r1**;$\mathbf{d}_f\mathbf{x}$).

Proof:

I1[]•((vax(**r**(**x1**,a)–**r**(**x1**,a0))–uax(**r**(**x1**,a)–**r**(**x1**,a2)))
∗**D1**[**r**(**x1**,a),**r**(**x1**,a+da)]∗(**f**|**x1**;$\mathbf{d}_f\mathbf{r}$)|**x1**;$\mathbf{d}_f\mathbf{r}$)
=lim S[bn(CX∩Prn)](**r**(**x1**,(i+1)∗pt/n) – **r**(**x1**,(i∗pt/n)))
•T[**D1**[**r**(**x1**,a(Prni)),**r**(**x1**,a(Prni)+da)]∗(**f**(**r**(**x**,a))|**x1**;$\mathbf{d}_f\mathbf{r}$)]
= lim S[bn(CX∩Prn)](**r**(**x1**,(i+1)∗pt/n) – **r**(**x1**,(i∗pt/n)))
•**qd**(**r**(**x1**,a(Prni)+da)–**r**(**x1**,a(Prni)))
∗(**f**(**r**(**x1**,a(Prni)+da))–**f**(**r**(**x1**,a(Prni))))]
= lim S[bn(CX∩Prn)](**f**(**r**(**x1**,a(Prni)+da))–**f**(**r**(**x1**,a(Prni))))]

Theoretical Physics: Integration

= lim **f**(**r**(**x1**,a(Prn(j+1))))−**f**(**r**(**x1**,a(Prn(j))))
+ **f**(**r**(**x1**,a(Prn(j+2))))−**f**(**r**(**x1**,a(Prn(j+1))))
+ ...
+ **f**(**r**(**x1**,a(Prn(k))))−**f**(**r**(**x1**,a(Prn(k−1))))
where Prn(j) contains **r**(**x1**,a0) and Prn(k) contains **r**(**x1**,a2)
= **f**(**r**(**x1**,a2))−**f**(**r**(**x1**,a0))

The dual is proven similarly.

qed

This is the **fundamental theorem of integral calculus extended to directional gradients over traces and tracks**.

Again, **I1**[CR]•(**f**|**x1**;**dr**) may be considered a value (for fixed a2) or a function of a2 (for variable a2). To emphasize the directional integral as function, it is sometimes written as

g(**x1**,a) =.**I1**[CR(a)]•(**f**|**x1**;**dr**)
= **I1**[a0,a]•(**f**|**x1**;**dr**) (1593)

g(**x1**,a) = **I1**[CR(a)]∧(**f**|**x1**;**dr**)
= **I1**[a0,a]∧(**f**|**x1**;**dr**) (1594)

G(**x1**,a) = **I1**[CR(a)]∗(**f**|**x1**;**dr**) (1595)
= **I1**[a0,a]∗(**f**|**x1**;**dr**) (1596)

Now let **f**(**r**(**x1**,a)) have step discontinuity at **r**(**x1**,a1), a0<a1<a2, that is

f(**r**(**x1**,a)) = **f1**(**r**(**x1**,a)) + **f0**∗(uax(**r**(**x1**,a)−**r**(**x1**,a1))) (1597)

where **f1** enjoys interior cancellation and **f0** is constant. Then

D1[**r**(**x1**,a),**r**(**x1**,a+da)]∗(**f**|**x1**;$\mathbf{d_f r}$)
= **D1**[**r**(**x1**,a),**r**(**x1**,a+da)]∗(**f1**|**x1**;$\mathbf{d_f r}$)
+ **f0**∗**D1**∗(uax(**r**(**x1**,a)−**r**(**x1**,a1))|**x1**;$\mathbf{d_f r}$) (1598)

and

I1•[CR](**D1**[**r**(**x1**,a),**r**(**x1**,a+da)]∗(**f**|**x1**;$\mathbf{d_f r}$)|**x1**;$\mathbf{d_f r}$)
= **f1**(**r**(**x1**,a2)) − **f1**(**r**(**x1**,a0)) + **f0**. (1599)

The results extend to functions with *n* step discontinuities. For

f(**r**(**x1**,a)) = **f1**(**r**(**x1**,a)) + S[1,n](**f0i**∗(uax(**r**(**x1**,a)−**r**(**x1**,ai)))) (1600)

D1[**r**(**x1**,a),**r**(**x1**,a+da)]∗(**f**|**x1**;**d**$_{f}$**r**)
= **D1**[**r**(**x1**,a),**r**(**x1**,a+da)]∗(**f1**|**x1**;**d**$_{f}$**r**)
+ S[1,n](**f0i**∗**D1**∗(uax(**r**(**x1**,a)−**r**(**x1**,ai))|**x1**;**d**$_{f}$**r**)) (1601)

and

I1•[CR](**D1**[**r**(**x1**,a),**r**(**x1**,a+da)]∗(**f**|**x1**;**d**$_{f}$**r**)|**x1**;**d**$_{f}$**r**)
= **f1**(**r**(**x1**,a2)) − **f1**(**r**(**x1**,a0))+ S[1,n](**f0i**) (1602)

where **f1** has a continuous directional gradient.

This is the **fundamental theorem of integral calculus extended to vectorial functions with discontinuities over traces**.

Dually, let **f**(**x**(**r1**,a)) have step discontinuity at **x**(**r1**,a1), a0<a1<a2, that is

f(**x**(**r1**,a)) = **f1**(**x**(**r1**,a)) + **f0**∗(uar(**x**(**r1**,a)−**x**(**r1**,a1))) (1603)

where **f1** enjoys interior cancellation and **f0** is constant. Then

D1[**x**(**r1**,a),**x**(**r1**,a+da)]∗(**f**|**r1**;**d**$_{f}$**x**)
= **D1**[**x**(**r1**,a),**x**(**r1**,a+da)]∗(**f1**|**r1**;**d**$_{f}$**x**)
+ **f0**∗**D1**∗(uar(**x**(**r1**,a)−**x**(**r1**,a1))|**r1**;**d**$_{f}$**x**) (1604)

and

I1•[CX](**D1**[**x**(**r1**,a),**x**(**r1**,a+da)]∗(**f**|**r1**;**d**$_{f}$**x**)|**r1**;**d**$_{f}$**x**)
= **f1**(**x**(**r1**,a2)) − **f1**(**x**(**r1**,a0)) + **f0**. (1605)

The results extend to functions with n step discontinuities. For

f(**x**(**r1**,a)) = **f1**(**x**(**r1**,a)) + S[1,n](**f0i**∗(uar(**x**(**r1**,a)−**x**(**r1**,ai)))) (1606)

D1[**x**(**r1**,a),**x**(**r1**,a+da)]∗(**f**|**r1**;**d**$_{f}$**x**)
= **D1**[**x**(**r1**,a),**x**(**r1**,a+da)]∗(**f1**|**r1**;**d**$_{f}$**x**)
+ S[1,n](**f0i**∗**D1**∗(uar(**x**(**r1**,a)−**x**(**r1**,ai))|**r1**;**d**$_{f}$**x**)) (1607)

and

I1•[CX](**D1**[**x**(**r1**,a),**x**(**r1**,a+da)]∗(**f**|**r1**;**d**$_{f}$**x**)|**r1**;**d**$_{f}$**x**)
= **f1**(**x**(**r1**,a2)) − **f1**(**x**(**r1**,a0))+ S[1,n](**f0i**) (1608)

where **f1** has a continuous directional gradient.

This is the **fundamental theorem of integral calculus extended to vectorial functions with discontinuities over tracks**.

Theoretical Physics: Integration

Now let **f**(**r**(**x1**,a)) have continuous basic directional gradients except perhaps at **r1**(**x1**,a1) where

A(**x1**,a1) ≡ **D1**[**r**(**x1**,a1)]∗(**f**|**x1**;$\mathbf{d}_f\mathbf{r}$) – **D1**[**r**(**x1**,a1)]∗(**f**|**x1**;$\mathbf{d}_b\mathbf{r}$). (1609)

Then the forward and backward directional gradients define

D1[**r**(**x1**,a)]∗(**f1**|**x1**;$\mathbf{d}_f\mathbf{r}$)
≡ **D1**[**r**(**x1**,a)]∗(**f**|**x1**;$\mathbf{d}_b\mathbf{r}$)
+ uar(**r**(**x1**,a)–**r**(**x1**,a1))
∗(**D1**[**r**(**x1**,a)]∗(**f**|**x1**;$\mathbf{d}_f\mathbf{r}$) – **A**(**x1**,a1)).

Then **D1**[**r**(**x1**,a)]∗(**f1**|**x1**;$\mathbf{d}_f\mathbf{r}$) = **D1**[**r**(**x1**,a)]∗(**f1**|**x1**;**dr**) is continuous.

Consequently

D1[**r**(**x1**,a)]∗(**f1**|**x1**;**dr**) + ua(**r**(**x1**,a)–**r**(**x1**,a1))∗**A**(**x1**,a1)
= **D1**[**r**(**x1**,a)]∗(**f**|**x1**;$\mathbf{d}_b\mathbf{r}$)
+ uar(**r**(**x1**,a)–**r**(**x1**,a1))
∗[**D1**[**r**(**x1**,a)]∗(**f**|**x1**;$\mathbf{d}_f\mathbf{r}$) – **D1**[**r**(**x1**,a)]∗(**f**|**x1**;$\mathbf{d}_b\mathbf{r}$)] (1610)

provides a unified description of the gradient.

The unified description may be extended to a finite number of discontinuities,

D1[**r**(**x1**,a)]∗(**f1**|**x1**;**dr**) + S[1,n](uax(**r**(**x1**,a)–**r**(**x1**,ai))∗**Ai**)
= S[1,n](**D1**[(**x1**,a)]∗(**f**|**x1**;$\mathbf{d}_b\mathbf{r}$)
+ uax(**r**(**x1**,a)–**r**(**x1**,ai))
∗[**D1**[(**x1**,a)]∗(**f**|**x1**;$\mathbf{d}_f\mathbf{r}$) – **D1**[a]∗(**f**|**x1**;$\mathbf{d}_b\mathbf{r}$)]) (1611)

where **D1**[**r**(**x1**,a)]∗(**f1**|**x1**;**dr**) is continuous over **r**(**x1**,a).

Dual results for functions over the material track **x**(**r1**,a) may be similarly developed.

Integration of Sectional Gradients

Here one is given
a single observation of a function of the universe **f**(**r**(**x**,a1));
fourteen possibly distinct sectional gradients listed on page 292;
the mutually related sections at **r1**(**x1**,a1)
SECT(**r1**,**u1Ri**,**u2Ri**,**u3Ri**)
SECT(**x1**,**u1Xi**,**u2Xi**,**u3Xi**).

The constitutive derivatives of the gradients may be impulsive or discontinuous. Sectional gradients are functions of their sections; they may differ from continuous gradients.

Definition 57 (material and local partitions)

Given
a material partition vector,
pX1 = p1X1∗**u1X1** + p2X1∗**u2X1** + p3X1∗**u3X1**;
for
n>0 a given integer;
i,j,k= – $(n^2–1)$,...,–,1,0,1,2,3,...n^2
(i–1)/n≤t1<i/n for abs(t1)≤n;
(j–1)/n≤t2<j/n for abs(t2)≤n;
(k–1)/n≤t3<k/n for abs(t3)≤n;
the sets
PX1n(i,j,k) ≡ {**x**=t1∗p1X1∗**u1X1**
+ t2∗p2X1∗**u2X1**
+ t3∗p3X1∗**u3X1**}
and PX1n∞ otherwise. (1612)
are called **the nth partition of matter based on pX1**.

Given
a local partition vector,
pR1 = p1R1∗**u1R1** + p2X1∗**u2R1** + p3X1∗**u3R1**;
the sets
PR1n(i,j,k) ≡ {r=t1∗p1X1∗**u1R1**
+ t2∗p2R1∗**u2R1**
+ t3∗p3R1∗**u3R1**}
and PR1n∞ otherwise (1613)
are called **the nth partition of locations based on pR1**.

end of definition

Theoretical Physics: Integration

The individual sets in a partition are called **cells**, material or local according to their partitions.

Definition 58 (center of a cell in a partition)

Given

PX1n(i,j,k), a cell in partition PX1n;

then

$$\mathbf{x}(\text{PX1n}(i,j,k)) \equiv (i - ½)*\text{p1X1}*\mathbf{u1X1}/n + (j - ½)*\text{p2X1}*\mathbf{u2X1}/n + (k - ½)*\text{p3X1}*\mathbf{u3X1}/n \quad (1614)$$

$$\mathbf{x}(\text{PX1n}\infty) = \mathbf{0}.$$

Given

PR1n(i,j,k), a cell in partition PR1n;

then

$$\mathbf{r}(\text{PR1n}(i,j,k)) \equiv (i - ½)*\text{p1R1}*\mathbf{u1R1}/n + (j - ½)*\text{p2R1}*\mathbf{u2R1}/n + (k - ½)*\text{p3R1}*\mathbf{u3R1}/n \quad (1615)$$

$$\mathbf{r}(\text{Pn}\infty) = \mathbf{0}.$$

end of definition

For each PX1n(i,j,k) of PX1n let

$$\begin{aligned}\mathbf{mxn}(\text{PX1n}(i,j,k)) &\equiv ((\mathbf{x}(\text{PX1n}(i+1,j,k)) - \mathbf{x}(\text{PX1n}(i,j,k)))*\mathbf{u1X1} \\ &\quad + (\mathbf{x}(\text{PX1n}(i,j+1,k)) - \mathbf{x}(\text{PX1n}(i,j,k)))*\mathbf{u2X1} \\ &\quad + (\mathbf{x}(\text{PX1n}(i,j,k+1)) - \mathbf{x}(\text{PX1n}(i,j,k)))*\mathbf{u3X1})/n^2 \\ &= (\text{p1X1}*\mathbf{u1X1}+\text{p2X1}*\mathbf{u2X1}+\text{p3X1}*\mathbf{u3X1})/n^3\end{aligned}$$

with $\mathbf{mxn}(\text{PX1n}\infty) = \mathbf{0}$. (1616)

For each PR1n(i,j,k) of PR1n let

$$\begin{aligned}\mathbf{mrn}(\text{PR1n}(i,j,k)) &\equiv ((\mathbf{r}(\text{PR1n}(i+1,j,k)) - \mathbf{r}(\text{PR1n}(i,j,k)))*\mathbf{u1R1} \\ &\quad + (\mathbf{r}(\text{PR1n}(i,j+1,k)) - \mathbf{r}(\text{PR1n}(i,j,k)))*\mathbf{u2R1} \\ &\quad + (\mathbf{r}(\text{PR1n}(i,j,k+1)) - \mathbf{r}(\text{PR1n}(i,j,k)))*\mathbf{u3R1})/n^2 \\ &= (\text{p1R1}*\mathbf{u1R1}+\text{p2R1}*\mathbf{u2R1}+\text{p3R1}*\mathbf{u3R1})/n^3\end{aligned}$$

with $\mathbf{mrn}(\text{Pn}\infty) \equiv \mathbf{0}$. (1617)

For a measurable set MX in a given partition PX1n let *bn(MX∩PX1n)* be the number of cells for which MX∩PX1n(i,j,k) is non–empty; likewise for a measurable set MR in a given partition PR1n let *bn(MR∩PR1n)* be the number of cells for which MR∩PR1n(i,j,k) is non–empty.

The **measure of** MX **with respect to the partition PX1n** is defined as

mxn(MX∩PX1n)
≡ bn(M∩Pn)∗(**mxn**(PX1n(i,j,k)))
= (p1X1∗**u1X1**+p2X1∗**u2X1**+p3X1∗**u3X1**)
∗bn(MX∩PX1n)/n^3 (1618)

The **measure of** MR **with respect to the partition PR1n** is defined as

mrn(MR∩PR1n)
≡ bn(MR∩PR1n)∗(**mrn**(PR1n(i,j,k)))
= (p1R1∗**u1R1**+p2R1∗**u2R1**+p3R1∗**u3R1**)
∗bn(MR∩PR1n)/n^3. (1619)

Definition 59 (measure of a measurable sets MR or MX)

Given

pX1 =p1X1∗**u1X1**+p2X1∗**u2X1**+p3X1∗**u3X1**;
MX a measurable set of matter;
PX1n, a sequenced set of partitions based on **pX1**;
mxn(MX∩PX1n), the measure of MX with respect to PX1n;

then

mx(MX) ≡ lim(**mxn**(MX∩PX1n)) as n⟶∞ (1620)
≡ mx(MX)∗**pX1**. (1621)

Given

pR1 =p1∗**u1R1**+p2∗**u2R1**+p3∗**u3R1**;
MR a measurable set of locations;
PR1n, a sequenced set of partitions based on **pR1**;
mrn(MR∩PR1n), the measure of MR with respect to PR1n;

then

mr(MR) ≡ lim(**mrn**(MR∩PR1n)) as n⟶∞ (1622)
≡ mr(MR)∗**pR1**. (1623)

end of definition

Theoretical Physics: Integration

We shall often write

$$\mathbf{dX1} \equiv \mathbf{pX1}/n^3 \quad (1624)$$

$$\mathbf{dR1} \equiv \mathbf{pR1}/n^3. \quad (1625)$$

Surfaces of the measurable sets MX and MR are also defined with reference to the partition.

Definition 60 (surface of MX and MR)

Given

pX1 = (p1X1∗**u1X1**+p2X1∗**u2X1**+p3X1∗**u3X1**);
MX a measurable set of V3(**x**);
PX1n, a sequenced set of partitions of V3(**x**) based on **pR1**;
mxn(MX∩PX1n), the measure of MX with respect to PX1n;

then

if PX1n(i,j,k) ∩ MX = PX1n(i,j,k) (1626)
PX1n(i,j,k) is called an **cell interior to** MX
if PX1n(i,j,k) ∩ MX = null (1627)
PX1n(i,j,k) is called an **cell exterior to** MX
if PX1n(i,j,k) ∩ MX ⊂ PX1n(i,j,k) properly (1628)
PX1n(i,j,k) is called a **surface cell of** MX.

Given

pR1 = (p1R1∗**u1R1**+p2R1∗**u2R1**+p3R1∗**u3R1**);
MR a measurable set of V3(**r**);
PR1n, a sequenced set of partitions of V3(**r**) based on **pR1**;
mrn(MR∩PR1n), the measure of MR with respect to PR1n;

then

if PR1n(i,j,k) ∩ MR = PR1n(i,j,k) (1629)
PR1n(i,j,k) is called an **cell interior to** MR
if PR1n(i,j,k) ∩ MR = null (1630)
PR1n(i,j,k) is called an **cell exterior to** MR
if PR1n(i,j,k) ∩ MR ⊂ PR1n(i,j,k) properly (1631)
PR1n(i,j,k) is called a **surface cell of** MR.

end of definition

The surfaces SMX and SMR are defined as limits of the surface cells as the partitions, material or local, become more refined with increasing *n*.

Exterior and interior, positive and negative subsets of the surface are also defined in the usual way as in equation (727) and following.

SMXPE|**pX1** ≡ {**r** in SMX|(**r+pX1**/n^3 is in V3 – (MX∪SMX) (1632)
is called the **exterior positive surface of** MX.
SMXPI|**pX1** ≡ {**r** in SMX|(**r+pX1**/n^3 is in (MX – SMX) (1633)
is called the **interior positive surface of** MX.
SMXP0|**pX1** ≡ {**r** in SMX|(**r+pX1**/n^3 is in SMX (1634)
is called the **zero positive surface of** MX.
SMXNE|**pX1** ≡ {**r** in SMX|(**r–pX1**/n^3 is in V3 – (MX∪SMX) (1635)
is called the **exterior negative surface of** MX.
SMXNI|**pX1** ≡ {**r** in SMX|(**r–pX1**/n^3 is in (MX – SMX) (1636)
is called the **interior negative surface of** MX.
SMXN0|**pX1** ≡ {**r** in SMX|(**r–pX1**/n^3 is in SMX (1637)
is called the **zero negative surface of** MX.

SMRPE|**pR1** ≡ {**r** in SMR|(**r+pR1**/n3) is in V3 – (MR∪SMR) (1638)
is called the **exterior positive surface of** MR.
SMRPI|**pR1** ≡ {**r** in SMR|(**r+pR1**/n^3 is in (MR – SMR) (1639)
is called the **interior positive surface of** MR.
SMRP0|**pR1** ≡ {**r** in SMR|(**r+pR1**/n^3 is in SMR (1640)
is called the **zero positive surface of** MR.
SMRNE|**pR1** ≡ {**r** in SMR|(**r–pR1**/n^3 is in V3 – (MR∪SMR) (1641)
is called the **exterior negative surface of** MR.
SMRNI|**pR1** ≡ {**r** in SMR|(**r–pR1**/$n^{3)}$ is in (MR – SMR) (1642)
is called the **interior negative surface of** MR.
SMRN0|**pR1** ≡ {**r** in SMR|(**r–pR1**/$n^{3)}$ is in SMR (1643)
is called the **zero negative surface of** MR.

Theoretical Physics: Integration

Definition 61 (invergences, incurls, and ingradients)

Given

f(**x**(**r**,ai)), a measurable function over V3(**x**);

MX, a measurable set of V3(**x**);

pX1 = pX1∗**uX1**

= p1X1∗**u1X1**+p2X1∗**u2X1**+p3X1∗**u3X1**, a partition vector;

for

S[bn(MX∩PX1n)], the sum over which MX∩PX1n is non–empty

I3[MX]•(**f**(**x**(**r**,ai));**dX1**)

≡ lim(S[MX∩PX1n](**f**(**x**(PX1n(i,j,k)))•**mx**(PX1n(i,j,k))))

= lim(S[MX∩PX1n](**f**(**x**(PX1n(i,j,k)))•**pX1**/n^3))

= lim(S[MX∩PX1n](**f**(**x**(PX1n(i,j,k)))•**dX1**)) (1644)

I3[MX]∧(**f**(**x**(**r**,ai));**dX1**)

≡ lim(S[MX∩PX1n](**f**(**x**(PX1n(i,j,k)))∧**mx**(PX1n(i,j,k))))

= lim(S[MX∩PX1n](**f**(**x**(PX1n(i,j,k)))∧**pX1**/n^3))

= lim(S[MX∩PX1n](**f**(**x**(PX1n(i,j,k)))∧**dX1**)) (1645)

I3[MX]∗(**f**(**x**(**r**,ai));**dX1**)

≡ lim(S[MX∩PX1n](**f**(**x**(PX1n(i,j,k)))∗**mx**(PX1n(i,j,k))))

= lim(S[MX∩PX1n](**f**(**x**(PX1n(i,j,k)))∗**pX1**/n^3))

= lim(S[MX∩PX1n](**f**(**x**(PX1n(i,j,k)))∗**dX1**)) (1646)

as $n \longrightarrow \infty$.

Given

f(**r**(**x**,ai)), a measurable function over V3(**r**);

MR, a measurable set of V3(**r**);

pR1 = pR1∗**uR1**

= p1∗**u1R1**+p2∗**u2R1**+p3∗**u3R1**, a partition vector;

for

S[bn(MR∩PR1n)], the sum over which MR∩PR1n is non–empty

I3[MR]•(**f**(**r**)|**mr**;**dR1**)

≡ lim S[bn(MR∩PR1n)](**f**(**r**(PR1n(i,j,k)))•**m1**(PR1n(i,j,k)))

= lim S[bn(MR∩PR1n)](**f**(**r**(PR1n(i,j,k)))•**pR1**/n^3)

= lim(S[MX∩PX1n](**f**(**x**(PX1n(i,j,k)))•**dX1**)) (1647)

I3[MR]∧(**f**(**r**)|**mr**;**dR1**)

≡ lim S[bn(MR∩PR1n)](**f**(**r**(PR1n(i,j,k)))∧**m1**(PR1n(i,j,k)))

= lim S[bn(MR∩PR1n)](**f**(**r**(PR1n(i,j,k)))∧**pR1**/n^3)

= lim(S[bn(MX∩PX1n](**f**(**x**(PX1n(i,j,k)))∧**dX1**)) (1648)

I3[MR]∗(**f**(**r**)|**m**;**dR1**)

≡ lim S[bn(MR∩PR1n)](**f**(**r**(PR1n(i,j,k)))∗**m1**(PR1n(i,j,k)))

= lim S[bn(MR∩PR1n)](**f**(**r**(PR1n(i,j,k)))∗**pR1**/n^3)

= lim(S[bn(MX∩PX1n](**f**(**x**(PX1n(i,j,k)))∗**dX1**)) (1649)

as n⟶∞.

end of definition

Similarly, sectional integration of a transformation, **A**(**r**(**x**,a)), is defined as

I3[MX]•(**A**;**dX1**)

≡ lim(S[MX∩PX1n](**mx**(PX1n(i,j,k))•**T[A(x**(PX1n(i,j,k)))])). (1650)

Material sectional integrals are functions of the partition vector as shown by the decomposition:

I3[MX]•(**f**(**x**(**r**,ai));**dX1**) ≡ **I3**[MX](**f**(**x**(**r**,ai))•**pX1**) (1651)

I3[MX]∧(**f**(**x**(**r**,ai));**dX1**) ≡ **I3**[MX](**f**(**x**(**r**,ai))∧**pX1**) (1652)

I3[MX]∗(**f**(**x**(**r**,ai));**dX1**) ≡ **I3**[MX](**f**(**x**(**r**,ai))∗**pX1**) (1653)

where

I3[MX](**f**(**x**(**r**,ai))) ≡ lim S[MX∩PX1n](**f**(**x**(PX1n(i,j,k)))/n^3. (1654)

For functions with a basic, continuous, material gradient an **orthogonal partition** may be specified with **px** = px1∗**u1** + px2∗**u2** + px3∗**u3** from which a partition PX, a measure **mx** and finally invergences, incurls and ingradients with respect to **px** at observation ai are defined as

I3[MX]•(**f**(**x**(**r**,ai));**dx**)

≡ lim(S[MX∩PXn](**f**(**x**(PXn(i,j,k)))•**mx**(PXn(i,j,k))))

= lim(S[MX∩PXn](**f**(**x**(PXn(i,j,k)))•**px**/n^3))

= lim(S[MX∩PXn](**f**(**x**(PXn(i,j,k)))•**dx**)) (1655)

I3[MX]∧(**f**(**x**(**r**,ai));**dx**)

≡ lim(S[MX∩PXn](**f**(**x**(PXn(i,j,k)))∧**mx**(PXn(i,j,k))))

= lim(S[MX∩PXn](**f**(**x**(PXn(i,j,k)))∧**px**/n^3))

= lim(S[MX∩PXn](**f**(**x**(PXn(i,j,k)))∧**dx**)) (1656)

Theoretical Physics: Integration

$$\mathbf{I3}[\text{MX}]*(\mathbf{f}(\mathbf{x}(\mathbf{r},\text{ai}));\mathbf{dx})$$
$$\equiv \lim(\text{S}[\text{MXnPXn}](\mathbf{f}(\mathbf{x}(\text{PXn}(i,j,k)))*\mathbf{mx}(\text{PXn}(i,j,k))))$$
$$= \lim(\text{S}[\text{MXnPXn}](\mathbf{f}(\mathbf{x}(\text{PXn}(i,j,k)))*\mathbf{px}/n^3))$$
$$= \lim(\text{S}[\text{MXnPXn}](\mathbf{f}(\mathbf{x}(\text{PXn}(i,j,k)))*\mathbf{dx})) \quad (1657)$$
$$\text{as } n \longrightarrow \infty$$

where

$$\mathbf{dx} \equiv \mathbf{px}/n^3. \quad (1658)$$

Dually for local functions with basic, continuous, local gradients with an orthogonal partition PR based on **pr** = pr1∗**u1** + pr2∗**u2** + pr3∗**u3,** a measure **mr** and finally invergences, incurls and ingradients with respect to **pr** at observation ai are similarly defined.

Now consider the relationships induced[81] by **r**(**x**,ai) where

$$\mathbf{r}(\text{V3}(\mathbf{x},\text{ai})) = \text{V3}(\mathbf{r})$$
$$\mathbf{r}(\text{MX},\text{ai}) = \text{MR}$$
$$\mathbf{r}(\text{SMX},\text{ai}) = \text{SMR}$$
$$\mathbf{r}(\mathbf{x1},\text{ai}) = \mathbf{r1}$$
$$\mathbf{f}(\mathbf{r}(\mathbf{x},\text{ai})) = \mathbf{f}(\mathbf{x}(\mathbf{r},\text{ai}))$$
$$\mathbf{D3}[\mathbf{x1}]*(\mathbf{f}|\text{ai};\mathbf{dX1}) = \mathbf{D3}[\mathbf{r1}]*(\mathbf{f}|\text{ai};\mathbf{dR1})\cdot[\mathbf{UR1}^{-1}]\cdot[\mathbf{T}[\mathbf{UR1}^{-1}]]$$
$$\cdot\mathbf{D3}[\mathbf{x1}]*(\mathbf{r}|\text{ai};\mathbf{dX1}) \quad (1659)$$

and dually,

$$\mathbf{x}(\text{V3}(\mathbf{r},\text{ai})) = \text{V3}(\mathbf{x})$$
$$\mathbf{x}(\text{MR},\text{ai}) = \text{MX}$$
$$\mathbf{x}(\text{SMR},\text{ai}) = \text{SMX}$$
$$\mathbf{x}(\mathbf{r1},\text{ai}) = \mathbf{x1}$$
$$\mathbf{f}(\mathbf{r}(\mathbf{x},\text{ai})) = \mathbf{f}(\mathbf{x}(\mathbf{r},\text{ai}))$$
$$\mathbf{D3}[\mathbf{r1}]*(\mathbf{f}|\text{ai};\mathbf{dR1}) = \mathbf{D3}[\mathbf{x1}]*(\mathbf{f}|\text{ai};\mathbf{dX1})\cdot[\mathbf{UX1}^{-1}]\cdot[\mathbf{T}[\mathbf{UX1}^{-1}]]$$
$$\cdot\mathbf{D3}[\mathbf{r1}]*(\mathbf{x}|\text{ai};\mathbf{dR1}. \quad (1660)$$

Theorem 24 (the first partition theorem)

Given

f(**r**(**x**,ai)) with a basic continuous local gradient everywhere;
MR, a measurable set in V3(**r**);
pr = pr1 * **u1**+pr2 * **u2**+pr3 * **u3** a partition vector;
PRn, a set of partitions defined from **pr**;

for

I3[MR]•(**f**(**r**(**x**,ai));**dr**) ≡ **sr**(**f**)•**pr**;

then in another partition

PR1m defined by
pR1 = p1R1 * **u1R1**+p2R1 * **u2R1**+p3R1 * **u3R1**

I3[MR]•(**f**(**r**(**x**,ai));**dR1**) = **sr**(**f**)•**pR1** (1661)
I3[MR]∧(**f**(**r**(**x**,ai));**dR1**) = **sr**(**f**)∧**pR1** (1662)
I3[MR] * (**f**(**r**(**x**,ai));**dR1**) = **sr**(**f**) * **pR1**. (1663)

Given

f(**x**(**r**,ai)) with a basic continuous local gradient everywhere;
MX, a measurable set in V3(**x**);
px = px1 * **u1**+px2 * **u2**+px3 * **u3** a partition vector;
PXn, a set of partitions defined from **px**;

for

I3[MX]•(**f**(**x**(**r**,ai));**dx**) ≡ **sx**(**f**)•**px**;

then in another partition

PX1m defined by
pX1 = p1X1 * **u1X1**+p2X1 * **u2X1**+p3X1 * **u3X1**

I3[MX]•(**f**(**x**(**r**,ai));**dX1**) = **sx**(**f**)•**pX1** (1664)
I3[MX]∧(**f**(**x**(**r**,ai));**dX1**) = **sx**(**f**)∧**pX1** (1665)
I3[MX] * (**f**(**x**(**r**,ai));**dX1**) = **sx**(**f**) * **pX1**. (1666)

Proof:

Let **a** = p1R1 * **u1**/pr1 + p2R1 * **u2**/pr2 +p3R1 * **u3**/pr3.
By the dilation theorem
m(PRn(i,j,k)|**pR1**) = m(PRn(i,j,k)|**pr**)/(det[**UR1**] * det[**G**(**a**)])
= 1/(det[**UR1**] * det[**G**(**a**)] * n^3).
One subset under the PRn partition is thus occupied by
det[**UR1**] * det[**G**(**a**)] subsets[82]
under the PR1m partition.

Theoretical Physics: Integration

Setting m^3 = det[**UR1**]✱det[**G**(**a**)]✱n^3,
S[bn(MR∩PRn)](**f**(**r**(PRn(i,j,k)))•**m**(PRn(i,j,k)))
= S[bn(MR∩PRn)](**f**(**r**(PRn(i,j,k)))•**pr**/n^3)
= S[bn(MR∩PRn)](**f**(**r**(PRn(i,j,k)))•det[**UR1**]✱det[**G**(**a**)]
✱**pr**/m^3)
= S[bn(MR∩PRn)](det[**UR1**]✱det[**G**(**a**)]✱**f**(**r**(PRn(i,j,k)))
•**pr**/m^3)

Let **r1** be the location of PR1m(i',j',k') in PRn(i,j,k)
and
h1 ≡ **r1** − **r**(PRn(i,j,k)).
Then since **f** has a continuous gradient
f(**r**(PR1m(i',j',k'))) = **f**(**r**(PRn(i,j,k)) + **h1**)
≈ **f**(**r**(PRn(i,j,k)))
+ **h1**•**T**[**D3**[**r**(PRn(i,j,k))]✱(**f**|ai;**pr**)].
For k such particles
S[1,k](**f**(**r**(PRn(i,j,k)) + **hi**))
= k✱**f**(**r**(PRn(i,j,k)))
+ S[1,k](**hi**•**T**[**D3**[**r**(PRn(i,j,k))]✱(**f**|ai;**dr**)]).
As n⟶∞, the **hi**⟶**0** so that
f(**r**(PRn(i,j,k))) ≈ S[1,k](**f**(**r**(PRn(i,j,k))+**hi**))/k
= S[1,k]**f**(**r**(PR1m(i',j',k')))/k.
Set k=det[**UR1**]✱det[**G**(**a**)] and observe
S[bn(MR∩PRn)](S[1,k](**f**(**r**(PR(i',j',k')))))
= S[bn(MR∩PRm)](**f**(**r**(PR1m(i',j',k')))).
Consequently
S[bn(MR∩PRn)](**f**(**r**(PRn(i,j,k)))•**m**(PRn(i,j,k)))
≈ Sbn(MR∩PRm)](**f**(**r**(PR1m(i',j',k')))•**pr**/m^3)
Thus
s(**f**) ≡ lim (S[bn(MR∩PRn)](**f**(**r**(PRn(i,j,k)))/n^3))
= lim (S[bn(MR∩PRm)](**f**(**r**(PR1m(i',j',k')))/m^3))
Consequently,
I3[MX]•(**f**;**dR1**) = lim (S[bn(MR∩PRm)](**f**(**r**(PR1m(i',j',k')))
•**pR1**/m^3))
= **s**(**f**)•**pR1**
I3[MR]**∧**(**f**;**dR1**) = lim (S[bn(MR∩PRm)](**f**(**r**(PR1m(i',j',k')))
∧pR1/m^3))
= **s**(**f**)**∧pR1**
I3[MR]✱(**f**;**dR1**) = lim (S[bn(MR∩PRm](**f**(**r**(PR1m(i',j',k')))
✱**pR1**/m^3))
= **s**(**f**)✱**pR1**.

The dual is proven similarly. qed

Thus, for a given MR or MX, **sr**(**f**) and **sx**(**f**) are invariant with respect to partitions.

For **x**(MR) = MX, the relationship between **sr**(**f**) and **sx**(**f**) is given in the following theorem.

Theorem 25 (the second partition theorem)

Given

f(**r**(**x**,ai)) with a basic continuous local gradient everywhere;
MR, a measurable set in V3(**r**);
MX, a measurable set in V3(**x**);
pr = pr1 ∗**u1**+pr2∗**u2**+pr3∗**u3** a partition vector;
px = px1∗**u1**+px2∗**u2**+px3∗**u3** a partition vector;
PRn, a set of partitions defined from **pr**;
PXn, a set of partitions defined from **px**;

for

MX = **x**(MR);

then

I3[MX]•(**f**(**x**(**r**,ai));**dX1**) ≡ **sx**(**f**)•**pX1** = **sr**(**f**)•**pX1**
I3[MX]∧(**f**(**x**(**r**,ai));**dX1**) ≡ **sx**(**f**)∧**pX1** = **sr**(**f**)∧**pX1**
I3[MX]∗(**f**(**x**(**r**,ai));**dX1**) ≡ **sx**(**f**)∗**pX1** = **sr**(**f**)∗**pX1**.

(1667)

Dually

for

MR = **r**(MX);

then

I3[MR]•(**f**(**r**(**x**,ai));**dR1**) ≡ **sr**(**f**)•**pR1** = **sx**(**f**)•**pR1**
I3[MR]∧(**f**(**r**(**x**,ai));**dR1**) ≡ **sr**(**f**)∧**pR1** = **sx**(**f**)∧**pR1**
I3[MR]∗(**f**(**r**(**x**,ai));**dR1**) ≡ **sr**(**f**)∗**pR1** = **sx**(**f**)∗**pR1**.

(1668)

Proof:

Consider first the partition PRn(i,j,k) of V3(**r**) by the partition vector **pr=u1+u2+u3**
and an similar partition PXm(i,j,k) of V3(**x**) by the partition vector **px=u1+u2+u3**. Let

xn = **x**(**r**(PRn(i,j,k)),ai)

be found in PXm(i',j',k') and

dxn ≡ **xn** – **x**(PXm(i',j',k')).

Then

drn ≡ **r**(**x**(PXm(i',j',k')),ai) – **r**(PRn(i,j,k))
≡ **rn** – **r**(PRn(i,j,k))
= **dxn**•**T**[**D3**[**r**(PRn(i,j,k))]∗(**x**|ai;**dr**)]

and

dxn = **drn**•**T**[**D3**[**x**(PXm(i',j',k'))]∗(**r**|ai;**dx**)].

Now **rn** may not lie in PRn(i,j,k). So we ask how many of the PRn(i,j,k) subsets are mapped into PXm(i',j',k')? To answer this question, let

D3[**r**(PRn(i,j,k))]∗(**x**|ai;**dr**) ≡ b1∗**u1X1**∗**u1**
+ b2∗**u2X1**∗**u2** + b3∗**u3X1**∗**u3**
= **T**[**UX1**]•[**G**(**b**)]

for **b** = b1∗**u1** + b2∗**u2** + b3∗**u3**. Then formally constructing an identity transformation of the PR1n partition into V3(**x**), we see that **b** is a dilation vector and the dilation theorem applies, namely

(PXm(i',j',k')|**pr1**)/m(PXm(i',j',k')|**px1**)
= det[**UX1**]∗det[**G**(**b**)]
= det[**T**[**D3**[**r**(PRn(i,j,k))]∗(**x**|ai;**dr**)]]
= det[**D3**[**r**(PRn(i,j,k))]∗(**x**|ai;**dr**)].

In the limit then,

k(**r**(PRn(i,j,k))) = det[**D3**[**r**(PRn(i,j,k))]∗(**x**|ai;**dr**)]

subsets of PR1n in the vicinity of **r**(PRn(i,j,k)) are mapped into PXm(i',j',k') by **x**(**r**,ai).

Conversely 1/k(**r**(PRn(i,j,k))) = det[**D3**[**x**(PXm(i',j',k'))]∗(**r**|ai;**dx**)] subsets of PX1n in the vicinity of **x**(PXm(i',j',k')) are mapped into PRn(i,j,k) by **r**(**x**,ai).

Also,

f(**r**(PRn(i,j,k)),ai)
= **f**(**xn**,ai)
= **f**(**x**(PXm(i',j',k'))+**dxn**,ai)
= **f**(**x**(PXm(i',j',k'))) + **dxn**•**T**[**D3**[**x**(PXm(i',j',k'))]∗(**f**|ai;**dx**)]

and

f(**x**(PXm(i',j',k')),ai)
= **f**(**r**(PRn(i,j,k))+**drn**,ai)
≡ **f**(**r**(PRn(i,j,k))) + **drn**•**T**[**D3**[**r**(PRn(i,j,k))]∗(**f**|ai;**dr**)].

Thus,

S[bn(MR∩PRn)](**f**(**r**(PRn(i,j,k)))/n^3)
= S[bn(MR∩PRn)](**f**(**x**(PXm(i',j',k')))
+ **dxnk**•**T**[**D3**[**x**(PXm(i',j',k'))]∗(**f**|ai;**dx**)]/n^3)
= S[bn(MR∩PRn)](S[1,k(**r**(PRn(i,j,k)))]∗(**f**(**x**(PXm(i',j',k')))
+ **dxnk**•**T**[**D3**[**x**(PXm(i',j',k'))]∗(**f**|ai;**dx**)])/n^3)
= S[bn(MX∩PXm)](k(**r**(PRn(i,j,k)))∗**f**(**x**(PXm(i',j',k')))
+ S[1,k(**r**(PRn(i,j,k)))]
(**dxnk**•**T**[**D3**[**x**(PXm(i',j',k'))]∗(**f**|ai;**dx**)])/n^3)
= S[bn(MX∩PXm)](**f**(**x**(PXm(i',j',k')))/(n^3/k(**r**(PRn(i,j,k))))
+ S[1,k(**r**(PRn(i,j,k)))]
(**dxnk**•**T**[**D3**[**x**(PXm(i',j',k'))]∗(**f**|ai;**dx**)])/n^3)
= S[bn(MX∩PXm)](**f**(**x**(PXm(i',j',k')))/m^3)
+ S[1,k(**r**(PRn(i,j,k)))]
(**dxnk**•**T**[**D3**[**x**(PXm(i',j',k'))]∗(**f**|ai;**dx**)]/n^3)
≡ S[bn(MX∩PXm)](**f**(**x**(PXm(i',j',k')))/m^3).

Consequently,

sx(**f**) = **sr**(**f**).

Now applying the first partition theorem once

I3[MR]•(**f**;**dR1**) = **sr**(**f**)•**dR1**= I3[MR]•(**f**;**dr**) = **sr**(**f**)•**pr** = **sx**(**f**)•**pr**
I3[MX]•(**f**;**dX1**) = **sx**(**f**)•**dX1** =I3[MX]•(**f**;**dx**) = **sx**(**f**)•**px** = **sr**(**f**)•**px**

and again,

I3[MR]•(**f**;**dr**) = **sr**(**f**)•**pr** = I3[MR]•(**f**;**dR1**) = **sr**(**f**)•**pR1** = **sx**(**f**)•**pR1**
I3[MX]•(**f**;**dx**) = **sx**(**f**)•**px** = I3[MX]•(**f**;**dX1**) = **sx**(**f**)•**pX1** = **sr**(**f**)•**pX1**

Consequently,

I3[MX]•(**f**(**x**(**r**,ai));**dX1**) = **sx**(**f**)•**pX1** = **sr**(**f**)•**pX1**
I3[MX]∧(**f**(**x**(**r**,ai));**dX1**) = **sx**(**f**)∧**pX1** = **sr**(**f**)∧**pX1**
I3[MX]*(**f**(**x**(**r**,ai));**dX1**) = **sx**(**f**)***pX1** = **sr**(**f**)***pX1**.

The duals are proven similarly.

qed

Corollary

Given mutually related entities
MR, a measurable set in V3(**r**);
MX, a measurable set in V3(**x**);
SECT(**r**, **u1R1**,**u2R1**,**u3R1**);
SECT(**x**, **u1X1**,**u2X1**,**u3X1**);
pR1 = p1R1***u1R1**+p2R1***u2R1**+p3R1***u3R1**,
a partition vector;
pX1 = p1X1***u1R1**+p2X1***u2X1**+p3X1***u3X1**,
a partition vector;
PR1n, a set of partitions defined from **pR1**;
PX1n, a set of partitions defined from **pX1**;
f(**r**(**x**,ai)) with a basic continuous local gradient everywhere;

for

G(pR1/pX1) ≡ (p1R1/p1X1)***u1*****u1**
+ (p2R2/p2X2)***u2*****u2**
+ (p3R3/p3X3)***u3*****u3**

then

I3[MR]•(**f**(**r**(**x**,ai));**dR1**)
= I3[MX]•(**f**(**x**(**r**,ai));**dX1**)•**UX1**$^{-1}$•**G**(pR1/pX1)•**UR1**
/det[**UX1**] (1669)

I3[MX]•(**f**(**x**(**r**,ai));**dX1**)
= I3[MR]•(**f**(**r**(**x**,ai));**dR1**)•**UR1**$^{-1}$•**G**(pX1/pR1)•**UX1**
/det[**UR1**]. (1670)

Proof:

pX1•**UX1**⁻¹•**G**(pR1/pX1)•**UR1**/det[**UX1**]
= (p1X1∗**u1**∗**u1** +p2X1∗**u2**∗**u2** +p3X1∗**u3**∗**u3**)
•**G**(pR1/pX1)•**UR1**
= p1R1∗**u1R1** +p2R1∗**u2R1** +p3R1∗**u3R1**
= **pR1**.

Therefore,

I3[MX]•(**f**(**x**(**r**,ai));**dX1**)•**UX1**⁻¹•**G**(pR1/pX1)•**UR1**/det[**UX1**]
= **sx**(**f**)•**pX1**•**UX1**⁻¹•**G**(pR1/pX1)•**UR1**/det[**UX1**]
= **sx**(**f**)•**pR1**
= **sr**(**f**)•**pR1**
= I3[MR]•(**f**(**r**(**x**,ai));**dR1**).

The dual is proven similarly. qed

Theorem 26 (integration of sectional gradients)

Given at observation ai
MR, a measurable set in V3(**r**);
SMR the surface of MR;
pR1 = p1R1∗**u1R1**+p2R1∗**u2R1**+p3R1∗**u3R1**,
a partition vector;
PR1n, a set of partitions defined from **pR1**;
SMRPE|**pR1**, the positive exterior surface of MR;
SMRPI|**pR1**, the positive interior surface of MR;
f(**r**(**x**,ai)) with a basic continuous sectional gradient over MR;
then

I3[MR]•(**D3**[**r**]∗(**f**|ai;**dR1**)•[**UR1**]⁻¹•**T**[**UR1**]⁻¹;**dR1**)
=lim S[bn(SMRPE|**pR1**∩PR1n)](**f**(**r**(PR1n(i,j,k))+**dR1**))
– S[bn(SMRPI|**pR1**∩PR1n)](**f**(**r**(PR1n(i,j,k)))) (1671)
= **I**[SMRPI|**pR1**,SMRPE|**pR1**](**f**(**sr**)). (1672)

Given
MX, a measurable set in V3(**x**);
pX1 = p1X1∗**u1X1**+p2X1∗**u2X1**+p3X1∗**u3x1**,
a partition vector;
PX1n, a set of partitions defined from **pX1**;
SMXPE|**pX1**, the positive exterior surface of MX;
SMXPI|**pX1**, the positive interior surface of MX;
f(**x**(**r**,ai)) with a basic continuous sectional gradient over MX;

then

$$\mathbf{I}3[MX]\bullet(\mathbf{D3}[\mathbf{x}]*(\mathbf{f}|ai;\mathbf{dX1})\bullet[\mathbf{UX1}]^{-1}\bullet\mathbf{T}[\mathbf{UX1}]^{-1};\mathbf{dX1})$$
$$=\lim S[bn(SMXPE|\mathbf{pX1}\cap PX1n)](\mathbf{f}(\mathbf{r}(PX1n(i,j,k))+\mathbf{dX1}))$$
$$- S[bn(SMXPI|\mathbf{pX1}\cap PX1n)](\mathbf{f}(\mathbf{r}(PX1n(i,j,k)))) \quad (1673)$$
$$= \mathbf{I}[SMXPI|\mathbf{pX1},SMXPE|\mathbf{pX1}](\mathbf{f}(\mathbf{sx})). \quad (1674)$$

The proof is similar to the development of equation (780).

Corollary

$$\mathbf{I3}[MR]\bullet(\mathbf{D3}[\mathbf{r}]*(\mathbf{f}|ai;\mathbf{dR1})\bullet[\mathbf{UR1}]^{-1}\bullet\mathbf{T}[\mathbf{UR1}]^{-1};\mathbf{dR1})$$
$$= \mathbf{I3}[MX]\bullet(\mathbf{D3}[\mathbf{x}]*(\mathbf{f}|ai;\mathbf{dX1})\bullet[\mathbf{UX1}]^{-1}\bullet\mathbf{T}[\mathbf{UX1}]^{-1};\mathbf{dX1}) \quad (1675)$$
$$\mathbf{I}[SMRPI|\mathbf{pR1},SMRPE|\mathbf{pR1}](\mathbf{f}(\mathbf{sr}))$$
$$= \mathbf{I}[SMXPI|\mathbf{pX1},SMXPE|\mathbf{pX1}](\mathbf{f}(\mathbf{sx})). \quad (1676)$$

Proof:

$$\mathbf{I3}[MR]\bullet(\mathbf{D3}[\mathbf{r}]*(\mathbf{f}|ai;\mathbf{dR1})\bullet[\mathbf{UR1}]^{-1}\bullet\mathbf{T}[\mathbf{UR1}]^{-1};\mathbf{dR1})$$
$$= \mathbf{I}3[MR]\bullet(\mathbf{D3}[\mathbf{x}]*(\mathbf{f}(\mathbf{x}(\mathbf{r},a1))|a1;\mathbf{dX1})\bullet[\mathbf{UX1}^{-1}]\bullet[\mathbf{G}^{-1}(\mathbf{p})]\bullet\mathbf{UR1}$$
$$\bullet[\mathbf{UR1}]^{-1}\bullet\mathbf{T}[\mathbf{UR1}]^{-1};\mathbf{dR1}) \quad \text{from Theorem 13}$$
$$= \lim S\,[bn(Pn\cap MX)]\bullet(\mathbf{D3}[\mathbf{x}]*(\mathbf{f}(\mathbf{x}(\mathbf{r},a1))|a1;\mathbf{dX1})$$
$$\bullet[\mathbf{UX1}^{-1}]\bullet[\mathbf{G}^{-1}(\mathbf{p})]\bullet\mathbf{T}[\mathbf{UR1}]^{-1}$$
$$\bullet\mathbf{T}[\mathbf{UR1}]\bullet\mathbf{G}(\mathbf{p})\bullet\mathbf{T}[\mathbf{UX1}^{-1}]\bullet\mathbf{dX1})$$

from equation (1027)

$$= \lim S\,[bn(Pn\cap MX)]\bullet(\mathbf{D3}[\mathbf{x}]*(\mathbf{f}(\mathbf{x}(\mathbf{r},a1))|a1;\mathbf{dX1})$$
$$\bullet[\mathbf{UX1}^{-1}]\bullet\mathbf{T}[\mathbf{UX1}^{-1}]\bullet\mathbf{dX1})$$
$$= \mathbf{I3}[MX]\bullet(\mathbf{D3}[\mathbf{x}]*(\mathbf{f}|ai;\mathbf{dX1})\bullet[\mathbf{UX1}]^{-1}\bullet\mathbf{T}[\mathbf{UX1}]^{-1};\mathbf{dX1}).$$

This proves the first proposition.
The second is a direct consequence of the theorem. qed

Theoretical Physics: Integration

Theorem 27 (integration of sectional gradients over measurable sets)
Given at observation a1
MR be a measurable set in V3(**r**);
SMR the surface of MR;
sr, a location on SMR;
pR1 = p1R1∗**u1R1**+p2R1∗**u2R1**+p3R1∗**u3R1**, a partition vector;
PR1n, a set of partitions defined from **pR1**;
SMRPE|**pR1**, the positive exterior surface of MR;
SMRPI|**pR1**, the positive interior surface of MR;
f(**r**(**x**,ai)) a generalized function over V3(**r**);
for
I[SMR] defined as in equation (869)
then

I3[V3]•(**f**(**r**(**x**,a1))
∗(**D3**[**r**]∗(ur(**r**,a1)|MR;**dR1**) + **D3**[**r**]∗(vr(**r**,a1)|MR;**dR1**))
•[**UR1**]$^{-1}$•**T**[**UR1**]$^{-1}$|a1;**dR1**)
≡ **I**[SMR](**f**(**sr**,a1)∗(ur(**sr**+d1R1∗**u1R1**)|MR
+ vr(**sr**+d1R1∗**u1R1**)|MR−1
+ ur(**sr**+d2R1∗**u2R1**)|MR
+ vr(**sr**+d2R1∗**u2R1**)|MR−1
+ ur(**sr**+d3R1∗**u3R1**)|MR
+ vr(**sr**+d3R1∗**u3R1**)|MR−1)). (1677)

Dually
given at observation a1
MX be a measurable set in V3(**r**);
SMX the surface of MX;
sx, a particle on SMX
pX1 = p1X1∗**u1X1**+p2X1∗**u2X1**+p3X1∗**u3X1**, a partition vector;
PX1n, a set of partitions defined from **pX1**;
SMXPE|**pX1**, the positive exterior surface of MX;
SMXPI|**pX1**, the positive interior surface of MX;
f(**x**(**r**,ai)) a generalized function over V3(**x**)
for
I[SMX] defined as in equation (869)

then

$$
\begin{aligned}
&\mathbf{I3}[\mathrm{V3}]\bullet(\mathbf{f}(\mathbf{x}(\mathbf{r},\mathrm{a1}))\\
&\qquad\ast(\mathbf{D3}[\mathbf{x}]\ast(\mathrm{ux}(\mathbf{x},\mathrm{a1})|\mathrm{MX};\mathbf{dX1}) + \mathbf{D3}[\mathbf{x}]\ast(\mathrm{vx}(\mathbf{x},\mathrm{a1})|\mathrm{MX};\mathbf{dX1}))\\
&\qquad\qquad\bullet[\mathbf{UX1}]^{-1}\bullet\mathbf{T}[\mathbf{UX1}]^{-1}|\mathrm{a1};\mathbf{dX1})\\
&\equiv \mathbf{I}[\mathrm{SMX}](\mathbf{f}(\mathbf{sx},\mathrm{a1})\ast(\mathrm{ux}(\mathbf{sx}+\mathrm{d1X1}\ast\mathbf{u1X1})|\mathrm{MX}\\
&\qquad + \mathrm{vx}(\mathbf{sx}+\mathrm{d1X1}\ast\mathbf{u1X1})|\mathrm{MX}-1\\
&\qquad + \mathrm{ux}(\mathbf{sx}+\mathrm{d2X1}\ast\mathbf{u2X1})|\mathrm{MX}\\
&\qquad + \mathrm{vx}(\mathbf{sx}+\mathrm{d2X1}\ast\mathbf{u2X1})|\mathrm{MX}-1\\
&\qquad + \mathrm{ux}(\mathbf{sx}+\mathrm{d3X1}\ast\mathbf{u3X1})|\mathrm{MX}\\
&\qquad + \mathrm{vx}(\mathbf{sx}+\mathrm{d3X1}\ast\mathbf{u3X1})|\mathrm{MX}-1)). \qquad (1678)
\end{aligned}
$$

Proof:

$$
\begin{aligned}
&\mathbf{I3}[\mathrm{V3}]\bullet(\mathbf{f}(\mathbf{r}(\mathbf{x},\mathrm{a1}))\\
&\qquad\ast(\mathbf{D3}[\mathbf{r}]\ast(\mathrm{ur}(\mathbf{r},\mathrm{a1})|\mathrm{MR};\mathbf{dR1}) + \mathbf{D3}[\mathbf{r}]\ast(\mathrm{vr}(\mathbf{r},\mathrm{a1})|\mathrm{MR};\mathbf{dR1}))\\
&\qquad\qquad\bullet[\mathbf{UR1}]^{-1}\bullet\mathbf{T}[\mathbf{UR1}]^{-1}|\mathrm{a1};\mathbf{dR1})\\
&= \lim \mathrm{S}[\mathrm{bn}(\mathrm{V3}\cap\mathrm{PR1n})](\mathbf{f}(\mathbf{r}(\mathrm{PR1n}(i,j,k)))\\
&\qquad\ast(\mathbf{D3}[\mathbf{r}(\mathrm{PR1n}(i,j,k)))]\ast(\mathrm{ur}(\mathbf{r})|\mathrm{MR};\mathbf{dR1})\\
&\qquad\quad + \mathbf{D3}[\mathbf{r}(\mathrm{PR1n}(i,j,k))]\ast(\mathrm{vr}(\mathbf{r})|\mathrm{MR};\mathbf{dR1}))\\
&\qquad\qquad\bullet[\mathbf{UR1}]^{-1}\bullet\mathbf{T}[\mathbf{UR1}]^{-1}\\
&\qquad\qquad\quad\bullet\mathbf{m}(\mathrm{PR1n}(i,j,k))\\
&= \lim \mathrm{S}[\mathrm{bn}(\mathrm{SMR}\cap\mathrm{PR1n})](\mathbf{f}(\mathbf{sr}(\mathrm{PR1n}(i,j,k)))\\
&\qquad\ast(\mathbf{D3}[\mathbf{sr}(\mathrm{PR1n}(i,j,k))]\ast(\mathrm{ur}(\mathbf{r})|\mathrm{MR};\mathbf{dR1})\\
&\qquad\quad + \mathbf{D3}[\mathbf{sr}(\mathrm{PR1n}(i,j,k))]\ast(\mathrm{vr}(\mathbf{r})|\mathrm{MR};\mathbf{dR1}))\\
&\qquad\qquad\quad\bullet\mathbf{m}(\mathrm{PR1n}(i,j,k)))\\
&= \lim \mathrm{S}[\mathrm{bn}(\mathrm{SMR}\cap\mathrm{PR1n})](\mathbf{f}(\mathbf{sr}(\mathrm{PR1n}(i,j,k)))\\
&\qquad\ast(((\mathrm{ur}(\mathbf{sr}+\mathrm{d1R1}\ast\mathbf{u1R1})|\mathrm{MR}\\
&\qquad\quad+\mathrm{vr}(\mathbf{sr}+\mathrm{d1R1}\ast\mathbf{u1R1})|\mathrm{MR}-1)\ast\mathbf{u1R1}/\mathrm{d1R1})\\
&\qquad + ((\mathrm{ur}(\mathbf{sr}+\mathrm{d2R1}\ast\mathbf{u2R1})|\mathrm{MR}\\
&\qquad\quad + \mathrm{vr}(\mathbf{sr}+\mathrm{d2R1}\ast\mathbf{u2R1})|\mathrm{MR}-1)\ast\mathbf{u2R1}/\mathrm{d2R1})\\
&\qquad + ((\mathrm{ur}(\mathbf{sr}+\mathrm{d3R1}\ast\mathbf{u3R1})|\mathrm{MR}\\
&\qquad\quad + \mathrm{vr}(\mathbf{sr}+\mathrm{d3R1}\ast\mathbf{u3R1})|\mathrm{MR}-1)\ast\mathbf{u3R1}/\mathrm{d3R1}))\\
&\qquad\qquad\bullet[\mathbf{UR1}]^{-1}\bullet\mathbf{T}[\mathbf{UR1}]^{-1}\\
&\qquad\bullet(\mathrm{p1R1}\ast\mathbf{u1R1} + \mathrm{p2R1}\ast\mathbf{u2R1} + \mathrm{p3R1}\ast\mathbf{u3R1})/n^3\\
&= \lim \mathrm{S}[\mathrm{bn}(\mathrm{SMR}\cap\mathrm{PR1n})](\mathbf{f}(\mathbf{sr}(\mathrm{PR1n}(i,j,k)))\\
&\qquad\ast((\mathrm{ur}(\mathbf{sr}+\mathrm{d1R1}\ast\mathbf{u1R1})|\mathrm{MR}\\
&\qquad\quad + \mathrm{vr}(\mathbf{sr}+\mathrm{d1R1}\ast\mathbf{u1R1})|\mathrm{MR}-1)\ast(\mathrm{p1R1}/\mathrm{d1R1})/n^3\\
&\qquad + (\mathrm{ur}(\mathbf{sr}+\mathrm{d2R1}\ast\mathbf{u2R1})|\mathrm{MR}\\
&\qquad\quad + \mathrm{vr}(\mathbf{sr}+\mathrm{d2R1}\ast\mathbf{u2R1})|\mathrm{MR}-1)\ast(\mathrm{p2R1}/\mathrm{d2R1})/n^3\\
&\qquad + (\mathrm{ur}(\mathbf{sr}+\mathrm{d3R1}\ast\mathbf{u3R1})|\mathrm{MR}\\
&\qquad\quad + \mathrm{vr}(\mathbf{sr}+\mathrm{d3R1}\ast\mathbf{u3R1})|\mathrm{MR}-1)\\
&\qquad\qquad\ast(\mathrm{p3R1}/\mathrm{d3R1})/n^3))
\end{aligned}
$$

= lim S[bn(SMR∩PR1n)](**f**(**sr**(PR1n(i,j,k)))
∗(ur(**sr**+d1R1∗**u1R1**)|MR
+ vr(**sr**+d1R1∗**u1R1**)|MR−1
+ ur(**sr**+d2R1∗**u2R1**)|MR
+ vr(**sr**+d2R1∗**u2R1**)|MR−1
+ ur(**sr**+d3R1∗**u3R1**)|MR
+ vr(**sr**+d3R1∗**u3R1**)|MR−1))
≡ **I**[SMR](**f**(**sr**)∗(ur(**sr**+d1R1∗**u1R1**)|MR
+ vr(**sr**+d1R1∗**u1R1**)|MR−1
+ ur(**sr**+d2R1∗**u2R1**)|MR
+ vr(**sr**+d2R1∗**u2R1**)|MR−1
+ ur(**sr**+d3R1∗**u3R1**)|MR
+ vr(**sr**+d3R1∗**u3R1**)|M−1)).

The dual is proven similarly.

qed

Theorem 28 (integration of sectional gradients over measurable sets)

Given at observation a1
MR, a measurable set in V3(**r**);
SMR, the surface of MR;
sr, a location on SMR;
pR1 = p1R1∗**u1R1**+p2R1∗**u2R1**+p3R1∗**u3R1**,
a partition vector;
PR1n, a set of partitions defined from **pR1**;
SMRPE|**pR1**, the positive exterior surface of MR;
SMRPI|**pR1**, the positive interior surface of MR;
f(**r**(**x**,ai)) continuous,
bounded,
and asymptotic over V3 (**r**)
with a basic continuous sectional gradient over MR;

for
I[SMR] defined as in equation (869)
then

I3[V3]•((vr(**r**,a1)|MR −ur(**r**,a1)|MR)
∗**D3**[**r**]∗(**f**(**r**(**x**,a1)|a1;**dR1**))•[**UR1**]$^{-1}$•**T**[**UR1**]$^{-1}$|a1;**dR1**)
= **0**. (1679)

Dually
given at observation a1
MX be a measurable set in V3(**r**);
SMX the surface of MX;
sx, a location on SMX;
pX1 = p1X1 ∗ **u1X1**+p2X1 ∗ **u2X1**+p3X1 ∗ **u3X1**,
a partition vector;
PX1n, a set of partitions defined from **pX1**;
SMXPE|**pX1**, the positive exterior surface of MX;
SMXPI|**pX1**, the positive interior surface of MX;
f(**x**(**r**,ai)) continuous,
bounded,
and asymptotic over V3 (**x**)
with a basic continuous sectional gradient over MX;
for
I[SMX] defined as in equation (869)

then

I3[V3]•((vx(**x**,a1)|MX −ux(**x**,a1)|MX)
∗**D3**[**x**]∗(**f**(**x**(**r**,a1)|a1;**dX1**))•[**UX1**]$^{-1}$•**T**[**UX1**]$^{-1}$|a1;**dX1**)
= **0**. (1680)

Proof:
I3[V3]•((vr(**r**,a1)|MR −ur(**r**,a1)|MR)
∗**D3**[**r**]∗(**f**(**r**(**x**,a1)|a1;**dR1**))
•[**UR1**]$^{-1}$•**T**[**UR1**]$^{-1}$|a1;**dR1**)
= **I**3[SMR]•(**D3**[**r**]∗(**f**(**r**)|a1;**dR1**)
•[**UR1**]$^{-1}$•**T**[**UR1**]$^{-1}$|**a1**;**dR1**)
= lim S[bn(SMR∩PR1n)]
(**f**(**sr**(PR1n(i,j,k))+d1R1 ∗ **u1R1**)−**f**(**sr**(PR1n(i,j,k)))
+ **f**(**sr**(PR1n(i,j,k))+d2R1 ∗ **u2R1**)−**f**(**sr**(PR1n(i,j,k)))
+ **f**(**sr**(PR1n(i,j,k))+d3R1 ∗ **u3R1**)−**f**(**sr**(PR1n(i,j,k))))
= **0** since **f** is continuous.
The dual is proven similarly. qed

Theoretical Physics: Integration

Theorem 29 (integration of sectional gradients over measurable sets)

Given at observation a1
MR be a measurable set in V3(**r**);
SMR the surface of MR;
sr, a location on SMR;
pR1 = p1R1∗**u1R1+**p2R1∗**u2R1+**p3R1∗**u3R1**,
a partition vector;
PR1n, a set of partitions defined from **pR1**;
SMRPE|**pR1**, the positive exterior surface of MR;
SMRPI|**pR1**, the positive interior surface of MR;
f(**r**(**x**,ai)) continuous,
bounded,
and asymptotic over V3 (**r**)
with a basic continuous sectional gradient over MR;
for
I[SMR] defined as in equation (869)
then

I3[V3]•((vr(**r**,a1)|MR −ur(**r**,a1)|MR)
∗**D3[r]**∗(**f**(**r**(**x**,a1)|a1;**dR1**))•[**UR1**]$^{-1}$•**T**[**UR1**]$^{-1}$|a1;**dR1**)
= **0**. (1681)

Dually
given at observation a1
MX be a measurable set in V3(**r**);
SMX the surface of MX;
sx, a location on SMX;
pX1 = p1X1∗**u1X1+**p2X1∗**u2X1+**p3X1∗**u3X1**,
a partition vector;
PX1n, a set of partitions defined from **pX1**;
SMXPE|**pX1**, the positive exterior surface of MX;
SMXPI|**pX1**, the positive interior surface of MX;
f(**x**(**r**,ai)) continuous,
bounded,
and asymptotic over V3 (**x**)
with a basic continuous sectional gradient over MX;
for
I[SMX] defined as in equation (869)

then

I3[V3]•((vx(**x**,a1)|MX −ux(**x**,a1)|MX)
∗**D3**[**x**]∗(**f**(**x**(**r**,a1)|a1;**dX1**))•[**UX1**]$^{-1}$•**T**[**UX1**]$^{-1}$|a1;**dX1**)
= **0**. (1682)

Proof:

I3[V3]•((vr(**r**,a1)|MR −ur(**r**,a1)|MR)
∗**D3**[**r**]∗(**f**(**r**(**x**,a1)|a1;**dR1**))
•[**UR1**]$^{-1}$•**T**[**UR1**]$^{-1}$|a1;**dR1**)
= **I**3[SMR]•(**D3**[**r**]∗(**f**(**r**)|a1;**dR1**)
•[**UR1**]$^{-1}$•**T**[**UR1**]$^{-1}$|**a1**;**dR1**)
= lim S[bn(SMR∩PR1n)]
(**f**(**sr**(PR1n(i,j,k))+d1R1∗**u1R1**)−**f**(**sr**(PR1n(i,j,k)))
+ **f**(**sr**(PR1n(i,j,k))+d2R1∗**u2R1**)−**f**(**sr**(PR1n(i,j,k)))
+ **f**(**sr**(PR1n(i,j,k))+d3R1∗**u3R1**)−**f**(**sr**(PR1n(i,j,k))))
= **0** since **f** is continuous.

The dual is proven similarly. qed

Theorem 30 (integration of sectional gradients over measurable sets)

Given at observation a1
MR, a measurable set in V3(**r**);
SMR, the surface of MR;
sr, a location on SMR;
pR1 = p1R1∗**u1R1**+p2R1∗**u2R1**+p3R1∗**u3R1**,
a partition vector;
PR1n, a set of partitions defined from **pR1**;
SMRPE|**pR1**, the positive exterior surface of MR;
SMRPI|**pR1**, the positive interior surface of MR;
f(**r**(**x**,ai)) continuous,
bounded,
and asymptotic over V3 (**r**)
with a basic continuous sectional gradient over MR;

for
I[SMR] defined as in equation (869)

then

I3[V3]•((**D3**[**r**]∗(**f**(**r**,a1)∗ur(**r**)|MR;**dR1**)
+**D3**[**r**]∗(**f**(**r**,a1)∗vr(**r**)|MR;**dR1**))
•[**UR1**]$^{-1}$•**T**[**UR1**]$^{-1}$|a1;**dR1**)

= **I**[SMR](**f**(**sr**(PR1n(i,j,k))+d1R1∗**u1R1**)
∗(ur(**sr**+d1R1∗**u1R1**)|MR
+ vr(**sr**+d1R1∗**u1R1**)|MR)
– **f**(**sr**(PR1n(i,j,k)))
+ **f**(**sr**(PR1n(i,j,k))+d2R1∗**u2R1**)
∗(ur(**sr**+d2R1∗**u2R1**)|MR
+ vr(**sr**+d2R1∗**u2R1**)|MR)
–**f**(**sr**(PR1n(i,j,k)))
+ **f**(**r**(PR1n(i,j,k))+d3R1∗**u3R1**)
∗ur((**r**+d3R1∗**u3R1**)|MR
+ vr(**sr**+d3R1∗**u3R1**)|MR)
–**f**(**sr**(PR1n(i,j,k)))). (1683)

Dually
given at observation a1
MX, a measurable set in V3(**r**);
SMX, the surface of MX;
sx, a location on SMX;
pX1 = p1X1∗**u1X1**+p2X1∗**u2X1**+p3X1∗**u3X1**
a partition vector;
PX1n, a set of partitions defined from **pX1**;
SMXPE|**pX1**, the positive exterior surface of MX;
SMXPI|**pX1**, the positive interior surface of MX;
f(**x**(**r**,ai)) continuous,
bounded,
and asymptotic over V3 (**x**)
with a basic continuous sectional gradient over MX;
for
I[SMX] defined as in equation (869)
then

I3[V3]•((**D3**[**x**]∗(**f**(**x**,a1)∗ux(**x**)|MX;**dX1**)
+**D3**[**x**]∗(**f**(**x**,a1)∗vx(**x**)|MX;**dX1**))
•[**UX1**]$^{-1}$•**T**[**UX1**]$^{-1}$|a1;**dX1**)
= **I**[SMX](**f**(**sx**(PX1n(i,j,k))+d1X1∗**u1X1**)
∗(ux(**sx**+d1X1∗**u1X1**)|MX
+ vx(**sx**+d1X1∗**u1X1**)|MX)
– **f**(**sx**(PX1n(i,j,k)))
+ **f**(**sx**(PX1n(i,j,k))+d2X1∗**u2X1**)
∗(ux(**sx**+d2X1∗**u2X1**)|MX
+ vx(**sx**+d2X1∗**u2X1**)|MX)
–**f**(**sx**(PX1n(i,j,k)))

+ **f**(**x**(PX1n(i,j,k))+d3X1∗**u3X1**)
∗ux((**x**+d3X1∗**u3X1**)|MX
+ vx(**sx**+d3X1∗**u3X1**)|MX)
−**f**(**sx**(PX1n(i,j,k)))). (1684)

Proof:

I3[V3]•((**D3**[**r**]∗(**f**(**r**,a1)∗ur(**r**)|MR;**dR1**)
+**D3**[**r**]∗(**f**(**r**,a1)∗vr(**r**)|MR;**dR1**))
•[**UR1**]$^{-1}$•**T**[**UR1**]$^{-1}$|a1;**dR1**)
= lim S[bn(V3∩PR1n)]((**D**[**r**(PR1n(i,j,k))]∗(**f**(**r**)∗ur(**r**)|MR;**dR1**)
+ **D**[**r**(PR1n(i,j,k))]∗(**f**(**r**)∗vr(**r**)|MR;**dR1**))
•[**UR1**]$^{-1}$•**T**[**UR1**]$^{-1}$•**m**(PR1n(i,j,k)))
= lim(S[bn(V3∩PR1n)]((**f**(**r**(PR1n(i,j,k))+d1R1∗**u1R1**)
∗ur(**r**+d1R1∗**u1R1**)|MR
−**f**(**r**(PR1n(i,j,k)))
∗ur(**r**(PR1n(i,j,k)))|MR
+ **f**(**r**(PR1n(i,j,k))+d1R1∗**u1R1**)
∗vr(**r**+d1R1∗**u1R1**)|MR
−**f**(**r**(PR1n(i,j,k)))
∗vr(**r**(PR1n(i,j,k)))|MR)
∗**u1R1**/d1R1
+ (**f**(**r**(PR1n(i,j,k))+d2R1∗**u2R1**)
∗ur(**r**+d2R1∗**u2R1**)|MR
−**f**(**r**(PR1n(i,j,k)))
∗ur(**r**(PR1n(i,j,k)))|MR
+ **f**(**r**(PR1n(i,j,k))+d2R1∗**u2R1**)
∗vr(**r**+d2R1∗**u2R1**)|MR
−**f**(**r**(PR1n(i,j,k)))
∗vr(**r**(PR1n(i,j,k)))|MR)
∗**u2R1**/d2R1
+ (**f**(**r**(PR1n(i,j,k))+d3R1∗**u3R1**)
∗ur(**r**+d3R1∗**u3R1**)|MR
−**f**(**r**(PR1n(i,j,k)))
∗ur(**r**(PR1n(i,j,k)))|MR
+ **f**(**r**(PR1n(i,j,k))+d3R1∗**u3R1**)
∗vr(**r**+d3R1∗**u3R1**)|MR
−**f**(**r**(PR1n(i,j,k)))
∗vr(**r**(PR1n(i,j,k)))|MR)
∗**u3R1**/d3R1)
•[**UR1**]$^{-1}$•**T**[**UR1**]$^{-1}$
•(p1R1∗**u1R1** + p2R1∗**u2R1** +p3R1∗**u3R1**)/n^3
= lim S[bn(SMR∩PR1n)](**f**(**sr**(PR1n(i,j,k))+d1R1∗**u1R1**)
∗(ur(**sr**+d1R1∗**u1R1**)|MR
+ vr(**sr**+d1R1∗**u1R1**)|MR)
− **f**(**sr**(PR1n(i,j,k)))

+ **f**(**sr**(PR1n(i,j,k))+d2R1∗**u2R1**)
∗(ur(**sr**+d2R1∗**u2R1**)|MR
+ vr(**sr**+d2R1∗**u2R1**)|MR)
−**f**(**sr**(PR1n(i,j,k)))
+ **f**(**r**(PR1n(i,j,k))+d3R1∗**u3R1**)
∗ur(**r**+d3R1∗**u3R1**)|MR
+ vr(**sr**+d3R1∗**u3R1**)|MR)
−**f**(**sr**(PR1n(i,j,k)))
≡ **I**[SMR](**f**(**sr**(PR1n(i,j,k))+d1R1∗**u1R1**)
∗(ur(**sr**+d1R1∗**u1R1**)|MR
+ vr(**sr**+d1R1∗**u1R1**)|MR)
− **f**(**sr**(PR1n(i,j,k)))
+ **f**(**sr**(PR1n(i,j,k))+d2R1∗**u2R1**)
∗(ur(**sr**+d2R1∗**u2R1**)|MR
+ vr(**sr**+d2R1∗**u2R1**)|MR)
−**f**(**sr**(PR1n(i,j,k)))
+ **f**(**r**(PR1n(i,j,k))+d3R1∗**u3R1**)
∗ur((**r**+d3R1∗**u3R1**)|MR
+ vr(**sr**+d3R1∗**u3R1**)|MR)
−**f**(**sr**(PR1n(i,j,k)))).

The dual is proven similarly.

qed

Corollary:
For mutually related entities,

I3[V3]•(**f**(**r**(**x**,a1))
∗(**D3**[**r**]∗(ur(**r**,a1)|MR;**dR1**)
+ **D3**[**r**]∗(vr(**r**,a1)|MR;**dR1**))
•[**UR1**]$^{-1}$•**T**[**UR1**]$^{-1}$|a1;**dR1**)
= **I3**[V3]•(**f**(**x**(**r**,a1))
∗(**D3**[**x**]∗(ux(**x**,a1)|MX;**dX1**)
+ **D3**[**r**]∗(vx(**x**,a1)|MX;**dX1**))
•[**UX1**]$^{-1}$•**T**[**UX1**]$^{-1}$|a1;**dX1**) (1685)

I3[V3]•((vr(**r**,a1)|MR −ur(**r**,a1)|MR)
∗**D3**[**r**]∗(**f**(**r**(**x**,a1)|a1;**dR1**))
•[**UR1**]$^{-1}$•**T**[**UR1**]$^{-1}$|a1;**dR1**)
= **I3**[V3]•((vx(**x**,a1)|MX − ux(**x**,a1)|MX)
∗**D3**[**x**]∗(**f**(**x**(**r**,a1)|a1;**dX1**))
•[**UX1**]$^{-1}$•**T**[**UX1**]$^{-1}$|a1;**dX1**) (1686)

I3[V3]•((**D3**[**r**]∗(**f**(**r**,a1)∗ur(**r**)|MR;**dR1**)
+ **D3**[**r**]∗(**f**(**r**,a1)∗vr(**r**)|MR;**dR1**))
•[**UR1**]$^{-1}$•**T**[**UR1**]$^{-1}$|a1;**dR1**)
= **I3**[V3]•((**D3**[**x**]∗(**f**(**x**,a1)∗ux(**x**)|MX;**dX1**)
+ **D3**[**x**]∗(**f**(**x**,a1)∗vx(**x**)|MX;**dX1**))
•[**UX1**]$^{-1}$•**T**[**UX1**]$^{-1}$|a1;**dX1**). (1687)

Proof:

f(**sr**(**x**,a1) = **f**(**sx**(**r**,a1) = **f**(**sr**(**sx**,a1)

Many other integral operations are possible.

Under the conditions of the previous theorems:

I[V3]•(**f**(**r**,a1)∧(**D**[**r**]∗(ur(**r**)|MR;**dR1**) + **D**[**r**]∗(vr(**r**)|MR;**dR1**))
•[**UR1**]$^{-1}$•**T**[**UR1**]$^{-1}$|a1;**dR1**)
= **I**[SMR](**f**(**sr**,a1)
•((ur(**sr**+d1R1∗**u1R1**)|MR
+ vr(**sr**+d1R1∗**u1R1**)|MR−1)
∗**u1R1**∧(**u2R1**∧**u3R1**)
+ (ur(**sr**+d2R1∗**u2R1**)|MR
+ vr(**sr**+d2R1∗**u2R1**)|MR−1)
∗**u2R1**∧(**u3R1**∧**u1R1**)
+ (ur(**sr**+d3R1∗**u3R1**)|MR
+ vr(**sr**+d3R1∗**u3R1**)|MR−1)
∗**u3R1**∧(**u1R1**∧**u2R1**))) (1688)

I[V3]•(**f**(**r**,a1)∧(**D**[**r**]∗(ur(**r**)|MR;**dR1**)
+ **D**[**r**]∗(vr(**r**)|MR;**dR1**))|,a1;**dR1**)
= **I**[SMR](**f**(**sr**,a1)
•**C**((ur(**sr**+d1R1∗**u1R1**)|MR
+ vr(**sr**+d1R1∗**u1R1**)|MR−1)∗**u1R1**/d1R1
+ (ur(**sr**+d2R1∗**u2R1**)|MR
+ vr(**sr**+d2R1∗**u2R1**)|MR−1)∗**u2R1**/d2R1
+ (ur(**sr**+d3R1∗**u3R1**)|MR
+ vr(**sr**+d3R1∗**u3R1**)|MR−1)∗**u3R1**/d3R1)
•**pR1**) (1689)

I[V3]∧(f(r,a1)∧(D[r]∗(ur(r)|MR;dR1)
+ D[r]∗(vr(r)|MR;dR1))|a1;dR1)
= I[SMR](f(sr,a1)
•C((ur(sr+d1R1∗u1R1)|MR
+ vr(sr+d1R1∗u1R1)|MR–1)∗u1R1/d1R1
+ (ur(sr+d2R1∗u2R1)|MR
+ vr(sr+d2R1∗u2R1)|MR–1)∗u2R1/d2R1
+ (ur(sr+d3R1∗u3R1)|MR
+ vr(sr+d3R1∗u3R1)|MR–1)∗u3R1/d3R1)
•C(pR1)) (1690)

I[V3]∧(f(r)∗(D[r]∗(ur(r)|MR;dR1)
+ D[r]∗(vr(r)|MR;dR1))|a1;dR1)
= I[SMR](f(sr,a1)
∗((ur(sr+d1R1∗u1R1)|MR
+ vr(sr+d1R1∗u1R1)|MR–1)∗u1R1/d1R1
+ (ur(sr+d2R1∗u2R1)|MR
+ vr(sr+d2R1∗u2R1)|MR–1)∗u2R1/d2R1
+ (ur(sr+d3R1∗u3R1)|MR
+ vr(sr+d3R1∗u3R1)|MR–1)∗u3R1/d3R1)
•C(dR1)) (1691)

I[V3]∗(f(r,a1)∗(D[r]∗(ur(r)|MR;dR1)
+ D[r]∗(vr(r)|MR;dR1))|a1;dR1)
= I[SMR](f(sr)
∗((ur(sr+d1R1∗u1R1)|MR
+ vr(sr+d1R1∗u1R1)|MR–1)∗u1R1/d1R1
+ (ur(sr+d2R1∗u2R1)|MR
+ vr(sr+d2R1∗u2R1)|MR–1)∗u2R1/d2R1
+ (ur(sr+d3R1∗u3R1)|MR
+ vr(sr+d3R1∗u3R1)|MR–1)∗u3R1/d3R1)
∗dR1 (1692)

I[V3]∗(f(r,a1)•(D[r]∗(ur(r)|MR;dR1)
+ D[r]∗(vr(r)|MR;dR1))|a1;dR1)
= I[SMR](f(sr)
•((ur(sr+d1R1∗u1R1)|MR
+ vr(sr+d1R1∗u1R1)|MR–1)∗u1R1/d1R1
+ (ur(sr+d2R1∗u2R1)|MR
+ vr(sr+d2R1∗u2R1)|MR–1)∗u2R1/d2R1
+ (ur(sr+d3R1∗u3R1)|MR
+ vr(sr+d3R1∗u3R1)|MR–1)∗u3R1/d3R1)
∗dR1) (1693)

I[V3]∗(f(r,a1)∧(D[r]∗(ur(r)|MR;dR1)
+ D[r]∗(vr(r)|MR;dR1))|a1;dR1)
= I[SMR](f(sr)
∧((ur(sr+d1R1∗u1R1)|MR
+ vr(sr+d1R1∗u1R1)|MR−1)∗u1R1/d1R1
+ (ur(sr+d2R1∗u2R1)|MR
+ vr(sr+d2R1∗u2R1)|MR−1)∗u2R1/d2R1
+ (ur(sr+d3R1∗u3R1)|MR
+ vr(sr+d3R1∗u3R1)|MR−1)∗u3R1/d3R1)
∗dR1) (1694)

I[V3]∗(f(r,a1)∗(D[r]∗(ur(r)|MR;dR1)
+ D[r]∗(vr(r)|MR;dR1))|a1;dR1)
= I[SMR](f(sr)
∗((ur(sr+d1R1∗u1R1|MR)
+ vr(sr+d1R1∗u1R1|MR)−1)∗u1R1/d1R1
+ (ur(sr+d2R1∗u2R1|MR)
+ vr(sr+d2R1∗u2R1|MR)−1)∗u2R1/d2R1
+ (ur(sr+d3R1∗u3R1|MR)
+ vr(sr+d3R1∗u3R1|MR)−1)∗u3R1/d3R1)
∗dR1). (1695)

For ***f*** or *f* continuous, bounded and asymptotic over V3

I[V3]•((vr(r)|MR−ur(r)|MR)∗(D[r]∗(f(r,a1);dR1))
•[UR1]$^{-1}$•T[UR1]$^{-1}$|a1;dR1)
= 0 (1696)

I[V3]•((vr(r)|MR −ur(r)|MR)∗(D[r]∗f(r,a1);dR1))
•[UR1]$^{-1}$•T[UR1]$^{-1}$|a1;dR1)
= 0 (1697)

I[V3]•((vr(r)|MR −ur(r)|MR)∗(D[r]∧(f(r,a1);dR1))|a1;dR1)
= I[SMR](D[sr]∧(f(r,a1);dR1)•dR1) (1698)

I[V3]∧((vr(r)|MR −ur(r)|MR)∗(D[r]∧(f(r,a1);dR1))|a1;dR1)
= I[SMR](D[r]∧(f(r,a1);dR1)∧dR1) (1699)

I[V3]∧((vr(r)|MR−ur(r)|MR)∗(D[r]∗(f(r,a1);dR1))|a1;dR1)
= I[SMR](D[r]∗(f(r,a1);dR1)∧dR1) (1700)

I[V3]∗((vr(r)|MR −ur(r)|MR)∗(D[r]•(f(r,a1);dR1))|a1;dR1)
= I[SMR](D[r]•(f(r,a1);dR1)∗dR1) (1701)

I[V3]∗((vr(**r**)|MR −ur(**r**)|MR)∗(**D**[**r**]∧(**f**(**r**,a1);**dR1**))|a1;**dR1**)
= **I**[SMR](**D**[**r**]∧(**f**(**r**,a1);**dR1**)∗**dR1**) (1702)

I[V3]∗((vr(**r**)|MR −ur(**r**)|MR)∗(**D**[**r**]∗(f(**r**,a1);**dR1**))|a1;**dR1**)
= **I**[SMR](**D**[**r**]∗(f(**r**,a1);**dR1**)∗**dR1**). (1703)

Now at observations *a1* let

MX = SECT(**x1**, **u1X1**,**u2X1**,**u2X1**)

where

SECT(**x1**, **u1X1**,**u2X1**,**u2X1**)
= {**x**|**x**= **x1**+d1R1∗**u1R1**+d2R1∗**u2R1**+d3R1∗**u3R1**, diR1>0}

is the full section.

Then

SMXPE|**pX1** at infinity
SMXPI|**pX1** = **x1**
SMXP0|**pX1**

are the faces of the section.

For **f**(**x**(**r**,a1) a generalized function, by Theorem 27

I3[V3]•(**f**(**x**(**r**,a1))
∗(**D3**[**x**]∗(ux(**x**,a1)|SECT(**x1**, **u1X1**,**u2X1**,**u2X1**);**dX1**)
+ **D3**[**x**]∗(vx(**x**,a1)|SECT(**x1**, **u1X1**,**u2X1**,**u2X1**);**dX1**))
•[**UX1**]$^{-1}$•**T**[**UX1**]$^{-1}$|a1;**dX1**)
= −**f**(**x1**,a1)|SECT(**x1**, **u1X1**,**u2X1**,**u2X1**) (1704)

where **f**(**x1**,a1)|SECT(**x1**, **u1X1**,**u2X1**,**u2X1**) is the limiting value of **f** from SECT(**x1**, **u1X1**,**u2X1**,**u2X1**) at **x1**.

Suppose now *n* such sections
SECT(**x1**, **u1X1**,**u2X1**,**u2X1**),...,SECT(**x1**,**u1Xn**,**u2Xn**,**u2Xn**).

Then for each section

I3[V3]•(**f**(**x**(**r**,a1))
∗(**D3**[**x**]∗(ux(**x**,a1)|SECT(**x1**, **u1Xk**,**u2Xk**,**u2Xk**);**dXk**)
+ **D3**[**x**]∗(vx(**x**,a1)|SECT(**x1**, **u1Xk**,**u2Xk**,**u2Xk**);**dXk**))
•[**UXk**]$^{-1}$•**T**[**UXk**]$^{-1}$|a1;**dXk**)
= −**f**(**x1**,a1)|SECT(**x1**, **u1Xk**,**u2Xk**,**u2Xk**), k= 1...n. (1705)

In Theoretical Physics, except to satisfy the principle of non–collocation, these limits need not be equal. That is, at **r**(**x1**,a1) a generalized function may have *n* distinct limits in addition to its proper value.

Similar statements may be made for the local duals.

In Theoretical Physics, then, **f**(**r1**(**x1**,ai)) may be[83]

continuous, with continuous gradients
continuous, with varying sectional gradients,
discontinuous, with varying sectional gradients

For any given observation *ai* only a finite number[84] of discontinuities, whether in the function or its gradient, are considered.

At a single observation *a1,* a function may be represented almost everywhere as a combination of a continuous function and combinations of step functions.

f(**x**(**r**,a1))
= **f1**(**x**(**r**,a1))
+ S[1,n1](**I3**[SECT(**x1**,**u1X1i**,**u2X1i**,**u3X1i**)]
∗(ux(**x1**|SECT(**x1**,**u1X1i**,**u2X1i**,**u3X1i**);**dX1i**)
•**UX1i**$^{-1}$•**T**[**UX1i**]$^{-1}$•**T**[**C1i**]
+ ...
+ S[1,nm](**I3**[SECT(**xm**,**u1Xmi**,**u2Xmi**,**u3Xmi**)]
∗ux(**xm**|SECT(**xm**,**u1X1mi**,**u2X1mi**,**u3X1mi**);**dX1m**)
•**UX1i**$^{-1}$•**T**[**UX1i**]$^{-1}$•**T**[**Cmi**]
+ S[1,n01](**c01i**
∗ux(**x01**)|SECT(**x01**,**u1X01i**,**u2X02i**,**u3X03i**)
+ ...
+ S[1,n0m](**c0mi**
∗ux(**x0m**)|SECT(**x0m**,**u1X0mi**,**u2X0mi**,**u3X0mi**)
+ S[1,n11](**D3**[SECT(**x11**,**u1X11i**,**u2X11i**,**u3X11i**)]
∗(vx(**x11**|SECT(**x11**,**u1X11i**,**u2X11i**,**u3X11i**);**dXi1**)
•**C11j**]
+ ...
+ S[1,n1m]
(c1m∗**D3**[SECT(**x1m**,**u1X1mi**,**u2X1mi**,**u3X1mi**)]
∗(vx(**x1**|SECT(**x1m**,**u1X1mi**,**u2X1mi**,**u3X1mi**);**dXi1**)
•**C1mj**]
+ ... (1706)

where

f1 is continuous everywhere with continuous gradients;

and

Cij = lim (**D3**[**xi**]∗(**f**(**xi+dX1i**,a1)|a1;**dX1i**)
−**D3**[**xi**]∗(**f1**(**xi**,a1)|a1;**dX1i**))

c0ij= lim **f**(**x0i+dX1i**,a1)|SECT(**x0i**,**u1X1i**,**u2X1i**,**u3X1i**)
− **f1**(**x0i**(**r**,a1))

C1ij the impulse constant

are constants.

Dually,

f(**r**(**x,**a1))
= **f1**(**r**(**x,**a1))
+ S[1,n1](**I3**[SECT(**r1**,**u1R1i**,**u2R1i**,**u3R1i**)]
∗(ur(**r1**|SECT(**r1**,**u1R1i**,**u2R1i**,**u3R1i**);**dR1i**)
•**UR1i**$^{-1}$•**T**[**UR1i**]$^{-1}$•**T**[**C1i**]
+ ...
+ S[1,nm](**I3**[SECT(**rm**,**u1Rmi**,**u2Rmi**,**u3Rmi**)]
∗ur(**rm**|SECT(**rm**,**u1R1mi**,**u2R1mi**,**u3R1mi**);**dR1m**)
•**UR1i**$^{-1}$•**T**[**UR1i**]$^{-1}$•**T**[**Cmi**]
+ S[1,n01](**c01i**
∗ur(**r01**)|SECT(**r01**,**u1R01i**,**u2R02i**,**u3R03i**)
+ ...
+ S[1,n0m](**c0mi**
∗ur(**r0m**)|SECT(**r0m**,**u1R0mi**,**u2R0mi**,**u3R0mi**)
+ S[1,n11](**D3**[SECT(**r11**,**u1R11i**,**u2R11i**,**u3R11i**)]
∗(vr(**r11**|SECT(**r11**,**u1R11i**,**u2R11i**,**u3R11i**);**dRi1**)
•**C11j**]
+ ...
+ S[1,n1m]
(c1m∗**D3**[SECT(**r1m**,**u1R1mi**,**u2R1mi**,**u3R1mi**)]
∗(vr(**r1**|SECT(**r1m**,**u1R1mi**,**u2R1mi**,**u3R1mi**);**dRi1**)
•**C1mj**]
+ ...

(1707)

where

f1 is continuous everywhere with continuous gradients;

and

Cij = lim (**D3**[**ri**]∗(**f**(**ri+dR1i**,a1)|a1;**dR1i**)
−**D3**[**ri**]∗(**f1**(**ri**,a1)|a1;**dR1i**))

c0ij = lim **f**(**r0i+dR1i**,a1)|SECT(**r0i**,**u1R1i**,**u2R1i**,**u3R1i**)
− **f1**(**r0i**(**r**,a1))

C1ij the impulse constant

are constants.

Interchanges

In modern Physics it is common to manipulate compounded differential and/or integral operators by interchanging operators. In Theoretical Physics, however, local and indexed derivatives may *not* be freely interchanged.

In particular, the widely used

D3•(**D3**∗**f**) = **D3**∗(D3•**f**)

and

D(**D3**∗**f**) = **D3**∗(D**f**)

are *generally false*.

Indeed, some interchanges are expressed ambiguously or without meaning. For instance,

D1[**r1**,**r2**]∗(D[a1](**f**|p;d_na)|p;**d**q)
= D[a1](**D1**[**r1**,**r2**]∗(**f**|p;**d**q)|p;d_na)

for n or m indicating the sense, p symbolizing **r1**, **x1**, **r**(**x**,a1), **r**(**x**,a1+da) or **r**(**x1**,a1−da) and q symbolizing **r**|**x1** or **x**|**r1**, is an empty symbolism.

The study of conditions under which operators may be interchanged validly is vast. There are 21 generic types of possible interchanges as seen in the schema:

Operator#1/ Operator#2	D	**D1**	**D3**	I	**I1**	**I3**
D	DD	D**D1**	D**D3**	DI	D**I1**	D**I3**
D1		**D1D1**	**D1D3**	**D1I**	**D1I1**	**D1I3**
D3			**D3D3**	**D3I**	**D3I1**	**D3I3**
I				II	I**I1**	I**I3**
I1					**I1I1**	**I1I3**
I3						**I3I3**

Generic Types of Derivative/Integral Exchanges

Each generic interchange may have up to 8 specific interchanges according to the schema

Operator#1	Operator#2	scalar or vector
•	*	**f**
•	∧	**f**
∧	∧	**f**
*	•	**f**
*	∧	**f**
•	*	f
∧	*	f
*	*	f

Specific Exchanges of Derivatives and Integrals

Each specific type has its own condition for validity. Within each specific type of interchange there also exist variations.

Appendix 4 gives the conditions and restrictions under which interchanges hold in Theoretical Physics.[85]

The Bridge Theorem

The relationship between material and local references is clarified in the following important theorem.

Theorem 31 (bridge theorem: first form)

Given

three ordered observations designated

$\mathbf{r}(\mathbf{x},a1-da)$, $\mathbf{r}(\mathbf{x},a1)$, and $\mathbf{r}(\mathbf{x},a1+da)$

corresponding to

$\mathbf{x}(\mathbf{r},a1-da)$, $\mathbf{x}(\mathbf{r},a1)$, and $\mathbf{x}(\mathbf{r},a1+da)$;

three functions of these observations having similar units

$\mathbf{f}(\mathbf{r}(\mathbf{x},a1-da))$, $\mathbf{f}(\mathbf{r}(\mathbf{x},a1))$, and $\mathbf{f}(\mathbf{r}(\mathbf{x},a1+da))$;

for

$\mathbf{r1}(\mathbf{x1},a1)$ the location of the reference particle $\mathbf{x1}(\mathbf{r1},a1)$;

$\mathbf{r0} \equiv \mathbf{r}(\mathbf{x1},a1-da)$;

$\mathbf{r2} \equiv \mathbf{r}(\mathbf{x1},a1+da)$;

$\mathbf{x0} \equiv \mathbf{x}(\mathbf{r1},a1-da)$;

$\mathbf{x2} \equiv \mathbf{x}(\mathbf{r1},a1+da)$;

$\mathbf{s0} \equiv \mathbf{r}(\mathbf{x0},a1)$;

$\mathbf{s2} \equiv \mathbf{r}(\mathbf{x2},a1)$;

$\mathbf{y0} \equiv \mathbf{x}(\mathbf{r0},a1)$;

$\mathbf{y2} \equiv \mathbf{x}(\mathbf{r2},a1)$;

then

$$\mathbf{f}(\mathbf{r}(\mathbf{x1},a1+da)) = \mathbf{f}(\mathbf{r1}(\mathbf{x},a1+da)) + \mathbf{f}(\mathbf{r2}(\mathbf{x},a1+da)) - \mathbf{f}(\mathbf{r1}(\mathbf{x},a1+da)) \tag{1708}$$

$$\mathbf{f}(\mathbf{r}(\mathbf{x1},a1)) = \mathbf{f}(\mathbf{r2}(\mathbf{x},a1)) + \mathbf{f}(\mathbf{r1}(\mathbf{x},a1)) - \mathbf{f}(\mathbf{r2}(\mathbf{x},a1)) \tag{1709}$$

$$\mathbf{f}(\mathbf{x}(\mathbf{r1},a1+da)) = \mathbf{f}(\mathbf{x1}(\mathbf{r},a1+da)) + \mathbf{f}(\mathbf{x2}(\mathbf{r},a1+da)) - \mathbf{f}(\mathbf{x1}(\mathbf{r},a1+da)) \tag{1710}$$

$$\mathbf{f}(\mathbf{x}(\mathbf{r1},a1)) = \mathbf{f}(\mathbf{x2}(\mathbf{r},a1)) + \mathbf{f}(\mathbf{x1}(\mathbf{r},a1)) - \mathbf{f}(\mathbf{x2}(\mathbf{r},a1)) \tag{1711}$$

and

$$\mathbf{f}(\mathbf{r}(\mathbf{x1},a1)) = \mathbf{f}(\mathbf{r0}(\mathbf{x},a1)) + \mathbf{f}(\mathbf{r1}(\mathbf{x},a1)) - \mathbf{f}(\mathbf{r0}(\mathbf{x},a1)) \tag{1712}$$

$$\mathbf{f}(\mathbf{r}(\mathbf{x1},a1-da)) = \mathbf{f}(\mathbf{r1}(\mathbf{x},a1-da)) + \mathbf{f}(\mathbf{r0}(\mathbf{x},a1-da)) - \mathbf{f}(\mathbf{r1}(\mathbf{x},a1-da)) \tag{1713}$$

Theoretical Physics: Bridge Theorem

$$\begin{aligned}
&\mathbf{f}(\mathbf{x}(\mathbf{r1},\mathrm{a1})) \\
&\qquad = \mathbf{f}(\mathbf{x0}(\mathbf{r},\mathrm{a1})) + \mathbf{f}(\mathbf{x1}(\mathbf{r},\mathrm{a1})) - \mathbf{f}(\mathbf{x0}(\mathbf{r},\mathrm{a1})) && (1714) \\
&\mathbf{f}(\mathbf{x}(\mathbf{r1},\mathrm{a1-da})) \\
&\qquad = \mathbf{f}(\mathbf{x1}(\mathbf{r},\mathrm{a1-da})) + \mathbf{f}(\mathbf{x0}(\mathbf{r},\mathrm{a1-da})) - \mathbf{f}(\mathbf{x1}(\mathbf{r},\mathrm{a1-da})). && (1715)
\end{aligned}$$

Proof:

By the principle of non–collocation

$$\begin{aligned}
\mathbf{r2}(\mathbf{x},\mathrm{a1+da}) &= \mathbf{r}(\mathbf{x1},\mathrm{a1+da}) = \mathbf{r2}(\mathbf{x1},\mathrm{a1+da}), \\
\mathbf{r1}(\mathbf{x},\mathrm{a1}) &= \mathbf{r}(\mathbf{x1},\mathrm{a1}) = \mathbf{r1}(\mathbf{x1},\mathrm{a1}), \\
\mathbf{x2}(\mathbf{r},\mathrm{a1+da}) &= \mathbf{x}(\mathbf{r1},\mathrm{a1+da}) = \mathbf{x2}(\mathbf{r1},\mathrm{a1+da}), \\
\mathbf{x1}(\mathbf{r},\mathrm{a1}) &= \mathbf{x}(\mathbf{r1},\mathrm{a1}) = \mathbf{x1}(\mathbf{r1},\mathrm{a1}), \\
\mathbf{r0}(\mathbf{x},\mathrm{a1-da}) &= \mathbf{r}(\mathbf{x1},\mathrm{a1-da}) = \mathbf{r0}(\mathbf{x1},\mathrm{a1-da}), \\
\mathbf{x0}(\mathbf{r},\mathrm{a1-da}) &= \mathbf{x}(\mathbf{r1},\mathrm{a1-da}) = \mathbf{x0}(\mathbf{r1},\mathrm{a1-da}).
\end{aligned}$$

qed

Except for compatibility of units, no restrictions are placed on the functions, **f**(a1–da), **f**(a1), and **f**(a1+da). The theorem holds even for stalled processes[86].

Now consider the three functions above as three instances of a single function, **f**(**r**(**x**,a))=**f**(**x**(**r**,a)) over the real index *a*. This consideration leads to a second form of the bridge theorem.

Theorem 32 (bridge theorem: second form)

Given

r1(**x1**,a1) the location of the reference particle **x1**(**r1**,a1);

f(**r**(**x**,a));

for

$\mathbf{diff}_n(\mathbf{f}(\mathbf{r}(\mathbf{x},a))|q)$ the change in **f** from **f**(**r1**(**x1**,a1)) in either a backward or forward sense symbolized by n with q as reference

then

$$\begin{aligned}
\mathbf{diff}_f(\mathbf{f}|\mathbf{x1}) &= \mathbf{diff}_f(\mathbf{f}|\mathbf{r1}) + \mathbf{diff}_f(\mathbf{f}(\mathbf{r}|\mathrm{a1+da})) \\
&= \mathbf{diff}_f(\mathbf{f}|\mathbf{r2}) + \mathbf{diff}_f(\mathbf{f}(\mathbf{r}|\mathrm{a1})) && (1716) \\
\mathbf{diff}_f(\mathbf{f}|\mathbf{r1}) &= \mathbf{diff}_f(\mathbf{f}|\mathbf{x1}) + \mathbf{diff}_f(\mathbf{f}(\mathbf{x}|\mathrm{a1+da})) \\
&= \mathbf{diff}_f(\mathbf{f}|\mathbf{x2}) + \mathbf{diff}_f(\mathbf{f}(\mathbf{x}|\mathrm{a1})) && (1717) \\
\mathbf{diff}_b(\mathbf{f}|\mathbf{x1}) &= \mathbf{diff}_b(\mathbf{f}|\mathbf{r1}) + \mathbf{diff}_b(\mathbf{f}(\mathbf{r}|\mathrm{a1-da})) \\
&= \mathbf{diff}_b(\mathbf{f}|\mathbf{r0}) + \mathbf{diff}_b(\mathbf{f}(\mathbf{r}|\mathrm{a1})) && (1718) \\
\mathbf{diff}_b(\mathbf{f}|\mathbf{r1}) &= \mathbf{diff}_b(\mathbf{f}|\mathbf{x1}) + \mathbf{diff}_b(\mathbf{f}(\mathbf{x}|\mathrm{a1-da})) \\
&= \mathbf{diff}_b(\mathbf{f}|\mathbf{x0}) + \mathbf{diff}_b(\mathbf{f}(\mathbf{x}|\mathrm{a1})) && (1719)
\end{aligned}$$

which imply

$$\mathbf{diff}_f(\mathbf{f}(\mathbf{r}|\mathrm{a1{+}da})) = -\mathbf{diff}_f(\mathbf{f}(\mathbf{x}|\mathrm{a1{+}da})) \tag{1720}$$
$$\mathbf{diff}_b(\mathbf{f}(\mathbf{r}|\mathrm{a1{-}da})) = -\mathbf{diff}_b(\mathbf{f}(\mathbf{x}|\mathrm{a1{-}da})) \tag{1721}$$
$$\mathbf{diff}_f(\mathbf{f}|\mathbf{x1}) - \mathbf{diff}_f(\mathbf{f}|\mathbf{x2}) = \mathbf{diff}_f(\mathbf{f}(\mathbf{r}|\mathrm{a1{+}da})) + \mathbf{diff}_f(\mathbf{f}(\mathbf{x}|\mathrm{a1})) \tag{1722}$$
$$\mathbf{diff}_b(\mathbf{f}|\mathbf{x1}) - \mathbf{diff}_b(\mathbf{f}|\mathbf{x0}) = \mathbf{diff}_b(\mathbf{f}(\mathbf{r}|\mathrm{a1{-}da})) + \mathbf{diff}_b(\mathbf{f}(\mathbf{x}|\mathrm{a1})) \tag{1723}$$
$$\mathbf{diff}_f(\mathbf{f}|\mathbf{r1}) - \mathbf{diff}_f(\mathbf{f}|\mathbf{r2}) = \mathbf{diff}_f(\mathbf{f}(\mathbf{x}|\mathrm{a1{+}da})) + \mathbf{diff}_f(\mathbf{f}(\mathbf{r}|\mathrm{a1})) \tag{1724}$$
$$\mathbf{diff}_b(\mathbf{f}|\mathbf{r1}) - \mathbf{diff}_b(\mathbf{f}|\mathbf{r0}) = \mathbf{diff}_b(\mathbf{f}(\mathbf{x}|\mathrm{a1{-}da})) + \mathbf{diff}_b(\mathbf{f}(\mathbf{x}|\mathrm{a1})). \tag{1725}$$

Proof:

$\mathbf{diff}_f(\mathbf{f}|\mathbf{x1}) = \mathbf{f}(\mathbf{r}(\mathbf{x1},\mathrm{a1{+}da})) - \mathbf{f}(\mathbf{r}(\mathbf{x1},\mathrm{a1}))$.

Using the expansion of $\mathbf{f}(\mathbf{r}(\mathbf{x1},\mathrm{a1{+}da}))$ from the first form,

$$\begin{aligned}\mathbf{diff}_f(\mathbf{f}|\mathbf{x1}) &= \mathbf{f}(\mathbf{r1}(\mathbf{x},\mathrm{a1{+}da})) + \mathbf{f}(\mathbf{r2}(\mathbf{x},\mathrm{a1{+}da}))\\ &\qquad - \mathbf{f}(\mathbf{r1}(\mathbf{x},\mathrm{a1{+}da})) - \mathbf{f}(\mathbf{r}(\mathbf{x1},\mathrm{a1}))\\ &= \mathbf{f}(\mathbf{r1}(\mathbf{x},\mathrm{a1{+}da})) - \mathbf{f}(\mathbf{r1}(\mathbf{x},\mathrm{a1}))\\ &\qquad + \mathbf{f}(\mathbf{r2}(\mathbf{x},\mathrm{a1{+}da})) - \mathbf{f}(\mathbf{r1}(\mathbf{x},\mathrm{a1{+}da}))\\ &= \mathbf{diff}_f(\mathbf{f}|\mathbf{r1}) + \mathbf{diff}_f(\mathbf{f}(\mathbf{r}|\mathrm{a1{+}da})).\end{aligned}$$

Again using the expansion of $\mathbf{f}(\mathbf{r}(\mathbf{x1},\mathrm{a1}))$ from the first form,

$$\begin{aligned}\mathbf{diff}_f(\mathbf{f}|\mathbf{x1}) &= \mathbf{f}(\mathbf{r}(\mathbf{x1},\mathrm{a1{+}da})) - \mathbf{f}(\mathbf{r2}(\mathbf{x},\mathrm{a1}))\\ &\qquad - \mathbf{f}(\mathbf{r1}(\mathbf{x},\mathrm{a1})) + \mathbf{f}(\mathbf{r2}(\mathbf{x},\mathrm{a1}))\\ &= \mathbf{f}(\mathbf{r2}(\mathbf{x},\mathrm{a1{+}da})) - \mathbf{f}(\mathbf{r2}(\mathbf{x},\mathrm{a1}))\\ &\qquad + \mathbf{f}(\mathbf{r2}(\mathbf{x},\mathrm{a1})) - \mathbf{f}(\mathbf{r1}(\mathbf{x},\mathrm{a1}))\\ &= \mathbf{diff}_f(\mathbf{f}|\mathbf{r2}) + \mathbf{diff}_f(\mathbf{f}(\mathbf{r}|\mathrm{a1})).\end{aligned}$$

which proves the material form.

Also,

$\mathbf{diff}_f(\mathbf{f}|\mathbf{r1}) = \mathbf{f}(\mathbf{x}(\mathbf{r1},\mathrm{a1{+}da})) - \mathbf{f}(\mathbf{x}(\mathbf{r1},\mathrm{a1}))$

Using the expansion of $\mathbf{f}(\mathbf{x}(\mathbf{r1},\mathrm{a1{+}da}))$ from the first form,

$$\begin{aligned}\mathbf{diff}_f(\mathbf{f}|\mathbf{r1}) &= \mathbf{f}(\mathbf{x1}(\mathbf{r},\mathrm{a1{+}da})) + \mathbf{f}(\mathbf{x2}(\mathbf{r},\mathrm{a1{+}da}))\\ &\qquad - \mathbf{f}(\mathbf{x1}(\mathbf{r},\mathrm{a1{+}da})) - \mathbf{f}(\mathbf{x1}(\mathbf{r},\mathrm{a1}))\\ &= \mathbf{f}(\mathbf{x1}(\mathbf{r},\mathrm{a1{+}da})) - \mathbf{f}(\mathbf{x1}(\mathbf{r},\mathrm{a1}))\\ &\qquad + \mathbf{f}(\mathbf{x2}(\mathbf{r},\mathrm{a1{+}da})) - \mathbf{f}(\mathbf{x1}(\mathbf{r},\mathrm{a1{+}da}))\\ &= \mathbf{diff}_f(\mathbf{f}|\mathbf{x1}) + \mathbf{diff}_f(\mathbf{f}(\mathbf{x}|\mathrm{a1{+}da})).\end{aligned}$$

Again using the expansion of $\mathbf{f}(\mathbf{x}(\mathbf{r1},\mathrm{a1}))$ from the first form,

$$\begin{aligned}\mathbf{diff}_f(\mathbf{f}|\mathbf{r1}) &= \mathbf{f}(\mathbf{x2}(\mathbf{r},\mathrm{a1{+}da})) - \mathbf{f}(\mathbf{x2}(\mathbf{r},\mathrm{a1}))\\ &\qquad + \mathbf{f}(\mathbf{x2}(\mathbf{r},\mathrm{a1})) - \mathbf{f}(\mathbf{x1}(\mathbf{r},\mathrm{a1}))\\ &= \mathbf{diff}_f(\mathbf{f}|\mathbf{x2}) + \mathbf{diff}_f(\mathbf{f}(\mathbf{x}|\mathrm{a1}))\end{aligned}$$

which proves the local form.

Similarly for the backward sense

$\mathbf{diff}_b(\mathbf{f}|\mathbf{x1}) = \mathbf{f}(\mathbf{r}(\mathbf{x1},\mathrm{a1})) - \mathbf{f}(\mathbf{r}(\mathbf{x1},\mathrm{a1{-}da}))$.

Using the expansion of $\mathbf{f}(\mathbf{r}(\mathbf{x1},\mathrm{a1}))$ from the first form,

$$\begin{aligned}\mathbf{diff}_b(\mathbf{f}|\mathbf{x1}) &= \mathbf{f}(\mathbf{r0}(\mathbf{x},\mathrm{a1})) + \mathbf{f}(\mathbf{r1}(\mathbf{x},\mathrm{a1}))\\ &\qquad - \mathbf{f}(\mathbf{r0}(\mathbf{x},\mathrm{a1})) - \mathbf{f}(\mathbf{r}(\mathbf{x1},\mathrm{a1{-}da}))\\ &= \mathbf{f}(\mathbf{r0}(\mathbf{x},\mathrm{a1})) - \mathbf{f}(\mathbf{r0}(\mathbf{x},\mathrm{a1{-}da}))\\ &\qquad + \mathbf{f}(\mathbf{r1}(\mathbf{x},\mathrm{a1})) - \mathbf{f}(\mathbf{r0}(\mathbf{x},\mathrm{a1}))\\ &= \mathbf{diff}_b(\mathbf{f}|\mathbf{r0}) + \mathbf{diff}_b(\mathbf{f}(\mathbf{r}|\mathrm{a1})).\end{aligned}$$

Theoretical Physics: Bridge Theorem

Again using the expansion of $\mathbf{f}(\mathbf{r}(\mathbf{x1},a1-da))$ from the first form,

$$\begin{aligned}\mathbf{diff_b}(\mathbf{f}|\mathbf{x1}) &= \mathbf{f}(\mathbf{r}(\mathbf{x1},a1)) - \mathbf{f}(\mathbf{r1}(\mathbf{x},a1-da)) \\ &\qquad - \mathbf{f}(\mathbf{r0}(\mathbf{x},a1-da)) + \mathbf{f}(\mathbf{r1}(\mathbf{x},a1-da)) \\ &= \mathbf{f}(\mathbf{r1}(\mathbf{x},a1)) - \mathbf{f}(\mathbf{r1}(\mathbf{x},a1-da)) \\ &\qquad + \mathbf{f}(\mathbf{r1}(\mathbf{x},a1-da)) - \mathbf{f}(\mathbf{r0}(\mathbf{x},a1-da)) \\ &= \mathbf{diff_b}(\mathbf{f}|\mathbf{r1}) + \mathbf{diff_b}(\mathbf{f}(\mathbf{r}|a1-da))\end{aligned}$$

and

$$\mathbf{diff_b}(\mathbf{f}|\mathbf{r1}) = \mathbf{f}(\mathbf{x}(\mathbf{r1},a1)) - \mathbf{f}(\mathbf{x}(\mathbf{r1},a1-da)).$$

Using the expansion of $\mathbf{f}(\mathbf{x}(\mathbf{r1},a1))$ from the first form,

$$\begin{aligned}\mathbf{diff_f}(\mathbf{f}|\mathbf{r1}) &= \mathbf{f}(\mathbf{x0}(\mathbf{r},a1)) + \mathbf{f}(\mathbf{x1}(\mathbf{r},a1)) \\ &\qquad - \mathbf{f}(\mathbf{x0}(\mathbf{r},a1)) - \mathbf{f}(\mathbf{x}(\mathbf{r1},a1-da)) \\ &= \mathbf{f}(\mathbf{x0}(\mathbf{r},a1)) - \mathbf{f}(\mathbf{x0}(\mathbf{r},a1-da)) \\ &\qquad + \mathbf{f}(\mathbf{x1}(\mathbf{r},a1)) - \mathbf{f}(\mathbf{x0}(\mathbf{r},a1)) \\ &= \mathbf{diff_b}(\mathbf{f}|\mathbf{x0}) + \mathbf{diff_b}(\mathbf{f}(\mathbf{x}|a1)).\end{aligned}$$

Finally using the expansion of $\mathbf{f}(\mathbf{x}(\mathbf{r1},a1-da))$ from the first form,

$$\begin{aligned}\mathbf{diff_b}(\mathbf{f}|\mathbf{r1}) &= \mathbf{f}(\mathbf{x}(\mathbf{r1},a1)) - \mathbf{f}(\mathbf{x1}(\mathbf{r},a1-da)) \\ &\qquad - \mathbf{f}(\mathbf{x0}(\mathbf{r},a1-da)) + \mathbf{f}(\mathbf{x1}(\mathbf{r},a1-da)) \\ &= \mathbf{diff_b}(\mathbf{f}|\mathbf{x1}) + \mathbf{diff_f}(\mathbf{f}(\mathbf{x}|a1-da)).\end{aligned}$$

qed

Again no restrictions of continuity are placed on the function **f**; consequently the bridge theorem holds for any **f**, scalar or vector, whether or not continuous[87]. The theorem holds even for stalled processes.

The differences of the bridge theorem are called the **materially referenced diff(f|x)**, the **locally referenced diff(f|r)**, and the **net diff(f(r|a))**, differences respectively. The latter is so named because

$$\begin{aligned}\mathbf{diff_f}(\mathbf{f}(\mathbf{r}|a1+da)) &= -\mathbf{diff_f}(\mathbf{f}(\mathbf{x}|a1+da)) \\ &= \mathbf{diff_f}(\mathbf{f}|\mathbf{x1}) - \mathbf{diff_f}(\mathbf{f}|\mathbf{r1})\end{aligned} \qquad (1726)$$

which refers functional values for $\mathbf{x}(\mathbf{r2},a1+da)$ to $\mathbf{x}(\mathbf{r1},a1+da)$, that is the *net* change of values, $\mathbf{f}(\mathbf{x1}(\mathbf{r2},a1+da)) - \mathbf{f}(\mathbf{r1}(\mathbf{x2},a1+da))$, at observation a1+da. For stalled processes the net change is zero.

It should noted, however, that while

$$\mathbf{diff_f}(\mathbf{f}(\mathbf{r}|a1+da)) + \mathbf{diff_f}(\mathbf{f}(\mathbf{x}|a1+da)) = \mathbf{0},$$

$$\begin{aligned}&\mathbf{diff_f}(\mathbf{f}(\mathbf{r}|a1)) + \mathbf{diff_f}(\mathbf{f}(\mathbf{x}|a1)) \\ &\qquad = \mathbf{f}(\mathbf{r2}(\mathbf{y2},a1)) - 2*\mathbf{f}(\mathbf{r1}(\mathbf{x1},a1)) + \mathbf{f}(\mathbf{s2}(\mathbf{x2},a1)).\end{aligned} \qquad (1727)$$

In the backward sense,

$$\mathbf{diff_b}(\mathbf{f}(\mathbf{r}|a1-da)) = -\mathbf{diff_b}(\mathbf{f}(\mathbf{x}|a1-da)) = \mathbf{diff_b}(\mathbf{f}|\mathbf{x1}) - \mathbf{diff}(\mathbf{f}|\mathbf{r1}) \quad (1728)$$

and

$$\mathbf{diff_b}(\mathbf{f}(\mathbf{r}|a1)) + \mathbf{diff_b}(\mathbf{f}(\mathbf{x}|a1)) = -\mathbf{f}(\mathbf{r0}(\mathbf{y0},a1)) + 2*\mathbf{f}(\mathbf{r1}(\mathbf{x1},a1)) - \mathbf{f}(\mathbf{s0}(\mathbf{x0},a1)). \quad (1729)$$

Thus the same material difference generates two distinct local and two distinct net differences just as the same local difference generates two distinct material and two distinct net differences.

The importance of the bridge theorem lies precisely in the relationship it expresses between an indexed change in a function (which implies more than one observation) with a fixed local change (which implies only one observation).

This perspective can be appreciated by applying the second form to **r**(**x1**,a), the changing location of the particle **x1**. Since

$$\text{diff}(\mathbf{r}(\mathbf{x1},a)|\mathbf{r}) = \mathbf{0},$$

$$\mathbf{diff}(\mathbf{r}|\mathbf{x1}) = \mathbf{diff}(\mathbf{r}|a1+da) = \mathbf{diff}(\mathbf{r}(\mathbf{x}|a1)) \quad (1730)$$

that is ,

$$\mathbf{r}(\mathbf{x1},a1+da) - \mathbf{r}(\mathbf{x1},a1) = \mathbf{r2}(\mathbf{x},a1+da) - \mathbf{r1}(\mathbf{x},a1+da) = \mathbf{r}(\mathbf{x2},a1) - \mathbf{r}(\mathbf{x1},a1)$$

where the first difference involves two observations while the latter two involve a single observation.

The length of the difference vector may be similarly analyzed; that is, for

$$d(\mathbf{r}(\mathbf{x1},a)) \equiv \text{abs}(\mathbf{diff}(\mathbf{r}|\mathbf{x1})),$$

then

$$d(\mathbf{r}(\mathbf{x1},a)) = d(\mathbf{r}(\mathbf{x},a1+da)) = d(\mathbf{r}(\mathbf{x},a1)). \quad (1731)$$

Theoretical Physics: Bridge Theorem

Now further, given an ordered set of real indexed observations with corresponding function, $\mathbf{f}(\mathbf{r}(\mathbf{x},a))$, results depending on continuity can be written.

First suppose ***f*** continuous[88] at $\mathbf{r1}(\mathbf{x1},a1)$ with derivatives which perhaps may be discontinuous in either ***r***, ***x***, or *a*. Then, for example, from the second form

$$\lim \mathbf{diff_f}(\mathbf{f}|\mathbf{x1})/da$$
$$= \lim \mathbf{diff_f}(\mathbf{f}|\mathbf{r1})/da + \lim \mathbf{diff_f}(\mathbf{f}(\mathbf{r}|a1+da))/da, \quad \text{as } da \longrightarrow 0$$

implies

$$D[a1](\mathbf{f}|\mathbf{x1};d_f a) = D[a1](\mathbf{f}|\mathbf{r1};d_f a) + D[\mathbf{r1},\mathbf{r2}](\mathbf{f}|a1+da;da)$$

where the derivatives are taken as basic, and

$$D[a1](\mathbf{f}|\mathbf{x1};d_f a) \equiv \lim (\mathbf{f}(\mathbf{r}(\mathbf{x1},a1+da)) - \mathbf{f}(\mathbf{r}(\mathbf{x1},a1)))/da$$
$$D[a1](\mathbf{f}|\mathbf{r1};d_f a) \equiv \lim (\mathbf{f}(\mathbf{x}(\mathbf{r1},a1+da)) - \mathbf{f}(\mathbf{x}(\mathbf{r1},a1)))/da$$
$$D[\mathbf{r1},\mathbf{r2}](\mathbf{f}|a1+da;da) \equiv \lim (\mathbf{f}(\mathbf{r2}(\mathbf{x},a1+da)) - \mathbf{f}(\mathbf{r1}(\mathbf{x},a1+da)))/da.$$

The last derivative, which clearly differs from the first two, would be more accurately symbolized as

$$\lim D[\mathbf{r1},\mathbf{r}(\mathbf{x1},a1+da)](\mathbf{f}|a1+da;da) \text{ as } da \longrightarrow 0.$$

Since the bridge theorem holds for all differences, it likewise holds as a limit where these exist. Assuming now these limits exist (which implies more than one observation)[89] at least in a forward or backward sense, the second form of the bridge theorem can be used to obtain derivatives.

Corollary (the first indexed corollaries to the bridge theorem.)

Given

$\mathbf{r1}(\mathbf{x1},a1)$, the location of the reference particle $\mathbf{x1}(\mathbf{r1},a1)$;

$\mathbf{f}(\mathbf{r}(\mathbf{x},a))$ continuous almost everywhere

with forward and backward indexed derivatives at $\mathbf{r1}(\mathbf{x1},a1)$;

then

$$D[a1](\mathbf{f}|\mathbf{x1};d_f a) = D[a1](\mathbf{f}|\mathbf{r1};d_f a) + D[\mathbf{r1},\mathbf{r2}](\mathbf{f}|a1+da;da) \qquad (1732)$$
$$D[a1](\mathbf{f}|\mathbf{x1};d_f a) = D[a1](\mathbf{f}|\mathbf{r2};d_f a) + D[\mathbf{r1},\mathbf{r2}](\mathbf{f}|a1;da) \qquad (1733)$$
$$D[a1](\mathbf{f}|\mathbf{r1};d_f a) = D[a1](\mathbf{f}|\mathbf{x1};d_f a) + D[\mathbf{x1},\mathbf{x2}](\mathbf{f}|a1+da;da) \qquad (1734)$$
$$D[a1](\mathbf{f}|\mathbf{r1};d_f a) = D[a1](\mathbf{f}|\mathbf{x2};d_f a) + D[\mathbf{x1},\mathbf{x2}](\mathbf{f}|a1;da) \qquad (1735)$$

D[a1](**f**|**x1**;d_ba) = D[a1](**f**|**r1**;d_ba) + D[**r0**,**r1**](**f**|a1−da;da) (1736)
D[a1](**f**|**x1**;d_ba) = D[a1](**f**|**r0**;d_ba) + D[**r0**,**r1**](**f**|a1;da) (1737)
D[a1](**f**|**r1**;d_ba) = D[a1](**f**|**x1**;d_ba) + D[**x0**,**x1**](**f**|a1−da;da) (1738)
D[a1](**f**|**r1**;d_ba) = D[a1](**f**|**x0**;d_ba) + D[**x0**,**x1**](**f**|a1;da). (1739)

Thus

D[**r1**,**r2**](**f**|a1+da;da) = −D[**x1**,**x2**](**f**|a1+da;da) (1740)
D[**r0**,**r1**](**f**|a1−da;da) = −D[**x0**,**x1**](**f**|a1−da;da). (1741)

D[a1](**f**|**x1**;d_fa) + D[**x1**,**x2**](**f**|a1+da;da)
= D[a1](**f**|**x2**;d_fa) + D[**x1**,**x2**](**f**|a1;da) (1742)
D[a1](**f**|**r1**;d_fa) + D[**r1**,**r2**](**f**|a1+da;da)
= D[a1](**f**|**r2**;d_fa) + D[**r1**,**r2**](**f**|a1;da) (1743)
D[a1](**f**|**x1**;d_ba) + D[**x0**,**x1**](**f**|a1−da;da)
= D[a1](**f**|**x0**;d_ba) + D[a1][**x0**,**x1**](**f**|a1;da) (1744)
D[a1](**f**|**r1**;d_ba) + D[**r0**,**r1**](**f**|a1−da;da)
= D[a1](**f**|**r0**;d_ba) + D[**r0**,**r1**](**f**|a1;da). (1745)

Proof:

The derivatives are limiting forms of the bridge theorem second form in ratio to the index of observations.

The second form of the bridge theorem may also be applied to directional gradients.

Corollary (the first material and local corollaries to the bridge theorem)

Given

r1(**x1**,a1), the location of the reference particle **x1**(**r1**,a1);
f(**r**(**x**,a)) continuous almost everywhere
with forward and backward indexed derivatives at **r1**(**x1**,a1);

for

n symbolizing forward or backward;
q symbolizing **r**|**x1**, **r**|**x0**, **x**|**r1**, or **x**|**r2**;
qd(q), the directional reciprocal vector;

then

D1[**r1**,**r2**]∗(**f**|**x1**;$\mathbf{d}_n$(q))
= **D1**[**x1**,**x2**]∗(**f**|**r1**;$\mathbf{d}_n$(q)) + **D1**[**r1**,**r2**]∗(**f**|a1+da;$\mathbf{d}_n$(q)) (1746)

Theoretical Physics: Bridge Theorem

$$\mathbf{D1}[\mathbf{r1},\mathbf{r2}]\ast(\mathbf{f}|\mathbf{x1};\mathbf{d}_n(q)) = \mathbf{D1}[\mathbf{y2},\mathbf{x1}]\ast(\mathbf{f}|\mathbf{r2};\mathbf{d}_n(q)) + \mathbf{D1}[\mathbf{r1},\mathbf{r2}]\ast(\mathbf{f}|a1;\mathbf{d}_n(q)) \tag{1747}$$

$$\mathbf{D1}[\mathbf{x1},\mathbf{x2}]\ast(\mathbf{f}|\mathbf{r1};\mathbf{d}_n(q)) = \mathbf{D1}[\mathbf{r1},\mathbf{r2}]\ast(\mathbf{f}|\mathbf{x1};\mathbf{d}_n(q)) + \mathbf{D1}[\mathbf{x1},\mathbf{x2}]\ast(\mathbf{f}|a1+da;\mathbf{d}_n(q)) \tag{1748}$$

$$\mathbf{D1}[\mathbf{x1},\mathbf{x2}]\ast(\mathbf{f}|\mathbf{r1};\mathbf{d}_n(q)) = \mathbf{D1}[\mathbf{r1},\mathbf{s2}]\ast(\mathbf{f}|\mathbf{x2};\mathbf{d}_n(q)) + \mathbf{D1}[\mathbf{x1},\mathbf{x2}]\ast(\mathbf{f}|a1;\mathbf{d}_n(q)) \tag{1749}$$

$$\mathbf{D1}[\mathbf{r0},\mathbf{r1}]\ast(\mathbf{f}|\mathbf{x1};\mathbf{d}_n(q)) = \mathbf{D1}[\mathbf{x0},\mathbf{x1}]\ast(\mathbf{f}|\mathbf{r1};\mathbf{d}_n(q)) + \mathbf{D1}[\mathbf{r0},\mathbf{r1}]\ast(\mathbf{f}|a1-da;\mathbf{d}_n(q)) \tag{1750}$$

$$\mathbf{D1}[\mathbf{r0},\mathbf{r1}]\ast(\mathbf{f}|\mathbf{x1};\mathbf{d}_n(q)) = \mathbf{D1}[\mathbf{x1},\mathbf{y0}]\ast(\mathbf{f}|\mathbf{r0};\mathbf{d}_n(q)) + \mathbf{D1}[\mathbf{r0},\mathbf{r1}\ast(\mathbf{f}|a1;\mathbf{d}_n(q)) \tag{1751}$$

$$\mathbf{D1}[\mathbf{x0},\mathbf{x1}]\ast(\mathbf{f}|\mathbf{r1};\mathbf{d}_n(q)) = \mathbf{D1}[\mathbf{r0},\mathbf{r1}]\ast(\mathbf{f}|\mathbf{x1};\mathbf{d}_n(q)) + \mathbf{D1}[\mathbf{x0},\mathbf{x1}]\ast(\mathbf{f}|a1-da;\mathbf{d}_n(q)) \tag{1752}$$

$$\mathbf{D1}[\mathbf{x0},\mathbf{x1}]\ast(\mathbf{f}|\mathbf{r1};\mathbf{d}_n(q)) = \mathbf{D1}[\mathbf{s0},\mathbf{r1}]\ast(\mathbf{f}|\mathbf{x0};\mathbf{d}_n(q)) + \mathbf{D1}[\mathbf{x0},\mathbf{x1}]\ast(\mathbf{f}|a1;\mathbf{d}_n(q)). \tag{1753}$$

Also

$$\mathbf{D1}[\mathbf{r1},\mathbf{r2}]\ast(\mathbf{f}|a1+da;\mathbf{d}_n(q)) = -\mathbf{D1}[\mathbf{x1},\mathbf{x2}]\ast(\mathbf{f}|a1+da;\mathbf{d}_n(q)) \tag{1754}$$

$$\mathbf{D1}[\mathbf{r0},\mathbf{r1}]\ast(\mathbf{f}|a1-da;\mathbf{d}_n(q)) = -\mathbf{D1}[\mathbf{x0},\mathbf{x1}]\ast(\mathbf{f}|a1da;\mathbf{d}_n(q)). \tag{1755}$$

Proof:

Directional gradients are outer products of indexed derivatives with directional reciprocal vectors.

From these equations divergences and curls can be written explicitly, as for instance:

$$\begin{aligned} D1[\mathbf{r1},\mathbf{r2}]\bullet(\mathbf{f}|\mathbf{x1};\mathbf{dr}) &= tr[\mathbf{D1}[\mathbf{r1},\mathbf{r2}]\ast(\mathbf{f}|\mathbf{r1};\mathbf{dr}) + \mathbf{D1}[\mathbf{r1},\mathbf{r2}]\ast(\mathbf{f}|a1+da;\mathbf{dr})] \\ &= tr[\mathbf{D1}[\mathbf{r1},\mathbf{r2}]\ast(\mathbf{f}|\mathbf{r1};\mathbf{dr})] + tr[\mathbf{D1}[\mathbf{r1},\mathbf{r2}]\ast(\mathbf{f}|a1+da;\mathbf{dr})] \end{aligned} \tag{1756}$$

$$\begin{aligned} \mathbf{D1}[\mathbf{r1},\mathbf{r2}]\wedge(\mathbf{f}|\mathbf{x1};\mathbf{dr}) &= \mathbf{c}[\mathbf{D1}[\mathbf{r1},\mathbf{r2}]\ast(\mathbf{f}|\mathbf{r1};\mathbf{dr}) + \mathbf{D1}[\mathbf{r1},\mathbf{r2}]\ast(\mathbf{f}|a1+da;\mathbf{dr})] \\ &= \mathbf{c}[\mathbf{D1}[\mathbf{r1},\mathbf{r2}]\ast(\mathbf{f}|\mathbf{r1};\mathbf{dr})] + \mathbf{c}[\mathbf{D1}[\mathbf{r1},\mathbf{r2}]\ast(\mathbf{f}|a1+da;\mathbf{dr})]. \end{aligned} \tag{1757}$$

The indexed derivatives can be given an intrinsic representation as, for instance by the chain rule,

$$D[a1](\mathbf{f}|\mathbf{x1};d_f a) = D[a1](sfx;d_f a)\ast D[a1](\mathbf{f}|\mathbf{x1};d_f sfx) \tag{1758}$$

which, in turn, can be used to show the relationship between the indexed and material/local corollaries. For instance,

$$\begin{aligned}
&\mathbf{D1}[\mathbf{r1},\mathbf{r2}]\ast(\mathbf{f}|\mathbf{x1};\mathbf{dr})\\
&\qquad = D[a1](\mathbf{f}|\mathbf{x1};d_f a)\ast\mathbf{qd}(D[a1](\mathbf{r}|\mathbf{x1};d_f a))\\
&\qquad = D[a1](sfx;d_f a)\ast D[a1](\mathbf{f}|\mathbf{x1};d_f sfx)\\
&\qquad\qquad\qquad \ast\mathbf{qd}(\mathbf{utfx})/D[a1](sfx;d_f a)\\
&\qquad = D[a1](\mathbf{f}|\mathbf{x1};d_f sfx)\ast\mathbf{qd}(\mathbf{utfx}).
\end{aligned} \tag{1759}$$

The relationship between indexed and material/local derivatives of Theoretical Physics is further elucidated using the relationship between indexed derivatives and directional gradients.

Corollary (the second indexed corollaries to the bridge theorem)
Given
r1(**x1**,a1), the location of the reference particle **x1**(**r1**,a1);
f(**r**(**x**,a)) continuous almost everywhere
with forward and backward indexed derivatives at **r1**(**x1**,a1);
then

$$\begin{aligned}
&D[a1](\mathbf{f}|\mathbf{x1};d_f a)\\
&\qquad = D[a1](\mathbf{f}|\mathbf{r1};d_f a)\\
&\qquad\quad + D[a1](\mathbf{r}|\mathbf{x1};d_f a)\bullet\mathbf{T}[\mathbf{D1}[\mathbf{r1},\mathbf{r2}]\ast(\mathbf{f}|a1+da;\mathbf{d_f r})]
\end{aligned} \tag{1760}$$

$$\begin{aligned}
&D[a1](\mathbf{f}|\mathbf{x1};d_f a)\\
&\qquad = D[a1](\mathbf{f}|\mathbf{r2};d_f a)\\
&\qquad\quad + D[a1](\mathbf{r}|\mathbf{x1};d_f a)\bullet\mathbf{T}[\mathbf{D1}[\mathbf{r1},\mathbf{r2}]\ast(\mathbf{f}|a1;\mathbf{d_f r})]
\end{aligned} \tag{1761}$$

$$\begin{aligned}
&D_f[a1](\mathbf{f}|\mathbf{r1};d_f a)\\
&\qquad = D[a1](\mathbf{f}|\mathbf{x1};d_f a)\\
&\qquad\quad + D[a1](\mathbf{x}|\mathbf{r1};d_f a)\bullet\mathbf{T}[\mathbf{D1}[\mathbf{x1},\mathbf{x2}]\ast(\mathbf{f}|a1+da;\mathbf{d_f x})]
\end{aligned} \tag{1762}$$

$$\begin{aligned}
&D[a1](\mathbf{f}|\mathbf{r1};d_f a)\\
&\qquad = D[a1](\mathbf{f}|\mathbf{x2};d_f a)\\
&\qquad\quad + D[a1](\mathbf{x}|\mathbf{r1};d_f a)\bullet\mathbf{T}[\mathbf{D1}[\mathbf{x1},\mathbf{x2}]\ast(\mathbf{f}|a1;\mathbf{d_f x})]
\end{aligned} \tag{1763}$$

$$\begin{aligned}
&D[a1](\mathbf{f}|\mathbf{x1};d_b a)\\
&\qquad = D[a1](\mathbf{f}|\mathbf{r1};d_b a)\\
&\qquad\quad + D[a1](\mathbf{r}|\mathbf{x1};d_b a)\bullet\mathbf{T}[\mathbf{D1}[\mathbf{r0},\mathbf{r1}]\ast(\mathbf{f}|a1-da;\mathbf{d_b r})]
\end{aligned} \tag{1764}$$

Theoretical Physics: Bridge Theorem

D[a1](**f**|**x1**;d_ba)
    = D[a1](**f**|**r0**;d_ba)
        + D[a1](**r**|**x1**;d_ba)•**T**[**D1**[**r0**,**r1**]∗(**f**|a1;$\mathbf{d}_b$**r**)]   (1765)
D[a1](**f**|**r1**;d_ba)
    = D[a1](**f**|**x1**;d_ba)
        + D[a1](**x**|**r1**;d_ba)•**T**[**D1**[**x0**,**x1**]∗(**f**|a1−da;$\mathbf{d}_b$**x**)]   (1766)
D[a1](**f**|**r1**;d_ba)
    = D[a1](**f**|**x2**;d_ba)
        + D[a1](**x**|**r1**;d_ba)•**T**[**D1**[**x0**,**x1**]∗(**f**|a1;$\mathbf{d}_b$**x**)]   (1767)
from which

D[a1](**r**|**x1**;d_fa)•**T**[**D1**[**r1**,**r2**]∗(**f**|a1+da;$\mathbf{d}_f$**r**)]
    = −D[a1](**x**|**r1**;d_fa)•**T**[**D1**[**x1**,**x2**]∗(**f**|a1+da;$\mathbf{d}_f$**x**)]   (1768)
D[a1](**r**|**x1**;d_ba)•**T**[**D1**[**r0**,**r1**]∗(**f**|a1−da;$\mathbf{d}_b$**r**)]
    = −D[a1](**x**|**r1**;d_ba)•**T**[**D1**[**x0**,**x1**]∗(**f**|a1−da;$\mathbf{d}_b$**x**)].   (1769)

Proof:
D[a1](**r**|**x1**;d_fa) = D[**r1**,**r2**](**r**|a1+da;d_fa).
Thus for a non–stalled process,
D[a1](**r**|**x1**;d_fa)•**qd**(D[**r1**,**r2**](**r**|a1+da;d_fa)) = 1.
Consequently,
D[**r1**,**r2**](**f**|a1+da;da)
    = D[a1](**r**|**x1**;d_fa)
        •**qd**(D[**r1**,**r2**](**r**|a1+da;d_fa))∗D[**r1**,**r2**](**f**|a1+da;da)
    = D[a1](**r**|**x1**;d_fa)•**T**[**D1**[**r1**,**r2**]∗(**f**|a1+da;$\mathbf{d}_f$**r**)].
Results follow from the first indexed corollaries to the bridge theorem.
For **f** stalled,
D[a1](**r**|**x1**;d_fa) = **0**.
Thus,
D[a1](**f**|**x1**;d_fa) = D[a1](**f**|**r1**;d_fa).
The remaining propositions are proven similarly.   qed

Corollary (the second material and local corollaries to the bridge theorem)
Given
r1(**x1**,a1), the location of the reference particle **x1**(**r1**,a1);
f(**r**(**x**,a)) continuous almost everywhere
with forward and backward indexed derivatives at **r1**(**x1**,a1);
for
n symbolizing forward or backward;
q symbolizing **r**|**x1**, **r**|**x0**, **x**|**r1**, or **x**|**r2**;
qd(q), the directional reciprocal vector;

then

$$\mathbf{D1}[\mathbf{r1},\mathbf{r2}]*(\mathbf{f}|\mathbf{x1};\mathbf{d}_{n}(q)) = \mathbf{D1}[\mathbf{x1},\mathbf{x2}]*(\mathbf{f}|\mathbf{r1};\mathbf{d}_{n}(q)) + D[\mathbf{r1},\mathbf{r2}](\mathbf{f}|a1+da;da)*\mathbf{qd}(D[a1](q;d_{n}a)) \tag{1770}$$

$$\mathbf{D1}[\mathbf{r1},\mathbf{r2}]*(\mathbf{f}|\mathbf{x1};\mathbf{d}_{n}(q)) = \mathbf{D1}[\mathbf{y2}\ ,\mathbf{x1}]*(\mathbf{f}|\mathbf{r2};\mathbf{d}_{n}(q)) + D[\mathbf{r1},\mathbf{r2}](\mathbf{f}|a1;da)*\mathbf{qd}(D[a1](q;d_{n}a)) \tag{1771}$$

$$\mathbf{D1}[\mathbf{x1},\mathbf{x2}]*(\mathbf{f}|\mathbf{r1};\mathbf{d}_{n}(q)) = \mathbf{D1}[\mathbf{r1},\mathbf{r2}]*(\mathbf{f}|\mathbf{x1};\mathbf{d}_{n}(q)) + D[\mathbf{x1},\mathbf{x2}](\mathbf{f}|a1+da;da)*\mathbf{qd}(D[a1](q;d_{n}a)) \tag{1772}$$

$$\mathbf{D1}[\mathbf{x1},\mathbf{x2}]*(\mathbf{f}|\mathbf{r1};\mathbf{d}_{n}(q)) = \mathbf{D1}[\mathbf{s2},\mathbf{r1}]*(\mathbf{f}|\mathbf{x2};\mathbf{d}_{n}(q)) + D[\mathbf{x1},\mathbf{x2}](\mathbf{f}|a1;da)*\mathbf{qd}(D[a1](q;d_{n}a)) \tag{1773}$$

$$\mathbf{D1}[\mathbf{r0},\mathbf{r1}]*(\mathbf{f}|\mathbf{x1};\mathbf{d}_{n}(q)) = \mathbf{D1}[\mathbf{x0},\mathbf{x1}]*(\mathbf{f}|\mathbf{r1};\mathbf{d}_{n}(q)) + D[\mathbf{r0},\mathbf{r1}](\mathbf{f}|a1-da;da)*\mathbf{qd}(D[a1](q;d_{n}a)) \tag{1774}$$

$$\mathbf{D1}[\mathbf{r0},\mathbf{r1}]*(\mathbf{f}|\mathbf{x1};\mathbf{d}_{n}(q)) = \mathbf{D1}[\mathbf{y0},\mathbf{x1}]*(\mathbf{f}|\mathbf{r0};\mathbf{d}_{n}(q)) + D[\mathbf{r0},\mathbf{r1}](\mathbf{f}|a1;da)*\mathbf{qd}(D[a1](q;d_{n}a)) \tag{1775}$$

$$\mathbf{D1}[\mathbf{x0},\mathbf{x1}]*(\mathbf{f}|\mathbf{r1};\mathbf{d}_{n}(q)) = \mathbf{D1}[\mathbf{r0},\mathbf{r1}]*(\mathbf{f}|\mathbf{x1};\mathbf{d}_{n}(q)) + D[\mathbf{x0},\mathbf{x1}](\mathbf{f}|a1-da;da)*\mathbf{qd}(D[a1](q;d_{n}a)) \tag{1776}$$

$$\mathbf{D1}[\mathbf{x0},\mathbf{x1}]*(\mathbf{f}|\mathbf{r1};\mathbf{d}_{n}(q)) = \mathbf{D1}[\mathbf{s0},\mathbf{r1}]*(\mathbf{f}|\mathbf{x0};\mathbf{d}_{n}(q)) + D[\mathbf{x0},\mathbf{x1}](\mathbf{f}|a1;da)*\mathbf{qd}(D[a1](q;d_{n}a)) \tag{1777}$$

$$D[\mathbf{r1},\mathbf{r2}](\mathbf{f}|a1+da;da)*\mathbf{qd}(D[a1](q;d_{n}a)) = -D[\mathbf{x1},\mathbf{x2}](\mathbf{f}|a1+da;da)*\mathbf{qd}(D[a1](q;d_{n}a)) \tag{1778}$$

$$D[\mathbf{r0},\mathbf{r1}](\mathbf{f}|a1-da;da)*\mathbf{qd}(D[a1](q;d_{n}a)) = -D[\mathbf{x0},\mathbf{x1}](\mathbf{f}|a1-da;da)*\mathbf{qd}(D[a1](q;d_{n}a)). \tag{1779}$$

Proof:

Results follow from the decomposition of the directional gradient into an outer product.

The first corollaries refer the bridge theorem to the observational index; the second corollaries to the intrinsic features of the particle's motion.

There is still another context for the bridge theorem––the open neighborhood of **r1**(**x1**,a1) in V3.

Theoretical Physics: Bridge Theorem

Corollary (the third indexed corollaries to the bridge theorem)

Given

r1(**x1**,a1), the location of the reference particle **x1**(**r1**,a1);

f(**r**(**x**,a)) continuous almost everywhere
with continuous gradient in section **URn** at **r1**(**x1**,a1)
into which D[a1](**r**|**x1**;d_na) extends;

for **n** signifying either forward or backward;

then

$$D[a1](\mathbf{f}|\mathbf{x1};d_f a) = D[a1](\mathbf{f}|\mathbf{r1};d_f a) + D[a1](\mathbf{r}|\mathbf{x1};d_f a)\bullet\mathbf{T}[\mathbf{URf}(a1+da)^{-1}]\bullet[\mathbf{URf}(a1+da)]^{-1}\bullet\mathbf{T}[\mathbf{D3}[\mathbf{r1}]*(\mathbf{f}|a1+da;\mathbf{dRf})] \tag{1780}$$

$$D[a1](\mathbf{f}|\mathbf{x1};d_f a) = D[a1](\mathbf{f}|\mathbf{r2};d_f a) + D[a1](\mathbf{r}|\mathbf{x1};d_f a)\bullet\mathbf{T}[\mathbf{URf}(a1)^{-1}]\bullet[\mathbf{URf}(a1)]^{-1}\bullet\mathbf{T}[\mathbf{D3}[\mathbf{r1}]*(\mathbf{f}|a1;\mathbf{dRf})] \tag{1781}$$

$$D[a1](\mathbf{f}|\mathbf{r1};d_f a) = D[a1](\mathbf{f}|\mathbf{x1};d_f a) + D[a1](\mathbf{x}|\mathbf{r1};d_f a)\bullet\mathbf{T}[\mathbf{UXf}(a1+da)^{-1}]\bullet[\mathbf{UXf}(a1+da)]^{-1}\bullet\mathbf{T}[\mathbf{D3}[\mathbf{x1}]*(\mathbf{f}|a1+da;\mathbf{dXf})] \tag{1782}$$

$$D[a1](\mathbf{f}|\mathbf{r1};d_f a) = D[a1](\mathbf{f}|\mathbf{x2};d_f a) + D[a1](\mathbf{x}|\mathbf{r1};d_f a)\bullet\mathbf{T}[\mathbf{UXf}(a1)^{-1}]\bullet[\mathbf{UXf}(a1)]^{-1}\bullet\mathbf{T}[\mathbf{D3}[\mathbf{x1}]*(\mathbf{f}|a1;\mathbf{dXf})] \tag{1783}$$

$$D[a1](\mathbf{f}|\mathbf{x1};d_b a) = D[a1](\mathbf{f}|\mathbf{r1};d_b a) + D[a1](\mathbf{r}|\mathbf{x1};d_b a)\bullet\mathbf{T}[\mathbf{URb}(a1-da)^{-1}]\bullet[\mathbf{URb}(a1-da)]^{-1}\bullet\mathbf{T}[\mathbf{D3}[\mathbf{r1}]*(\mathbf{f}|a1-da;\mathbf{dRb})] \tag{1784}$$

$$D[a1](\mathbf{f}|\mathbf{x1};d_b a) = D[a1](\mathbf{f}|\mathbf{r0};d_b a) + D[a1](\mathbf{r}|\mathbf{x1};d_b a)\bullet\mathbf{T}[\mathbf{URb}(a1)^{-1}]\bullet[\mathbf{URb}(a1)]^{-1}\bullet\mathbf{T}[\mathbf{D3}[\mathbf{r1}]*(\mathbf{f}|a1;\mathbf{dRb})] \tag{1785}$$

$$D[a1](\mathbf{f}|\mathbf{r1};d_b a) = D[a1](\mathbf{f}|\mathbf{x1};d_b a) + D[a1](\mathbf{x}|\mathbf{r1};d_b a)\bullet\mathbf{T}[\mathbf{UXb}(a1-da)^{-1}]\bullet[\mathbf{UXb}(a1-da)]^{-1}\bullet\mathbf{T}[\mathbf{D3}[\mathbf{x1}]*(\mathbf{f}|a1-da;\mathbf{dXb})] \tag{1786}$$

$$D[a1](\mathbf{f}|\mathbf{r1};d_b a) = D[a1](\mathbf{f}|\mathbf{x0};d_b a) + D[a1](\mathbf{x}|\mathbf{r1};d_b a)\bullet\mathbf{T}[\mathbf{UXb}(a1)^{-1}]\bullet[\mathbf{UXb}(a1)]^{-1}\bullet\mathbf{T}[\mathbf{D3}[\mathbf{x1}]*(\mathbf{f}|a1;\mathbf{dXb})] \tag{1787}$$

which imply

D[a1](**r**|**x1**;d_fa)•**T**[**URf**(a1+da)$^{-1}$]•[**URf**(a1+da)]$^{-1}$
•**T**[**D3**[**r1**]∗(**f**|a1+da;**dRf**)]
= –D[a1](**x**|**r1**;d_fa)•**T**[**UXf**(a1+da)$^{-1}$]•[**UXf**(a1+da)]$^{-1}$
•**T**[**D3**[**x1**]∗(**f**|a1+da;**dXf**)] (1788)

D[a1](**r**|**x1**;d_ba)•**T**[**URb**(a1–da)$^{-1}$]•[**URb**(a1–da)]$^{-1}$
•**T**[**D3**[**r1**]∗(**f**|a1–da;**dRb**)]
= –D[a1](**x**|**r1**;d_ba)•**T**[**UXb**(a1–da)$^{-1}$]•[**UXb**(a1–da)]$^{-1}$
•**T**[**D3**[**x1**]∗(**f**|a1–da;**dXb**)]. (1789)

Proof:

Results follow from the first indexed corollaries and Theorem 14.

Corollary (the third material and local corollaries to the bridge theorem)

Given

r1(**x1**,a1), the location of the reference particle **x1**(**r1**,a1);
f(**r**(**x**,a)) continuous almost everywhere
with continuous gradient in section **URn** at **r1**(**x1**,a1)
into which D[a1](**r**|**x1**;d_na) extends
for **n** signifying either forward or backward;
then

D1[**r1**,**r2**]∗(**f**|**x1**;$\mathbf{d_n}$(q))
= **D1**[**x1**,**x2**]∗(**f**|**r1**;$\mathbf{d_n}$(q))
+ **D3**[**r1**]∗(**f**|a1+da;**dRf**)•**T**[**URf**(a1+da)$^{-1}$]•[**URf**(a1+da)$^{-1}$]
•**D1**[**r1**,**r2**]∗(**r**|**x1**;$\mathbf{d_n}$(q)) (1790)

D1[**r1**,**r2**]∗(**f**|**x1**;$\mathbf{d_n}$(q))
= **D1**[**y2**,**x1**]∗(**f**|**r2**;$\mathbf{d_n}$(q))
+ **D3**[**r1**]∗(**f**|a1;**dRf**)•**T**[**URf**(a1)$^{-1}$]•[**URf**(a1)$^{-1}$]
•**D1**[**r1**,**r2**]∗(**r**|**x1**;$\mathbf{d_n}$(q)) (1791)

D1[**x1**,**x2**]∗(**f**|**r1**;$\mathbf{d_n}$(q))
= **D1**[**r1**,**r2**]∗(**f**|**x1**;$\mathbf{d_n}$(q))
+ **D3**[**x1**]∗(**f**|a1+da;**dXf**)•**T**[**UXf**(a1+da)$^{-1}$]•[**UXf**(a1+da)$^{-1}$]
•**D1**[**x1**,**x2**]∗(**x**|**r1**;$\mathbf{d_n}$(q)) (1792)

D1[**x1**,**x2**]∗(**f**|**r1**;$\mathbf{d_n}$(q))
= **D1**[**s2**,**r1**]∗(**f**|**x2**;$\mathbf{d_n}$(q))
+ **D3**[**x1**]∗(**f**|a1;**dXf**)•**T**[**UXf**(a1)$^{-1}$]•[**UXf**(a1)$^{-1}$]
•**D1**[**x1**,**x2**]∗(**x**|**r1**;$\mathbf{d_n}$(q)) (1793)

$$\begin{aligned}
&\mathbf{D1[r0,r1]\ast(f|x1;d_n}(q)) \\
&\quad = \mathbf{D1[x0,x1]\ast(f|r1;d_n}(q)) \\
&\quad\quad + \mathbf{D3[r1]\ast(f|}a1-da;\mathbf{dRb})\cdot\mathbf{T[URb}(a1-da)^{-1}]\cdot[\mathbf{URb}(a1-da)^{-1}] \\
&\quad\quad\quad \cdot\mathbf{D1[r0,r1\ast(r|x1;d_n}(q))
\end{aligned} \tag{1794}$$

$$\begin{aligned}
&\mathbf{D1[r0,r1]\ast(f|x1;d_n}(q)) \\
&\quad = \mathbf{D1[x1,y0]\ast(f|r0;d_n}(q)) \\
&\quad\quad + \mathbf{D3[r1]\ast(f|}a1;\mathbf{dRb})\cdot\mathbf{T[URb}(a1)^{-1}]\cdot[\mathbf{URb}(a1)^{-1}] \\
&\quad\quad\quad \cdot\mathbf{D1[r0,r1]\ast(r|x1;d_n}(q))
\end{aligned} \tag{1795}$$

$$\begin{aligned}
&\mathbf{D1[x0,x1]\ast(f|r1;d_n}(q)) \\
&\quad = \mathbf{D1[r0,r1]\ast(f|x1;d_n}(q)) \\
&\quad\quad + \mathbf{D3[x1]\ast(f|}a1-da;\mathbf{dXb})\cdot\mathbf{T[UXb}(a1-da)^{-1}]\cdot[\mathbf{UXb}(a1-da)^{-1}] \\
&\quad\quad\quad \cdot\mathbf{D1[x0,x1]\ast(x|r1;d_n}(q))
\end{aligned} \tag{1796}$$

$$\begin{aligned}
&\mathbf{D1[x0,x1]\ast(f|r1;d_n}(q)) \\
&\quad = \mathbf{D1[r1,s0]\ast(f|x0;d_n}(q)) \\
&\quad\quad + \mathbf{D3[x1]\ast(f|}a1;\mathbf{dXb})\cdot\mathbf{T[UXb}(a1)^{-1}]\cdot[\mathbf{UXb}(a1)^{-1}] \\
&\quad\quad\quad \cdot\mathbf{D1[x0,x1]\ast(x|r1;d_n}(q))
\end{aligned} \tag{1797}$$

which imply

$$\begin{aligned}
&\mathbf{D3[r1]}(\mathbf{f}|a1+da;\mathbf{dRf})\cdot\mathbf{T[URf}(a1+da)^{-1}]\cdot[\mathbf{URf}(a1+da)^{-1}] \\
&\quad\quad\quad \cdot\mathbf{D1[r1,r2]\ast(r|x1;d_n}(q)) \\
&\quad = -\mathbf{D3[x1]}(\mathbf{f}|a1+da;\mathbf{dXf})\cdot\mathbf{T[UXf}(a1+da)^{-1}]\cdot[\mathbf{UXf}(a1+da)^{-1}] \\
&\quad\quad\quad \cdot\mathbf{D1[x1,x2]\ast(x|r1;d_n}(q))
\end{aligned} \tag{1798}$$

$$\begin{aligned}
&\mathbf{D3[r1]}(\mathbf{f}|a1-da;\mathbf{dRb})\cdot\mathbf{T[URb}(a1-da)^{-1}]\cdot[\mathbf{URb}(a1-da)^{-1}] \\
&\quad\quad\quad \cdot\mathbf{D1[r0,r1]\ast(r|x1;d_n}(q)) \\
&\quad = -\mathbf{D3[x1]}(\mathbf{f}|a1-da;\mathbf{dXb})\cdot\mathbf{T[UXb}(a1-da)^{-1}]\cdot[\mathbf{UXb}(a1-da)^{-1}] \\
&\quad\quad\quad \cdot\mathbf{D1[x0,x1]\ast(x|r1;d_n}(q)).
\end{aligned} \tag{1799}$$

Proof:

Results follow from the previous corollary and the decomposition of the directional gradient into an outer product.

The Continuous Case of the Bridge Theorem.

The corollaries to the bridge theorem hold even when **f** and its derivatives enjoy only a limited continuity at **r1**(**x1**,a1). Where **f** and its derivatives are simply continuous simpler forms may be written. In the simply continuous case forward and backward derivatives are equal, that is,

$$\begin{aligned}
D[a1](\mathbf{r}|\mathbf{x1};d_f a) &= D[a1](\mathbf{r}|\mathbf{x1};d_b a) \\
&\equiv D[a1](\mathbf{r}|\mathbf{x1};da) \\
D[a1](\mathbf{x}|\mathbf{r1};d_f a) &= D[a1](\mathbf{x}|\mathbf{r1};d_b a) \\
&\equiv D[a1](\mathbf{x}|\mathbf{r1};da)
\end{aligned}$$

$$D[a1](\mathbf{f}|\mathbf{x1};d_f a) = D[a1](\mathbf{f}|\mathbf{x1};d_b a)$$
$$\equiv D[a1](\mathbf{f}|\mathbf{x1};da)$$
$$D[a1](\mathbf{f}|\mathbf{x2};d_f a) = D[a1](\mathbf{f}|\mathbf{x0};d_b a)$$
$$\equiv D[a1](\mathbf{f}|\mathbf{x2};da)$$
$$D([a1]\mathbf{f}|\mathbf{r1};d_f a) = D[a1](\mathbf{f}|\mathbf{r1};d_b a)$$
$$\equiv D[a1](\mathbf{f}|\mathbf{r1};da)$$
$$D[\mathbf{r1},\mathbf{r2}](\mathbf{f}|a1;da) = D[\mathbf{r0},\mathbf{r1}](\mathbf{f}|a1;da)$$
$$D[\mathbf{x1},\mathbf{x2}](\mathbf{f}|a1;da) = D[\mathbf{x0},\mathbf{x1}](\mathbf{f}|a1;da)$$
$$D[\mathbf{r1},\mathbf{r2}](\mathbf{f}|a1+da;da) = D[\mathbf{r0},\mathbf{r1}](\mathbf{f}|a1-da;da)$$
$$D[\mathbf{x1},\mathbf{x2}](\mathbf{f}|a1+da;da) = D[\mathbf{x0},\mathbf{x1}](\mathbf{f}|a1-da;da)$$
$$\mathbf{D1}[r1,r2]*(\mathbf{f}|\mathbf{x1};\mathbf{d}_n(q)) = \mathbf{D1}[r0,r1]*(\mathbf{f}|\mathbf{x1};\mathbf{d}_n(q))$$
$$\equiv \mathbf{D1}[a1](\mathbf{f}|\mathbf{x1};\mathbf{d}_n(q))$$
$$\mathbf{D1}[x1,x2]*(\mathbf{f}|\mathbf{r1};\mathbf{d}_n(q)) = \mathbf{D1}[x0,x1]*(\mathbf{f}|\mathbf{r1};\mathbf{d}_n(q))$$
$$\equiv \mathbf{D1}[a1]*(\mathbf{f}|\mathbf{r1};\mathbf{d}_n(q))$$
$$\mathbf{D1}[r1,r2]*(\mathbf{f}|a1+da;\mathbf{d}_n(q)) = \mathbf{D1}[r0,r1*(\mathbf{f}|a1-da;\mathbf{d}_n(q))$$
$$\equiv \mathbf{D1}[a1]*(\mathbf{f}(\mathbf{r}|a1+da);\mathbf{d}_n(q))$$
$$\mathbf{D1}[x1,x2]*(\mathbf{f}|a1+da;\mathbf{d}_n(q)) = \mathbf{D1}[x0,x1*(\mathbf{f}|a1-da;\mathbf{d}_n(q))$$
$$\equiv \mathbf{D1}[a1]*(\mathbf{f}(\mathbf{x}|a1+da);\mathbf{d}_n(q))$$
$$\mathbf{D1}[r1,r2]*(\mathbf{f}|a1;\mathbf{d}_n(q)) = \mathbf{D1}[r0,r1]*(\mathbf{f}|a1;\mathbf{d}_n(q))$$
$$\equiv \mathbf{D1}[a1]*(\mathbf{f}(\mathbf{r}|a1);\mathbf{d}_n(q))$$
$$\mathbf{D1}[x1,x2]*(\mathbf{f}|a1;\mathbf{d}_n(q)) = \mathbf{D1}[x0,x1]*(\mathbf{f}|a1;\mathbf{d}_n(q))$$
$$\equiv \mathbf{D1}[a1]*(\mathbf{f}(\mathbf{x}|a1);\mathbf{d}_n(q))$$
$$\mathbf{D1}[r1,r2*(\mathbf{f}|a1+da;\mathbf{d}_f\mathbf{r}) = \mathbf{D1}[r0,r1*(\mathbf{f}|a1-da;\mathbf{d}_b\mathbf{r})$$
$$\equiv \mathbf{D1}[a1]*(\mathbf{f}(\mathbf{r}|a1+da);\mathbf{dr})$$
$$\mathbf{D1}[r1,r2]*(\mathbf{f}|a1;\mathbf{d}_f\mathbf{r}) = \mathbf{D1}[r0,r1]*(\mathbf{f}|a1;\mathbf{d}_b\mathbf{r})$$
$$\equiv \mathbf{D1}[a1]*(\mathbf{f}(\mathbf{r}|a1);\mathbf{dr})$$
$$\mathbf{D1}[x1,x2]*(\mathbf{f}|a1+da;\mathbf{d}_f\mathbf{x}) = \mathbf{D1}[x0,x1]*(\mathbf{f}|a1-da;\mathbf{d}_b\mathbf{x})$$
$$\equiv \mathbf{D1}[a1]*(\mathbf{f}(\mathbf{x}|a1+da);\mathbf{dx})$$
$$\mathbf{D1}[x1,x2]*(\mathbf{f}|a1;\mathbf{d}_f\mathbf{x}) = \mathbf{D1}[x0,x1]*(\mathbf{f}|a1;\mathbf{d}_b\mathbf{x})$$
$$\equiv \mathbf{D1}[a1]*(\mathbf{f}(\mathbf{r}|a1);\mathbf{dr}|\mathbf{x1})$$
$$\mathbf{D3}[\mathbf{r1}]*(\mathbf{f}|a1+da;\mathbf{dRf})\cdot\mathbf{T}[\mathbf{URf}(a1+da)^{-1}]\cdot[\mathbf{URf}(a1+da)]^{-1}$$
$$\equiv \mathbf{D3}[\mathbf{r1}]*(\mathbf{f}|a1+da;\mathbf{dr})$$
$$\mathbf{D3}[\mathbf{r1}]*(\mathbf{f}|a1;\mathbf{dRf})\cdot\mathbf{T}[\mathbf{URf}(a1)^{-1}]\cdot[\mathbf{URf}(a1)]^{-1}$$
$$\equiv \mathbf{D3}[\mathbf{r1}]*(\mathbf{f}|a1;\mathbf{dr})$$
$$\mathbf{D3}[\mathbf{x1}]*(\mathbf{f}|a1+da;\mathbf{dXf})\cdot\mathbf{T}[\mathbf{UXf}(a1+da)^{-1}]\cdot[\mathbf{UXf}(a1+da)]^{-1}$$
$$\equiv \mathbf{D3}[\mathbf{x1}]*(\mathbf{f}|a1+da;\mathbf{dx})$$
$$\mathbf{D3}[\mathbf{x1}]*(\mathbf{f}|a1;\mathbf{dXf})\cdot\mathbf{T}[\mathbf{UXf}(a1)^{-1}]\cdot[\mathbf{UXf}(a1)]^{-1}$$
$$\equiv \mathbf{D3}[\mathbf{x1}]*(\mathbf{f}|a1;\mathbf{dx}).$$

Theoretical Physics: Bridge Theorem

Corollary (the first indexed corollaries to the bridge theorem, simply continuous case)

D[a1](**f**|**x1**;da) = D[a1](**f**|**r1**;da) + D[a1](**f**(**r**|a1+da);da) (1800)
D[a1](**f**|**x1**;da) = D[a1](**f**|**r1**;da) – D[a1](**f**(**x**|a1+da);da) (1801)
D[a1](**f**|**x1**;da) = D[a1](**f**|**r2**;da) + D[a1](**f**(**r**|a1);da) (1802)
D[a1](**f**|**x2**;da) = D[a1](**f**|**r1**;da) – D[a1](**f**(**x**|a1);da) (1803)

implying

D[a1](**f**(**r**|a1+da);da) = –D[a1](**f**(**x**|a1+da);da) (1804)

D[a1](**f**|**x1**;da) + D[a1](**f**(**x**|a1+da);da)
= D[a1](**f**|**x2**;da) + D[a1](**f**(**x**|a1);da) (1805)
D[a1](**f**|**r1**;da) + D[a1](**f**(**r**|a1+da);da)
= D[a1](**f**|**r2**;da) + D[a1](**f**(**r**|a1);da). (1806)

Corollary (the first material and local corollaries to the bridge theorem, simply continuous case)

D1[a1]∗(**f**|**x1**;$\mathbf{d}_n$(q))
= **D1**[a1]∗(**f**|**r1**;$\mathbf{d}_n$(q))
+ **D1**[a1]∗(**f**(**r**|a1+da);$\mathbf{d}_n$(q)) (1807)
D1[a1]∗(**f**|**x1**;$\mathbf{d}_n$(q))
= **D1**[a1]∗(**f**|**r1**;$\mathbf{d}_n$(q))
–**D1**[a1]∗(**f**(**x**|a1+da);$\mathbf{d}_n$(q)) (1808)
D1[a1]∗(**f**|**x1**;$\mathbf{d}_n$(q))
= **D1**[a1]∗(**f**|**r2**;$\mathbf{d}_n$(q))
+ **D1**[a1]∗(**f**(**r**|a1);$\mathbf{d}_n$(q)) (1809)
D1[a1]∗(**f**|**x2**;$\mathbf{d}_n$(q))
= **D1**[a1]∗(**f**|**r1**;$\mathbf{d}_n$(q))
– **D1**[a1]∗(**f**(**x**|a1);$\mathbf{d}_n$(q)) (1810)

for q symbolizing **r**|**x1** or **x**|**r1**, implying

D1[a1]∗(**f**(**r**|a1+da);$\mathbf{d}_n$(q)) = – **D1**[a1]∗(**f**(**x**|a1+da);$\mathbf{d}_n$(q)). (1811)

Corollary (the second indexed corollaries to the bridge theorem, simply continuous case)

$$D[a1](\mathbf{f}|\mathbf{x1};da) = D[a1](\mathbf{f}|\mathbf{r1};da) + D[a1](\mathbf{r}|\mathbf{x1};da)\bullet\mathbf{T}[\mathbf{D1}[a1]\ast(\mathbf{f}(\mathbf{r}|a1+da);\mathbf{dr})] \quad (1812)$$

$$D[a1](\mathbf{f}|\mathbf{x1};da) = D[a1](\mathbf{f}|\mathbf{r1};da) - D[a1](\mathbf{x}|\mathbf{r1};da)\bullet\mathbf{T}[\mathbf{D}[a1]\mathbf{1}\ast(\mathbf{f}(\mathbf{x}|a1+da);\mathbf{dx})] \quad (1813)$$

$$D[a1](\mathbf{f}|\mathbf{x1};da) = D[a1](\mathbf{f}|\mathbf{r2};da) + D[a1](\mathbf{r}|\mathbf{x1};da)\bullet\mathbf{T}[\mathbf{D}[a1]\mathbf{1}\ast(\mathbf{f}(\mathbf{r}|a1);\mathbf{dr})] \quad (1814)$$

$$D[a1](\mathbf{f}|\mathbf{x2};da) = D[a1](\mathbf{f}|\mathbf{r1};da) - D[a1](\mathbf{x}|\mathbf{r1};da)\bullet\mathbf{T}[\mathbf{D}[a1]\mathbf{1}\ast(\mathbf{f}(\mathbf{x}|a1);\mathbf{dx})] \quad (1815)$$

from which

$$D[a1](\mathbf{r}|\mathbf{x1};da)\bullet\mathbf{T}[\mathbf{D1}[a1]\ast(\mathbf{f}(\mathbf{r}|a1+da);\mathbf{dr})] = -D[a1](\mathbf{x}|\mathbf{r1};da)\bullet\mathbf{T}[\mathbf{D1}[a1]\ast(\mathbf{f}(\mathbf{x}|a1+da);\mathbf{dx})]. \quad (1816)$$

Corollary (second material and local corollaries to the bridge theorem, simply continuous case)

$$\mathbf{D1}[a1]\ast(\mathbf{f}|\mathbf{x1};\mathbf{d}_{\mathbf{n}}(q)) = \mathbf{D1}[a1]\ast(\mathbf{f}|\mathbf{r1};\mathbf{d}_{\mathbf{n}}(q)) + D[a1](\mathbf{f}(\mathbf{r}|a1+da);da)\ast\mathbf{qd}(D[a1](q;da)) \quad (1817)$$

$$\mathbf{D1}[a1]\ast(\mathbf{f}|\mathbf{x1};\mathbf{d}_{\mathbf{n}}(q)) = \mathbf{D1}[a1]\ast(\mathbf{f}|\mathbf{r1};\mathbf{d}_{\mathbf{n}}(q)) - D[a1](\mathbf{f}(\mathbf{x}|a1+da);da)\ast\mathbf{qd}(D[a1](q;da)) \quad (1818)$$

$$\mathbf{D1}[a1]\ast(\mathbf{f}|\mathbf{x1};\mathbf{d}_{\mathbf{n}}(q)) = \mathbf{D1}[a1]\ast(\mathbf{f}|\mathbf{r2};\mathbf{d}_{\mathbf{n}}(q)) + D[a1](\mathbf{f}(\mathbf{r}|a1);da)\ast\mathbf{qd}(D[a1](q;da)) \quad (1819)$$

$$\mathbf{D1}[a1]\ast(\mathbf{f}|\mathbf{x2};\mathbf{d}_{\mathbf{n}}(q)) = \mathbf{D1}[a1]\ast(\mathbf{f}|\mathbf{r1};\mathbf{d}_{\mathbf{n}}(q)) - D[a1](\mathbf{f}(\mathbf{x}|a1);da)\ast\mathbf{qd}(D[a1](q;da)) \quad (1820)$$

implying

$$D[a1](\mathbf{f}(\mathbf{r}|a1+da);da)\ast\mathbf{qd}(D[a1](q;da)) = -D[a1](\mathbf{f}(\mathbf{x}|a1+da);da)\ast\mathbf{qd}(D[a1](q;da)) \quad (1821)$$

where q symbolizes **r**|**x1** or **x**|**r1**.

Theoretical Physics: Bridge Theorem

Corollary (the third indexed corollaries to the bridge theorem, simply continuous case)

$$D[a1](\mathbf{f}|\mathbf{x1};da) = D[a1](\mathbf{f}|\mathbf{r1};da) + D[a1](\mathbf{r}|\mathbf{x1};da)\cdot\mathbf{T}[\mathbf{D3}[\mathbf{r1}](\mathbf{f}|a1+da;\mathbf{dr})] \tag{1822}$$

$$D[a1](\mathbf{f}|\mathbf{x1};da) = D[a1](\mathbf{f}|\mathbf{r1};da) - D[a1](\mathbf{x}|\mathbf{r1};da)\cdot\mathbf{T}[\mathbf{D3}[\mathbf{x1}](\mathbf{f}|a1+da;\mathbf{dx})] \tag{1823}$$

$$D[a1](\mathbf{f}|\mathbf{x1};da) = D[a1](\mathbf{f}|\mathbf{r2};da) + D[a1](\mathbf{r}|\mathbf{x1};da)\cdot\mathbf{T}[\mathbf{D3}[\mathbf{r1}](\mathbf{f}|a1;\mathbf{dr})] \tag{1824}$$

$$D[a1](\mathbf{f}|\mathbf{x2};da) = D[a1](\mathbf{f}|\mathbf{r1};da) - D[a1](\mathbf{x}|\mathbf{r1};da)\cdot\mathbf{T}[\mathbf{D3}[\mathbf{x1}](\mathbf{f}|a1;\mathbf{dx})] \tag{1825}$$

which imply

$$D[a1](\mathbf{r}|\mathbf{x1};da)\cdot\mathbf{T}[\mathbf{D3}[\mathbf{r1}](\mathbf{f}|a1+da;\mathbf{dr})] = -D[a1](\mathbf{x}|\mathbf{r1};da)\cdot\mathbf{T}[\mathbf{D3}[\mathbf{x1}](\mathbf{f}|a1+da;\mathbf{dx})]. \tag{1826}$$

Corollary (the third material and local corollaries to the bridge theorem, simply continuous case)

$$\mathbf{D1}[a1]*(\mathbf{f}|\mathbf{x1};\mathbf{d}_n(q)) = \mathbf{D1}[a1]*(\mathbf{f}|\mathbf{r1};\mathbf{d}_n(q)) + \mathbf{D3}[\mathbf{r1}](\mathbf{f}|a1+da;\mathbf{dr})\cdot\mathbf{D1}[a1]*(\mathbf{r}|\mathbf{x1};\mathbf{d}_n(q)) \tag{1827}$$

$$\mathbf{D1}[a1]*(\mathbf{f}|\mathbf{x1};\mathbf{d}_n(q)) = \mathbf{D1}[a1]*(\mathbf{f}|\mathbf{r1};\mathbf{d}_n(q)) - \mathbf{D3}[\mathbf{x1}](\mathbf{f}|a1+da;\mathbf{dx})\cdot\mathbf{D1}[a1]*(\mathbf{x}|\mathbf{r1};\mathbf{d}_n(q)) \tag{1828}$$

$$\mathbf{D1}[a1]*(\mathbf{f}|\mathbf{x1};\mathbf{d}_n(q)) = \mathbf{D1}[a1]*(\mathbf{f}|\mathbf{r2};\mathbf{d}_n(q)) + \mathbf{D3}[\mathbf{r1}](\mathbf{f}|a1;\mathbf{dr})\cdot\mathbf{D1}[a1]*(\mathbf{r}|\mathbf{x1};\mathbf{d}_n(q)) \tag{1829}$$

$$\mathbf{D1}[a1]*(\mathbf{f}|\mathbf{x2};\mathbf{d}_n(q)) = \mathbf{D1}[a1]*(\mathbf{f}|\mathbf{r1};\mathbf{d}_n(q)) - \mathbf{D3}[\mathbf{x1}](\mathbf{f}|a1;\mathbf{dx})\cdot\mathbf{D1}[a1]*(\mathbf{x}|\mathbf{r1};\mathbf{d}_n(q)) \tag{1830}$$

which imply

$$\mathbf{D3}[\mathbf{r1}](\mathbf{f}|a1+da;\mathbf{dr})\cdot\mathbf{D1}[a1]*(\mathbf{r}|\mathbf{x1};\mathbf{d}_n(q)) = -\mathbf{D3}[\mathbf{x1}](\mathbf{f}|a1+da;\mathbf{dx})\cdot\mathbf{D1}[a1]*(\mathbf{x}|\mathbf{r1};\mathbf{d}_n(q)) \tag{1831}$$

where q symbolizes **r|x1** or **x|r1**.

Theoretical Physics: Bridge Theorem

The bridge theorem also applies to transformations. To illustrate the continuous case, let **M** ≡ **u1*m1 + u2*m2 + u3*m3** where **mi** are the row vectors of **M**. Then,

D[a1](**M**|**x1**;da)
 = **u1***D[a1](**m1**|**x1**;da)
  + **u2***D[a1](**m2**|**x1**;da)
  + **u3***D[a1](**m3**|**x1**;da) (1832)

where D[a1](**mi**|**x1**;da) may now be analyzed by the bridge theorem.

The units of the bridge theorem are those of ***f***; those of the indexed corollaries are those of ***f***; those of the material/local corollaries are those of *f/L*.

Example: Let **r**(**x**,a) = **x•M**(a) = **x•**(**m1**(a)***u1** + **m2**(a)***u2** + **m3**(a)***u3**), where **M** is a continuous indexed set of transformations with continuous indexed derivatives and the **mi** are the column vectors of **M** and det[**M**(a)] does not equal zero. Then

x(**r**,a) = **r•M**$^{-1}$(a) ≡ **r•**(**n1**(a)***u1** + **n2**(a)***u2** + **n3**(a)***u3**).

Let f(**r**(**x**,a)) = **r•x** = **x•M**(a)**•x** = **r•T[M**$^{-1}$(a)]**•r**. By direct computation,

r1(**x1**,a1) = **x1•M**(a1)
r2(**x1**,a1+da) = **x1•M**(a1+da)
x1(**r1**,a1) = **r1•M**$^{-1}$(a1)
x2(**r1**,a1+da) = **r1•M**$^{-1}$(a1+da)
s2(**x2**,a1) = **x2•M**(a1)
y2(**r2**,a1) = **r2•M**$^{-1}$(a1)
r1(**x2**,a1+da) = **x2•M**(a1+da)
f(**r**(**x1**,a1+da)) = **x1•M**(a1+da)**•x1** = **r2•x1**
f(**r**(**x1**,a1)) = **x1•M**(a1)**•x1** = **r1•x1**
f(**r1**(**x**,a1+da)) = **r1•T[M**$^{-1}$(a1+da)]**•r1** = **r1•x2**
f(**r2**(**x**,a1)) = **r2•T[M**$^{-1}$(a1)]**•r2** = **y2•x1** f(**r2**(**x**,a1+da))
 = **r2•T[M**$^{-1}$(a1+da)]**•r2**
 = **r2•x1**
f(**r1**(**x**,a1)) = **r1•T[M**$^{-1}$(a1)]**•r1** = **r1•x1**
f(**x**(**r1**,a1+da)) = **x2•M**(a1+da)**•x2** = **r1•x2**
f(**x**(**r1**,a1)) = **x1•M**(a1)**•x1** = **r1•x1**
f(**x1**(**r**,a1+da)) = **x1•M**(a1+da)**•x1** = **r2•x1**
f(**x2**(**r**,a1)) = **x2•M**(a1)**•x2** = **s2•x2**
f(**x2**(**r**,a1+da)) = **x2•M**(a1+da)**•x2** = **r1•x2**
f(**x1**(**r**,a1))= **x1•M**(a1)**•x1** = **r1•x1**

from which the first form of the bridge theorem can be verified;

Theoretical Physics: Bridge Theorem

$\mathrm{diff}_f(f|\mathbf{x1}) = \mathbf{x1}\bullet(\mathbf{M}(a1+da)-\mathbf{M}(a1))\bullet\mathbf{x1} \qquad = \mathbf{r2}\bullet\mathbf{x1} - \mathbf{r1}\bullet\mathbf{x1}$
$\mathrm{diff}_f(f|\mathbf{r1}) = \mathbf{r1}\bullet(\mathbf{x2} - \mathbf{x1})$
$\mathrm{diff}_f(f(\mathbf{r}|a1+da)) = \mathbf{r2}\bullet\mathbf{x1} - \mathbf{r1}\bullet\mathbf{x2}$
$\mathrm{diff}_f(f|\mathbf{r2}) = \mathbf{r2}\bullet(\mathbf{x1} - \mathbf{y2})$
$\mathrm{diff}_f(f(\mathbf{r}|a1)) = \mathbf{r2}\bullet\mathbf{y2} - \mathbf{r1}\bullet\mathbf{x1}$
$\mathrm{diff}_f(f(\mathbf{x}|a1+da)) = \mathbf{r1}\bullet\mathbf{x2} - \mathbf{r2}\bullet\mathbf{x1}$
$\mathrm{diff}_f(f|\mathbf{x2}) = (\mathbf{r1} - \mathbf{s2})\bullet\mathbf{x2}$
$\mathrm{diff}_f(f(\mathbf{x}|a1)) = \mathbf{s2}\bullet\mathbf{x2} - \mathbf{r1}\bullet\mathbf{x1}$
from which the second form of the bridge theorem can be verified.

Likewise, given the requisite continuity, with $\mathbf{r1}(\mathbf{x1},a1)$ as reference,
$D[a1](\mathbf{r}|\mathbf{x1};da) = D[a1](\mathbf{x1}\bullet\mathbf{M}(a)|\mathbf{x1};da) = \mathbf{x1}\bullet D[a1](\mathbf{M};da)$

$$
\begin{aligned}
D[\mathbf{r1},\mathbf{r2}](\mathbf{r}|a1;da) &\longleftarrow (\mathbf{r2}(\mathbf{y2},a1) - \mathbf{r1}(\mathbf{x1},a1))/da \\
&= (\mathbf{y2}\bullet\mathbf{M}(a1) - \mathbf{x1}\bullet\mathbf{M}(a1))/da \\
&\longrightarrow D(\mathbf{x1},\mathbf{y2})[a1](\mathbf{x}|a1;da)\bullet\mathbf{M}(a1) \\
&= \mathbf{x1}\bullet D[a1](\mathbf{M};da) \\
D[\mathbf{x1},\mathbf{x2}](\mathbf{r}|a1;da) &\longleftarrow (\mathbf{r}(\mathbf{x2},a1) - \mathbf{r1}(\mathbf{x1},a1))/da \\
&= (\mathbf{x2}\bullet\mathbf{M}(a1) - \mathbf{x1}\bullet\mathbf{M}(a1))/da \\
&\longrightarrow D[a1](\mathbf{x}|\mathbf{r1};da)\bullet\mathbf{M}(a1) \\
&= \mathbf{r1}\bullet D[a1](\mathbf{M}^{-1};da)\bullet\mathbf{M}(a1) \\
D[\mathbf{r1},\mathbf{r2}](\mathbf{r}|a1+da;da) &\longleftarrow (\mathbf{r2}(\mathbf{x1},a1+da) - \mathbf{r1}(\mathbf{x2},a1+da))/da \\
&= (\mathbf{x1}\bullet\mathbf{M}(a1+da) - \mathbf{x2}\bullet\mathbf{M}(a1+da))/da \\
&\longrightarrow -D(\mathbf{x}|\mathbf{r1};da)\bullet\mathbf{M}(a1+da) \\
&= (\mathbf{x1}\bullet\mathbf{M}(a1+da) - \mathbf{x1}\bullet\mathbf{M}(a1))/da \\
&\longrightarrow \mathbf{x1}\bullet D[a1](\mathbf{M};da) \\
D[\mathbf{x1},\mathbf{x2}](\mathbf{r}|a1+da;da) &\longleftarrow (\mathbf{r1}(\mathbf{x2},a1+da) - \mathbf{r2}(\mathbf{x1},a1+da))/da \\
&= (\mathbf{x2}\bullet\mathbf{M}(a1+da) - \mathbf{x1}\bullet\mathbf{M}(a1+da))/da \\
&\longrightarrow D(\mathbf{x}|\mathbf{r1};da)\bullet\mathbf{M}(a1+da) \\
&= (\mathbf{x1}\bullet\mathbf{M}(a1) - \mathbf{x1}\bullet\mathbf{M}(a1+da))/da \\
&\longrightarrow -\mathbf{x1}\bullet D[a1](\mathbf{M};da) \\
D[a1](\mathbf{r}|\mathbf{x2};da) &= \mathbf{x2}\bullet D[a1](\mathbf{M};da) \\
&= \mathbf{r1}\bullet[\mathbf{M}^{-1}(a1+da)]\bullet D[a1](\mathbf{M};da) \\
D[a1](\mathbf{x}|\mathbf{r1};da) &= \mathbf{r1}\bullet D[a1](\mathbf{M}^{-1};da) \\
D[\mathbf{r1},\mathbf{r2}](\mathbf{x}|a1;da) &\longleftarrow (\mathbf{x}(\mathbf{r2},a1) - \mathbf{x1}(\mathbf{r1},a1))/da \\
&= (\mathbf{r2}\bullet\mathbf{M}^{-1}(a1) - \mathbf{r1}\bullet\mathbf{M}^{-1}(a1))/da \\
&\longrightarrow D[a1](\mathbf{r}|\mathbf{x1},da)\bullet\mathbf{M}^{-1}(a1) \\
D[\mathbf{x1},\mathbf{x2}](\mathbf{x}|a1;da) &\longleftarrow (\mathbf{x2}(\mathbf{r},a1) - \mathbf{x1}(\mathbf{r1},a1))/da \\
&= (\mathbf{r1}\bullet\mathbf{M}^{-1}(a1+da) - \mathbf{r1}\bullet\mathbf{M}^{-1}(a1))/da \\
&\longrightarrow \mathbf{r1}\bullet D[a1](\mathbf{M}^{-1};da)
\end{aligned}
$$

D[**r1**,**r2**(**x**|a1+da;da) ⟵ (**x1**(**r2**,a1+da) − **x2**(**r1**,a1+da))/da
= (**r2**•**M**$^{-1}$(a1+da) − **r1**•**M**$^{-1}$(a1+da))/da
⟶ D[a1](**r**|**x1**,da)•**M**$^{-1}$(a1+da)
= (**r1**•**M**$^{-1}$(a1) − **r1**•**M**$^{-1}$(a1+da))/da
⟶ −**r1**•D[a1](**M**$^{-1}$;da)
D[**x1**,**x2**(**x**|a1+da;da) = **r1**•D[a1](**M**$^{-1}$;da)
D[a1](**x**|**r2**;da) = **r2**•D[a1](**M**$^{-1}$;da) = **x1**•**M**(a1+da)•D[a1](**M**$^{-1}$;da)
qd(D[a1](**r**|**x1**;da)) = **qd**(**x1**)•[**T**[D[a1](**M**;da)]]$^{-1}$
qd(D[a1](**x**|**r1**;da)) = **qd**(**r1**)•[**T**[D[a1](**M**$^{-1}$;da)]]$^{-1}$
D[a1](sr;da) = sqrt((**x1**•D[a1](**m1**(a);da))^2
+ (**x1**•D[a1](**m2**(a);da))^2
+ (**x1**•D[a1](**m3**(a);da))^2)
D[a1](sx;da) = sqrt((**r1**•D[a1](**n1**(a);da))^2
+ (**r1**•D[a1](**n2**(a);da))^2
+ (**r1**•D[a1](**n3**(a);da))^2)
D[a1](sx;dsr) = D[a1](sx;da)/D[a1](sr;da)
D[a1](sr;dsx) = D[a1](sr;da)/D[a1](sx;da)
utr = **x1**•D[a1](**M**;da)/D[a1](sx;da)
qd(**utr**) = D[a1](sx;da)∗**qd**(**x1**)•[**T**[D[a1](**M**;da)]]$^{-1}$
utr∗**qd**(**utr**) = **T**[D[a1](**M**;da)]•[**x1**∗**qd**(**x1**)]•[**T**[D[a1](**M**;da)]]$^{-1}$
utx = **r1**•D[a1](**M**$^{-1}$;da)/D[a1](sx;da)
q(**utx**) = D[a1](sx;da)∗**q**(**r1**)•[**T**[D[a1](**M**$^{-1}$;da)]]$^{-1}$
utx∗**q**(**utx**) = **T**[D[a1](**M**$^{-1}$;da)]•[**r1**∗(**qd**(**r1**))]•[**T**[D[a1](**M**$^{-1}$;da)]]$^{-1}$
D1[r1,r2]∗(**r**|**x1**;**d**$_f$**r**) = **utr**∗**q**(**utr**)
D1[x1,x2]∗(**x**|**r1**;**d**$_f$**x**) = **utx**∗**q**(**utx**)
D1[r1,r2]∗(**r**|**x1**;**d**$_n$(q)) = D[a1](**r**|**x1**;da)∗**qd**(D[a1](q;da))
D1[x1,x2]∗(**x**|**r1**;**d**$_n$(q)) = D[a1](**x**|**r1**;da)∗**qd**(D[a1](q;da))
D3[x1]∗(**r**|a1;**dx**) = **T**[**M**(a1)]
D3[x1]∗(**r**|a1+da;**dx**) = **T**[**M**(a1+da)]
D3[y2]∗(**r**|a1;**dx**) = **T**[**M**(a1)]
D3[x2]∗(**r**|a1+da;**dx**) = **T**[**M**(a1)]
D3[r1]∗(**x**|a1;**dr**) = **T**[**M**$^{-1}$(a1)]
D3[r1]∗(**x**|a1+da;**dr**) = **T**[**M**$^{-1}$(a1+da)]
D3[r2]∗(**x**|a1;**dr**) = **T**[**M**$^{-1}$(a1)]
D3[r2]∗(**x**|a1+da;**dr**) = **T**[**M**$^{-1}$(a1)]
D[a1](f|**x1**;da) = D[a1](**r**|**x1**;da)•**x1**
= **x1**•D[a1](**M**;a)•**x1**
D[a1](f|**r1**;da) = **r1**•D[a1](**x**|**r1**;da)
= **r1**•**T**(D[a1](**M**$^{-1}$;da))•**r1**
= **r1**•D[a1](**M**$^{-1}$;da)•**r1**
D[a1](f|**x2**;da)= D[a1](**r**|**x2**;da)•**x2**
= **r1**•**M**$^{-1}$(a1+da)•D[a1](**M**;da)•**T**[**M**$^{-1}$(a1+da)]•**r1**

Theoretical Physics: Bridge Theorem

D[a1](f|r2;da) = r2•D[a1](x|r2;da)
D[r1,r2](f|a1;da) = D[a1](r•x(r,a1);da)
⟵(r2•y2 − r1•x1)/da
= (r2•y2 − r2•x1 + r2•x1 − r1•x1)/da
= (r2•(y2 − x1) + (r2 − r1)•x1)/da
⟶ −r2•D[a1](x|r2;a) + D[a1](r|x1;a)•x1
D[x1,x2](f|a1;da) = D[a1](r•x(r,a1);da)
⟵(s2•x2 − r1•x1)/da
= (s2•x2 − r1•x2 + r1•x2 − r1•x1)/da
= ((s2 − r1)•x2 + r1•(x2 − x1))/da
⟶ −D[a1](r|x2;da)•x2+ r1•D[a1](x|r1;da)
D[r1,r2](f|a1+da;da) = D[a1](r•x(r,a1+da);da)
= −r2•D[a1](x|r1;da) + D[a1](r|x1;da)•x2
= −r1•D[a1](x|r1;da) + D[a1](r|x1;da)•x1
D[x1,x2](f|a1+da;da) = D[a1](r•x(x,a1+da);da)
= r2•D[a1](x|r1;da) − D[a1](r|x1;da)•x2
= r1•D[a1](x|r1;da) − D[a1](r|x1;da)•x1
D1[r1,r2]*(f|x1;d(q)) = x1•D1[r1,r2]*(r|x1;d(q))
D1[x1,x2]*(f|r1;d(q)) = r1•D1[x1,x2]*(x|r1;d(q))
D1[s2,r1]*(f|x2;d(q)) = x2•D1[s2,r1]*(r|x2;d(q))
D1[y2,x1]*(f|r2;d(q)) = r2•D1[y2,x1]*(x|r2;d(q))
D1[r1,r2]*(f|a1;d(q)) = −r2•D1[y2,x1]*(x|r2;d(q))
+ x1•D1[r1,r2]*(r|x1;d(q))
D1[x1,x2]*(f|a1;d(q)) = −x2•D1[s2,r1]*(r|x2;d(q))
+ r1•D1[x1,x2]*(x|r1;d(q))
D1[r1,r2]*(f|a1+da;d(q)) = −r1•D1[x1,x2]*(x|r1;d(q))
+ x1•D1[r1,r2]*(r|x1;d(q))
D1[x1,x2]*(f|a1+da;d(q)) = −x1•D1[r1,r2]*(r|x1;d(q))
+ r1•D1[x1,x2]*(x|r1;d(q))
D3[r1]*(f|a1;dr) = D3[r1]*((r•x)|a1;dr)
= r1•T[D3[r1]*(x|a1;dr)] + x1•D3[r1]*(r|a1;dr)
= x1•M(a1)•M^{-1}(a1) + x1
= 2*x1
D3[r1]*(f|a1+da;dr)= r1•T[D3[r1]*(x|a1+da;dr)]
+ x2•D3[r1]*(r|a1+da;dr)
= r1•M^{-1}(a1+da) + x2
= 2*x2
D3[x1]*(f|a1;dx) = 2*r1
D3[x1]*(f|a1+da;dx) = 2*r2

D[a1](**r**|**x1**;da)•**T**[**D1**[**r1**,**r2**]∗(f|a1;**dr**)]
= – D[a1](**r**|**x1**;da)•**qd**(D[a1](**r**|**x1**;da))∗D[a1](**x**|**r2**;da)•**r2**
+ D[a1](**r**|**x1**;da)•**qd**(D[a1](**r**|**x1**;da))∗D[a1](**r**|**x1**;da)•**x1**
= –**r2**•D[a1](**x**|**r2**;da) + **x1**•D[a1](**r**|**x1**;da)
D[a1](**r**|**x1**;da)•**T**[**D1**[**r1**,**r2**]∗(f|a1+da;**dr**)]
= –**r1**•D[a1](**x**|**r1**;da) + **x1**•D[a1](**r**|**x1**;da)
D[a1](**x**|**r1**;da)•**T**[**D1**[**x1**,**x2**]∗(f|a1;**dx**)]
= –**x2**•D[a1](**r**|**x2**;da) + **r1**•D[a1](**x**|**r1**;da)
D[a1](**x**|**r1**;da)•**T**[**D1**[**x1**,**x2**]∗(f|a1+da;**dx**)]
= –**x1**•D[a1](**r**|**x1**;da) + **r1**•D[a1](**x**|**r1**;da)
D[**r1**,**r2**](f|a1;da)∗**q**(D[a1](q;da))
= (–**r2**•D[a1](**x**|**r2**;da)
+ D[a1](**r**|**x1**;da•**x1**))∗**q**(D[a1](q;da))
= –**r2**•**D1**[**y2**,**x1**]∗(**x**|**r2**;**d**(q))+ **x1**•**D1**[**r1**,**r2**]∗(**r**|**x1**;**d**(q))
D[**r1**,**r2**](f|a1+da;da)∗**q**(D[a1](q;da))
= –**r1**•**D1**[**x1**,**x2**]∗(**x**|**r1**;**d**(q))+ **x1**•**D1**[**r1**,**r2**]∗(**r**|**x1**;**d**(q))
D[**x1**,**x2**](f|a1;da)∗**q**(D[a1](q;da))
= –**x2**•**D1**[**s2**,**r1**]∗(**r**|**x2**;**d**(q))+ **r1**•**D1**[**x1**,**x2**]∗(**x**|**r1**;**d**(q))
D[**x1**,**x2**](f|a1+da;da)∗**q**(D[a1](q;da))
= –**x1**•**D1**[**r1**,**r2**]∗(**r**|**x1**;**d**(q))+ **r1**•**D1**[**x1**,**x2**]∗(**x**|**r1**;**d**(q))
D[a1](**r**|**x1**;da)•**T**[**D3**[**r1**]∗(f|a1;**dr**)] = 2∗**x1**•D[a1](**M**;da)•**x1**
D[a1](**r**|**x1**;da)•**T**[**D3**[**r1**]∗(f|a1+da;**dr**)] = 2∗**x1**•D[a1](**M**;da)•**x2**
D[a1](**x**|**r1**;da)•**T**[**D3**[**x1**]∗(f|a1;**dx**)] = 2∗**r1**•D[a1](**M**$^{-1}$;da)•**r1**
D[a1](**x**|**r1**;da)•**T**[**D3**[**x1**]∗(f|a1+da;**dx**)] = 2∗**r1**•D[a1](**M**$^{-1}$;da)•**r2**
D3[**r1**]∗(f|a1;**dr**)•**D1**[**r1**,**r2**]∗(**r**|**x1**;**d**(q)) = 2∗**x1**•**D1**[**r1**,**r2**]∗(**r**|**x1**;**d**(q))
D3[**r1**]∗(f|a1+da;**dr**)•**D1**[**r1**,**r2**]∗(**r**|**x1**;**d**(q)) = 2∗**x2**•**D1**[**r1**,**r2**]∗(**r**|**x1**;**d**(q))
D3[**x1**]∗(f|a1;**dx**)•**D1**[**x1**,**x2**]∗(**x**|**r1**;**d**(q)) = 2∗**r1**•**D1**[**x1**,**x2**]∗(**x**|**r1**;**d**(q))
D3[**x1**]∗(f|a1+da;**dx**)•**D1**[**x1**,**x2**]∗(**x**|**r1**;**d**(q)) = 2∗**r2**•**D1**[**x1**,**x2**]∗(**x**|**r1**;**d**(q))
for q symbolizing **r**|**x1** or **x**|**r1**.

The reader can verify the relevant bridge corollaries as joint limits of a1+da⟶a1, **x2**⟶**x1** and **r2**⟶**r1**.
The relationship

D[a1]([**M**•**M**$^{-1}$];da)=[**0**]=**M**•D[a1](**M**$^{-1}$;da)+D[a1](**M**;da)•**M**$^{-1}$

will be useful for verifying the third corollary.

From this example the reader can appreciate the rich texture of ideas which Theoretical Physics brings to the service of Physics.

Fallacies Related to Partial Derivatives.

When in equation (1822), a=t is referred to as time; $D[t1](f\vert\mathbf{x};dt)$ as a whole derivative, df/dt; and $D[a1](f\vert\mathbf{r};dt)$ as a partial derivative, ∂f/∂t––the third indexed corollary of the bridge theorem can be matched to the Eulerian relationship, namely,

$$df/dt = \partial f/\partial t + (d\mathbf{r}/dt)\bullet\boldsymbol{\nabla}f.$$

We have arrived at a crucial stage in the development of Theoretical Physics. Modern Physics relies fundamentally on the Eulerian relationship as justification for the use of partial derivatives to describe the relationship between Eulerian (local) and Lagrangian (material) references. Nevertheless the Eulerian relationship is frequently false. The importance of this statement needs to be underscored. To the extent that modern Physics relies on the Eulerian relationship as justification for using partial derivatives falsely, modern Physics cannot be classified as science in the strict meaning of the term.

The third indexed corollary to the bridge theorem in the simply continuous case and the Eulerian relationship refer to six different derivatives:

Derivative: Symbol	**Name**	**Definition: limit as** da ⟶ 0
$D[a1](\mathbf{f}\vert\mathbf{x1};da)$	material	$(\mathbf{f}(\mathbf{r}(\mathbf{x1},a+da))-\mathbf{f}(\mathbf{r}(\mathbf{x1},a)))/da$
$D[a1](\mathbf{f}\vert\mathbf{r1};da)$	local	$(\mathbf{f}(\mathbf{x}(\mathbf{r1},a+da))-\mathbf{f}(\mathbf{x}(\mathbf{r1},a)))/da$
$\mathbf{D3}[\mathbf{r1}]*(\mathbf{f}\vert a1+da;\mathbf{dr})$	gradient	$(\mathbf{f}(\mathbf{x}(\mathbf{r1}+dri*\mathbf{ui},a1+da))$ $-\mathbf{f}(\mathbf{r1}(\mathbf{x1},a1+da))*\mathbf{ui})/dri$
$d\mathbf{f}/dt$	whole	$(\mathbf{f}(\mathbf{r}+\mathbf{h},t+dt)-\mathbf{f}(\mathbf{r},t))/dt$
$\partial\mathbf{f}/\partial t$	partial	$(\mathbf{f}(\mathbf{r},t+dt)-\mathbf{f}(\mathbf{r},t))/dt$
$\nabla\mathbf{f}$	gradient	$(\mathbf{f}(\mathbf{r}+dri*\mathbf{ui})-\mathbf{f}(\mathbf{r}))*\mathbf{ui}/dri$

The Eulerian relationship

Difficulties arise because the latter three derivatives (of the calculus) can be interpreted in different ways, only one of which equates to the former three derivatives (of Theoretical Physics.)

In the Eulerian relationship the factor, d**r**/dt, is ambiguous. Although the function **r**(t) = **r**(**x1**,t) is unique by the principle of non–collocation, there are many, even continuous, functions **r**:R⟶V3 which map into the same trajectory **r**(t). The Eulerian relationship must then always be associated with a definite "particle velocity," which however is not expressed as such in the Eulerian relationship. The Eulerian relationship is false unless d**r**/dt refers only to D[t1](**r**|**x1**;dt).

The gradient, **∇f**, in the Eulerian relationship can also be interpreted ambiguously as

D3[**r1**]∗(f|a1+da;**dr**),
or **D3**[**r1**]∗(f|a1;**dr**),
or **D3**[**r1**]∗(f|a1−da;**dr**),
or **D3**[**r2**]∗(f|a1+da;**dr**),
or even **D1**[**r1**,**r2**]∗(f|**x1**;**dr**).

While these distinctions become less significant for regions of simple continuity where they may coalesce to the same result, at locations of diminished discontinuity, the Eulerian relationship easily leads to error.

The greater difficulties with the Eulerian relationship, however, arise from identifying whole derivatives with material derivatives and partial derivatives with local derivatives.

In the calculus ambiguities surround partial derivatives. These are resolved by designating variables as either dependent or independent. For example, consider u=f(x,y) and y=g(x,z).

If x and y are the independent variables, then

$$\partial u/\partial x=\partial f/\partial x$$

and

$$\partial z/\partial x = -(\partial g/\partial x)/(\partial g/\partial z).$$

If x and z are the independent variables, then

$$\partial u/\partial x = \partial f/\partial x + (\partial f/\partial y)*(\partial g/\partial x)$$

and

$$\partial y/\partial x= \partial g/\partial x.$$

The "correct" result thus depends on an arbitrary and external labeling of the variables.

The designation of a variable as dependent or independent, as used in the calculus, is thus only an artifice of Mathematics to indicate the desired chaining rules of differentiation. Logically correct partial differentiation compatible with Theoretical Physics occurs only if the selected rule (usually unmentioned) coincides with the differentiation of compound functions of Theoretical Physics[90].

Theoretical Physics: Fallacies of Partial Derivatives

The whole derivative of a function f(**r**,t) is defined from the difference **f**(**r**+**h**,t+h)–**f**(**r**,t). Since the notion, **h⟶0** is not well defined, the usual definition of whole derivative is not well–defined either. This condition is referred to as "both **r** and t independent." Thus the whole derivative in its general meaning is unsuitable for Theoretical Physics.

If we hope to validate the Eulerian relationship we are forced to modify the definition of whole derivatives to functions

f(**r**,t)=**f**(**r**(t))=**f**(t),

or **f**(**r**,t)=**f**(**r**(t),t).

These cases are called: "t independent, **r** dependent."

In the first case, where **r**(t) is some trace in V3, the definition of whole derivative based on (**f**(**r**(t+h))–**f**(**r**(t)))/h is meaningful. However, in this case the partial derivative is undefined.

In the second case, **f**(**r**,t)=**f**(**r**(t),t) the whole derivative may be meaningfully based on the ratio (**f**(**r**(t+h),t+h)–**f**(**r**(t),t))/h. If then the partial derivative is formed, for example, from (**f**(**r**(t),t+h)–**f**(**r**(t),t))/h and only when so defined can we hope to put d**f**/dt=∂**f**/∂t+(d**r**/dt)•**∇**f. Even so, **∇**f needs to be defined precisely as **D3**[**r1**]∗(f|t1+h;**dr**) for the relationship to be true. A discrepancy appears immediately when the Eulerian relationship is expanded from scalar to vectorial functions.

Consequently every element of the Eulerian relationship carries its weight of ambiguity which, easily misinterpreted, leads to error.

In summary, then, the Eulerian relationship succeeds only for *continuous cases of a scalar function* where

df/dt is forced to D[t1](f|**x1**;dt),
∂f/∂t is forced to D[t1](f|**r1**;dt),
d**r**/dt is forced to D[t1](**r**|**x1**;dt),
and
∇f is forced to **D3**[**r1**]∗(f|t1+h;**dr**).

Theoretical Physics: Fallacies of Partial Derivatives

Example: To illustrate the differences between derivatives suitable for Theoretical Physics and the usual whole/partial derivatives of calculus, consider the derivatives of the following simple one–dimensional function r(x,t)=x∗(t–t0), t0 constant.

The derivatives of the calculus for this function are:

Derivatives of the calculus	Condition	Result
1. dr/dt	t,x independent	undefined
2. dr/dt	x dependent	x+(t–t0)∗dx/dt
3. ∂r/∂t	t,x independent	x
4. ∂r/∂t	x dependent	x+(t–t0)∗∂x/∂t
5. dr/dx	t,x independent	undefined
6. dr/dx	t dependent	t–t0+x/(dx/dt)
7. ∂r/∂x	t independent	t–t0
8. ∂r/∂x	t dependent	t–t0+x/(∂x/∂t)
9. dr/dr		1
10. ∂r/∂r		1

The derivatives of Theoretical Physics for this function are:

Derivatives of Theoretical Physics	Result	Common identity, True/False
1. D(r|x;dt)	x	dr/dt, False
2. D(r|t;dx)	t–t0	∂r/∂x, True
3. D(r|r;dt)	0	∂r/∂t, False
4. D(r|r;dr)	0	∂r/∂r, False
5. D(r|r;dx)	0	∂r/∂x, False
6. D(r|t;dr)	1	∂r/∂r, True
7. D(r|x;dr)	1	dr/dr, True

Theoretical Physics: Fallacies of Partial Derivatives

Derivatives of Theoretical Physics	Result	Common identity, True/False
8. D(r|t;dt)	∞	dr/dt, False
9. D(r|x;dx)	∞	dr/dx, False
10. D(x|r;dt)	−x/(t−t0)	∂x/∂t, False
11. D(x|t;dr)	1/(t−t0)	∂x/∂r, True
12. D(x|x;dt)	0	dx/dt, False
13. D(x|x;dx)	0	dx/dx, False
14. D(x|x;dr)	0	dx/dr, False
15. D(x|t;dx)	1	∂x/∂x, True
16. D(x|r;dx)	1	∂x/∂x, True
17. D(x|t;dt)	∞	dx/dt, False
18. D(x|r;dr)	∞	∂x/∂r, False

The derivatives of the calculus may be interpreted in many ways, but unless they conform to the material and local derivatives of Theoretical Physics, they yield false relationships.

The falsity of identifying the locally referenced derivative of Theoretical Physics with the partial derivative of the calculus can be starkly exhibited as:

$$1 = \partial r/\partial r \quad \text{vs} \quad D[\mathbf{r1}](r|\mathbf{r};dr) = 0.$$

Since asserting the identity $\{\partial r/\partial r = D[\mathbf{r1}](r|\mathbf{r};dr)\}$ creates a logical contradiction, it is a false identity. When this false identity is taken as true, as is often the case in modern Physics, any statement of Physics that can be possibly constructed with sufficient mathematical manipulation can be proven logically true. While the prudent and judicious use of contradiction–bearing statements may be useful for creating models, the results can never rise to science because they cannot be understood. Explanation of physical observations so based is useless for Physics because reality is what it is and is not what it is not.

Theoretical Physics: Fallacies of Partial Derivatives

To establish Theoretical Physics on a solidly consistent foundation, a conditional notation for derivatives has been adopted which at first may have seemed gratuitous, but now may be better appreciated.

The clarity introduced by the notation can be seen by applying equation (1822) to **r**(**x1**,a) for the continuous case.

D[a1](**r**|**x1**;da)
= D[a1](**r**|**r1**;da) + D[a1](**r**|**x1**;da)•**T**[**D3**[**r1**]∗(**r**|a1+da;**dr**)].

Since D[a1](**r**|**r1**;da)=**0** and since **D3**[**r1**]∗(**r**|a1+da;**dr**) is the identity transformation, the equation is reduced to the tautology,

D[a1](**r**|**x1**;da) = D[a1](**r**|**x1**;da).

The Eulerian relationship,

$$d\mathbf{r}/dt = \partial\mathbf{r}/\partial t + (d\mathbf{r}/dt)\bullet\nabla\mathbf{r}$$

however, is not so clearly a tautology. In fact, to set

$$\partial\mathbf{r}/\partial t \equiv \mathbf{0}$$

can be most distressing for those who interpret Newton's second law as

$$\mathbf{f} = m*\partial\mathbf{r}^2/\partial t^2.$$

The ambiguities associated with the Eulerian relationship are a major reason why many propositions of Modern Physics are badly posed mathematically.[91]

The inconvenience to Modern Physics is great, inasmuch as it follows that all argumentation based on partial differential equations, which is practically all of modern Physics, is suspect.

Summary[92]

Physics is the study of reality observable as extended, moving, or forcing. As a study, Physics has as its goal the analysis and unification of physical observations.

Theoretical Physics is the elaboration of intellectual instruments (ideas) into a set of statements[93], logically true, which can be applied to the observations of Physics. Ideas are needed as means for study because, while extended, moving, or forcing objects can be sensed, only ideas can be understood. But ideas can only be understood if they are logically consistent in a framework of other ideas and propositions.

Our primitive observations of physical objects discover certain general attributes of both the objects and by reflection also of the process of observation. Theoretical Physics must create ideas to match these attributes. The structures of Mathematics provide a storehouse of consistent ideas from which Theoretical Physics may be generated. Thus is born the first problem of Theoretical Physics: how to transform Mathematics into Theoretical Physics.

Summary

The association between Physics, Theoretical Physics, and Mathematics is indicated in the following table:

Physics	Attributes	Theoretical Physics
The study of	consistent ideas	Mathematics & symbology constrain and expand mathematical entities by physical units constrain and expand mathematical operators by physical references
observable reality	relative location of observer and observed object observation limited in scope dynamic	three dimensional set of vectors and an arbitrary origin. particle as topological material and local references. See also *Theoretical Physics The Third Problem.*
as extended,	priority having parts, i.e., measurable.	material reference particle as topological
moving, and forcing.		See *Theoretical Physics The Second Problem.*

Physics, Theoretical Physics, Mathematics

Symbols label ideas. An elaboration of these as well as some useful mathematical structures are provided for practice and reference.

Some new mathematical ideas are developed, notably
division in the set of vectors
new types of gradients
new types of integrals.

Summary

The answer to the first problem is given by both constraining mathematical ideas and equations (some mathematical operations are not allowed) and also expanding them, much as a generic idea is expanded into specific ideas.

The ideas of Mathematics are first expanded by attaching to each a label called a physical unit. The units operate as constraints which forbid certain otherwise valid mathematical operations because the equations of Theoretical Physics must agree by units.

On the other hand, for each idea of Mathematics the labeling generates for Theoretical Physics a panoply of related ideas, one for each label.

Each idea of Mathematics is further transformed into Theoretical Physics by expanding it with and constraining it to references. The relationships between the references engender a family of related ideas, each different as an idea from its mathematical parent. In particular the mathematical ideas of differentiation and integration blossom into a copious family of new ideas of differentiation and integration. The relationships between these new ideas are explicated in this book.

Physical references are provided by:

1. the origin of the three dimensional set of vectors, and
2. a hypothetical universe at rest.

The origin references the observer's location and orientation; the universe at rest references objects in the universe.

A **particle**, (**x**), is defined by means of the topology of three dimensional set of vectors as a limiting operation which corresponds to increasingly refined physical observation.

Change is referenced by the two ways of observation, called the material reference and the local reference.

The two ways of observation are linked by the bridge theorem.

The Eulerian relationship, widely used by physicists to describe the linkage, is shown to lead easily to falsity because of its ambiguities. It follows that Physics based on partial differential equations is likely to be logically ill–founded. Since most of modern Physics is based on partial differential equations and the Eulerian relationship, almost all of modern Physics is badly posed mathematically.

Summary

The first problem of Theoretical Physics is mostly solved. The solution enables mutual enhancement of both Physics and Mathematics.

Finally, some ways are indicated in which Physics itself is oriented to the larger reality.

Epilogue

Having found no few errors himself, the author is not so deceived as to imagine that no others still lurk in the text. He would be greatly pleased to have them pointed out to him. Generous readers may do so by writing the publisher (address in front–piece) or they may post their findings electronically at

http://theoretical.Physics.books@gmail.com

Endnotes

[1] Not every logically consistent proposition is true, though every true proposition is logically consistent.

[2] Language is not always helpful in distinguishing ideas closely interrelated with cognition. When, looking at an object, we say: "I see a man," we are coalescing a more elaborate saying: "I understand that the object I am looking at is a man."

Surely the perception itself is always truly perceived and so taken for granted. Knowing *what* is perceived, however, requires more than perception; it requires understanding––an intellectual process. The perceiver needs first to have conceived the idea "man" before asserting a concordance between the perceived object and his idea. If the concordance matches reality, the understanding is correct and the statement true. So truth is the conformity of the intellect to an objective reality.

For both sensory perception and intellectual understanding, it is objective reality which reigns. The senses always conform; the intellect only upon the formation of a conforming idea.

The burden of theoretical Physics is to form ideas useful for understanding what is observed. The burden of the observing physicist is to select ideas of Theoretical Physics which conform to the observation and thus understand and explain the observation. Pseudo ideas, those implying contradictions, only lead to false understanding and false explanations.

The process of cognition is greatly muddied in modern thinking. An outstanding discourse, perhaps unsurpassed, was made by Thomas Aquinas. Cf *Summa Theologiae*, passim. One might also usefully consult Hoenen, Peter, *Reality and Judgment according to St. Thomas, Regnant*, 1952.

[3] Readers who find the following unfamiliar should consult a standard text discussing the invoked mathematical structures.

[4] The real number system is not strictly necessary, since the quotient numbers with their usual topology suffice for Theoretical Physics.

[5] Many authors confuse functions as descriptions of relationships with the relationships themselves. For those failing to make the distinction, examples of different functions describing the same relationship becomes confusing.

[6] The case with derivatives is similar to the case of plane geometry. There are an infinite number of plane geometries, depending on the angular number one wishes to assign to the interior angles of a triangle. The familiar Euclidean geometry assigns uniquely the number pi to this angular number. Similarly, for any one continuous relationship from domain to range, there are an infinite number of functions with different derivatives describing the same relationship. For example, the functions

$$f0(x) = c, \text{ constant},$$
$$f1(x) \equiv \lim f(x+dx) = c + dx * \sin(a) \text{ as } dx \longrightarrow 0,$$

for any angle a, describe the same relationship. The derivative of f1 is clearly sin(a) while the derivative of f0 is 0. Clearly f0(x) and f1(x) are different functions. Yet for every x, f1(x) = f0(x). The function f0 is merely one of an infinity of such functions with identical values.

Endnotes

Choosing the angular number equal to zero defines the unique function with a Newton/Leibniz (basic) derivative. It is easily seen that in the set of all functions of the same relationship, the one with a Newton/Leibniz derivative has a measure of zero.

[7] The basic or classical Newton/Leibniz derivative is used in the text much the way Euclidean geometry is commonly used. Propositions in geometry which do not rely on the parallel postulate are considered to apply to the larger set of plane geometries while those which rely on the postulate are considered restricted to Euclidean geometry. In the same way, propositions which do not rely on the basic approximation are considered to apply to the larger set of functions with derivatives, while those which do are considered to apply only to functions with a basic derivative.

Just as Euclidean geometry finds great utility in practical matters, so too does the basic derivative.

[8] The four derivatives are not a little difficult to express using the conventional notation d/dx or ∂/∂x. The difficulty provides a first example of how an inadequate symbology hinders our ability to think clearly. Many examples follow.

[9] The confusion of the two conventions causes ambiguity. For instance,

$$f(t+(-1)*dt) \approx f(t) + (-1)*dt*D[t](f|u;(+1)*dt)$$

gives

$$f(t) \approx f(t-dt) + dt*D[t](f|u;(+1)*dt)$$

so that

$$\begin{aligned} D[t](f|u;(+1)*dt) &= D[t](f|u;(-1)*dt) \\ &= D[t](f|u;\ d_f t) \\ &= D[t](f|u;\ d_b t) \end{aligned}$$

which becomes a contradiction where the derivative is discontinuous.

[10] Functions with regions discontinuous beyond denumerability are excluded because then the idea of differentiability makes no sense. Thus the attribute of differentiability implies a function continuous in a region around the point of interest, though not necessarily at that point.

[11] See *Differentiation and Integration of Compound Functions* by JRBreton ISBN 978-0-9844299-4-3, 110 pages, Library of Congress Control Number: 2011935109 for a discussion on specific derivatives of compound functions.

[12] See *Differentiation and Integration of Compound Functions* by JRBreton ISBN 978-0-9844299-4-3, 110 pages, Library of Congress Control Number: 2011935109 for a discussion on specific integrals of compound functions.

[13] The proofs for these initial results are assumed well known and thus omitted. Later development will make the proofs apparent.

[14] The reader may profitably consult *Step Functions and Product Rules* ISBN 978-0-9844299-2-9 which functions as an Appendix of this volume.

Endnotes

[15] The multiplication of two step functions needs careful attention. For instance, if one puts

u(x−x1)*u(x−x1) = u(x−x1)

as a point by point multiplication
then

D[x](u(t−x1)*u(t−x1);dt) = D[x](u(t−x1);dt) = i(x−x1).

while if one considers u(x−x1)*u(x−x1) as the product of two functions

D[x](u(t−x1)*u(t−x1);dt) = 2*u(x−x1)*i(x−x1)

which clearly differ. This is again an example of two different functions having the same functional values.

Functional multiplication is used in the text for which reason step functions are always shown explicitly. Consequently, for step functions, u(x−x1)*u(x−x1) differs from u(x−x1).

For a full discussion of these issues see *Step Functions and Product Rules* ISBN 978-0-9844299-2-9.

[16] The derivation might also have proceeded thus:

D[x](f(t)*u(t−x1);d$_f$t)
= lim (f(x+dx)*u(x+dx−x1) − f(x)*u(x−x1))/dx
= lim (f(x+dx)*u(x+dx−x1)
− f(x)*u(x+dx−x1)
+ f(x)*u(x+dx−x1)
− f(x)*u(x−x1))/dx
= lim (f(x)*D[x](u(t−x1);d$_f$t))
+u(x+dx−x1)*(f(x+dx)− f(x))/dx
= v(x−x1)*D[x]f(t);d$_f$t +f(x)*D[x](u(t−x1);d$_f$t)

which appears markedly different from the text for x=x1.

The presentation may deceive however since

lim f(x)/dx + u(x+dx−x1)*(f(x+dx)−f1+f1−f(x1))/dx = lim f(x+dx)/dx

as in the text.

These functions do *not* obey the product rule for limits, even for *f* continuous at *x1*:

lim f(x1+dx)*(u(x1+dx−x1)−u(x1−x1))/dx
≠ lim f(x1+dx)*lim (u(x1+dx−x1)−u(x1−x1))/dx,

but rather

lim f(x1+dx)*(u(x1+dx−x1)−u(x1−x1))/dx
= f(x1)*((lim u(x1+dx−x1)−u(x1−x1))/dx, + lim (f(x1+dx)−f(x))/dx.

The limiting value of u(x1+dx−x1)−u(x1−x1))/dx is not bounded at x1.

[17] The integral of a derivative consists of two limiting processes, one pertaining to the derivative, the other to the integral. The processes are not independent but are regulated by the integral expression such that the integral increment always dominates the differential increment. For example

I[x0,x1](D[x](f;d$_f$t);dx)
= lim dx*(D[x0](f;d$_f$t)+D[x0+dx](f;d$_f$t)+...+D[x1](f;d$_f$t))
= lim dx*(lim (f(x0+dt)−f(t0)/dt)
+lim (f(x0+dx+dt)−f(t0+dx)/dt)
+...
+lim (f(x1+dt)−f(x)/dt))

Endnotes

= lim (dx/dt)∗(lim (f(x0+dt)−f(t0)
+f(x0+dx+dt)−f(t0+dx)
+...
+f(x1+dt)−f(x1))

provided 0<dt<dx. For *f* continuous, *dt* can be arbitrarily close to *dx.*

[18] Some results are possible for *f* discontinuous at xi
I[](u(x−xi)∗D[x](f(t);d_ft);dx)= −lim (f(xi+dx)
I[](u(x−xi)∗D[x](f(t);d_bt);dx)= lim (−f(xi+dx−dt)
I[](v(x−xi)∗D[x](f(t);d_bt);dx)= −lim (f(xi)−f(xi−dt))−f(xidx−dt))
I[](v(x−xi)∗D[x](f(t);d_ft);dx)= −f(xi)
The result then depends on the functional value and the limits of the function at the discontinuity.

[19] Clearly not all functional pairs obey the product rule. For a thorough discussion on how the step functions u(x) and v(x) themselves relate to the product rule, see *Step Functions and Product Rules* ISBN 978-0-9844299-2-9, Library of Congress Control Number: 2010935175, published separately.

[20] For proof, see *Step Functions and Product Rules*, published separately.

[21] The text *describes* V3. An *axiomatic* foundation may be readily found in texts treating the set of vectors.

[22] R itself may be considered a set of vectors. Then 0 serves as origin. Ordinary addition and multiplication serve as vector addition and scalar multiplication (the angle *a* defining inner and cross products always equals 0; thus for the real numbers x1 and x2, x1∧x2=0 always.) Integration and differentiation in this set of vectors are identical to ordinary integration and differentiation. R is a one dimensional set of vectors.

[23] Propositions proved for any arbitrary origin are called **relative** in contrast to propositions which depend on a specific origin (called **absolute**). Newton, for example, insisted on an absolute orientation for Physics, a necessity disputed by Poincaré and Einstein.

[24] "Hand rules", left or right, are sometimes used to specify vector products. The "hand rules" come from making the following directional assignment:

index finger ⟶ **u1**
middle finger ⟶ **u2** or −**u2**
thumb ⟶ **u3**

Absent a given orientation, however, hand rules can be confusing inasmuch as any left–hand rule can be made into a right–hand rule by altering the orientation. The definition of the text, for instance, can be made into either a right–hand rule or a left–hand rule. Hand rules, then, are superfluous.

The alternative definition has

u1∧(−**u2**) = **u3** = **u2∧u1**
(−**u2**)**∧u3** = **u1** = **u3∧u2**
−(**u3∧u1**) = **u2** = **u1∧u3**

Endnotes

[25] When b=0, **r** and **a** are orthogonal. The minimum solution then is **0**, the origin, while the set of solutions is described as any vector **r** such that **ur•ua**=0.

[26] Since sin(angle(**a**,**r**)) = (**ur∧ua**)•**ub**,

r = b∗**ur**/(a∗(**ur∧ua**)•**ub**)

expresses the same solution for **r∧a**=**b**.

[27] The matrix function **T** is sometimes applied to vectors to indicate pre– or post–multiplication. We shall forgo this usage inasmuch as the position of the vector with respect to the matrix resolves all ambiguities.

The convention used here is this: **r•A** refers to a row vector pre–multiplied in a matrix sense with the matrix **A** which yields in turn a row vector. This is a complete operation in itself.

The notation **A•r** refers to a post–matrix multiplication of **A** with the column vector **r**. This is not a complete operation in itself, but must be part of another expression which completes it. What is sometimes written as **T**[**A•r**] is here written **r•T**[**A**].

[28] The proof of this assertion considers the projection of t1*$\mathbf{r1}$ onto the plane. The projection may be represented by a linear combination of the two given variables **n2** and **n3** which in turn result in two equations in two unknowns. A quick check of **rn** may be made for **n2** and **n3** orthogonal.

[29] The solution is unique as a diagonal matrix. Indeed starting with the assumption **A** = **G**(**r**), a diagonal matrix, the equation

r0•**G**(**r**) = **r1**

has a unique solution, namely

r = r11∗**u1**/r01 + r12∗**u2**/r02 + r13∗**u3**/r03
G(**r**) = r11∗**u1**∗**u1**/r01 + r12∗**u2**∗**u2**/r02 + r13∗**u3**∗**u3**/r03
= **G**(**r1•u1∗u1/r0•u1 + r1•u2∗u2/r0•u2 + r1•u3∗u3/r0•u3**)

Whereas **r0•T**[**A**] = **r1** is a system of 3 equations with 9 unknowns, **r0•G**(**r**) = **r1** is a system of 3 equations with 3 unknowns.

[30] The set

{**r**(dr)|**r**(dr) = **r0** + dr∗**u**(**r**–**r2**), dr in R}

is identical with the line

{**r**(x)|**r**(x) = **r0** + f(x)∗**r1**, x in R}

provided **r1** = **r2**–**r0** and f(x)∗r1 = dr. The functions **r**(x) and **r**(dr) differ in their domains:

r(x): R⟶V3
r(dr): V3⟶V3.

[31] Directional limits arise from a topology borrowed from R. Let

r(x) = **r0** + (x – x1)∗**ur** ≡ **r0** + dx∗**ur**.

Then

f(**r**(x)) = **f**(**r0** + dx∗**ur**).

If then **f**(**r**(x))⟶**f1** as dr⟶0, it also follows by equating dr and dx that **f**(**r**)⟶**f1** as dr∗**ur**⟶**0**.

In effect the topology of R is mapped onto **r**(x) = **r0** + dr∗**ur**.

Endnotes

[32] To transform a first quadrant vector **a**=a1***u1**+a2***u2**+a3***u3** to another first quadrant vector **b**=b1***u1**+b2***u2**+b3***u3**, the matrix **q**(**a**)***b**/3 may be used since **a**•**q**(**a**)***b**/3 = **b**. This transform is constrained to ai***ui** ⟶ bi***ui**.
To transform a sectional vector **aR1**=a1***u1R1**+a2***u2R1**+a3***u3R1** to another vector **bR1**=b1***u1R1**+b2***uR12**+b3***u3R1** in the same section, we may use the matrix

(**UR1**$^{-1}$)•**q**(**a**)***b**•**UR1**/3.

This transform is constrained to ai***uiR1** ⟶ bi***uiR1**.

[33] Unless D[x](**r**(t);dt) = **0**. In the case of a stalled process, the direction is indeterminate.

[34] The unit tangent, **ut**, can be used to construct an intrinsic system of coordinates for the curve. Since abs(D[s(x)](**r**(s);ds))=1 for all s, it follows that for all s(x),

D[s(x)](D[s(x)](**r**(s1);ds1)•D[s(x)](**r**(s1);ds1);ds)
= 0
= 2*D[s(x)](**r**(s);ds)•D[s(x)](D[s(x)](**r**(s1);ds1);ds)

that is, D[s(x)](**r**(s);ds)=**ut** and D[s(x)](D[s(x)](**r**(s1);ds1);ds) are orthogonal.
Define
D[s(x)](D[s(x)](**r**(s1);ds1);ds) ≡ **c**(s(x)) = abs(**c**(s(x)))***uc**. The magnitude,

abs(**c**(s(x)) ≡ curv(s(x)),

is called the **curvature**
and its reciprocal,

1/curv(s(x)) ≡ rad(s(x))

is called the **radius of curvature** of **r**(t) at x.
Noting that **uc** and **ut** are orthogonal, a third vector, **ub**, called the bi–normal **vector**, can be constructed from their cross product, that is,

ub = **ut** ∧ **uc**.

All vectors associated with a curve can be written as varying linear combinations of these mutually orthogonal vectors.
The intrinsic derivatives of this coordinate set produce a set of self–referring equations.
Consider D[s(x)](**ut**;ds) = curv(s(x))***uc**. Since abs(**ub**(s))=1 for all s, D[s(x)](**ub**;ds) is a vector in the plane of **ut** and **uc**. Now, **ub** and **ut** are orthogonal, so that
D[s(x)](**ub**•**ut**;ds) = D[s(x)](**ub**;ds)•**ut**) + **ub**•curv]***uc** = 0.
Again **ub**•**uc**=0, so that D[s(x)](**ub**;ds) must also be orthogonal to **ut**.
In other words
D[s(x)](**ub**;ds) is parallel to **uc**, so we may write,

D[s(x)](**ub**;ds) ≡ tor***uc**.

Tor is called the **torsion** of the curve, **r**(t) at x.
Next,
D[s(x)](**uc**;ds) = D[s(x)](**ub**∧**ut**;ds)
= **ub**∧D[s(x)](**ut**;ds) + D[s(x)](**ub**;ds)∧**ut**
= **ub**∧curv***uc** + tor***uc**∧**ut**
= – curv***ut** – tor***ub**

Thus

D[s(x)](**ut**;ds) = curv***uc**
D[s(x)](**uc**;ds) = –(curv***ut** + tor***ub**)
D[s(x)](**ub**;ds) = tor***uc**.

Endnotes

It follows then that all higher derivatives can be written in terms of the same intrinsic coordinate system (**ut**, **uc**, **ub**) and the higher derivatives of curv(s) and tor(s).

It can be shown that

D[s(x)](**ut**;ds)•D[s(x)](**ub**;ds) = curv*tor,
D[s(x)](**uc**;ds)•D[s(x)](**ub**;ds) = 0,
D[s(x)](**ut**;ds)•D[s(x)](**uc**;ds) = 0,

The curvature and torsion of a curve depend on position and are functions of *s*. These two functions are called the **intrinsic equations** of the curve. If two curves have the same intrinsic equations they are identical, except possibly for orientation.

Example. **r**(x) = **r1***u(x–1) + x***r2**, **r1** and **r2** fixed and u the unit step function, describes a straight line with a "jump" at x=1. Taking for reference x=0,

s(x) = abs(x***r2**), x<1
= abs(**r1** + x***r2**) , x>1

D[x](**r**(t);dt) = **r1***i(x–1) + **r2** = D[x](s(t);dt)***ut**

and

D[x](D[x](**r**;dt);dt) = **r1***D[x](i(t–1);dt)

that is, **ut** is constant except at the jump, so that except at x=1,

curv(s) = tor(s) = 0.

At x=1, the arc–length, s, is an indeterminate value between

1*r2 and abs(**r1** + 1***r2**).

Additionally,
D[s(x)](**ut**;ds)
= D[x](**ut**;dt)/D[x](s;dt)
= D[x](D[x](**r**;dt);dt)/(D[x](s;dt))^2
= **r1***D[x](i(t–1);dt)/(abs(**r1***i(x–1) + **r2**))^2
= curv(s(1))***uc**

so that
curv(s(1)) = r1*D(i(x–1)/(abs(**r1***i(x–1) + **r2**))^2
Thus at x=1, curv takes on multiple values.
Since **r**(x) is a plane curve, tor is undefined.

---

[35] In geometry three kinds of transformations are often studied because they preserve certain geometrical properties. These geometric transformations are defined as follows:

Translations: **r2** = **r1** + **c**, **c**, a constant vector;
Dilations: **r2** = c***r1**, c a constant scalar;
Rotations: **r2** = **r1**•**R**, **R** a constant transformation whose determinant equals 1.

These transformations are all one to one. Considered as algebraic groups, the geometric transformations are closed, that is, a translation of a translation is itself a translation, etc. In particular, the group of rotations have an interesting structure. Note that outer multiplication **r1*r2** is not a geometric transformation since it is many to one. Diagonal transformations, **G**(**r**), whose determinants equal one are rotations.

Endnotes

For reference we give the rotation through an angle g of a vector **r1** around an axial direction **ua**, producing a new vector **r2**.

$$\mathbf{r2} = \mathbf{r1}\cdot[\mathbf{ua}*\mathbf{ua} + \cos(g)*[\mathbf{I}-\mathbf{ua}*\mathbf{ua}] - \sin(g)*\mathbf{C}(\mathbf{ua})]$$

Thus by the suitable choice of two parameters: an axial direction and a rotation angle all the rotations of a given vector can be determined.

We also give the solution for the transformation which carries one direction into another, that is in our symbology, for the rotation Rot:**ur1**⟶**ur2**. Let

$$\mathbf{ur1}=a11*\mathbf{u1}+a12*\mathbf{u2}+a13*\mathbf{u3}$$

and

$$\mathbf{ur2}=a21*\mathbf{u1}+a22*\mathbf{u2}+a23*\mathbf{u3}$$

Because **ur•ur**=1, only two parameters need be specified for a given direction. There are numerous ways to specify the desired rotation.

It is often helpful to specify the problem in terms of angles defining the rotation. For instance, a rotation around **u3** need only specify a rotation angle, b, in the plane determined by **u1** and **u2**. For this special case:

$$\mathbf{ur2}=\mathbf{ur1}\cdot\begin{bmatrix}\cos(b) & \sin(b) & 0\\ -\sin(b) & \cos(b) & 0\\ 0 & 0 & 1\end{bmatrix}$$

Likewise rotation around **u1** yields

$$\mathbf{ur2}=\mathbf{ur1}*\begin{bmatrix}0 & 0 & 1\\ 0 & \cos(b) & \sin(b)\\ 0 & -\mathrm{sinb} & \cos(b)\end{bmatrix}$$

and around the **u2** axis,

$$\mathbf{ur2}=\mathbf{ur1}\cdot\begin{bmatrix}\cos(b) & 0 & \sin(b)\\ 0 & 1 & 0\\ -\sin(b) & 0 & \cos(b)\end{bmatrix}$$

For an arbitrary rotation consider

$$\mathbf{ur1} = (\mathbf{ur1}\cdot\mathbf{u1})*\mathbf{u1}+(\mathbf{ur1}\cdot\mathbf{u2})*\mathbf{u2}+(\mathbf{ur1}\cdot\mathbf{u3})*\mathbf{u3}$$

where **ur1•ui** is the cosine of the angle between the given direction **ur1** and the orientation axes **ui**, i=1,2,3. Since we can write any of these angles in terms of the other two, let us choose **ur1•u3**=cos(a1).

Then

$$\mathbf{ur1}\cdot\mathbf{u1}=\sin(a1)*\cos(b1) \text{ and } \mathbf{ur1}\cdot\mathbf{u2}=\sin(a1)*\sin(b1)$$

defines another angle b1 which together with the angle a1 defines the direction **ur1**.

Call the direction to which **ur1** is rotated **ur2**. Then **ur2** is likewise defined by two analogously defined angles, a3 and b3. In fact, with **ur1** as the starting direction, the rotation can be defined by the two angles a2 = a3–a1 and b2 = b3–b1. In terms of these incremental angles, let

$$x=\cos(a2)+\cot(a1)*\sin(a2) \text{ and } y=\cos(a2)-\tan(a1)*\sin(a1).$$

Then

$$\mathbf{ur2}=\mathbf{ur1}\cdot\begin{bmatrix}x*\cos(b2) & x*\sin(b2) & 0\\ -x*\sin(b2) & x*\cos(b2) & 0\\ 0 & 0 & y\end{bmatrix}$$

In this form it can be easily seen that a rotation can always be expressed as

$$\mathbf{ur2} = \mathbf{ur1} \bullet \mathbf{G}(\mathbf{a}) - \mathbf{ur1} \wedge \mathbf{b}$$

where $\mathbf{a}$=x∗cos(b2)∗**u1**+x∗cos(b2)∗**u2**+y∗**u3**, and $\mathbf{b}$=x∗sin(b2)∗**u3**. Similar equations can be written for formulations referenced to **u1** or **u3**.

[36] The curl is usually defined for a function **f** = f1∗**u1**+f2∗**u3**+f3∗**u3** as

$$\begin{aligned}\mathbf{D}\wedge\mathbf{f} = &(\partial f3/\partial h2-\partial f2/\partial h3) * \mathbf{u1}\\ &+ (\partial f1/\partial h3-\partial f3/\partial h1) * \mathbf{u2}\\ &+ (\partial f2/\partial h1-\partial f1/\partial h2) * \mathbf{u3}\end{aligned}$$

The symbology of the usual definition is somewhat confusing since, for instance, f3 can be interpreted as independent of h2. However putting f3 ≡ **f•u3** and interpreting

$$\partial f3/\partial h2 \equiv ((\mathbf{f}(\mathbf{r}+h2*\mathbf{u2})-\mathbf{f}(\mathbf{r}))/h2)\bullet\mathbf{u3},$$

then the curl of the text becomes the negative of the usual one.

[37] Two things are necessary for the proposed definition to be successful. First it must be a consistent elaboration of the previous definition of gradient. Secondly, it must be an unambiguous choice from what might be more than one possibility.

One possible definition might equate **D**∗(**c**∗f) = **c**∗(**D**∗f(**r**)) to (**D**∗f(**r**))∗**c** generally. For this possible definition, we would be forced to define **D**∗(**c**∗f(**r**)) in terms of diagonal transformations. Then

$$\mathbf{D}*(\mathbf{c}*f(\mathbf{r})) = \mathbf{G}(\mathbf{c})\bullet\mathbf{G}(\mathbf{D}*f) = \mathbf{G}(\mathbf{D}*f)\bullet\mathbf{G}(\mathbf{c})$$

Using this definition, it follows that the gradient of any vector function **f** would then be defined as

$$\begin{aligned}\mathbf{D}*\mathbf{f}(\mathbf{r}) &= \mathbf{D}*(f1(\mathbf{r})*\mathbf{u1}+f2(\mathbf{r})*\mathbf{u2}+f3(\mathbf{r})*\mathbf{u3})\\ &= \mathbf{G}(\mathbf{u1})\bullet\mathbf{G}(\mathbf{D}*(f1))+\mathbf{G}(\mathbf{u2})\bullet\mathbf{G}(\mathbf{D}*(f2))+\mathbf{G}(\mathbf{u3})\bullet\mathbf{G}(\mathbf{D}*(f3))\\ &\equiv Df11*\mathbf{u1}*\mathbf{u1}+Df22*\mathbf{u2}*\mathbf{u2}+Df33*\mathbf{u3}*\mathbf{u3}\end{aligned}$$

where D∗(fi) = Dfi1∗**u1**+Dfi2∗**u2**+Dfi3∗**u3**, i=1,2,3. Thus of the nine derivatives contemplated in the **D**∗**f**(**r**) of the text, only three would finally figure in this possible definition.

Note that with this definition **D**∗**r**=**u1**∗**u1**+**u2**∗**u2**+**u3**∗**u3**=**I** also.

[38] A formula cast specifically for continuous gradients yields specific results for continuous gradients, not directional or sectional gradients, etc.

No small care is required in applying these concepts––ambiguity easily leads to error. The mathematical confusion surrounding these concepts is clarified in Theoretical Physics.

[39] Notation needs to be carefully respected here. For

$$\mathbf{f} = f1*\mathbf{u1}+ f2*\mathbf{u2}+f3*\mathbf{u3},$$

$$\mathbf{C}(\mathbf{f}) = \begin{bmatrix} 0 & -f3 & f2 \\ f3 & 0 & -f1 \\ -f2 & f1 & 0 \end{bmatrix}$$

so that

$$\begin{aligned}D\bullet(-f3*\mathbf{u2} + f2*\mathbf{u3}) &\longleftarrow (f3(\mathbf{r}+h2*\mathbf{u2})-f3(\mathbf{r})/h2 + f2(\mathbf{r}+h3*\mathbf{u3})-f2(\mathbf{r})/h3)\\ &= (\mathbf{f}(\mathbf{r}+h2*\mathbf{u2})-\mathbf{f}(\mathbf{r}))\bullet\mathbf{u3}/h2 - (\mathbf{f}(\mathbf{r}+h3*\mathbf{u3})-\mathbf{f}(\mathbf{r}))\bullet\mathbf{u2}/h3.\end{aligned}$$

Consequently

$$\begin{aligned}\mathbf{u1}\bullet(\mathbf{D}\bullet(\mathbf{C}(\mathbf{f}))) &= D\bullet(-f3*\mathbf{u2} + f2*\mathbf{u3})\\ &= \mathbf{u1}\bullet(\mathbf{D}\wedge\mathbf{f}).\end{aligned}$$

Endnotes

[40] The preceding development of derivatives and integrals from real functions to vector functions over curves relies to some extent on a property of order. All such integrals may be defined more generally over measurable sets (which need not be intervals) by methods which abandon order.

[41] What properties should a measure have? It is an open question. We put down some desirable ones:

1. For M1 and M2 disjoint and for all n > m, finite
   m(M1 ∪ M2) = **m**(M1) + **m**(M2),
2. For constant c1, **c1** and c2, **c2** with f1, **f1** and f2, **f2** bounded on M,
   mn(Pn(i,j,k)) = **m**(Pn(i,j,k))
3. I•[M](c1 ∗ f1+c2 ∗ f2|m;r) = c1 ∗ I•[M](f1|m;r) + c2 ∗ I•[M](f2|m;r)

If M is the intersection of all open sets containing **r0** and **f** is bounded at **r0**

4. I•[M](**f**|**m**;**r**) ≡ I•[**r0**](**f**|**m**;**r**) = lim (S[M∩Pn]**f**•**m**(Pn(i,j,k))) = 0
5. **I**∧[M](**f**|**m**;**r**) ≡ **I**∧[**r0**](**f**|**m**;**r**) = lim (**S**[M∩Pn]**f**∧**m**(Pn(i,j,k))) = **0**
6. I∗[M](f|m;r) ≡ I∗[r0](f|m;r) = lim (S[M∩Pn]f∗m(Pn(i,j,k))) = [0]
7. I∗[M](1|m;r) = **m**(M)
8. A translation property
9. A dilation property
10. For M symmetrical around **r0**, **m**(M) = m ∗ u(r0).

The first three: the disjoint property, mesh invariance, and linearity are required in any measure.

For an example of a measure–like function which fits only one of the three requirements consider defining

mn ≡ S[M∩Pn](d/n^2) = S[M∩Pn]((sqrt(3))/n^3)

and

umn ≡ **u**(S[M∩Pn](**r**(M∩Pn(i,j,k)|Pn))

a pseudo–measure of M with respect to the partition Pn as

mn(M|Pn) ≡ mn ∗ umn

and a pseudo–measure of M as

m(M) = lim(**mn**(M|Pn)) as n⟶∞.
≡ m(M) ∗ u(M).

For M=P1(1,1,1) under this pseudo–measure

m1(P1(1,1,1)) = (sqrt(3) ∗ (u1+u2+u3)/(sqrt(3) = **u1+u2+u3**
m2(P1(1,1,1)) = ((sqrt(3)/8) ∗ 8
∗ u(u1/4 + u2/4 + u3/4
+ u1/4 + u2/4 + 3 ∗ u3/4
+ u1/4 + 3 ∗ u2/4 + u3/4
+ 3 ∗ u1/4 + u2/4 + u3/4
+ 3 ∗ u1/4 + 3 ∗ u2/4 + u3/4
+ 3 ∗ u1/4 + u2/4 + 3 ∗ u3/4
+ u1/4 + 3 ∗ u2/4 + 3 ∗ u3/4
+ 3 ∗ u1/4 + 3 ∗ u2/4 + 3 ∗ u3/4)
= sqrt(3) ∗ u(16 ∗ (u1 + u2 + u3)/4)
= sqrt(3) ∗ u(4 ∗ (u1 + u2 + u3)
= sqrt(3) ∗ u((u1 + u2 + u3)
= sqrrt3) ∗ (u1 + u2 + u3)/sqrt(3)
= **u1+u2+u3**
= **m1**(P1(1,1,1)).

For M=P2(1,1,1)
m1(P2(1,1,1)) = ((sqrt)3)/(8)∗**u**(**u1**/2 + **u2**/2 + **u3**/2)
= ((sqrt(3)/8)∗(**u1** + **u2** + **u3**)/sqrt(3)
= (**u1** + **u2** + **u3**)/8
= **m1**(P1(1,1,1))/8.

For M=P1(0,0,0))
m1(P1(0,0,0)) = −(**u1**+**u2**+**u3**) = −**m1**(P1(1,1,1)).
For M=Pn(i,j,k) such that M∩Pn∞ is empty
m1(Pn(i,j,k)) = ((sqrt(3)/(n^3))
∗**u**((i−1/2)∗**u1** + (j−1/2)∗**u2** + (k−1/2)∗**u3**)
= ((sqrt(3)/(n^3))
∗((2∗i−1)∗**u1** + (2∗j−1)∗**u2** + (2∗k−1)∗**u3**)
/sqrt($(2∗i−1)^2+(2∗j−1)^2+(2∗k−1)^2$)
for i,j,k=−(n^2−1),...,−1,0,1,2,3,...n^2.
If **m1**(Pn(i,j,k)) = c1∗**u1** + c2∗**u2** + c3∗**u3**,
then
m1(Pn(−i+1,−j+1,−k+1) = −c1∗**u1** − c2∗**u2** − c3∗**u3**.

If **m1**(Pn(i,j,k)) = (d/n^2)∗**u**(**r**(Pn(i,j,k))),
then
m(Pn(1,j,k)) = (d/n^2)∗**u**(r1(Pn(i,j,k))∗**u1**
+ r2(Pn(i,j,k))∗**u2**
+ r3(Pn(i,j,k))∗**u3**)
= **m1**(Pn(i,j,k))
since m2n = S[Pn(i,j,k))∩P2n](d/($(2∗n)^2$)) = 8∗d/($(2∗n)^2$) = d/n^2
and
u(S[Pn(i,j,k))∩P2n](Pn(i,j,k)) = **u**((2∗i−1)∗**u1** + (2∗j−1)∗**u2** (2∗k−1)∗**u3**)
+ 2∗i−1)∗**u1** + (2∗j−1)∗**u2** + (2∗k+1)∗**u3**)
+ 2∗i−1)∗**u1** + (2∗j+1)∗**u2** (2∗k−1)∗**u3**)
+ 2∗i+1)∗**u1** + (2∗j−1)∗**u2** + (2∗k−1)∗**u3**)
+ 2∗i+1)∗**u1** + (2∗j+1)∗**u2** + (2∗k−1)∗**u3**)
+ 2∗i+1)∗**u1** + (2∗j−1)∗**u2** + (2∗k+1)∗**u3**)
+ 2∗i−1)∗**u1** + (2∗j+1)∗**u2** + (2∗k+1)∗**u3**)
+ 2∗i+1)∗**u1** + (2∗j+1)∗**u2** + (2∗k+1)∗**u3**)
= **u**(16∗i∗**u1** + 16∗j∗**u2** + 16∗k∗**u3**)
= **u**(2∗i∗**u1** + 2∗j∗**u2** + 2∗k∗**u3**)
= **u**(r1(Pn(i,j,k))∗**u1**
+ r2(Pn(i,j,k))∗**u2**
+ r1(Pn(i,j,k))∗**u1**).
Clearly this pseudo−measure has mesh stability.
Moreover for M symmetrical around **r0**,
m(M) = m∗**u**(**r0**);
for a dilated M,
m(MA) = a1∗a2∗a3∗m(M)∗**m**(M)•**G**(**a**)/(sqrt(**m**(M)•**G**(**a**)•**G**(**a**)•**m**(M))
so that for MA = {a∗**r**|**r** in M},
m(MA) = a^3∗**m**(M).
However, for M1 and M2 disjoint, **m**(M1UM2) does not equal **m**(M1)+**m**(M2).
Moreover integration under this pseudo−measure is nonlinear.

Endnotes

[42] The volume of an individual cell in the nth partition is

p1 ∗ p2 ∗ p3 ∗ det[**UR1**]/n^3

and the volume of M is

vol(M) = bn(M∩Pn) ∗ p1 ∗ p2 ∗ p3 ∗ det[**UR1**]/n^3

Thus

bn(M∩Pn)/n^3 = vol(M)/(p1 ∗ p2 ∗ p3 ∗ det[**UR1**])

becomes a handy way to calculate *bn*.

[43] This definition may be made more pointed by introducing inner and outer bounding sets using inclusion and showing the difference between the two sets approaches a set of measure 0.

[44] Another variation

I[M](**f**(**r**)|m;dV) ≡ lim (S[bn(M∩Pn)](**f**(**r**(Pn(i,j,k))) ∗ vol(**m**(Pn(i,j,k)))))
= lim (S[bn(M∩Pn)](**f**(**r**(Pn(i,j,k)))/n^3)
∗ (p1R1 ∗ p2R1 ∗ p3R1) ∗ (**u1R1 ∧ u2R1**)•**u3R1**

becomes the usual volume integration.

[45] More rigorously, let ml and mu be integers such that

ml≤det[**UR1**] ∗ det[**G**(**p1**)]/(det[**UR2**] ∗ det[**G**(**p2**)])≤mu.

Then with increasing n, mu−ml⟶0.

[46] Limits of limits involve a rule linking the two limits whereby the first limit dominates the second.
The increment associated with a sectional gradient is

dR1 = lim **u1R1**/d1R1 + **u2R1**/d2R1 + **u3R1**/d3R1
≡ **urR1**/rR1.

The increment associate with the partition vector **pR1** is

pR1/n^3 = (p1R1 ∗ **u1R1** + p2R1 ∗ **u2R1** + p3R1 ∗ **u3R1**)/n^3

which is the measure **mn**(Pn(i,j,k)) of a partition cell in Pn.

For matching the limiting processes for the integral of a gradient the following rule is applied. To match the increment of integration with the increment for the gradient requires

0 <diR1<piR1/n^3

and specifically

pR1/n^3= **urR1**/rR1
= **dR1**

Then

rR1 = sqrt(((**u1R1**/d1R1 + **u2R1**/d2R1 + **u3R1**/d3R1)•**u1**)^2
+ ((**u1R1**/d1R1 + **u2R1**/d2R1 + **u3R1**/d3R1)•**u2**)^2
+ ((**u1R1**/d1R1 + **u2R1**/d2R1 + **u3R1**/d3R1)•**u3**)^2).

While giving **pR1** the same direction as **dR1**, the rule matches the fineness of the partition cells (a function of n whose measure approaches zero with increasing n) to **dR1** which grows without bound as the diRi approach zero.

[47] For functions for which cancellation holds, non−vector integration gives similar results.

For ***f*** continuous over V3,

I[V3](u(**r−ser**) ∗ **D**[**r,r+dR1**]•(**f**(**r**);**dR1**)|M;dR1)
= −**f**(**ser**)•I[V3](**D**[**r,r+dR1**] ∗ (u(**r−ser**)|M;**dR1**);dR1)

I[V3](u(**r−ser**) ∗ **D**[**r,r+dR1**] ∧ (**f**(**r**);**dR1**)|M;**dR1**)
= −**f**(**ser**) ∧ I[V3](**D**[**r,r+dR1**] ∗ (u(**r−ser**)|M;**dR1**);dR1)

Endnotes

I[V3](u(**r**−**ser**)∗**D**[**r**,**r**+**dR1**]∗(**f**(**r**);**dR1**)|M;**dR1**)
= −**f**(**ser**)∗I[V3](**D**[**r**,**r**+**dR1**]∗(u(**r**−**ser**)|M;**dR1**);dR1)
I[V3](u(**r**−**sir**)∗**D**[**r**,**r**+**dR1**]•(**f**(**r**);**dR1**)|M;dR1)
= −**f**(**sir**)•I[V3](**D**[**r**,**r**+**dR1**]∗(u(**r**−**sir**)|M;**dR1**);dR1)
I[V3](u(**r**−**sir**)∗**D**[**r**,**r**+**dR1**]∧(**f**(**r**);**dR1**)|M;dR1)
= −**f**(**ser**)∧I[V3](**D**[**r**,**r**+**dR1**]∗(u(**r**−**ser**)|M;**dR1**);dR1)
I[V3](u(**r**−**ser**)∗**D**[**r**,**r**+**dR1**]∗(**f**(**r**);**dR1**)||M;dR1)
= −**f**(**sir**)∗I[V3](**D**[**r**,**r**+**dR1**]∗(u(**r**−**sir**)|M;**dR1**);dR1)
I[V3](v(**r**−**ser**)∗**D**[**r**,**r**+**dR1**]•(**f**(**r**);**dR1**)|M;dR1)
= **f**(**ser**)•I[V3](**D**[**r**,**r**+**dR1**]∗(v(**r**−**ser**)|M;**dR1**);dR1)
I[V3](v(**r**−**ser**)∗**D**[**r**,**r**+**dR1**]∧(**f**(**r**);**dR1**)|M;dR1)
= **f**(**ser**)∧I[V3](**D**[**r**,**r**+**dR1**]∗(v(**r**−**ser**)|M;**dR1**);dR1)
I[V3](v(**r**−**ser**)∗**D**[**r**,**r**+**dR1**]∗(**f**(**r**);**dR1**)|M;dR1)
= **f**(**ser**)∗I[V3](**D**[**r**,**r**+**dR1**]∗(v(**r**−**ser**)|M;**dR1**);dR1)
I[V3](v(**r**−**sir**)∗**D**[**r**,**r**+**dR1**]•(**f**(**r**);**dR1**)|M;dR1)
= **f**(**sir**)•I[V3](**D**[**r**,**r**+**dR1**]∗(v(**r**−**sir**)|M;**dR1**);dR1)
I[V3](v(**r**−**sir**)∗**D**[**r**,**r**+**dR1**]∧(**f**(**r**);**dR1**)|M;dR1)
= **f**(**sir**)∧I[V3](**D**[**r**,**r**+**dR1**]∗(v(**r**−**sir**)|M;**dR1**);dR1)
I[V3](v(**r**−**sir**)∗**D**[**r**,**r**+**dR1**]∗(**f**(**r**);**dR1**)|M;dR1)
= **f**(**sir**)∗I[V3](**D**[**r**,**r**+**dR1**]∗(v(**r**−**sir**)|M;**dR1**);dR1).

Analogous results hold for v(**r**−**ser**) and at **sir**.

[48] In SECT(**r1**,**u1R1**,**u2R1**,**u3R1**) for sectional gradient **D**[**r**]∗(u(**r**−**sr1**)|M;**dR1**) a given **sr** in SM may lie in SMP|**uiR1**, SM0|**ujR1**, SMN|**ukR1**, that is,

SMP|**uiR1** ≡ {**sr** in SM; **sr**+diR1∗**uiR1** in V3−(M∪SM)}, i = 1,2,3
SM0|**uiR1** ≡ {**sr** in SM; **sr**+diR1∗**uiR1** in SM}, i = 1,2,3
SMN|**uiR1** ≡ {**sr** in SM; **sr**+diR1∗**uiR1** in (M − SM)}, i = 1,2,3.

Thus at **sr**, the sectional gradient may be impulsive in only one sectional direction, or two, or all three.

The integrations then become, for illustration,

I[V3]•**D**[**r**]∗(u(**r**−**sr**)|M;**dR1**)•[**UR1**]$^{-1}$•**T**[**UR1**]$^{-1}$;**dR1**)
= 1, if only one **sr** + diR1**uiR1** is in SMP|**uiR1**
= 2 if two **sr** + diR1**uiR1** are in SMP|**uiR1**
= 3 if three **sr** + diR1**uiR1** are in SMP|**uiR1**.

[49] A differentiable surface is defined by two parameters **sr** = **r**(x1,x2) where

D(**rs**|x2;x1) and D(**sr**|x1;x2)

are basic derivatives at **rs1**.

[50] The coordinate system (**us1**, **us2**, **us3**) is the coordinate system (**un**,**ut1**,**ut2**) rotated such that (**us1**+**us2**+**us3**)/sqrt(3) = **un**.

[51] Equation (913) forms the kernel of the Gauss divergence theorem. The divergence theorem is expressed by a Fubini integration using a volumetric non−vector measure v ≈ **u1**•(**u2**∧**u3**) and a second non−vector measure for an integration related to surface. The theorem states

I[M](D•**f**;dv) = I[SM](**f**•**un**;ds)

where **un** is the outward orthogonal on the surface designated by SM.

Endnotes

The divergence theorem fails if,

- **f**(**r**) is discontinuous in M
- **f**(**r**) is unbounded in M
- D•**f**; is not basic or is discontinuous
- M is non–measurable
- SM is non–differentiable.

The equations of the text, besides allowing analysis of cases for which the divergence theorem fails, may also be used to formulate analogs to the divergence theorem for curls and gradients.

[52] Green's theorem

I[M]((f*D•D*g−g*D•D*f);dv) = I[SM](f*D*g−g*D*f)•un;ds)

suffers the same restrictions as the divergence theorem.

[53] As in the apex of a pyramid.

[54] A differentiable surface is defined by unique orthogonals at every point in the surface. Non–differentiable surfaces, however, may have more than one local orthogonal. For instance where the surface at **rs1** is the conjunction of many planes, a section can be assigned to each plane with **u1Ri** and **u2R1** coplanar and **u3Ri** orthogonal. The surface then at **rs1** would have multiple orthogonals.

[55] Analogous to the real one–dimensional case,

i(a*x−x0) = D[t](u(a*z−x0);d(a*z)/a) = i(x−x0)/abs(a).

[56] Physical units as labels are not the common units of measurement used in, for example, the metric system. The primary observables correspond to vectorial quantities, whereas the metric system is a scalar system of units.

[57] The physical units of the exponential function are always neutral. The mathematical function exp(x) differs from the function of Theoretical Physics, similarly symbolized, as follows:

exp(x) = 1 + x + x*x/2 + ... (mathematical)
exp(x) = 1 + x/(1*u(x)) + x*x/(2*u(x)*1*u(x)) + ... (physical)

where u(x) denotes the units of x associated with the factorial divisor.

Manipulations of exponential (log) functions are often misleading. Consider

D[x1](exp(f(x));dx) = D[x1](f;dt)*exp(f(x1)).

The units of D[x1](exp(f(x));dx) are 1/x, but the units of D[x1](f;dx)*exp(f(x1)) are apparently u(f)/x. However,

D[x1](exp(f(x);dx) = D[x1](1 + f(x)/(1*u(f)) + f(x)*f(x)/2! +...);dx)
= 0 + D[x1](f;dx)/(1*u(f)) + 2*f(x1)*D[x1](f;dx)/2! +...
= (D[x1](f;dx)/(1*u(f)))*exp(f(x1)).

The factor (D[x1](f;dx)/(1*u(f))) has units 1/x. Consequently, some care is required in using exp and log as functions of Theoretical Physics, even with apparently innocuous constants.

The physical unit of abs(**r**) is identical to the physical dimension of **r** itself.

From its definition, it is obvious that **ur** has a neutral physical dimension. It represents pure directionality.

From the point of view of physical units, the operators I and D act as inverses, that is

$$I(D(g;dx);dx) = D(I(g;dx);dx) \text{ both have units of g}$$

since

$$(g/x)*x = (g*x)/x \text{ have units of g.}$$

The usual Lebesgue measure in (R,R,R) has a unit L*L*L. This is useful for the Fubini theorem which maps I:(R,R,R)⟶R. The more fundamental idea of local integration, which preserves the inversion of **D***, has dimension L.

Inconsistency of physical units often causes errors in summations. The sum of two velocities is a third velocity, whereas the summation of velocity over an interval of time is not velocity but length.

[58] The "whole" universe can never be observed physically, as if from outside. If a physical observer observes the "whole" universe from outside, from what part of the universe does he make his observation?

Similarly, the "smallest" particle can never be observed physically. If a physical observer observes the "smallest" particle, how then can he observe its parts?

These fantasies are physically self–contradictory.

[59] A notable example comes from the early modern physicist, Isaac Newton in his Philosophiae Naturalis Principia Mathematica. After first positing his axioms, Newton then discusses the earth's motion around the sun. He treats the earth–moon system as a particle: a resolution on the order of 10^9 meters. The result is very good, but not altogether satisfactory. So Newton is led by his critics to consider oceanic tides. Here the result is somewhat less satisfactory: the discussion assumes a resolution on the order 10^4 meters. Newton continues the Principia by considering more and more detailed physical phenomena (unreflectingly assuming more and more refined resolution.) Finally towards the end of the Principia, he finally contemplates the "ultimate" particle, (that is, resolution equal to zero) of which he says: "Who can doubt but that the ultimate particle is hard (*durum*)." The hardness of the ultimate particle is an undeclared axiom of Newtonian Physics, which logically should have been stated in the beginning with his other axioms. All of the "particles" considered in the Principia are treated as "hard", although it is clear that physically none of the "particles" actually discussed (earth–sun; earth–moon, moon–oceans, etc.) are, in fact, "hard". The ultimate particle is involved, according to Newton, in the transmission of light.

A little more than a century later, Augustin Jean Fresnel proved the Newtonian conclusions about light false.

Newton appears not to have appreciated the distinction between Mathematics and Theoretical Physics. His confusion is manifested in many ways: he gives Euclidean geometry, absolute position, absolute rest, absolute motion, absolute time, etc., a physical significance (inferred, but unobservable) instead of recognizing these as intellectual (ideas) instruments useful in understanding physical reality.

Endnotes

[60] The fictitious reference of a universe at rest is merely an intellectual convenience and not strictly necessary. Any method of identifying physical observables as they move and change is acceptable. The reference to a universe at rest is convenient because it avoids the necessity of establishing initial conditions of motion.

[61] In contrast to integers, real numbers in Mathematics allow infinite "resolution." Likewise, while recognizing that physical observations will necessarily be discrete, Theoretical Physics provides ideas for infinitely resolvable observations. Similarly, while recognizing that physical observations will necessarily be localized, Theoretical Physics provides ideas for universal observations.

[62] A third way is also possible. The derivatives may be interpreted with reference to a surface, material or local. Such derivatives might then be symbolized with **D2∗**. The development of these interesting derivatives, not given here, would parallel the development given in the text.

[63] Given **r1**(**x1**,a1) and **r**(**x1**,a1+da) = **r2**, the text chooses to define **x2** and **y2** as those particles which occupy **r1** and **r2** at a1+da and a1 respectively. The alternate, starting with **x1**(**r1**,a1) and **x**(**r1**,a1+da) = **x2**, would have chosen **r2** and **s2** as the locations of **x1** and **x2** at a1+da and a1 respectively. The consequences of this arbitrary choice are not great inasmuch as the arbitrariness comes from the necessity of choosing either the material or the local reference, but not both together, as starting points. As illustrated, either starting point leads to the same description:

{**s2**(**x2**,a1); **r1**(**x1**,a1); **r1**(**x2**,a1+da); **r2**(**x1**,a1+da); **r2**(**y2**,a1)}

or

{**y2**(**r2**,a1); **x1**(**r1**,a1); **x2**(**r1**,a1+da); **x1**(**r2**,a1+da); **x2**(**s2**,a1)}.

[64] For **d**(**r**|**x1**) the direction depends on the order of the index a. By convention a positive order is always assumed; that is

d(**r**|**x1**) = **r**(**x1**,a1+da)−**r**(**x1**,a1) ⟶ da>0.

The same convention is used for **d**(**x**|**r1**), namely

d(**x**|**r1**) = **x**(**r1**,a1+da)−**x**(**r1**,a1) ⟶ da>0.

[65] Forward/backward are ideas derived from the ordering of observations. Independent of the observational index, forward/backward become ambiguous ideas when applied to V3. Given a direction, **ur**, as forward, then −**ur** may be said to be the backward direction. But calling −**ur** forward makes **ur** the backward direction. Common language leads to a facile application of the ideas of forward or backward to V3 independent of the observational index, but then they have no unambiguous meaning.

In the text, the sense of forward or backward in V3 is taken not only take from the observational index, but also from the backward incremental direction as −(−**ur**). This convention has the advantage, in the continuous case, of giving both forward and backward increments the same direction. For Theoretical Physics this convention has the forward direction defined from a particle's motion *away from* the reference, in contrast to the backward direction which is defined from a particle's motion *toward* the reference.

Endnotes

The simpler alternative of defining both backward and forward increments from motions away from the reference leads to unnecessary complexity and awkwardness.

[66] The symbol **r**(**x**,a1) allows not two but an infinity of choices for **r** and **x**. To remove the ambiguity one of the three values is selected:

r0 = **r**(**y0**,a1), **r1** = **r**(**x1**,a1) or **r2** = **r**(**y2**,a1).

Similarly the symbol **x**(**r**,a1) is allowed to take on values

x0(**s0**,a1), **x1**(**r1**,a1) or **x2**(**s2**,a1).

The variables **r0**, **x0**, **r2**, and **x2** arise from the set of observations **r**(**x1**,a) and are thus functions of the observational variable, *a*.

Note that while **r**(**x**(**r1**,a1),a) equals **r**(**x1**,a), it does not equal **r**(**x**(**r1**,a),a1) unless constrained.

[67] The reader should note that **dX1** has a orthogonal representation, namely

dX1 = g∗**ug** = g∗(c1∗**u1**+c2∗**u2**+c3∗**u3**),

which may violate the condition ci≥0. When so represented, g∗**ug**•[**UX1**]$^{-1}$ may not fall in the first quadrant.

[68] For a material section SECT(**x1**,**u1X1**,**u2X1**,**u3X1**) there are many possible mappings into arbitrary local sections. Of these many mappings the observation **r**(**x**,a1) designates one at **r1**(**x1**,a1), but one for which **D3**[**r1**]∗(**r**|a1;**dX1**) may not exist. Consequently, one material subsection is chosen such that the gradient does exist. But the existence of sectional gradients does not assure sectional differentiability. Consequently the subsection must be further divided until sectional differentiability is assured. But again, sectional differentiability over SECT(**x1**,**u1X1**,**u2X1**,**u3X1**) does not assure compatible sectional differentiability in the image local section SECT(**r1**,**u1R1**,**u2R1**,**u3R1**). Hence both sections are further restricted until mutual sectional differentiability is achieved.

With these restrictions, the neighborhood of **r1**(**x1**,a1) is partitioned into a possibly large number of sections in each of which a sectional gradient, possibly different from others, exists. The uniqueness of this decomposition is not dealt with formally here, although a criterion for discussing uniqueness is found in the text.

[69] Generally, sectional mappings need not be linear. For non–linear mappings the matrices **UR1** and **UX1** are functions of **d**. Analogous to real derivatives, interest is confined to intersections of sections with neighborhoods in which the function may be linearized. Then the matrices **UR1** and **UX1** are no longer functions of **d**.

[70] These restrictions make the number and size of the sections depend on **r**(**x**,a1) at **r1**(**x1**,a1).

For any given material neighborhood–section SECT(**x1**,**u1X1**,**u2X1**,**u3X1**)
r(**x1** + d1X1∗**u1X1**) ≈ **r1** + (**r**(**x1**+d1X1∗**u1X1**) − **r1**) ≡ **r1** + d1R1∗**u1R1**
r(**x1** + d2X1∗**u1X1**) ≈ **r1** + (**r**(**x1**+d2X1∗**u2X1**) − **r1**) ≡ **r1** + d2R1∗**u2R1**
r(**x1** + d3X1∗**u1X1**) ≈ **r1** + (**r**(**x1**+d3X1∗**u3X1**) − **r1**) ≡ **r1** + d3R1∗**u3R1**.

The neighborhood–section SECT(**x1**,**u1X1**,**u2X1**,**u3X1**) is thus said to generate neighborhood–section SECT(**r1**,**u1R1**,**u2R1**,u3R1) under the mapping **r**(**x**,a1) at **r1**(**x1**,a1).

Endnotes

Dually, a local neighborhood–section SECT($\mathbf{r1},\mathbf{u1R1},\mathbf{u2R1},\mathbf{u3R1}$) is said to generate a material neighborhood–section SECT($\mathbf{x1},\mathbf{u1X1},\mathbf{u2X1},\mathbf{u3X1}$) under the mapping $\mathbf{x}(\mathbf{r},a1)$ at $\mathbf{x1}(\mathbf{r1},a1)$.

However, such sections may not be locally differentiable.

Now restrict the material neighborhood–sections only to those whose local images under $\mathbf{r}(\mathbf{x},a1)$ are locally differentiable.

Let SECT($\mathbf{x1},\mathbf{u1X1},\mathbf{u2X1},\mathbf{u3X1}$) be such a material neighborhood–section.

Let $\mathbf{x1} + \mathbf{dX1} = \mathbf{x1} + d1X1\ast\mathbf{u1X1}+d2X1\ast\mathbf{u2X1}+d3X1\ast\mathbf{u3X1}$ be a vector in the section.

Then given local differentiability

$$
\begin{aligned}
\mathbf{r}(\mathbf{x1} + \mathbf{dX1})-\mathbf{r1})/dX1 &\equiv \mathbf{r1} + dr\ast\mathbf{ur}\\
&\approx (\mathbf{r}(\mathbf{x1}+d1X1\ast\mathbf{u1X1})-\mathbf{r1})/d1X1\\
&\qquad +(\mathbf{r}(\mathbf{x1}+d2X1\ast\mathbf{u2X1})-\mathbf{r1})/d2X1\\
&\qquad +(\mathbf{r}(\mathbf{x1}+d3X1\ast\mathbf{u3X1})-\mathbf{r1})/d3X1\\
&\approx (\mathbf{r1} + d1R1\ast\mathbf{u1R1})-\mathbf{r1})/d1X1\\
&\qquad +(\mathbf{r1} + d2R1\ast\mathbf{u2R1})-\mathbf{r1})/d2X1\\
&\qquad +(\mathbf{r1} + d3R1\ast\mathbf{u3R1})-\mathbf{r1})/d3X1\\
&= d1R1\ast\mathbf{u1R1}/d1X1 + d2R1\ast\mathbf{u2R1}/d2X2 + d3R1\ast\mathbf{u3R1}/d3X1.
\end{aligned}
$$

Now let SECT($\mathbf{r1},\mathbf{u1R1},\mathbf{u2R1},\mathbf{u3R1}$) be the local section generated by SECT($\mathbf{x1},\mathbf{u1X1},\mathbf{u2X1},\mathbf{u3X1}$). For

$$\mathbf{dR1} = d1R1\ast\mathbf{u1R1}+d2R1\ast\mathbf{u2R1}+d3R1\ast\mathbf{u3R1}$$

$$
\begin{aligned}
&(\mathbf{x}(\mathbf{r1} + d1R1\ast\mathbf{u1R1}) - \mathbf{x1})/d1R1\\
&\quad + (\mathbf{x}(\mathbf{r1} + d2R1\ast\mathbf{u2R1}) - \mathbf{x1})/d2R1\\
&\quad + (\mathbf{x}(\mathbf{r1} + d3R1\ast\mathbf{u3R1}) - \mathbf{x1})/d3R1\\
&\qquad = (\mathbf{x1} + d1X1\ast\mathbf{u1X1} - \mathbf{x1})/d1R1\\
&\qquad\qquad + (\mathbf{x1} + d2X1\ast\mathbf{u2X1} - \mathbf{x1})/d2R1\\
&\qquad\qquad + (\mathbf{x1} + d3X1\ast\mathbf{u3X1} - \mathbf{x1})/d3R1\\
&\qquad = d1X1\ast\mathbf{u1X1}/d1R1\\
&\qquad\qquad + d2X1\ast\mathbf{u2X1}/d2R1\\
&\qquad\qquad + d3X1\ast\mathbf{u3X1}/d3R1.
\end{aligned}
$$

Consequently the assumption of differentiability for $\mathbf{r}(\mathbf{x},a1)$ implies the existence of reciprocal sectional gradients, but not necessarily mutual local differentiability. In particular $\mathbf{r}(\mathbf{x1} + \mathbf{dX1})$ may not equal $\mathbf{r1} + \mathbf{dR1}$.

Sections in Theoretical Physics may be considered sets with consistent directional derivatives.

[71] Note this fallacious argument.

Given

$$
\begin{aligned}
\mathbf{D3}[\mathbf{r1}]\ast(\mathbf{r}|a;\mathbf{dX1}) &\equiv \mathbf{a1}\ast\mathbf{u1X1} + \mathbf{a2}\ast\mathbf{u2X1} + \mathbf{a3}\ast\mathbf{u3X1}\\
&= \lim \mathbf{r}(\mathbf{x1}+d1X1\ast\mathbf{u1X1})-\mathbf{r}(\mathbf{x1}))\ast\mathbf{u1X1}/d1X1\\
&\qquad + (\mathbf{r}(\mathbf{x1}+d2X1\ast\mathbf{u2X1})-\mathbf{r}(\mathbf{x1}))\ast\mathbf{u2X1}/d2X1\\
&\qquad + (\mathbf{r}(\mathbf{x1}+d3X1\ast\mathbf{u3X1})-\mathbf{r}(\mathbf{x1}))\ast\mathbf{u3X1}/d3X1
\end{aligned}
$$

in a locally differentiable sector where

$$
\begin{aligned}
(\mathbf{r}(\mathbf{x1} + \mathbf{dX1}) - \mathbf{r}(\mathbf{x1}))/dX1 &= \mathbf{r}(\mathbf{x1}+d1X1\ast\mathbf{u1X1})-\mathbf{r}(\mathbf{x1}))/d1X1\\
&\qquad + (\mathbf{r}(\mathbf{x1}+d2X1\ast\mathbf{u2X1})-\mathbf{r}(\mathbf{x1}))/d2X1\\
&\qquad + (\mathbf{r}(\mathbf{x1}+d3X1\ast\mathbf{u3X1})-\mathbf{r}(\mathbf{x1}))/d3X1\\
&= \mathbf{a1} + \mathbf{a2} + \mathbf{a3};
\end{aligned}
$$

then the increment

$$(\mathbf{r}(\mathbf{x1} + \mathbf{dX1}) - \mathbf{r}(\mathbf{x1})) \equiv \mathbf{dR1} = dX1\ast(\mathbf{a1}+\mathbf{a2}+\mathbf{a3}).$$

In the definition of a differentiable sector, the diX1 are determined by the prior choice of the increment **dX1**, whereas in the definition of the sectional gradient the diX1 are not so constrained. For **dX1** = d1X1∗**u1X1**, for example,

(**r**(**x1** + **dX1**) − **r**(**x1**)) ≡ **dR1** = d1X1∗**a1**

in contrast to the above.

Moreover,

d1X1∗**u1X1**•**T**[**D3**[**r1**](∗(**r**|a;**dX1**)]
  = d1X1∗(**a1** + **u1X1**•**u2X1**∗**a2** + **u1X1**•**u3X1**∗**a3**).

In contrast,

d1X1∗**u1X1**•[**UX1**$^{-1}$]•**T**[**UX1**$^{-1}$]•**T**[**D3**[**r1**](∗(**r**|a;**dX1**)]
  = d1X1∗**u1**
    •(d1R1∗**u1**∗**u1R1**/d1X1
    + d2R1∗**u2**∗**u2R1**/d2X1
    + d3R1∗**u3**∗**u3R1**/d3X1
  = d1R1∗**u1R1**
  = d1X1∗**a1**.

[72] The reader should note carefully the fallacy involved in the following reasoning:

dRXk = **dXk**•**T**[**D3**[**r1**]∗(**r**(**x**,a)|a1;**dXk**)]
  = (−**dXj**•**T**[**D3**[**r1**]∗(**r**(**x**,a)|a1;**dXk**)]
  = (−**dXj**•**T**[−**D3**[**r1**]∗(**r**(**x**,a)|a1;**dXj**)]
  = (**dXj**•**T**[**D3**[**r1**]∗(**r**(**x**,a)|a1;**dXj**)]
  = **dRXj**.

[73] This result relies on the assumption of null matter. One can now appreciate an alternative approach in which only physical matter is assigned to one or more subsets in V3 in a mapping from the observed universe to V3 which preserves topology. Then **r**(**x**):V3⟶V3 is no longer bijective on V3, but on some subset of V3. While the identities in **r** and **x** remain, **D3**∗(**r**|a1;**dx**) may no longer be defined everywhere. Where it is defined but discontinuous, **D3**∗(**r**|a1;**dX1**) may then be defined to accommodate finite discontinuities as in the text. The result is then to treat **x** rather like a generalized function in contrast to the text which treats **x** like **r**.

The two approaches are easily reconciled by defining **y**(**x**):V3⟶V3 as properly material particles excluding null matter. Then **y** is injective onto a subset of V3 as above and also a generalized function as in the text.

[74] We are assuming the continuous properties of **f** coincide with the continuous properties of **x**, so that the mutually related sections SECT(**r1**,**u1R1**,**u2R1**,**u3R1**) and SECT(**x1**,**u1X1**,**u2X1**,**u3X1**) apply equally to **x** and **f**. Where this is not true, the sections need to be adjusted by intersection. For instance if SECT(**r1**,**u1R1**,**u2R1**,**u3R1**) and SECT(**x1**,**u1X1**,**u2X1**,**u3X1**) apply to physical material, **x**, but **f** has a discontinuous derivative in SECT(**x1**,**u1X1**,**u2X1**,**u3X1**), then SECT(**x1**,**u1X1**,**u2X1**,**u3X1**) can be decomposed into, say, SECT(**x1**,**u1Xf1**,**u2Xf1**,**u3Xf1**) and SECT(**x1**,**u1Xf2**,**u2Xf2**,**u3Xf2**) with corresponding section SECT(**r1**,**u1RXf1**,**u2RXf1**,**u3RXf1**) and SECT(**r1**,**u1RXf2**,**u2RXf2**,**u3RXf2**). Analysis can then proceed on each subsection.

Endnotes

[75] When using this particular result in the more general framework of sectional gradients, due care for signs must be taken. If the section bridges quadrants, or must be restricted because of discontinuities of functions of the primary variables then ambiguities can be avoided by reverting to sectional notation.

[76] The sets MR and MX may be open, closed, or mixed closed and open.
Since **r**(**x**,a1) is a bijection, it maps open sets in V3(**r**) into corresponding open sets in V3(**x**). The inverse functions perform likewise.
Closures are formed as limits from elements in the open sets. Because both V3(**r**) and V3(**x**) are defined as continuous, these limits exist. In V3(**x**), however, the closure may contain null matter.

[77] Here also we begin arbitrarily with **r**(**x**,a1) defining subsets of V3(**r**) with corresponding subsets in V3(**x**). For Theoretical Physics the dual is equally acceptable.

[78] Given a specific local trace other integrations may be defined as
I[a0,an](**f**(**r**(**x**,a+da)|**r**=**r**(**x1**,a);da)
≡ lim (S[bn(Pn∩[a0,an])](**f**(**r**(**x**,ai+da)|**r**=**r**(**x1**,ai));da))
I[a0,an](**f**(**r**(**x**,a–da)|**r**=**r**(**x1**,a);da)
≡ lim (S[bn(Pn∩[a0,an])](**f**(**r**(**x**,ai–da)|**r**=**r**(**x1**,ai));da))
I[a0,an](**f**(**r**(**y2**,a)|**r**=**r**(**x1**,a+da);da)
≡ lim (S[bn(Pn∩[a0,an])](**f**(**r**(**y2**,ai)|**r**=**r**(**x1**,ai+da));da))
I[a0,an](**f**(**r**(**y0**,a)|**r**=**r**(**x1**,a–da);da)
≡ lim (S[bn(Pn∩[a0,an])](**f**(**r**(**y0**,ai)|**r**=**r**(**x1**,ai–da));da)).
Likewise given a specific material track. material integrations may be defined as
I[a0,an](**f**(**x**(**r**,a+da)|**x**=**x**(**r1**,a);da)
≡ lim (S[bn(Pn∩[a0,an])](**f**(**x**(**r**,ai+da)|**x**=**x**(**r1**,ai));da))
I[a0,an](**f**(**x**(**r**,a–da)|**x**=**x**(**r1**,a);da)
≡ lim (S[bn(Pn∩[a0,an])](**f**(**x**(**r**,ai–da)|**x**=**x**(**r1**,ai));da))
I[a0,an](**f**(**x**(**s2**,a)|**x**=**x**(**r1**,a+da);da)
≡ lim (S[bn(Pn∩[a0,an])](**f**(**x**(**s2**,ai)|**x**=**x**(**r1**,ai+da));da))
I[a0,an](**f**(**x**(**s0**,a)|**x**=**x**(**r1**,a–da);da)
≡ lim (S[bn(Pn∩[a0,an])](**f**(**x**(**s0**,ai)|**x**=**x**(**r1**,ai–da));da)).

[79] The integration of a derivative may lead to ambiguities. For a discussion of these ambiguities and symbolism to clarify them, the author has an unpublished appendix of considerable length for consultation.

[80] This restriction is non–consequential when interest is centered on [a0,a2]. If F is the set of all functions coincident with **f** on [a0,a2] and allowing interior cancellation within [a0,a2], then any **f1** of F violating interior cancellation elsewhere may be replaced by some **f2** of F which enjoys interior cancellation everywhere. Integration over [a0,a2] is unaffected by the choice of either **f1** or **f2**.

[81] In Theoretical Physics, while integrals and gradients may be separately defined over V3(**x**) and V3(**r**), the relationship **r**(**x**,a) induces for each local function a corresponding material function and vice–versa.
Let

PX1Rn(i,j,k) = {**r**(**x**,a1)|PX1n(i,j,k)}.

Then since **r**(**x**,a) is a bijection, the collection PX1Rn of the subsets PX1Rn(i,j,k) is an induced partition of V3(**r**).

Set
r(PX1Rn(i,j,k)) = **r**(**x**(PX1n(i,j,k)))
MXR = {**r**(**x**,a1)|MX}
mxr(PX1Rn(i,j,k)) = **mx**(PX1n(i,j,k)).
Then defining
I3[MXR]•(**f**(**r**(**x**,ai));**dX1R**)
≡ lim(S[MXR∩PX1Rn](**f**(**r**(PX1Rn(i,j,k)))•**mxr**(PX1Rn(i,j,k))))
= I3•[MX](**f**(**x**(**r**,ai));**dX1**)
I3[MXR]∧(**f**(**r**(**x**,ai));**dX1R**)
≡ lim(S[MXR∩PX1Rn](**f**(**r**(PX1Rn(i,j,k)))∧**mxr**(PX1Rn(i,j,k))))
= **I3**[MX]∧(**f**(**x**(**r**,ai));**dX1**)
I3[MXR]∗(**f**(**r**(**x**,ai));**dX1R**)
≡ lim(S[MXR∩PX1Rn](**f**(**r**(PX1Rn(i,j,k)))∗**mxr**(PX1n(i,j,k))))
= **I3**[MX]∗(**f**(**x**(**r**,ai));**dX1**).

For the corresponding images over V3(**x**) let
PR1Xn(i,j,k) = {**x**(**r**,a1)|**r**îPR1n(i,j,k)}.
Then since **x**(**r**,a) is a bijection, the collection PR1Xn of the subsets PR1Xn(i,j,k) is an induced partition of V3(**x**).

Set
x(PR1Xn(i,j,k)) = **x**(**r**(PR1n(i,j,k)))
MRX = {**x**(**r**,a1)|**r**îMR}
mrx(PR1Xn(i,j,k)) = **mr**(PR1n(i,j,k)).
Then defining
I3[MRX]•(**f**(**x**(**r**,ai));**dRX**)
≡ lim(S[MRX∩PR1Xn](**f**(**x**(PR1Xn(i,j,k)))•**mrx**(PR1Xn(i,j,k))))
= I3•[MR](**f**(**r**(**x**,ai));**dR1**)
I3[MRX]∧(**f**(**x**(**r**,ai));**dRX**)
≡ lim(S[MRX∩PR1Xn](**f**(**x**(PR1Xn(i,j,k)))∧**mrx**(PR1Xn(i,j,k))))
= **I3**[MR]∧(**f**(**r**(**x**,ai));**dR1**)
I3[MRX]∗(**f**(**x**(**r**,ai));**dRX**)
≡ lim(S[MRX∩PR1Xn](**f**(**x**(PR1Xn(i,j,k)))∗**mrx**(PR1Xn(i,j,k))))
= **I3**∗[MR](**f**(**r**(**x**,ai));**dR1**).

The reader should note that while the partition cells defined by **pX1** in V3(**x**) all have the same shape, their local images under **r**(**x**,a1) may vary from cell to cell. Likewise partition cells defined by **pR1** in V3(**r**) have an identical shapes, but their material images under **x**(**r**,a1) may vary.

[82] More rigorously, let ml and mu be integers such that
ml≤det(**UR1**)∗det(**G**(**a**))≤mu.
Then with increasing n, mu−ml⟶0.

[83] The relationship between location and matter expressed by **r**(**x**,ai) also coerces a relationship between the discontinuities of related material and local functions.

[84] Using methods promoted by H. Lebesgue, the finite restriction may be extended to integration of a denumerable number of discontinuities.

[85] Consult with the author for the unpublished Appendix.

[86] For stalled processes, **r0** = **r1** = **r2** and **x0** = **x1** = **x2**.

Endnotes

[87] However, if at **r1**(**x1**,a1) **f** suffers a discontinuity,

f(**r1**(**x1**,a1)) ≡ **f01**

may not equal

f11 ≡ lim **f**(**r1**(**x1**,a1+da)) as da⟶0.

The difference **diff**$_f$(**f**|**x1**) = **f**(**r**(**x1**,a1+da)) – **f01**) would then differ from **f**(**r**(**x1**,a1+da)) – **f11**). The second form holds for the former difference, not the latter.

[88] The bridge theorem in both forms applies to functions even discontinuous at the reference. For discontinuous functions, however, derivatives may not exist as defined in the text. For example, if

lim (**f**(**r**(**x1**,a1+da)) = **f1** ≠ **f**(**r**(**x1**,a1)))

then

lim (**f**(**r**(**x1**,a1+da)) – **f**(**r**(**x1**,a1)))/da

has no limit.
However,

lim (**f**(**r**(**x1**,a1+da)) – **f1**)/da

would then exist and so serve as the basis for a proper definition of D[a1](**f**|**x1**;d$_f$a).
At reference **r1**(**x1**,a1) a discontinuous function might have different values for

f(**r**(**x1**,a1))
lim (**f**(**r**(**x1**,a1+da))
lim (**f**(**x**(**r1**,a1+da))
lim (**f**(**r1**(**x**,a1+da))
lim (**f**(**r**(**x1**,a1–da))
lim (**f**(**x**(**r1**,a1–da))
lim (**f**(**r1**(**x**,a1–da)).

In assuming a function continuous, the development in the text takes all the above as equal for the sake of simplicity. The extension to discontinuous functions is easily accomplished by redefining the derivatives in terms of appropriate limiting values of the functions at the reference.

[89] The first indexed corollary to the bridge theorem holds even when **f** is a step function at **r1**(**x1**,a1).
For instance let
f(**r1**(**x1**,a1)) = **f**(**x1**(**r1**,a1) ≡ **f11**∗ua(a–a1)
f(**r2**(**x1**,a1+da)) = **f**(**x1**(**r2**,a1+da)) ≡ **f21**∗ua(a–a1)
f(**r1**(**x2**,a1+da)) = **f**(**x2**(**r1**,a1+da)) ≡ **f12**∗ua(a–a1)
f(**r2**(**y2**,a1)) = **f**(**y2**(**r2**,a1)) ≡ **f22y**∗ua(a–a1)
f(**s2**(**x2**,a1)) = **f**(**x2**(**s2**,a1)) ≡ **fs22**∗ua(a–a1)
f(**r0**(**x1**,a1–da)) = **f**(**x1**(**r0**,a1–da) ≡ **f01**∗ua(a–a1))
f(**r1**(**x0**,a1–da)) = **f**(**x0**(**r1**,a1–da) ≡ **f10**∗ua(a–a1))
f(**r0**(**y0**,a1)) = **f**(**y0**(**r0**,a1) ≡ **f00y**∗ua(a–a1))
f(**s0**(**x0**,a1)) = **f**(**x0**(**s0**,a1) ≡ **fs00**∗ua(a–a1))
where **fij** are constant vectors and ua is the unit observational step function.
Then,
D[a1](**f**|**x1**;d$_f$a) = (**f21**–**f11**)∗D[a1](ua(a–a1)|**x1**;da)
D[a1](**f**|**r1**;d$_f$a) = (**f12**–**f11**)∗D[a1](ua(a–a1)|**r1**;da)
D[a1](**f**|**x2**;d$_f$a) = (**f12**–**fs22**)∗D[a1](ua(a–a1)|**x2**;da)
D[a1](**f**|**r2**;d$_f$a) = (**f21**–**f22y**)∗D[a1](ua(a–a1)|**r2**;da)
D[**r1**,**r2**](**f**|a1+da;da) = (**f21**–**f12**)∗D[a1](ua(a–a1)|a1+da;da)

D[**r1**,**r2**](**f**|a1;da) = (**f22y–f11**)∗D[a1](ua(a–a1)|a1;da)
D[**x1**,**x2**](**f**|a1+da;da) = (**f12–f21**)∗D[a1](ua(a–a1)|a1+da;da)
D[**x1**,**x2**](**f**|a1;da) = (**fs22–f11**)∗D[a1](ua(a–a1)|a1;da)
D[a1](**f**|**x1**;d_ba) = –(**f11–f01**)∗D[a1](ua(a–a1)|**x1**;da)
D[a1](**f**|**r1**;d_ba) = –(**f11–f10**)∗D[a1](ua(a–a1)|**r1**;da)
D[a1](**f**|**x0**;d_ba) = –(**fs00–f10**)∗D[a1](ua(a–a1)|**x0**;da)
D[a1](**f**|**r0**;d_ba) = –(**f00y–f01**)∗D[a1](ua(a–a1)|**r0**;da)
D[**r0**,**r1**](**f**|a1–da;da) = –(**f10–f01**)∗D[a1](ua(a–a1)|a1–da ;da)
D[**r0**,**r1**](**f**|a1;da) = –(**f11–f00y**)∗D[a1](ua(a–a1)|a1 ;da)
D[**x0**,**x1**](**f**|a1–da;da) = –(**f01–f10**)∗D[a1](ua(a–a1)|a1–da ;da)
D[**x0**,**x1**](**f**|a1;da) = –(**f11–fs00**)∗D[a1](ua(a–a1)|a1 ;da).

The first indexed corollaries then become transparently
(**f21–f11**)∗D[a1](ua(a–a1)|**x1**;d_fa) = (**f12–f11**)∗D[a1](ua(a–a1)|**r1**;d_fa)
+ (**f21–f12**)∗D[a1](ua(a–a1)|a1+da;d_fa)
(**f21–f11**)∗D[a1](ua(a–a1)|**x1**;d_fa) = (**f21–f22y**)∗D[a1](ua(a–a1)|**r2**;d_fa)
+ (**f22y–f11**)∗D[a1](ua(a–a1)|a1;d_fa)
(**f12–f11**)∗D[a1](ua(a–a1)|**r1**;d_fa) = (**f21–f11**)∗D[a1](ua(a–a1)|**x1**;d_fa)
+ (**f12–f21**)∗D[a1](ua(a–a1)|a1+da;d_fa)
(**f12–f11**)∗D[a1](ua(a–a1)|**r1**;d_fa) = (**f12–fs22**)∗D[a1](ua(a–a1)|**x2**;d_fa)
+ (**fs22–f11**)∗D[a1](ua(a–a1)|a1;d_fa)
–(**f11–f01**)∗D[a1](va(a–a1)|**x1**;d_ba) = –(**f11–f10**)∗D[a1](va(a–a1)|**r1**;d_ba)
– (**f10–f01**)∗D[a1](va(a–a1)|a1–da;d_ba)
–(**f11–f01**)∗D[a1](va(a–a1)|**x1**;d_ba) = –(**f00y–f01**)∗D[a1](va(a–a1)|**r0**;d_ba)
– (**f11–f00y**)∗D[a1](va(a–a1)|a1;d_ba)
–(**f11–f10**)∗D[a1](va(a–a1)|**r1**;d_ba) = –(**f11–f01**)∗D[a1](va(a–a1)|**x1**;d_ba)
– (**f01–f10**)∗D[a1](va(a–a1)|a1–da;d_ba)
–(**f11–f10**)∗D[a1](va(a–a1)|**r1**;d_ba) = –(**fs00–f10**)∗D[a1](va(a–a1)|**x0**;d_ba)
– (**f11–fs00**)∗D[a1](va(a–a1)|a1;d_ba).

[90] See *Step Functions and Product Rules* by JRBreton, ISBN 978-0-9844299-2-9, Library of Congress Control Number: 2010935175 which function as Appendix 2. to this volume.

[91] Perhaps this explains why Poincaré insisted that every fundamental physical theory be cast only in Lagrangian (material) form. Then the difficulties with the Eulerian relationship (and partial derivatives) would be avoided. Cf. Darrigol, O. "Henri Poincaré's Criticism of Electrodynamics" in Studies in History and Philosophy of Modern Physics, Vol 26B, Number 1, Apr 1995, pg 12.

[92] One topic has been slighted and an important problem neglected in this book.
The slighted topic is the so-called "scientific method" propounded by Francis Bacon and other English empiricists. In this method the development of Physics is required to proceed by critical experiments as the basis for deciding between opposing hypotheses.
The "scientific method" has been carefully examined many times with the conclusion that it is unsound as a scientific method. Of course, Physics must be based on observation. The "scientific method" however, implies that only observation, not understanding, defines Physics. This implication subverts Physics as a science. It redefines Physics as engineering.

Endnotes

It is not hard to find historical instances where the "scientific method" would have led to false conclusions.

Conclusions are statements of understanding, intellectual statements, for which the "scientific method" provides no basis.

While it is usually beneficial to enlarge the scope of observation, so to proceed, absent intellectual enlightenment, is usually wasteful both of time and money.

The neglected problem is the reflection that observation is itself a physical process. The important problem of propagation is considered in *Theoretical Physics: The Third Problem* by JRBreton ISBN 978-0-9844299-5-0.

93

Reflection: On solutions. The labor of gathering data is obviously the domain of Physics rather than Theoretical Physics. The collected data has then to be understood in terms of the appropriate ideas and equations of Theoretical Physics. Equations, however, look to solutions; and solutions may be obtained in different ways. Methods of solution are the domain of Mathematics. Thus the practice of physicists forms a conjunction between Mathematics, Theoretical Physics and Physics.

Two methods have been uncovered for finding solutions in general which are given various names inasmuch as they occur in so many different contexts: explicit/implicit; closed/open; open/feedback. To illustrate, consider the solutions to the quadratic equation: $a*x^2+b*x+c=0$, which has solution–types:

closed: $x=(b\pm\text{sqrt}(b^2-4*a*c))/(2*a)$

open: $x=-c/(a*x+b)$.

Both types have their advantages and disadvantages. For instance, in the limit as $a\longrightarrow 0$, it is clear that the open solution–type is preferable. It yields the correct solution, $x=-c/b$, whereas the closed solution–type becomes either indeterminate or unbounded. However, as $b\longrightarrow 0$, it is equally clear that the closed solution is preferable inasmuch as the open solution has no measurable region of convergence, and thus for numerical computation is useless.

Ideally the search for solutions always respects the truth about physical reality. However, inasmuch as Physics is the practice of moral persons, the practice of Physics has sometimes become dysfunctional. Therein lies perhaps the most celebrated controversy of Physics.

With this brief background, consider now the helio/terrestrial–centric controversy about the solar system.

The ancient astronomer, Ptolemy, in his *Almagest* (A.D. 140), has provided a mathematical treatment of the observed movement of the planets with reference to the astral background. Because of its eminent success, the *Almagest* became a foremost scientific text for more than a millennium and has even been proclaimed the catalyst which induced the blossoming of western science. The Almagest is an open solution to a planetary orbit using perturbations, called epicycles, of circular orbits. As a method of solution it is flawless, yielding correct solutions to the accuracy of measurement. Indeed, as measurement became increasingly more accurate during the course of a millennium, Ptolemaic solutions could be and were refined with complete satisfaction. Ptolemaic solutions provided the basis for the eminently successful Gregorian calendar in use today.

The success of this effort of mathematical solution, however, induced popularizers to confuse the ideas presented in the Almagest with material realities, which error in turn led to increasingly bizarre concepts of the solar system.

The priest–astronomer, Copernicus, in a posthumous work (A.D. 1543), advanced a closed solution based on a change of reference (origin) from surface of the earth to center of the sun. The brilliant Copernican effort, the fruit of a lifetime, challenged the bizarre concepts of the solar system which flowered under Ptolemaic tutelage. Unfortunately, at all points of experimental verification, the Ptolemaic calculations proved more exact than the Copernican predictions. In fact, each small change Copernicus made to simplify his method made the correspondence of his efforts with observed data more erroneous. Cf. Derek J. de S. Price, "Contra–Copernicus: A Critical Re–Estimation of the Mathematical Planetary Theory of Ptolemy, Copernicus and Kepler, in Critical Problems in the History of Science, Madison, 1959, pp 197–218.

More unfortunately, these brilliant efforts were caught up in the religious quarreling of the time, and have spun off a subsequent mythology which still tarnishes both religion and science.

From the viewpoint of Theoretical Physics, the choice of origin is purely arbitrary; an open solution is adequate, provided convergence in rapid enough numerically, and indeed may be superior to an inexact closed solution.

The ideas of Theoretical Physics are spiritual entities, not physical entities; the ideas of Theoretical Physics serve the needs of Physics by providing logical explanations for the observations; of two adequate explanations, the simpler is preferable.

Had Bacon's notion of "scientific method" been applied in the helio/terrestrial–centric controversy, Copernicus would have been discarded in the 16th century in favor of Ptolemy. Fortunately, the appeal of scientific "unity" prompted refinements to the Copernican ideas until efficient, refined, but still inexact, closed solutions achieved the practical accuracies of the open solution.

Index of Tables

Positive Definite and Basic Conventions....................26
Categories of Integration....................31
Derivatives involving Step Functions....................41
Integrals involving Step Functions....................45
Inequalities of Inner Products....................62
Solutions of Algebraic Vector Equations....................69
Symbols for Limits in V3....................106
Some Specific Positive Quadrant Results....................130
Rank of Matrices in V3....................145
Gradients of Sums and Products....................146
Step Functions at Surfaces....................196
Integrals of Sums and Products....................217
Symbols for Local Gradients....................219
Symbols for Local Ingradients....................219
Labels for Physical Units....................221
Some Physical Units....................225
Material References 230
Symbols for Global or Local References....................232
Derivatives of Theoretical Physics....................235
Designations for the Forward Convention....................239
Designations for the Backward Convention....................240
Panoramic Reference Designations....................240
Sectional Gradients of Theoretical Physics....................292
Mathematical Derivatives vs Derivatives of Theoretical Physics....................293
Decomposition of Indexed Derivatives....................310
Decomposition of Indexed Derivatives....................311
Generic Types of Derivative/Integral Exchanges....................406
Specific Exchanges of Derivatives and Integrals....................407
The Eulerian Relationship....................433
Physics, Theoretical Physics, Mathematics....................440

Index of Tables

Definitions

Physics....7
Theoretical Physics....11
Restricted Function....20
Compound Function....20
Continuity of a Real Function....21
Measurability of a Real Function....22
Basic Derivative from above and below....24
Reciprocal Vectors....63
Continuous Local Differentiability....131
Basic Sectional Gradient, Divergence, and Curl....135
Sectional Local Differentiability....137
Directional Integrals....149
Integrals along CX with respect to x....150
Integrals along CR with respect to r....152
Integrals over a Curve in terms of Arc-lengths....153
Directional Integrals....156
Invergences, Incurls and Ingradients along CX....157
Invergences, Incurls and Ingradients along CR....158
Step Functions along Curves....162
Basic Gradients of Step Functions....165
Partitions of V3....174
Location of a Cell in a Partition....174
Measure of a Measurable set M....175
Surface of M....177
Volume Step Functions in V3....195
Directional Derivative and Gradients Orthogonal to a Surface....207
Sectional Gradients Orthogonal to a Surface....208
Particle....226
Linearity....249
Continuously Mutually and Locally Differentiable Sections....250
Volume Step Functions in V3....326
Observational Integrals....332
Integrals of a Measurable Set in a Trace or in a Track....347
Material and Local Partitions....375
Center of a Cell in a Partition....376
Measure of a Measurable Sets MR or MX....377

Index

Surface of MX and MR..378
Invergences, Incurls, and Ingradients..380

Index of Theorems

Chain rule....112
Relationship between simply continuous and sectional gradients....141
Sectional gradients and directional derivatives....142
Interior cancellation over curves....161
Measure and volumes....176
Measure of a translated set....178
Measure of a dilated set....179
Partition for functions....182
Interior cancellation....185
Gradients of primary functions....252
Merging sections....263
Relationship between mutually related gradients....301
Relationship between local derivatives and gradients....305
I[](**f**(**r**(**x**,a))∗D[a](ua(**r**(**x**,a)–**r**(**x**,a1));d_fa);da)....334
I[](ua(**r**(**x**,a)–**r**(**x**,a1))∗D[a](**f**(**r**(**x**,t));d_ft);da)....336
I[](D[a](**f**(**r**(**x**,a))∗ua(**r**(**x**,a)–**r**(**x**,a1));d_fa);da)....338
I[](D[a](**f**(**r**(**x**,t));da);da)....339
I1[]•(**f**(**r**(**x1**,a))∗**D1**[**r**(**x1**,a),**r**(**x1**,a+da)]∗(uax(**r**(**x**,a)–**r**(**x**,a1))|**x1**;d_f**r**)|**x1**;d_f**r**)
....349
I1[]•(uax(**r**(**x1**,a)–**r**(**x1**,a1))∗**D1**[**r**(**x1**,a),**r**(**x1**,a+da)]∗(**f**(**r**(**x**,a))|**x1**;d_f**r**)|**x1**;d_f**r**)
....351
I1[]•**D1**[**r**(**x1**,a),**r**(**x1**,a+da)]∗(**f**(**r**(**x1**,a))∗(uax(**r**(**x**,a)–**r**(**x**,a1))|**x1**;d_f**r**)|**x1**;d_f**r**)
....354
I1[]•(**D1**[**r**(**x1**,a),**r**(**x1**,a+da)]∗(**f**(**r**(**x1**,a))|**x1**;d_f**r**)|**x1**;d_f**r**)....356
Extended fundamental theorem....371
First partition theorem....383
Second partition theorem....385
Integration of sectional gradients....388
I3[V3]•(**f**(**r**(**x**,a1))∗(**D3**[**r**]∗(ur(**r**,a1)|MR;**dR1**) + **D3**[**r**]∗(vr(**r**,a1)|**MR**;**dR1**))
....390
I3[V3]•((vr(**r**,a1)|MR–ur(**r**,a1)|MR)∗**D3**[r]∗(**f**(**r**(**x**,a1)|a1;**dR1**))
•[**UR1**]$^{-1}$•**T**[**UR1**]$^{-1}$|a1;**dR1**)....394

Index

$\int$3[V3]•((**D3**[**r**]∗(**f**(**r**,a1)∗ur(**r**)|MR;**dR1**)+**D3**[**r**]∗(**f**(**r**,a1)∗vr(**r**)|MR;**dR1**))
•[**UR1**]$^{-1}$•**T**[**UR1**]$^{-1}$|a1;**dR1**).....395
Bridge theorem: first form..409
Bridge theorem: second form..410

Reflections

Physics as reductionist........6
Physics as existential adoration........8
Logical consistency........10
The ideas of physics are non-physical........10
Physics differs from Engineering........10
God has revealed himself as a triple unity........51
Divine Providence and resolution........228
Jesus, divinity, and physical matter........231
Duality in Theoretical Physics and in Jesus Christ........232
On solutions........468

Alphabetical Index

A

absolute value....18
adoration....51
algebra....17
  vector....**51**
Almagest....**468**
almost everywhere....344, 417, 420
alphabet....13
ambiguity....15, 434, 438
analysis....9
angle....54
  reflex....99
Aquinas, Thomas....**228,** 445
arbitrary....434
arc–length....101
  as compound variable....**102**
  local reference....**237**
  material reference....237
area....225
arithmetic....
  science of....11
asymptotic....**42**
attribute....439
axiom of choice....**42**

B

bijective....**230**
bounded....**42**
bridge theorem....**409,** 441
  and transformations....**427**
  continuous case....**422**
  first form....**409**
  second form....**410**

Index

C

calculus........**18,** 433, 437
canonical........**143**
cell........376
  observational........331
  volume........456
chain rule........28, 417
  along a curve........**113**
  and arc–length........**113**
  in V3........**133**
change........**9,** 229, 230, 235, 284, 413, 441
combination........
  of physical units........**222**
complement........17
connected........**90**
consistent........11, 223, 438
continuity........
  attribute of a relationship........22
  of a real function........**21**
  of topological spaces........**21**
continuous........
  and Eulerian relationship........435
contour........**134**
contradiction........**11,** 437
  in modern physics........11
convention........
  basic........**25, 53**
  for matrix multiplication........449
  positive definite........**25, 53**
coordinates........
  intrinsic........450
Copernicus........**469**
corner........211
corollaries to the bridge theorem........
  first indexed........**414**
  second indexed........**417**
  third indexed........**420**
cosine matrix........**76**

curl....................**107,** 289, 344, 416, 453
orthogonal....................
to surface....................**134**
positive quadrant....................**124**
sectional....................
basic....................**135**
curl matrix operator....................**123, 136**
curl vector operator....................**76**
curvature....................**450**
curve....................89
intrinsic equation....................451
intrinsic feature....................102

D

definition....................14, 15
formal....................16
generic....................**26**
informal....................15
of a real integral....................**30**
operational....................21
specific....................**26**
denumerable....................17
derivative....................22
along a branch....................
continuous....................111
along CR with respect to r....................107, **111**
along CX....................
backward....................110
continuous....................**110**
differentiable....................**110**
forward....................110
along CX with respect to x....................107
as attribute of functions....................22
as function....................23
basic....................23, **107,** 414
along a branch....................**111**
along CR with respect to r....................111
along CX with respect to x....................**110**

Index

continuous **27**
continuous everywhere 27
directional 107, 108
basic **108**
forward **108**
in Theoretical Physics **286**
from above 24
from below 24
Newton/Leibniz 446
of a compound function **27**
of restricted functions **27**
of sums and products **115**
of the nth order 26
orthogonal ....
directional **207**
partial 433
whole 433, 435
derivatives ....
of real functions 25
positive quadrant ....
of compound functions **129**
diagonal vector operator **75, 123,** 270
difference ....
net **412**
differences 286
differentiable surface ....
classification **205**
differentiation 22
in the vector space **107**
dilation **214**
dimension ....
of the set of vectors **59**
direction 24, 51, 225
negative 24
of origin **59**
of straight line **88**
positive 24

Index

directional
    curl
        basic **117**
    divergence
        basic **117**
    gradient
        basic **117**
directional cosines **61**
directional derivative
    basic
        backward 108
    continuous 109
        everywhere **109**
directional gradients
    sectional
        of a matrix **137**
discontinuities 18, 31
divergence **107,** 289, 344, 416
    directional
        along CR **118**
    matrix
        restricted **118**
    positive quadrant **123**
    sectional
        basic **135**
divergence theorem **210**
division 440
domain 18
    metric 21
duality
    in Theoretical Physics **231**
dummy variable 333

Index

E

Einstein, Albert....448
energy....225
Eulerian relationship....**433,** 435, 438, 441
  ambiguities....435, 441
experiment....11
extended fundamental theorem of integral calculus....173
extension....**9**
extensions to the fundamental theorem of integral calculus....**188**

F

fallacies....319
  related to partial derivatives....**433**
field....17
force....10, 221
forcing....**7, 9**
function....18
  branched values....**106**
  compound....**20**
  continuous....
    in r along CR....**106**
    in the positive quadrant....**98**
    in x along CX....**104**
    omni-directionally....**93**
  continuous directionally....**95**
  differentiable....**23, 109**
  directional limit....
    backward....94
    forward....94
  doublet....48
  generalized....**36**
  impulse....**37,** 225
  in Theoretical Physics....
    derivative....**284**
  linear....**19**
  look-alike....18
    categories of differentiation and integration....**30**

measurable..**22**
mirror..**46**
real...**18**
reference...**230**
restricted...**20**
scalar..**70**
sectionally continuous..**101**
stalled...**106,** 113
vector..**70,** 225
fundamental theorem of integral calculus..34
along CR...**153**
extended..
in Theoretical Physics...**342**
extended to curves..**161**
extended to directional gradients...372
extended to vectorial functions with discontinuities..........................**373**
for directional integrals...**150**
for integrals over CX...**151**
fundamental theorem of integral calculus for sectional gradients.............
in V3..**187**

G

geometry..
differentiable...
restricted...**119**
Euclidean...445
God..6, 8, 10, 51, 231
gradient...107
along a curve..
and arc–length..**119**
and directional derivatives...**125**
basic...
along CR...**118**
of step functions..**165**
continuous..**265**
in section...**138**

directional
in Theoretical Physics **289**
of directional step function **166**
in Theoretical Physics
of primary functions **252**
inverse 274
locally continuous 132
everywhere **132**
of a vector **127**
orthogonal **207**
positive quadrant 124
continuous in orientation **125**
differentiable **125**
restricted
as outer product **119**
continuous **119**
sectional
basic **135**
orthogonal to a surface **208**
reduced to outer product **144**
simply continuous **132, 143, 274**
gradients
mutually related **301**
orthogonal
at surface **134**
simply continuous
everywhere **132**
Green's theorem **211**

H

hand rules **448**

Index

I

idea....10
 invalid....**15**
 mathematical....12
 new....15
 true....445
 valid....15
ideas....**9, 11,** 12, **13,** 439
 coherent....**11**
 derivative....**222**
image....**18**
incurl....149, 344, 380
 directional....156
  along CR....**158**
  forward....**156**
 in V3....**180**
 sectional....
  along CX....**157**
index....22
 of integration....**29**
ingradient....149, 344, 380
 directional....156
  along CR....**158**
  along CX....**157**
  forward....**156**
 in V3....**180**
integral....28
 along CR with respect to r....**152**
 along CX with respect to x....**150**
 as a function....30
 as operator....29
 directional....**149**
 involving step function....45
 of a derivative....447
 of a measurable set in a trace or track....347
 of compound functions....**32**
 of impulse functions....47
 of restricted functions....**32**

of sums and products........ **155**
of the nth order........ **32**
integrals........
of Theoretical Physics........ **331**
integration........ 28
directional........ 149
of a transformation........ **159, 349**
interior cancellation........
over curves........ **161**
observational........
in Theoretical Physics........ **331**
of a transformation........ **181**
of directional gradients........
over traces and tracks........ 349
of sectional gradient over M........ **392, 394**
of sectional gradients........ **388**
in Theoretical Physics........ **375**
over the observational index........ **334**
with respect to the observational index........
in Theoretical Physics........ 333
interchange........ 406
interior cancellation........ **185**
interval........ 29
closed........ **29**
open........ **29**
intrinsic........ 120, 245, 417
invergence........ 149, 344
directional........ 156
along CR........ **158**
along CX........ **157**
forward........ **156**
in V3........ **180**

J

Jesus........ 231, 232

Index

K

knowledge....9

L

label....221
Lebesgue....33
length....413
  as metric....90
  of a direction....**53**
  of origin....**58**
limit....18
  along a curve....
    backward....**104**
    forward....**103**
  directional....**94**
  from above....**19**
  from below....**19**
  in positive quadrant....**97**
  in quadrant....**95**
  in section....**98, 100**
  of positive quadrant....**96**
  of section....**99**
  omni-directional....**92**
limit point....**18**
line....
  linear....**87**
  tangent....**101**
linearity....**249**
lines....
  of a plane....**88**
local differentiability....
  continuous....**131**
  of sums and products....**146**
  sectional....**137**
local gradient....
  and local derivatives....**305**
local reference....**231,** 441

locally differentiable
  in the positive quadrant **131**
location 10, 225
logically consistent 11, 12, 439

M

magnitude 51
map
  straight **87**
material reference **231,** 441
Mathematics 17, 439
matrix 73
  curl operator **73**
  determinant **73**
  diagonal operator **73**
  inverse **73**
  null space
    dimension **78**
  of directional cosines **139**
  rank **78**
  trace **73**
  transpose **73**
matter 231
measurability
  attribute of functions **22**
measurable **9**
measure
  and volume **176**
  directed 29
  for integration 28
  of a material or local set **377**
  of a measurable set **175**
  of a set of observations
    in Theoretical Physics **332**
  positive 29, 205
metric 21
  positive 18
motion 221

moving **7, 9**
multiplication ....
 scalar **52**
 step functions 447
mutable 12

N

neutral 225
notation ....
 conditional 438
null matter **229**
number 225

O

object 12, 439
 mutable **9**
 physical 13
observation **7,** 9, 227, **232**
 primary 222
observational index ....
 backward sense **240**
 forward sense **239**
open set ....
 in V3 .... **89**
 restricted **90**
operator 14, 23, 28
order 228, 454
origin **58,** 441
 and observational reference 52
 in the set of vectors 52
 of the set of vectors ....
  as arbitrary **59**
orthogonal partition 381
outer product ....
 rank **79**

P

particle....441
  and physical observation....**227**
  as limit....227
  as operational concept....227
  definition of....**226**
  of classical mechanics....330
  property....227
partition....
  cell....174
    location....**174**
  material and local....**375**
  observational....**331**
  of a trace....**345**
  of a track....**346**
  of V3....**174**
partition theorem....182
  first....**383**
  second....**385**
partition vector....**175**
perception....221, 445
physical attribute....221
physical unit....
  as label....221
  derived....**223**
  extension....221
  multiplication....223
  neutral....**223**
  primary....223
physicist....228
physicists....6
physics....6, **7, 9, 439**
  as science....9
  modern....433, 438
  Newtonian....459
  reality....6
plane....**88**
plane geometry....445

Poincare, Henri....448
power....225
primary functions....**234,** 316
principle of non–collocation....**234,** 252, 403, **410**
principle of non–collocation,....**433**
process....**103, 107, 439**
product....
  of functions....**19**
product rule....448
  differential....**38**
  integral....**44,** 172
properties....
  of measures....454
property....230
proposition....11, 14
  logically consistent....445
Ptolemy....**468**

Q

quadrant....95
  positive....**95**

R

radius of curvature....**450**
range....**18**
rank....
  in Theoretical Physics....**237**
  of a gradient....**133**
  of a transformation....**145**
real number system....
  as a set of vectors....448
  complete....17
  measurable sets....**17**
  order....17
  topology....**17**
reality....**3,** 6, 7, 10, 11, 439, 442
  physical....51

reciprocal........225
reciprocal vector........
directional........**116**
orthogonal........64
reference........**22, 229,** 441
Eulerian........**433**
fixed-local........**231**
material........**231**
universe at rest........**229**
relationship........18
between continuous and sectional gradients........**141**
between gradients........138
sectional........**139**
sectional gradients and directional derivatives........**142**
religion........6
religion and science........469
resolution........228
restriction........22, 28
curved........**89**
of integration........**29**
rule........
matched integration........**186**
unit........**223**

S

science........10, 11, 433, 437
of physics........12
science and religion........469
scientific method........467
section........98
face........**98**
relabeling........**258**
sections........
in Theoretical Physics........
restricted to neighborhoods........251
merging........**259**
set........11, 14
set function........227, 231

Index

set of vectors
  division **63**
  division in 17
  functions of 70
  limit vector 90
  limits **92**
  measurable sets 90
    restricted **90**
  neighborhood
    open **89**
sets
  open **17**
solutions
  for matrix equations **82**
  sparse **84**
stall
  material **243**
step function **36,** 225
  in V3 **189**
  over curves **162**
  point 207
  topological relationships **36**
  volume
    in V3 **195**
step functions
  observational
    in Theoretical Physics **320**
  point
    in V3 **190**
straight
  line **87**
  plane **87**
sum
  of functions **19**
sum of two vectors **51**
summation
  as discrete integration **33**

surface
continuous **91**
derivatives 207
exterior negative **178,** 379
exterior positive **178,** 379
interior negative **178,** 379
interior positive **178,** 379
linear 87
non-differentiable 211
of a measurable set 177
in Theoretical Physics **378**
zero negative **178,** 379
zero positive **178,** 379
symbol **13,** 14, 15, 440
for derivatives 22
for the set of vectors, V3 **51**
local context **14**
symbols
as arbitrary 13
rules for **13**
syntax
in Theoretical Physics
for operators **285**

T

tautology 438
Theoretical Physics 10, **11,** 12, 15, 439
step functions **320**
theory 10
of measure 10
of numbers 10
time 225
topology 17, 441
of V3 **89**
set of vectors
restricted **89**
torsion **450**
trace 135, **236**

track..**236**
transformation..
  as restricted..143
  dilation..451
  identity..**61**
  rotation..451
  translation..451
truth..**445**

U

unification..9
union..17
unit..12, **221,** 343
unit..
  physical..441
units..
  of V3..224
unity..12
universe at rest..**229,** 441

V

variable..12
  compound..**20,** 27
  dependent..434
  dummy..28
vector..
  addition..**51**
  bi-normal..**450**
  calculus..**87**
  difference..**51**
  direction..**52**
  function..
    compound..**72**
    restricted..**71**
  length..**52**
  measure..
    in V3..**174**

multiplication....51, 54
cross product....55
inner product....**54**
outer product....**55**
orthogonal....**54**
parallel....52
negatively....**52**
positively....**52**
reciprocal....**63,** 122
reciprocal direction....**122**
reciprocal length....**122**
triple product....**55**
unit....**53**
vector equation....
entire set of solutions....**66**
minimum solution....66, **68**
solutions....**65**
vector length....
properties....**52**
vector operations....
referred to the origin....**60**
vector product....
as transformation....55
vectors....
algebraic operations....**51**
velocity....**230**
idea of....10
vertex....99
volume....222, 225, 230
integration....456

W

words....
as symbols....**15**